ELSEVIER'S DICTIONARY OF GEOGRAPHICAL INFORMATION SYSTEMS

ELSEVIER'S DICTIONARY OF GEOGRAPHICAL INFORMATION SYSTEMS

in

English, German, French and Russian

compiled by

B. DELIJSKA
Sofia, Bulgaria

2002
ELSEVIER
Amsterdam – Boston – London – New York – Oxford – Paris
San Diego – San Francisco – Singapore – Sydney – Tokyo

ELSEVIER SCIENCE B.V.
Sara Burgerhartstraat 25
P.O. Box 211, 1000 AE Amsterdam, The Netherlands

First edition 2002

Library of Congress Cataloging in Publication Data
A catalog record from the Library of Congress has been applied for.

British Library Cataloguing in Publication Data
Elsevier's dictionary of geographical information systems :
  In English, German, French and Russian
  1.Geographical information systems - Dictionaries -
  Polyglot
  I.Delijska, Boriana II.Dictionary of geographical
  Information systems
  910.2'85

  ISBN 0444509917

ISBN: 0-444-50991-7

PREFACE

The dictionary contains 4,040 terms with about 1,700 cross-references that are commonly used in the theory and practice of geographical information systems. The terms were selected according to their significance or frequency of use. The terminology covers the areas of geoinformatics, geostatistics, computer cartography, geospatial databases, computer graphics, geodesy, photogrammetry, remote sensing, hardware and software for introduction, processing and visualization of geospatial data, etc. Geographical information systems are modern, widely distributed and strategic tools in many human activities (land survey and regulation, ecology, forestry, agronomy, demography, hydrology, urban planning, transport, etc.).

The dictionary consists of two parts. In the first part, the *Basic Table*, the English terms are listed alphabetically and numbered consecutively. The English term is followed by its German, French and Russian equivalents. Synonyms of the English terms are also given as cross-references to the main entries in their proper alphabetical order. The second part, the *Indexes,* contains separate alphabetical indexes of the German, French and Russian terms. The reference number(s) of every term stand(s) for the number of the English term(s) in the basic table.

The author hopes that *Elsevier's Dictionary of Geographical Information Systems* will be a valuable tool for specialists, scientists and students and for everyone who is interested in this area.

Dr. Boriana Delijska

BIBLIOGRAPHY

Aronoff, S., *Geographic Information Systems: A Management Perspective* (WDL Publications,Ottawa, Canada, 1993)

The ESRI Press Dictionary of GIS Terminology (http://www.ashleyshostingservices.com/The-ESRI-Press-Dictionary-of-GIS-Terminology.html)

European Terminology Database (http://europa.eu.int/eurodicautom/login.jsp)

Le grand dictionnaire terminologique (http://www.granddictionnaire.com/_fs_global_01.htm)

P.Manoilov, G.Manoilov, B. Delijska, *Elsevier's Dictionary of Computer Graphics* (Elsevier, Amsterdam, 2000)

D.Parr, *GIS Glossary of Terms, URISA* (1460 Renaissance Drive, Suite 305, Park Ridge, IL 60068, 2000)

Notes and Study Materials for GIS and Geographer's Craft (http://www.Colorado.EDU/geography/ gcraft/notes/notes.html)

Баранов Ю.Б., Берлянт А.М., Капралов Е.Г., Кошкарев А.В., Серапинас Б.Б.,Филиппов Ю.А., *Геоинформатика. Толковый словарь основных терминов* (М.: ГИС – Ассоциация, 1999)

Цветков В.Я., *Геоинформационные системы и технологии* (Москва, "Финансы и статистика", 1998)

EXPLANATION OF SPECIAL SIGNS

1. The italics *d*, *f* and *r* in the basic table stand respectively for the German, French and Russian equivalents of the main English terms.

2. The gender of nouns is indicated as follows:

f	feminine
fpl	feminine plural
m	masculine
mpl	masculine plural
n	neuter
npl	neuter plural
pl	plural
m/f	masculine or feminine

3. The symbol *v* designates a verb.

4. The symbol *adj* designates an adjective.

5. Synonyms and abbreviations are separated by semicolons.

6. Two kinds of brackets are used:

    [ ] the information can be either included or left out;

    ( ) the information does not form an integral part of the expression, but helps to clarify it.

Basic Table

A

* AAT → 147

1 abscissa
d Abszisse *f*
f abscisse *f*
r абсцисса *f*

2 absolute accuracy
d absolute Exaktheit *f*; absolute Genauigkeit *f*
f exactitude *f* absolue
r абсолютная точность *f*

3 absolute altitude; absolute elevation; absolute height
d absolute Höhe *f*
f altitude *f* absolue
r абсолютная высота *f*; абсолютная [высотная] отметка *f*

4 absolute coordinates
d absolute Koordinaten *fpl*
f coordonnées *fpl* absolues
r абсолютные координаты *fpl*

* **absolute elevation → 3**

* **absolute height → 3**

5 absolute location
d absolute Lokalisierung *f*
f localisation *f* absolue
r абсолютное расположение *n*; абсолютное размещение *n*

6 absolute measurement
d absolute Messung *f*
f mesurage *m* absolu
r абсолютное измерение *n*

* **absolute method → 1168**

7 absolute pointing device
d absoluter Positionsanzeiger *m*; absolutes Zeigegerät *n*; absolute Steuervorrichtung *f*
f dispositif *m* de pointage absolu; dispositif de désignation absolu
r абсолютное указательное устройство *n*

8 absorption
d Absorption *f*
f absorption *f*
r поглощение *n*

9 abstention
d stillschweigender Erbschaftsverzicht *m*; Unterlassung *f*; Enthaltung *f*; Stimmenthaltung *f*
f abstention *f*; renonciation *f* tacite à une succession
r воздержание *n*; неучастие *n* [в голосовании]

10 abstraction
d Abstraktion *f*
f abstraction *f*
r абстракция *f*

11 acceptance test
d Abnahmeprüfung *f*; Übergabeprüfung *f*; Empfangstest *m*
f test *m* de réception; essai *m* de réception; essai de recette
r приёмочное испытание *n*; приёмочный тест *m*

12 acceptance threshold
d Aufnahmeschwelle *f*
f seuil *m* de réception
r порог *m* приёма

13 access
d Zugang *m*; Zugriff *m*
f accès *m*
r доступ *m*

14 accessibility
d Erreichbarkeit *f*; Zugänglichkeit *f*; Durchlässigkeit *f*
f accessibilité *f*
r доступность *f*

15 accessibility index
d Zugänglichkeitsindex *m*; Erreichbarkeitsindex *m*
f indice *m* d'accessibilité
r индекс *m* доступности; индекс достижимости

16 accidental error; casual error; erratic error; irregular error; random error
d zufälliger Fehler *m*
f erreur *f* aléatoire
r случайная погрешность *f*

17 accuracy; precision
(degree to which information on a map or in a digital database matches true or accepted values)
d Exaktheit *f*; Genauigkeit *f*
f exactitude *f*; justesse *f*
r точность *f*; соответствие *n*

* **accuracy of measurement → 2382**

18 **accuracy of survey**
  d Genauigkeit *f* der Aufnahme
  f précision *f* de levé
  r точность *f* съёмки

19 **accuracy standards**
  d Präzisionstandards *mpl*; Präzisionsnormen *fpl*
  f standards *mpl* d'exactitude; standards de précision
  r стандарты *mpl* точности

* **ACI** → 59

* **ACMS** → 271

* **ACPS** → 272

20 **acquisition of land; land acquisition**
  d Erwerb *m* von Grundstücken; Flächenentzug *m*; Grunderwerb *m*
  f acquisition *f* de terre; concession *f*
  r приобретение *n* земельной собственности; приобретение земли; отвод *m* земель

21 **acquisition time**
  d Erfassungszeit *f*; Einstellzeit *f*; Beschaffungszeit *f*; Akquisitionsintervall *n*
  f temps *m* d'acquisition
  r время *n* сбора

22 **across-track scanning; transverse scanning**
  d Querabtastung *f*
  f échantillonnage *m* sur des routes convergentes
  r сканирование *n* в поперечном направлении

* **ACS** → 241, 285

23 **active area**
  d aktiver Bereich *m*
  f zone *f* active
  r рабочее поле *n*

24 **active layer; current layer; active theme**
  d aktive Schicht *f*
  f couche *f* active; thème *m* actif
  r активный слой *m*; рабочий слой

25 **active sensor**
  d aktiver Sensor *m*
  f senseur *m* actif; capteur *m* actif; capteur de luminescence
  r активный сенсор *m*

* **active theme** → 24

26 **actuality; reality**
  d Aktualität *f*
  f actualité *f*; réalité *f*
  r действительность *f*; реальность *f*

27 **adaptive triangular mesh; ATM**
  d adaptive dreiseitige Masche *f*
  f maille *f* triangulaire adaptative
  r адаптивная треугольная сетка *f*

* **ADAR** → 84

28 **add** *v*
  d addieren; hinzufügen
  f additionner; sommer; ajouter
  r складывать; суммировать; прибавлять; добавить; добавлять

29 **address**
  (of a screen or raster point)
  d Adresse *f*
  f adresse *f*
  r адрес *m*

* **address** → 2242

30 **addressable point**
  d adressierbarer Punkt *m*
  f point *m* adressable
  r адресуемая позиция *f*

31 **address assignment; addressing**
  d Adresszuordnung *f*; Adressierung *f*
  f adressage *m*
  r присваивание *n* адреса; адресование *n*

32 **address baseline**
  d Adressenbasislinie *f*; Adressengrundlinie *f*; Adressenstandlinie *f*
  f ligne *f* de base d'adresse
  r адресная базисная линия *f*

33 **address coding**
  d Adressencodierung *f*
  f codage *m* d'adresse
  r кодирование *n* адреса

34 **address component; address part**
  d Adressenteil *m*; Anschriftsteil *m*
  f partie *f* adresse; zone *f* adresse
  r адресная часть *f*; часть кода, занимаемая адресом; часть кода, занимаемая номером памяти

35 **address correction**
  d Adressenkorrektur *f*; Adressenverbesserung *f*; Adressenrichtigstellung *f*
  f correction *f* d'adresse
  r уточнение *n* адреса

36 **address data processing**
  d Adressdatenverarbeitung *f*
  f traitement *m* de données d'adresse
  r обработка *f* адресных сведений

* **address geocoding** → 39

37 address grid
d Adressengitter *n*
f grille *f* d'adresses
r сетка *f* адресов

* **addressing** → 31

38 address[ing] system
d Adressiersystem *n*
f système *m* d'adressage
r система *f* адресации; адресная система;
система адресов

**39 address matching; admatching; address
geocoding**
d Adresskonvertierung *f*; Address-Matching *n*;
Adressgeocodierung *f*
f géocodage *m* par adresse postale
r адресная привязка *f*
позиционно-неопределённых наборов
данных

40 address parity
d Adressenparität *f*
f parité *f* d'adresse
r чётность *f* адреса

* **address part** → 34

41 address posting
d Adressenversetzung *f*; Adressenverbuchung *f*
f enregistrement *m* d'adresses
r регистрация *f* адресов

42 address range
d Adressenbereich *m*
f rangée *f* d'adresses
r диапазон *m* адресов

43 address space
d Adress[en]raum *m*
f espace *m* d'adresses
r адресное пространство *n*

* **address system** → 38

* **ADF** → 275

44 adiabatic process
d adiabatischer Prozess *m*
f processus *m* adiabatique
r адиабатический процесс *m*; адиабатный
процесс

45 adjacency
d Adjazenz *f*; Angrenzende *n*
f adjacence *f*
r прилегание *n*; смежность *f*

46 adjacency analysis
d Adjazenzanalyse *f*
f analyse *f* adjacence
r анализ *m* смежности

47 adjacent angle
d anliegender Winkel *m*; Nebenwinkel *m*
f angle *m* adjacent
r соседний угол *m*; смежный угол;
прилежащий угол

48 adjacent arcs
d adjazente Bögen *mpl*
f arcs *mpl* adjacents
r смежные дуги *fpl*

49 adjacent edge
d adjazente Kante *f*; benachbarte Kante
f arête *f* adjacente
r смежное ребро *n*

50 adjacent face
d adjazente Fläche *f*; benachbarte Fläche
f face *f* adjacente
r смежная грань *f*; смежная сторона *f*

51 adjacent facets
d angrenzende Facetten *fpl*; nachfolgende
Facetten; anliegende Facetten
f facettes *fpl* adjacentes
r смежные фацеты *mpl*

52 adjacent map sheet
d adjazentes Kartenblatt *n*
f feuille *f* d'une carte adjacente
r смежный лист *m* карты

* **adjacent points** → 2514

53 adjunction
d Adjunktion *f*
f adjonction *f*
r сопряжение *n*; присоединение *n*

* **adjusting** → 54

54 adjustment; adjusting; alignment
d Ausgleichung *f*; Justierung *f*
f ajustement *m*; ajustage *m*
r уравнивание *n*; настраивание *n*

55 adjustment of polygonal traverse
d Ausgleichung *f* des Polygonzugs
f compensation *f* de cheminement polygonal
r уравнивание *n* полигонального хода

* **admatching** → 39

56 administrative district
d Regierungsbezirk *m*; Amtsbezirk *m*;
Verwaltungsbezirk *m*

f district *m* administratif
r административная единица *f*,
административный округ *m*;
административный район *m*;
административная область *f*

57 administrative division
d verwaltende Abteilung *f*, administrative
Division *f*, Verwaltungsgliederung *f*
f division *f* administrative
r административное деление *n*

58 admissible error
d zulässiger Fehler *m*
f erreur *f* admissible
r допустимая погрешность *f*

* **ADS → 149**

* **advanced cartographic environment → 59**

**59 advanced cartographic equipment;
advanced cartographic environment; ACI**
d fortgeschrittene Kartografierungsausrüstung *f*
f équipement *m* cartographique avancé
r улучшенное картографическое
оборудование *n*

60 advanced cartographic system
d fortschrittliches System *n* der Kartografie
f système *m* cartographique avancé
r улучшенная картографическая система *f*

**61 advanced very high resolution radiometer;
AVHRS**
(a satellite sensing system operated by the US
National Oceanic and Atmospheric
Administration)
d Radiometer *m* mit sehr hoher Auflösung
f radiomètre *m* à très grand pouvoir séparateur;
radiomètre à résolution très grande
r усовершенствованный радиометр *m* очень
высокой разрешающей способностью

* **aerial map → 73**

62 aerial mosaic; air photo mosaic; mosaic
d Luftreihenaufnahme *f*
f assemblage *m* d'aérophotographies;
mosaïque *f* [de prises de vue aériennes];
assemblage d'images
r монтаж *m* последовательных аэросъёмок;
маршрутный монтаж

63 aerial perspective; air perspective
d Luftperspektive *f*
f perspective *f* aérienne
r воздушная перспектива *f*

* **aerial photo → 65**

64 aerial photogrammetry; phototopography
d Luftbildvermessung *f*
f photogrammétrie *f* aérienne
r аэрофотограмметрия *f*;
аэрофототопография *f*

**65 aerial photo[graphy]; aerophoto[graphy];
air photo**
d Aerofotografie *f*
f photo[graphie] *f* aérienne; aérophoto *f*
r аэрофотоснимок *m*

**66 aerial photography information system;
APIS**
d Informationssystem *n* der Aerofotografie
f système *m* d'information d'aérophotos
r информационная система *f*
аэрофотоснимков

67 aerial photography quad file; APQF
d Aerofotografie-Quadratdatei *f*
f fichier *m* d'aérophotos de quads
r файл *m* аэрофотоснимков элементарных
квадратных участков

68 aerial photoplan
d Luftfotoplan *m*
f aérophotoplan *m*
r аэрофотоплан *m*

* **aerial profile recorder → 86**

* **aerial radiometry → 69**

**69 aerial spectrophotometry; aerial
radiometry**
d Luftspektrofotometrie *f*
f spectrophotométrie *f* aérienne
r аэроспектрометрирование *n*

70 aerial survey
d Luftvermessung *f*, Luftbildaufnahme *f*
f levé *m* aérien
r аэросъёмка *f*

71 aerial triangulation; aerotriangulation
d Aerotriangulation *f*; Aerotriangulierung *f*;
Luftbildtriangulation *f*; Stereotriangulation *f*
f triangulation *f* aérienne; aérotriangulation *f*
r аэротриангуляция *f*

72 aerial view
d Luftbild *n*
f vue *f* aérienne
r взгляд *m* с воздуха

**73 aeronautical chart; aeronautical map;
aerial map**
d Luftfahrtkarte *f*; Fliegerkarte *f*

f carte *f* aéronautique
r аэронавигационная карта *f*

74 aeronautical data
d Luftfahrtdaten *pl*
f données *fpl* aéronautiques
r аэронавигационные данные *pl*

* **aeronautical map** → 73

75 aeronautical topographic chart
d topografische Luftfahrtkarte *f*
f carte *f* topographique aéronautique
r аэронавигационная топографическая
карта *f*

* **aerophoto** → 65

* **aerophotography** → 65

* **aerospace data** → 3121

* **aerotriangulation** → 71

* **affine map** → 76

76 affine map[ping]
d affine Abbildung *f*
f application *f* affine
r аффинное отображение *n*

77 affine transformation
d affine Transformation *f*
f transformation *f* affine
r аффинная трансформация *f*

78 aged data
d Altdaten *pl*; veraltete Daten *pl*
f données *fpl* vieilles
r устаревшие данные *pl*

79 age of data
d Datenalter *n*
f âge *m* des données
r возраст *f* данных

* **age pyramid** → 963

* **age-sex pyramid** → 963

80 agglomeration
(a geographical concentration of people and/or
activities)
d Agglomeration *f*; Ballung *f*
f agglomération *f*
r накапливание *n*; скопление *n*;
агломерация *f*

81 agglomeration area
d Ballungsgebiet *n*; Ballungsraum *m*

f zone *f* d'agglomération
r зона *f* агломерации

* **aggregate data** → 82

82 aggregate[d] data
d globale Daten *pl*
f données *fpl* obtenues par agrégation
r синтетические показатели *mpl*; совокупные
показатели; агрегированные данные *pl*

* **aggregate map** → 647

83 aggregation
d Aggregation *f*; Gesamtheit *f*
f agrégation *f*
r агрегирование *n*; агрегация *f*;
совокупность *f*

* **AHRS** → 240

* **AI** → 187

**84 airborne data acquisition and registration;
ADAR**
d Flugborddatenerfassung *f* und -Registrierung *f*
f saisie *f* et registration *f* des données aériennes
r сбор *m* и регистрация *f* данных с борта
самолёта

85 airborne data acquisition system
d im Flugzeug eingebautes
Datenerfassungssystem *n*;
Flugborddatenerfassungssystem *n*
f système *m* d'acquisition de données aériennes
r бортовая система *f* сбора данных

**86 airborne profile recorder; aerial profile
recorder; APR**
d Luftprofilregistriergerät *n*
f enregistreur *m* de profils aéroporté
r бортовой самолётный
радиопрофилометр *m*; радиовысотомер *m* с
самописцем

87 airborne scanner
d Flugzeugscanner *m*
f scanner *m* aérien
r бортовой сканер *m*

88 aircraft flight simulator
d Flugsimulator *m*
f simulateur *m* de vol aérien
r авиасимулятор *m*

* **air perspective** → 63

* **air photo** → 65

* **air photo mosaic** → 62

89 Albers equal-area conic projection
d flächentreue konische Albers-Projektion *f*
f projection *f* conique équivalente d'Albers
r равновеликая коническая проекция *f* Алберса

90 algorithm
d Algorithmus *m*
f algorithme *m*
r алгоритм *m*

91 algorithmic generalization
d algorithmische Generalisierung *f*
f généralisation *f* algorithmique
r алгоритмическая генерализация *f*

92 aliasing; stairstepping; jaggies
(of screen images)
d Aliasing *n*; Treppeneffekt *m*
f crénelage *m*; repliement *m*
r неровность *f*; ступенчатость *f*

93 alidade; sight-rule; sight-bar; elevation scale
(a part of a surveying instrument which consists of a sighting device and index, with accessories for reading and recording data)
d Alhidade *f*; Visierlineal *n*; Diopterlineal *n*; Kippregel *f*; Zeigerarm *m*
f alidade *f*
r алидада *f*

* alidade level → 1844

* alignment → 54, 1487

* alignment chart → 2546

94 allocation
d Zuordnung *f*; Zuweisung *f*; Allozierung *f*
f allocation *f*
r размещение *n*; распределение *n*

95 allocation of resources
d Einsatz *m* der Ressourcen; Faktoreneinsatz *m*
f allocation *f* de ressources
r размещение *n* ресурсов

* allotment → 2089

* almanac → 96

96 almanac [data]
(data on the general location and health of all satellites in the GPS constellation)
d Almanach-Daten *pl*
f données *fpl* d'almanach
r календарные данные *pl*; альманах *m*

97 along-track scanning

d Längenbtastung *f*
f échantillonnage *m* de cheminement longitudinal; échantillonnage [de routes] longitudinal; échantillonnage de trajectoire
r сканирование *n* в продольном направлении

* alphameric → 98

98 alphanumeric; alphameric
d alphanumerisch
f alphanumérique
r буквенно-цифровой

99 alphanumeric data
d alphanumerische Daten *pl*; Alphadaten *pl*
f données *fpl* alphanumériques
r буквенно-цифровые данные *pl*

100 alphanumeric grid
d alphanumerisches Gitter *n*
f grille *f* alphanumérique
r буквенно-цифровая сетка *f*

* alteration → 1137

101 alternate key
d Sekundärschlüssel *m*; Taste *f* mit doppelter Funktion
f touche *f* à double fonction
r альтернативный ключ *m*; дублирующий ключ

102 altimeter
d Höhenmesser *m*; Altimeter *m*
f altimètre *m*
r высотомер *m*

103 altimetric survey; survey of the relief; contour survey
d Höhenaufnahme *f*
f levé *m* altimétrique
r высотная съёмка *f*

104 altimetry; measurement of heights; measurement of altitudes
d Höhenmessung *f*; Höhenmesskunde *f*; Höhenbestimmung *f*
f altimétrie *f*
r альтиметрия *f*; высотометрия *f*; измерение *n* высоты

* altitude → 1811

* altitude difference → 1812

105 altitude matrix
d Höhenmatrix *f*
f matrice *f* d'altitudes
r матрица *f* высот

*   **altitude tint** → 1878

106 amalgamation
  d Unternehmensintegration *f*;
    Unternehmenskonzentration *f*;
    Amalgamation *f*
  f amalgamation *f*
  r амальгамирование *n*

*   **ambience** → 1286

107 ambient data; environmental data
  d Umweltdaten *pl*; Klimadaten *pl*
  f données *fpl* climatiques
  r данные *pl* о состоянии окружающей среды;
    данные об окружающей среде

108 ambiguity; vagueness
  d Mehrdeutigkeit *f*; Vagheit *f*; Ambiguität *f*;
    Phasenmehrdeutigkeit *f*
  f ambiguïté *f*
  r неясность *f*; неоднозначность *f*

*   **AM/FM system** → 278

*   **amortization** → 975

*   **AMS** → 242

*   **anaglyph** → 109

109 anaglyphic[al] map; anaglyph
  d Anaglyphe *f*; Anaglyphenbild *n*
  f anaglyphe *m*
  r анаглифическая карта *f*; анаглиф *m*

*   **anaglyphic map** → 109

110 analog map
  d analoge Karte *f*
  f carte *f* analogique
  r аналоговая карта *f*

111 analog plotter
  d Analogplotter *m*
  f traceur *m* analogique
  r аналоговый графопостроитель *m*

*   **analog** → 112

112 analog[ue]
  d analog
  f analogue; analogique
  r аналогический; аналоговый

113 analysis
  d Analyse *f*
  f analyse *f*
  r анализ *m*

114 analysis operations
  d Analyse-Operationen *fpl*
  f opérations *fpl* d'analyse
  r операции *fpl* анализа

115 analytical hill shading; digital shading
  d analytische Schummerung *f*
  f estompage *m* analytique; estompage
    numérique
  r автоматическая отмывка *f* возвышений

116 analytical map
  d analytische Karte *f*
  f carte *f* analytique
  r аналитическая карта *f*

117 analytical product
  (a product (usually a map) derived through
  analysis of graphic and/or attribute data)
  d analytisches Produkt *n*
  f produit *m* analytique
  r аналитический продукт *m*

*   **analytical thematic map** → 3645

118 anamorphic map
  d anamorphische Karte *f*
  f carte *f* anamorphique
  r анаморфированная карта *f*

119 anamorphose
  d Anamorphosis *f*
  f anamorphose *f*
  r анаморфоза *f*

120 anchor
  d Anker
  f ancre *f*; signet *m*; pointeur *m* de lien
  r якорь *m*; опора *f*

121 ancillary data
  d Zusatzdaten *pl*
  f données *fpl* ancillaires; données auxiliaires
  r вспомогательные данные *pl*;
    дополнительные данные

122 angle
  d Winkel *m*
  f angle *m*
  r угол *m*

*   **angle measure** → 2381

123 angle of altitude; angle of elevation;
    elevation angle
  d Erhebungswinkel *m*; Elevationswinkel *m*
  f angle *m* d'élévation; angle de hauteur; angle
    de site positif
  r угол *m* возвышения; угловая высота *f*

124 **angle of coverage; angle of field**
  d Bildwinkel *m*; Gesichtswinkel *m*
  f angle *m* de champ [visuel]
  r угол *m* поля зрения

  * **angle of deviation of the vertical** → 948

  * **angle of elevation** → 123

  * **angle of field** → 124

125 **angle of gradient; gradient angle**
  d Gradient[en]winkel *m*
  f angle *m* de gradient
  r угол *m* градиента

  * **angle of incidence** → 126

  * **angle of inclination** → 126

126 **angle of pitch; angle of incidence; incidence angle; angle of inclination; inclination angle; angle of slope; angle of tilt; tilt angle; slant angle; gradient of slope; slope gradient**
  d Einfallswinkel *m*; Neigungswinkel *m*; Inklinationswinkel *m*; Kippungswinkel *m*; Böschungswinkel *m*
  f angle *m* d'incidence; angle d'inclinaison [de talus]; angle de déclivité; angle de talus
  r угол *m* наклона; угол уклона; угол ската; угол откоса

  * **angle of slope** → 126

  * **angle of tilt** → 126

127 **angle of view; view[ing] angle; visual angle**
  d Blickwinkel *m*; Sehwinkel *m*; Betrachtungswinkel *m*
  f angle *m* visuel; angle de vision
  r угол *m* зрения; угол взгляда

  * **anglepreserving projection** → 686

128 **angular distance**
  d Winkelabstand *m*; Winkeldistanz *f*; Winkelentfernung *f*
  f distance *f* angulaire
  r угловое расстояние *n*

  * **angular field** → 1403

  * **angular field of vision** → 1403

  * **angular measure** → 2381

  * **animated GIF** → 129

129 **animated GIF [image]**
  d animiertes GIF-Bild *n*
  f image *f* GIF animée
  r анимированный GIF [рисунок] *m*

130 **animated map**
  d animierte Karte *f*
  f carte *f* animée
  r оживлённая карта *f*

131 **animation**
  d Animation *f*
  f animation *f*
  r оживление *n*; анимация *f*

  * **anisotrope** → 132

  * **anisotropic** → 132

132 **anisotropic[al]; anisotrope; nonisotropic**
  d anisotrop; nichtisotrop
  f anisotrope; non isotrope
  r анизотропный; неизотропный

  * **annotated** → 2065

133 **annotated feature**
  d annotierendes Feature *n*; annotierendes Merkmal *n*
  f caractéristique *f* annotée; trait *m* commenté
  r аннотированный признак *m*; признак, снабжённый комментариями

134 **annotation**
  d Annotation *f*
  f annotation *f*
  r аннотация *f*

135 **Antarctic circle**
  d südlicher Polarkreis *m*; antarktischer Polarkreis
  f cercle *m* polaire antarctique
  r Южный полярный круг *m*; антарктический полярный круг

136 **antialiasing; dejagging**
  (of a curve)
  d Kantenglättung *f*; Antialiasing *n*
  f anticrénelage *m*; lissage *m*
  r сглаживание *n* [границ кривых, наклонных линий и шрифтов]; плавное изменение *n*; уменьшение *n* ступенчатости; антиалиасинг *m*

137 **antipodes**
  d Antipoden *mpl*
  f antipodes *mpl*
  r антиподы *mpl*; жители *mpl* или страны *fpl* противоположных частей света

138 anti-spoofing; A-S
   d Anti-Täuschen *n* (*Verschlüsselung des*
      P-Codes zum y-Code zur Verhinderung von
      Störungen)
   f anti-déception *f*; anti-tromperie *f*;
      anti-arnaque *f*; anti-brouillage *m*;
      anti-leurrage *m*
   r анти-спуфинг *m*

   * **AOIPS** → 226

   * **apex** → 3928

   * **apex angles** → 3930

   * **aphylactic projection** → 145

   * **APIS** → 66

139 application
   d Anwendung *f*; Verwendung *f*
   f application *f*
   r приложение *n*; применение *n*;
      использование *n*

   * **application** → 140

140 application [software]
   d Anwendungssoftware *f*; Anwendung *f*;
      Applikation *f*; Aufgabe *f*
   f logiciel *m* appliqué; progiciel *m*
   r прикладное программное обеспечение *n*;
      приложение *n*

141 applied geodesy; engineering geodesy
   d angewandte Geodäsie *f*
   f géodésie *f* appliquée
   r прикладная геодезия *f*; инженерная
      геодезия

   * **appraisal** → 1321

   * **appraised value** → 1320

142 approach zone
   d Approach-Zone *f*
   f zone *f* approche
   r зона *f* доступности

143 approximate interpolation
   d angenäherte Interpolation *f*
   f interpolation *f* approchée
   r приближённая интерполяция *f*

144 approximation
   d Approximation *f*
   f approximation *f*
   r аппроксимация *f*; аппроксимирование *n*

   * **APQF** → 67

   * **APR** → 86

**145 arbitrary projection; aphylactic projection;
     compromise map projection**
   d willkürliche Projektion *f*; vermittelnde
      Abbildung *f*
   f projection *f* arbitraire
   r произвольная проекция *f*

146 arc
   d Bogen *m*
   f arc *m*
   r дуга *f*

147 arc attribute table; AAT
   d Bogenattributtabelle *f*; Arc-Attributtabelle *f*
   f table *f* d'attributs d'arcs
   r таблица *f* атрибутов дуг

148 arc coverage
   d Bogenversicherungsschutz *m*
   f couverture *f* d'arcs
   r покрытие *n* дуг

149 arc digitizing system; ADS
   d Bogendigitalisierungssystem *n*
   f système *m* de numérisation d'arcs; système de
      digitalisation d'arcs
   r система *f* оцифровки дуг

   * **arc endpoint** → 1271

150 architecture
   d Architektur *f*
   f architecture *f*
   r архитектура *f*

151 archive; backup
   d Archiv *n*
   f archives *fpl*; historique *m*
   r архив *m*

152 archiving
   d Archivierung *f*
   f archivage *m*
   r архивирование *n*; архивация *f*; архивное
      хранение *n*

   * **arc length** → 2136

**153 arc-node map data structure; link and node
     structure**
   d Bogen-Knoten-Kartendatenstruktur *f*
   f structure *f* arc-nœud de données
      cartographiques
   r линейно-узловая структура *f*
      картографических данных

154 arc-node model
   d Bogen-Knoten-Modell *n*; Arc-Node-Modell *n*

f modèle *m* arc-nœud
r линейно-узловое представление *n*;
векторно-топологическое представление

155 arc-node topology
d Bogen-Knoten-Topologie *f*;
Arc-Node-Topologie *f*
f topologie *f* arc-nœud
r линейно-узловая топология *f*; топология
дуг и узлов

156 Arctic circle
d nördlicher Polarkreis *m*
f cercle *m* polaire arctique
r Северный полярный круг *m*; арктический
полярный круг

157 area
d Fläche *f*; Flächeninhalt *m*
f aire *f*
r площадь *f*

158 area; zone
d Bereich *m*; Gebiet *n*
f zone *f*
r ареал *m*; область *f*; зона *f*

* **area boundary** → 387

* **area calculation** → 659

159 area centroid; centroid [point]
d Zentroid *n*; Flächenschwerpunkt *m*;
Schwerpunkt *m*
f centroïde *m*
r центроид *m*

**160 area chart; area graph; area diagram;
surface chart; surface graph**
d Flächendiagramm *n*
f diagramme *m* en aires; diagramme surfacique;
graphique *m* de surface
r диаграмма *f* с областями; площадная
диаграмма

161 area code; zone code
d Orts[netz]kennzahl *f*; Vorwahl *f*
f indicatif *m* de zone
r код *m* зоны

162 area decomposition; domain decomposition
d Gebietszerlegung *f*
f décomposition *f* de zone; décomposition de
domaine
r декомпозиция *f* области; разбиение *n*
области

* **area diagram** → 160

163 area distortion

d Flächenverzerrung *f*
f distorsion *f* de zone
r искажение *n* области

164 area edge; polygon edge
d Polygonkante *f*
f arête *f* de polygone
r грань *f* области; край *m* поля; ребро *n*
многоугольника; ребро полигона

**165 area [emission] source; AS; nonpoint
[emission] source; surface source**
d Emissionsquelle *f*; diffuse Quelle *f*
f zone *f* source [de pollution]; source *f* [de
pollution] étendue; source diffuse; source
dispersée
r совокупность *f* слабых источников
[загрязнения], рассредоточенных по
большой площади; рассредоточенный
источник *m* загрязнения

166 area feature; area object; polygonal object
d Flächenobjekt *n*
f objet *m* surfacique; objet polygonal; objet
zonal
r полигональный объект *m*; контурный
объект

* **area graph** → 160

167 area intersection
d Flächenverschneidung *f*
f intersection *f* de zones
r пересечение *n* зон

168 areal; areolar
d Sektor-; Flächen-; areolar
f d'aire; aréolaire
r секторный; ареальный; ареоларный

169 areal cover
d Flächendeckel *m*; regionale Abdeckung *f*
f couverture *f* régionale
r региональное покрытие *n*

* **areal element** → 1242

170 areal interpolation; spatial interpolation
d räumliche Interpolation *f*; spatiale
Interpolation
f interpolation *f* d'aire; interpolation spatiale
r пространственная интерполяция *f*

* **areal unit** → 3858

* **area object** → 166

171 area of a surface; surface area
d Oberflächenbereich *m*

f aire *f* de surface
r площадь *f* поверхности

* **area of influence** → 4032

172 area of operations
d Betriebsgebiet *n*
f zone *f* d'opérations
r область *f* операций

* **area patch** → 3545

173 area patch generalization
d Flächenstückgeneralisierung *f*
f généralisation *f* de morceaux de surface
r генерализация *f* кусков поверхности

174 area/perimeter calculation
d Bereich- und Perimeter-Berechnung *f*
f calcul *m* d'aire/périmètre
r вычисление *n* площади и периметра;
    расчёт *m* поверхности/периметра

175 area planning
d Raumplanung *f*
f planification *f* de zone
r районная планировка *f*

176 area ranking; ranking
d Rangordnung *f*; Rangfolge *f*
f classement *m* de zones; graduation *f* de zones
r систематизация *f* областей; ранжирование *n*
    областей

177 areas covered by a projection
d Abbildungsflächen *fpl*
f surfaces *fpl* de projection
r области *fpl* проекции

178 area search; territory search
d Bereichssuche *f*
f recherche *f* sur zone; recherche sélective
r поиск *m* в определённой области;
    групповой поиск

179 area selection
d Gebiet[s]auswahl *f*
f sélection *f* de zone
r выбор *m* области

* **area source** → 165

180 area symbols
d Flächensymbole *npl*
f poncifs *mpl*
r условные площадные знаки *mpl*

181 area triangulation
d Flächentriangulation *f*
f triangulation *f* d'aire

r площадная триангуляция *f*;
    микротриангуляция *f*

* **area unit** → 3858

182 area-weighted average resolution; AWAR
d Flächen-gewichtete Mittelauflösung *f*
f résolution *f* moyenne pondérée par aire
r средневзвешенная по площади
    разрешающая способность *f*

* **areolar** → 168

183 array
d Feld *n*; Array *n*
f tableau *m*; champ *m*
r массив *m*

184 array processor; matrix processor
d Array-Prozessor *m*; Vektorprozessor *m*
f processeur *m* matriciel; processeur de
    tableaux; processeur vectoriel
r матричный процессор *m*; векторный
    процессор

185 arrow
d Zählpfeil *m*; Pfeil *m*
f flèche *f*
r стрелка *f*

186 artificial boundary
d künstliche Grenze *f*
f limite *f* artificielle
r искусственная граница *f*

187 artificial intelligence; AI
d künstliche Intelligenz *f*;
    Maschinenintelligenz *f*
f intelligence *f* artificielle; IA
r искусственный интеллект *m*; ИИ

* **artificial terrain data** → 3566

* **A-S** → 138

* **AS** → 165

188 ascending; bottom-up; increasing
d aufsteigend; hochsteigend; ansteigend;
    wachsend; zunehmend
f ascendant; montant; levant; de bas en haut;
    croissant
r возрастающий; восходящий

189 ascending sort
    (of a field in a table)
d aufsteigendes Sortieren *n*; Sortieren in
    aufsteigender Reihenfolge
f triage *m* ascendant; triage *m* croissant
r сортировка *f* в восходящем порядке

190 **aspatial; non-spatial**
  d nichträumlich
  f non spatial
  r непозиционной; непространственный

191 **aspatial attribute**
  d nichträumliches Attribut *n*
  f attribut *m* non spatial
  r непространственный атрибут *m*

192 **aspatial data; non-spatial data**
  d nichträumliche Daten *pl*
  f données *fpl* non spatiaux
  r непространственные данные *pl*

193 **aspatial object; non-spatial object**
  d nichträumliches Objekt *n*
  f objet *m* non spatial
  r непространственный объект *m*

  * **aspect → 633**

  * **assessment → 1321**

194 **asset**
  (anything valuable enough to track or keep
  information on)
  d Anlage *f*; Gewinn *m*
  f actif *m*; bien *m*
  r ресурс *m*

195 **asset file**
  d Anlagendatei *f*
  f fichier *m* d'actifs
  r файл *m* учёта основного капитала

196 **assignment statement**
  (in database query)
  d Zuweisungsanweisung *f*; Ergibt-Anweisung *f*
  f instruction *f* d'affectation
  r оператор *m* присваивания

197 **associated data**
  d assoziierte Daten *pl*
  f données *fpl* associées
  r связанные данные *pl*

198 **associated number**
  d assoziierte Zahl *f*
  f nombre *m* associé
  r ассоциированное число *n*

199 **association**
  d Assoziation *f*
  f association *f*
  r связывание *n*; ассоциация *f*

200 **association for geographic information**
  d Vereinigung *f* für geografische Informationen
  f association *f* d'information géographique
  r ассоциация *f* географической информации

201 **astrogeodetic network**
  d astrogeodätisches Netz *n*
  f réseau *m* astrogéodesique
  r астрономо-геодезическая сеть *f*

202 **astrolabe**
  d Sternhöhenmesser *m*
  f astrolabe *m*
  r астролябия *f*

  * **astronomic → 203**

203 **astronomic[al]**
  d astronomisch
  f astronomique
  r астрономический

204 **astronomic azimuth; astronomic bearing**
  d astronomischer Azimut *m*
  f azimut *m* astronomique
  r астрономический азимут *m*

  * **astronomic bearing → 204**

205 **astronomic latitude; geographic latitude;**
    geodetic latitude; latitude
  d [astronomische] Breite *f*; geodätische Breite;
    geografische Breite; Polhöhe *f*
  f latitude *f* [astronomique]; latitude
    géographique; latitude géodésique
  r [астрономическая] широта *f*; геодезическая
    широта

206 **astronomic longitude; geographical**
    longitude; longitude
  d [astronomische] Länge *f*; geografische Länge
  f longitude *f* [astronomique]; longitude
    géographique
  r [астрономическая] долгота *f*

207 **astronomic mapping**
  d astronomische Kartierung *f*
  f cartographie *f* astronomique
  r астрономическое картографирование *n*

208 **astronomic meridian**
  d astronomischer Meridian *m*
  f méridien *m* astronomique
  r астрономический меридиан *m*

209 **astronomic parallel**
  d astronomischer Breitengrad *m*
  f parallèle *m* astronomique
  r астрономическая параллель *f*

210 **astronomic unit; AU**
  d astronomische Einheit *f*

f unité *f* astronomique
r астрономическая единица *f*

211 astronomic zenith distance
d astronomische Zenitdistanz *f*; astronomischer
Zenitwinkel *m*
f distance *f* astronomique zénithale
r астрономическое зенитное расстояние *n*

212 astronomy
d Astronomie *f*
f astronomie *f*
r астрономия *f*

213 asymmetrical information distribution
d asymmetrische Verteilung *f* von Information
f répartition *f* d'information asymétrique
r асимметричное распространение *n*
информации

214 asymmetry
d Asymmetrie *f*; Schiefheit *f*
f dissymétrie *f*; asymétrie *f*
r скошенность *f*; асимметрия *f*;
несимметрия *f*

215 asynchronous
d asynchron
f asynchrone
r асинхронный

216 asynchronous request
d asynchrone Anforderung *f*
f requête *f* asynchrone
r асинхронный запрос *m*

217 asynchronous transfer
d asynchrone Übertragung *f*
f transfert *m* asynchrone
r асинхронная передача *f*; асинхронный
перенос *m*

218 asynchronous transfer mode; ATM
d asynchroner Transfermodus *m*; asynchroner
Übermittlungsmodus *m*
f mode *m* de transfert asynchrone
r режим *m* асинхронной передачи

* **AT → 233**

219 atemporal data
d nichtzeitliche Daten *pl*
f données *fpl* non temporelles
r вневременные данные *pl*; данные, не
относящиеся к определённому времени

220 atemporal database
d nichtzeitliche Datenbasis *f*
f base *f* de données non temporelles
r база *f* вневременных данных

* **atlas → 1575**

221 atlas dummy
d Atlasentwurf *m*
f maquette *f* d'atlas
r макет *m* атласа

222 atlas for education
d Bildungsatlas *m*
f atlas *m* d'éducation
r учёбный атлас *m*

* **atlas shape → 223**

223 atlas size; atlas shape
d Landkartenformat *n*
f format *m* d'atlas
r формат *m* атласа

* **ATM → 27, 218**

224 atmosphere
d Atmosphäre *f*
f atmosphère *m*
r атмосфера *f*

225 atmospheric absorption
d atmosphärische Absorption *f*
f absorption *f* atmosphérique
r атмосферное поглощение *n*; атмосферная
абсорбция *f*

**226 atmospheric and oceanographic image
processing system; AOIPS**
d atmosphärisches und ozeanografisches
Bildverarbeitungsystem *n*
f système *m* de traitement d'images
atmosphériques et océanographiques
r система *f* обработки атмосферных и
океанографических снимок

227 atmospheric correction
d atmosphärische Korrektur *f*
f correction *f* atmosphérique
r атмосферная коррекция *f*

228 atmospheric effects
d atmosphärische Effekte *mpl*
f effets *mpl* atmosphériques
r атмосферные эффекты *mpl*

229 atmospheric noise model
d atmosphärisches Rauchenmodell *n*
f modèle *m* de bruit atmosphérique
r модель *f* атмосферных помех; модель
атмосферного шума

230 atmospheric path
d atmosphärischer Schallweg *m*

f trajectoire f atmosphérique; parcours m atmosphérique

r траектория f в атмосфере; путь m в атмосфере

231 atmospheric path length
d atmosphärische Weglänge f
f longueur f de parcours atmosphérique; longueur de trajectoire atmosphérique
r длина f траектории в атмосфере; длина пути в атмосфере; атмосферный пробег m

232 atmospheric window
d atmosphärisches Fenster n
f fenêtre f atmosphérique
r окно n прозрачности атмосферы; атмосферное окно

233 atomic time; AT
d Atomzeit f
f temps m atomique
r атомное время n

234 atomic value
d atomarer Wert m; Atomwert m
f valeur f atomique
r элементарная величина f

235 atomization
d Atomisierung f; Zerstäubung f
f atomisation f
r атомизация f

* **attach → 236**

* **attaching → 236**

236 attach[ment]; attaching
d Vereinigung f; Anliegerung f; Befestigung f
f attachement m
r присоединение n

237 attachment boundary
d Befestigungsgrenze f
f limite f d'attachement
r граница f присоединения

* **attachment point → 3677**

238 attenuation; damping; falloff; drop-off
d Dämpfung f; Abblendung f
f atténuation f; amortissement m; affaiblissement m
r затухание n; гашение n; успокоение n; ослабление n

239 attitude
d Attitüde f; Fluglage f
f attitude f
r позиция f; положение n

240 attitude and heading reference system; AHRS
d Fluglage- und Steuerkursreferenzsystem n
f système m de référence de cap et d'attitude
r система f отсчёта курса и положения; курсовая система

241 attitude control system; ACS (Landsat)
d Lageregelungssystem n
f système m de contrôle d'attitude; système de commande d'orientation; chaîne f de pilotage
r система f управления ориентацией; система ориентации; система пространственной стабилизации

242 attitude measurement sensor; AMS
d Fluglagemeßwertgeber m
f détecteur m d'orientation; élément m sensible d'orientation
r сенсор m замера положения; сенсор измерения позиции

* **attraction power → 243**

243 attractiveness; attraction power
d Attraktivität f; Anziehungskraft f
f attractivité f; pouvoir m d'attraction; puissance f d'attraction
r привлекательность f; притягательная сила f

244 attribute
d Attribut n
f attribut m; caractère m qualitatif
r атрибут m; реквизит m

245 attribute accuracy
d Attributsgenauigkeit f
f exactitude f d'attribut; précision f sémantique
r точность f атрибута

246 attribute attaching
d Attributsbefestigung f
f attachement m d'attribut
r присоединение n атрибута

247 attribute class
d Attributklasse f
f classe f d'attribut
r класс m атрибута

248 attribute code; attribute ID
d Attributkennzeichen n; Attributcode m
f code m d'attribut; marque f d'attribut
r код m атрибута

249 attribute-control code
d Attributsteuercode m
f code m de commande d'attribut
r код m управления признаком

250 attribute data; attribute information
  d Attributdaten *pl*; Sachdaten *pl*
  f données *fpl* d'attribut
  r атрибутивные данные *pl*

251 attribute definition
  d Attributdefinition *f*
  f définition *f* d'attribut
  r атрибутная дефиниция *f*

252 attribute domain
  d Attributdomäne *f*
  f domaine *m* d'attribut
  r домен *m* атрибута

  * **attribute ID → 248**

  * **attribute information → 250**

253 attribute linking
  d Attributverknüpfung *f*
  f liaison *f* d'attributs
  r связывание *n* атрибутов

  * **attribute matching → 262**

254 attribute object
  d Attributobjekt *n*
  f objet *m* attribut
  r объект *m* типа атрибута

255 attribute of relation
  d Relationsattribut *n*
  f attribut *m* de relation
  r атрибут *m* отношения (*в реляционных БД*)

256 attribute option
  d Attributoption *f*
  f option *f* d'attribut
  r опция *f* атрибута

257 attribute prompt
  d Attributanzeige *f*
  f consigne *f* d'attribut
  r подсказка *f* атрибута

258 attribute reference
  d Attributverweis *m*
  f référence *f* d'attribut
  r ссылка *f* атрибута

259 attribute table
  d Attributtabelle *f*
  f table *f* d'attributs; table attributaire
  r атрибутивная таблица *f*; таблица атрибутов

260 attribute tag
  d Attributetikett *n*
  f étiquette *f* d'attribut
  r тег *m* атрибута; атрибутный тег

261 attribute tag field
  d Feld *n* des Attributetiketts
  f champ *m* d'étiquette d'attribut
  r поле *n* атрибутного тега

262 attribute tagging; attribute matching
  d Attributetikettierung *f*
  f étiquetage *m* d'attributs
  r присваивание *n* атрибута
       пространственным объектам; связывание *n*
       объектов с атрибутами; атрибутирование *n*

263 attribute update
  d Attributaktualisierung *f*
  f mise *f* en jour d'attributs
  r актуализирвание *n* атрибутов

264 attribute value
  d Attributwert *m*
  f valeur *f* d'attribut; valeur attributaire
  r значение *n* атрибута

  * **attribute-value table → 3898**

265 attributive query
  d Attributabfrage *f*
  f requête *f* attributaire; requête alphanumérique
  r атрибутивный запрос *m*

  * **AU → 210**

  * **authalic projection → 1316**

  * **authorship in cartography → 771**

266 autocorrelation; autovariance
  d Autokorrelation *f*; Eigenkorrelation *f*;
       Selbstkorrelation *f*
  f autocorrélation *f*
  r автокорреляция *f*

267 autoincrement
  d Selbstinkrement *n*
  f auto-incrément *m*
  r автоматическое увеличивание *n* [на
       единицу]

268 autoindexing
  d Selbstindizierung *f*
  f indexation *f* automatique
  r автоматическое индексирование *n*

269 auto-join
  d automatische Teilnahme *f*; automatische
       Verbindung *f*
  f union *f* automatique
  r автоматическое соединение *n*

270 autolabeling
  d automatische Etikettierung *f*

f auto-étiquetage *m*
r автоматическое присваивание *n* меток;
автоматическая запись *f* меток;
автоматическое присваивание обозначений

* **automated cartographic system** → 285

* **automated cartography** → 660

271 **automated census mapping system; ACMS**
d automatisiertes System *n* der Kartografie von
Erhebungsdaten
f système *m* automatisé de cartographie de
recensement
r автоматизированная система *f*
картографирования переписи

272 **automated chart production system; ACPS**
d automatisierte Kartenproduktionssystem *n*
f système *m* automatisé de production
cartographique
r автоматизированная система *f*
картографического производства

273 **automated data processing**
d automatisierte Datenverarbeitung *f*
f traitement *m* de données automatique
r автоматизированная обработка *f* данных

274 **automated digitizing**
d automatische Digitalisierung *f*
f digitalisation *f* automatique
r автоматизированное оцифрование *n*;
автоматическое дигитализирование *n*

275 **automated direction finding; ADF**
d automatische Funkpeilung *f*
f repérage *m* des directions automatique
r автоматическое определение *n*
направления; автоматическое определение
азимута; пеленгование *n*

276 **automated feature recognition**
d automatische Merkmalserkennung *f*
f reconnaissance *f* automatique de
caractéristiques
r автоматическое распознавание *n* признаков

277 **automated generalization**
d automatische Generalisierung *f*; automatische
Verallgemeinerung *f*
f généralisation *f* automatique
r автоматическая генерализация *f*

* **automated mapping** → 660

278 **automated mapping/facilities management
system; AM/FM system**
d AM/FM-System *n*
f système *m* MF-MA

r система *f* автоматизированной картографии
и ГИС в управлении сетями предприятий
коммунального хозяйства

279 **automated name placement; automatic
name placement**
d automatische Namenplazierung *f*
f placement *m* de noms automatisé
r автоматизированное размещение *n*
надписей

* **automatic** → 280

280 **automatic[al]**
d automatisch
f automatique
r автоматический

281 **automatic closure**
d automatischer Abschluss *m*
f fermeture *f* automatique
r автоматическое замыкание *n*

282 **automatic contouring**
d automatische Umrisszeichnung *f*,
automatisches Konturen *n*
f isolignage *m* automatique; tracé *m*
automatique des courbes de niveau
r автоматическое вычерчивание *n* контуров;
автоматическое оконтуривание *n*

283 **automatic geocoding**
d automatische Geocodierung *f*
f géocodage *m* automatique
r автоматическое геокодирование *n*

284 **automatic image transmission; automatic
picture transmission**
d automatische Bildübertragung *f*
f transmission *f* d'images automatique
r автоматическая передача *f* изображений

285 **automatic mapping system; automated
cartographic system; ACS;
computer[-aided] mapping system; CAMS**
d automatisiertes System *n* der Kartografie;
automatisiertes Kartierungssystem *n*;
automatisiertes kartografisches System;
System zur automatischen Kartierung
f système *m* de cartographie automatisé;
système de cartographie automatique
r автоматическая картографическая
система *f*; автоматизированная
картографическая система; АКС

* **automatic name placement** → 279

286 **automatic network discovery**
d automatische Netzwerkentdeckung *f*

f découverte *f* de réseau automatique
r автоматический анализ *m* конфигурации
   сети

* **automatic picture transmission** → 284

287 **automatic topology error**
   d automatischer topologischer Fehler *m*
   f erreur *f* topologique automatique
   r автоматическая топологическая
      погрешность *f*

288 **automatic vehicle location; AVL**
   d automatische Fahrzeugortung *f*; automatische
      Standortbestimmung *f* von Fahrzeugen
   f localisation *f* automatique de véhicules; LAV
   r автоматическое определение *n*
      местоположения подвижного объекта

* **autonomous** → 2615

289 **autonomous positioning**
   d autonome Positionierung *f*
   f positionnement *m* autonome
   r автономное позиционирование *n*

* **autovariance** → 266

290 **auxiliary contour**
   d Hilfshöhenlinie *f*
   f ligne *f* de niveau auxiliaire
   r вспомогательная горизонталь *f*

291 **average filter; mean filter**
   d Mean-Filter *m*
   f filtre *m* moyen
   r усредняющий фильтр *m*

* **AVHRS** → 61

* **AVL** → 288

* **AWAR** → 182

292 **axis of projection; projection axis; ground
   line**
   d Projektionsachse *f*
   f axe *m* de projection
   r ось *f* проекции; ось проектирования

293 **azimuth**
   d Azimut *m*
   f azimut *m*
   r азимут *m*

294 **azimuthal angle**
   d Azimutalwinkel *m*; Azimutwinkel *m*
   f angle *m* azimutal
   r азимутальный угол *m*

295 **azimuthal correction**
   d azimutales Eindrehen *n*; azimutale Korrektur *f*
   f rotation *f* en azimut
   r азимутальная поправка *f*

* **azimuthal orthomorphic projection** → 3488

296 **azimuthal projection; zenithal projection**
   d Azimutalprojektion *f*
   f projection *f* azimutale
   r азимутальная проекция *f*

297 **azimuth degree**
   (a unit of direction, representing 1/360 of a
   full circle)
   d Azimutalgrad *m*
   f degré *m* azimutal
   r азимутальный градус *m*

298 **azimuth resolution**
   d Azimutalauflösung *f*
   f résolution *f* azimutale
   r азимутальная разрешающая способность *f*

B

299 back azimuth; reverse azimuth
d Gegenazimut *m*
f azimut *m* inverse
r обратный азимут *m*

* **back country** → 1838

300 background
d Hintergrund *m*; Untergrund *m*
f fond *m*; arrière-plan *m*; plan *m* d'arrière;
background *m*
r фон *m*; задний план *m*

301 background color
d Hintergrundfarbe *f*; Grundfarbe *f*
f couleur *f* de fond; couleur *f* d'arrière-plan
r цвет *m* фона

302 background map
d Untergrundkarte *f*
f carte *f* de fond
r фоновая карта *f*

* **back land** → 1838

* **backscatter** → 303

303 backscatter[ing]
d Rückstreuung *f*
f rétrodiffusion *f*; diffusion *f* en arrière
r обратное рассеяние *n*

304 backsight; backward sight
(a sight taken with a level to a point of known
elevation)
d Rückenansicht *f*; Rückblick *m*
f coup-arrière *m*; visée *f* arrière; visée inverse;
visée rétrograde
r обратное визирование *n*; взгляд *m* назад;
обзор *m* заднего вида; задний отсчёт *m*

* **backup** → 151

* **backward sight** → 304

305 ballistic camera; BC
d ballistische Meßkammer *f*; ballistische
Kamera *f*
f caméra *f* balistique
r фототеодолит *m*; баллистическая
фотокамера *f*; фотокамера для внешних
траекторных измерений

* **band** → 3511

306 banding
d Bandeinpassen *n*; Siebkettung *f*
f effet *m* de bande; étirement *m*
r нанесение *n* полос; обозначение *n*
полосами

* **band interleaved by line format** → 307

**307 band interleaved by line [image] format;
BIL format**
d Band-Zeilenüberlagerungsformat *n*; Format *n*
in aufeinanderfolgenden Zeilen
f format *m* d'images à lignes entrelacées
r формат *m* передачи изображений с
построчным хранением данных; формат
BIL

* **band interleaved by pixel format** → 308

**308 band interleaved by pixel [image] format;
BIP format**
d Band-Pixelüberlagerungsformat *n*; Format *n*
aufeinanderfolgenden Grauwerten pro Pixel
f format *m* d'images à pixels entrelacés
r формат *m* передачи изображений
последовательностью значений яркости
каждого пиксела; формат BIP

309 band separation
d Bandauflösung *f*; Weiche *f*
f séparation *f* de bandes
r разделение *n* полосы частот

**310 band sequential image format; BSQ image
format**
d BSQ-Bildformat *n* (*Bildformat in drei
getrennten Bildern*)
f format *m* d'images BSQ
r формат *m* изображений BSQ

311 bandwidth
d Bandbreite *f*
f largeur *f* de bande; bande *f* passante
r ширина *f* полосы [ленты]

312 bandwidth compression
d Bandbreitenkompression *f*;
Bandbreitenreduzierung *f*;
Bandbreitenverdichtung *f*
f compression *f* de largeur de bande;
compression des signaux
r сжатие *n* полосы частот; сжатие спектра

313 bandwidth expansion
d Bandbreitenerweiterung *f*
f extension *f* du spectre; étalement *m* de la
largeur de bande
r расширение *n* полосы частот

* **bar** → 3511

314 bar; stroke; dash
  d Strich *m*; Stab *m*
  f barre *f*; trait *m*; raie *f*
  r черта *f*; штрих *m*; тире *n*

315 bar chart; column graph; bar diagram; bar
    graph; column diagram; histogram
  d Streifendiagramm *n*; Balkengrafik *f*;
    Balkendiagramm *n*; Histogramm *n*
  f diagramme *m* à colonnes; diagramme à tuyaux
    d'orgue; histogramme *m*
  r столбчатая диаграмма *f*; столбиковая
    диаграмма; гистограмма *f*

* **bar diagram** → 315

* **bar graph** → 315

* **bar scale** → 1744

316 base/height ratio
  d Basishöhe-Verhältnis *n*
  f rapport *m* base-éloignement; rapport
    base/hauteur
  r отношение *n* основания к высоте

317 baseline
  d Basislinie *f*; Grundlinie *f*; Standlinie *f*
  f ligne *f* de base
  r базисная линия *f*; основная линия

* **base map** → 3706

* **base sheet** → 2659

318 base station; reference station
  d Referenzstation *f*
  f station *f* de référence; station de base
  r базовая станция *f*; референц-станция *f*;
    опорная станция

319 base table
  d Basistabelle *f*
  f table *f* de base
  r базовая таблица *f*

320 basic cartographic unit; basic map unit
  d kartografische Grundeinheit *f*;
    Grundkarteneinheit *f*
  f unité *f* cartographique de base
  r основная картографическая единица *f*

321 basic geodetic survey
  d Basisvermessung *f*;
    Grundvermessungsdienst *m*
  f levé *m* géodésique de base; service *m*
    géodésique de base

  r основные геодезические работы *fpl*

322 basic land unit
  d Grundlandeinheit *f*
  f unité *f* de terrain de base
  r основная земельная единица *f*

* **basic map unit** → 320

323 basic spatial unit
  d Grundraumeinheit *f*
  f unité *f* spatiale de base
  r базовая пространственная единица *f*

324 bathymetric[al] map; bathymetric chart
  d bathymetrische Karte *f*
  f carte *f* bathymétrique
  r батиметрическая карта *f*

* **bathymetric chart** → 324

* **bathymetric map** → 324

325 bathymetric survey
  d bathymetrische Vermessung *f*
  f levé *m* bathymétrique; levé du fond
  r батиметрическая съёмка *f*

326 bathymetry
  d Bathymetrie *f*
  f bathymétrie *f*
  r батиметрия *f*; батиметрические данные *pl*;
    измерение *n* глубин

327 baud rate
  d Baudrate *f*; Baud-Zahl *f*
  f débit *m* en bauds
  r скорость *f* [передачи данных] в бод

328 Bayesian methods
  d Bayes-Methoden *fpl*
  f méthodes *fpl* de Bayesian
  r байесовские методы *mpl*

* **BC** → 305

* **bearing** → 329

329 bearing [angle]; direction angle; grid
    azimuth; grid bearing; Y-azimuth
  d Richtungswinkel *m*
  f angle *m* directeur; gisement *m*
  r дирекционный угол *m*; направляющий угол

* **bearing point** → 1112

330 bed
  d Lagerung *f*; Ladefläche *f*
  f banc *m*; berceau *m*
  r пласт *m*; русло *n*; основание *n*

* **bedding plane** → 2777

* **bed geometry** → 524

331 beginning arc
d Anfangsbogen *m*
f arc *m* initial
r начальная дуга *f*

* **beginning point** → 3468

332 behavioral matrix
d Verhaltensmatrix *f*
f matrice *f* de comportement
r матрица *f* поведения

* **below-sea-level contour** → 979

* **belt-bed plotter** → 1186

* **Beltrami mapping** → 1551

* **Beltrami's mapping** → 1551

333 benchmark; survey mark
d Benchmark *n*; Bezugsmarke *f*; Prüfmarke *f*;
  Abrisspunkt *m*; Vergleichspunkt *m*
f repère *m* [de nivellement]; banc *m* d'essai;
  marque *f* de jalonnement; borne-repère *f*
r нивелирный пункт *m*; репер *m*

334 benchmarking
d Leistungsvergleich *m*; Benchmarking *n*
f étalonnage *m* de performances; évaluation *f* de
  performance; analyse *f* comparative;
  étalonnage concurrentiel; référenciation *f*
r разметка *f*; установление *n* контрольных
  точек; эталонное тестирование *n*;
  получение *n* контрольных характеристик

335 best linear unbiased estimator; BLUE
d beste lineare nichtverzerrende
  Schätzfunktion *f*
f fonction *f* d'estimation linéaire sans biais;
  meilleur estimateur *m* linéaire sans biais
r наилучшая линейная несмещённая
  функция *f* оценки

336 beta index
(the ratio between the number of links (edges)
and the number of nodes in a network)
d Beta-Index *m*
f indice *m* bêta
r бета-индекс *m*

337 Bézier curve
d Bézier-Kurve *f*
f courbe *f* de Bézier
r кривая *f* Безье

338 Bézier polygon
d Bézier-Vieleck *n*
f polygone *m* de Bézier
r многоугольник *m* Безье

339 Bézier surface
d Bézier-Fläche *f*
f surface *f* de Bézier
r поверхность *f* Безье

340 Bhattacharya distance
(the distance between classes distribution
centers)
d Bhattacharya-Distanz *f*
f distance *f* de Bhattacharya
r расстояние *n* Бхаттачарья

341 bibliographic data
d bibliografische Daten *pl*
f données *fpl* bibliographiques
r библиографические данные *pl*

342 bibliographic indexing
d bibliografische Indizierung *f*
f indexage *m* bibliographique
r библиографическое индексирование *n*

343 bicubic interpolator
d bikubischer Interpolator *m*
f interpolateur *m* bicubique
r бикубический интерполятор *m*

344 bicubic surface
d bikubische Fläche *f*
f surface *f* bicubique
r бикубическая поверхность *f*

* **BIL format** → 307

345 bilinear interpolation
d bilineare Interpolation *f*
f interpolation *f* bilinéaire
r билинейная интерполяция *f*

346 bilinear interpolator
d bilinearer Interpolator *m*
f interpolateur *m* bilinéaire
r билинейный интерполятор *m*

347 binary code
d binärer Code *m*
f code *m* binaire
r двоичный код *m*

348 binary grid
d binäres Gitter *n*
f grille *f* binaire
r двоичная сетка *f*

349 binary large object; BLOB
    (of data)
    d großes binäres Objekt *n*; Blob *n*
    f grand objet *m* binaire
    r большой двоичный объект *m*; "блоб" *m*

350 binary line generalization; BLG
    d binäre lineare Generalisierung *f*
    f généralisation *f* linéaire binaire
    r двоичная линейная генерализация *f*

351 binary overlay
    d binäres Overlay *n*
    f calque *m* binaire
    r двоичный оверлей *m*

352 binary raster
    d binärer Raster *m*
    f rastre *m* binaire
    r двоичный растр *m*; бинарный растр

353 binary raster data
    d binäre Rasterdaten *pl*; Deckerfolie *f*; Folie *f*
    f données *fpl* de trame binaires
    r двоичные растровые данные *pl*

354 binary space partitioning
    d Halbebenenunterteilung *f*
    f partition *f* binaire d'espace
    r двоичное пространственное разбиение *n*

    * **bind → 2199**

    * **binding → 2199**

    * **biogeography → 536**

    * **biom → 355**

355 biom[e]
    d Biom *n*
    f biome *m*
    r биом *m*

356 bioregion
    d Bioregion *f*
    f biorégion *f*
    r биорегион *m*

    * **biotop → 1797**

    * **BIP format → 308**

357 bit
    d Bit *n*
    f bit *m*
    r бит *m*

    * **bitmap → 3015**

    * **bitmap image → 3015**

    * **bitmap model → 3009**

    * **bitmap object → 3019**

358 bit plane
    (a gridded memory in graphics device used for storing information for display)
    d Bit-Slice *n*; Bit-Scheibe *f*
    f tranche *f* de bits
    r плоскость *f* элементов отображения; разрядная матрица *f*; битовая плоскость *f*; битовый слой *m*

359 black-and-white aerial photograph; monochrome aerial photograph
    d Schwarzweiß-Aerofotografie *f*
    f aérophoto *f* noire et blanche
    r чёрно-белый аэрофотоснимок *m*; монохромный аэрофотоснимок

360 blank line
    d Leerzeile *f*
    f ligne *f* blanche; ligne vide
    r пустая строка *f*

    * **BLG → 350**

361 blind digitizing
    d blindes Digitalisieren *n*; Blindendigitalisierung *f*; nichtvisuelle Digitalisierung *f*
    f numérisation *f* cachée
    r слепое оцифрование *n*; анонимное дигитализирование *n*

    * **BLOB → 349**

    * **block → 3150**

362 block-based address assignment
    d block-basierte Adresszuordnung *f*
    f adressage *m* par blocs
    r поблочное присваивание *n* адреса

363 block boundary
    d Blockgrenze *f*
    f limite *f* d'arrondissement
    r граница *f* квартала

    * **block-diagram → 1443**

    * **blockette → 3517**

    * **block group → 489**

    * **blocking → 2222**

364 **block name**
 d Blockname *m*
 f nom *m* de bloc
 r имя *n* блока; имя переписного участка

365 **block numbering**
 d Blocknummerierung *f*
 f numérotage *m* de bloc; numérotage de quartier
 r нумерование *n* квартала; нумерация *f* квартала; нумерация переписного участка

366 **block numbering area; BNA**
 d Blockkennnummeringsgebiet *n*; Bereich *m* mit Blocknummerierung
 f zone *f* de numérotage de blocs
 r зона *f* нумерации переписного участка

 * **block relationship file** → 502

 * **block scheme** → 1443

 * **BLUE** → 335

367 **blunder; rough error**
 (error, usually due to human operators, whose magnitude cannot be modelled by regular statistical models)
 d Rohfehler *m*
 f erreur *f* grossière
 r грубая погрешность *f*

 * **blurred** → 1495

 * **BNA** → 366

 * **body** → 3339

368 **bog; mire; swamp; marsh**
 d Sumpf *m*; Morast *m*; Moor *n*
 f tourbière *f*; marais *m*
 r трясина *f*; болото *n*; грязь *f*

369 **Boolean expression; logical expression**
 d Boolescher Ausdruck *m*
 f expression *f* booléenne; expression de Boole
 r булево выражение *n*

370 **Boolean image**
 d Boolesches Bild *n*
 f image *f* booléenne
 r булево изображение *n*

371 **Boolean operations**
 d Boolesche Operationen *fpl*
 f opérations *fpl* booléennes
 r булевы операции *fpl*

372 **Boolean operator; logical operator**
 d Boolescher Operator *m*
 f opérateur *m* booléen

 r булев оператор *m*

373 **Boolean retrieval**
 d Boolesche Suche *f*
 f recherche *f* booléenne
 r булев поиск *m*

374 **border; frame**
 d Kante *f*; Rahmen *m*
 f bordure *f*; bord *m*
 r рамка *f*; полоса *f*; кайма *f*

375 **border arcs; boundary arcs**
 (the arcs that create the outer edge boundary of a polygon coverage)
 d Randbögene *mpl*
 f arcs *mpl* de frontière
 r дуги *fpl* границы; граничные дуги

 * **bordering** → 728

376 **border of a map; map border; margin of a map; map margin; frame of a map; map frame**
 d Kartenrand *m*
 f filet *m* de cadre; filet à border
 r рамка *f* листа карты

 * **bottom contour** → 979

 * **bottom note** → 1454

 * **bottom-up** → 188

 * **bound** → 377

377 **boundary; bound**
 d Grenze *f*; Rand *m*
 f limite *f*; frontière *f*
 r граница *f*

 * **boundary arcs** → 375

378 **boundary changes**
 d Grenzänderungen *fpl*
 f échanges *fpl* de limite
 r изменения *npl* границы

379 **boundary correspondence**
 d Ränderzuordnung *f*
 f correspondance *f* des frontières
 r соответствие *n* границ; соотнесенность *f* границ

380 **boundary edge; bounding edge**
 d Randkante *f*
 f arête *f* frontière; côte *f* frontière
 r граничное ребро *n*

381 boundary effect
d Auswirkung *f* von Grenzen;
Grenzbeeinflussung *f*
f influence *f* de la limite; effet *m* au joint
r краевой эффект *m*

382 boundary extraction
d Grenzextraktion *f*; Konturerkennung *f*
f extraction *f* de frontière
r извлечение *n* границы

383 boundary file
d Boundary-File *n*
f fichier *m* de limites
r файл *m* данных граничных условий

384 boundary-following algorithm
d Grenzfolgenalgorithmus *m*
f algorithme *m* de filage de limite
r алгоритм *m* слежения границы; алгоритм
слежения контура

385 boundary line; demarcation line
d Grenzlinie *f*; Begrenzungslinie *f*
f ligne *f* de frontière
r демаркационная линия *f*

* **boundary mark → 386**

* **boundary marker → 386**

* **boundary marking → 957**

386 boundary monument; monument;
boundary sign; boundary mark[er]
(a ground located structure marking an
accurately surveyed position on a boundary
line separating two defined regions)
d Umgrenzungsmarker *m*; Grenzmarke *f*;
Grenzzeichen *n*; Grenzen-Kartenzeichen *n*
f balise *f* du terrain; balise de délimitation;
borne *f* frontière; signe *m* de démarcation;
monument *m*
r граничный геодезический знак *m*;
граничный грунтовой репер *m*; маркер *m*
границ

387 boundary of an area; area boundary; zone
boundary
d Gebietsgrenze *f*; Rand *m* eines Bereichs; Rand
eines Gebiets
f limite *f* de la superficie
r граница *f* области; граница территории;
граница зоны

388 boundary of a surface
d Rand *m* einer Fläche
f bord *m* d'une surface
r граница *f* поверхности

389 boundary of particular area
d Rand *m* eines partikulären Bereichs
f frontière *f* d'un domaine particulier; frontière
d'une zone particulière
r граница *f* отдельной области

390 boundary point; frontier point
d Randpunkt *m*; Grenzpunkt *m*
f point *m* frontière; point limite
r граничная точка *f*; краевая точка

391 boundary representation; B-Rep
d Randdarstellung *f*
f représentation *f* de limites
r представление *n* границ

392 boundary set
d Randsatz *m*
f ensemble *m* de frontières
r множество *n* граней

* **boundary sign → 386**

393 boundary survey
d Grenzvermessung *f*
f mensuration *f* pour le rétablissement des
limites
r межевание *n*

* **bounded box → 395**

* **bounding box → 395**

394 bounding coordinates; bounds of
coordinate system
d begrenzte Koordinaten *fpl*
f coordonnées *fpl* délimitées
r ограничивающие координаты *fpl*

* **bounding edge → 380**

395 bounding rectangle; bounding box;
bounded box
d abgeschlossener Objektbereich *m*;
Grenzbereich *m*; begrenztes Kästchen *n*;
begrenztes Rechteck *n*
f zone *f* de délimitation; cadre *m* d'objet; boîte *f*
délimitée; case *f* délimitée
r ограничивающий прямоугольник *m*;
ограничивающая область *f* объектов;
кадр *m* объектов; ограничивающий ящик *m*

* **bounds of coordinate system → 394**

396 Bowditch rule
(for traverse adjustment)
d Bowditch-Regel *f*
f règle *f* de Bowditch
r правило *n* Боудитча

397 boxcar classification; boxcar interpretation
  d Boxcar-Klassifizierung *f*
  f classification *f* d'échantillonneur monocanal
  r классификация *f* блоков узкополосных фильтров; классификация интеграторов с узкополосным фильтром

  * **boxcar interpretation** → 397

  * **box plot** → 1443

398 branches of thematic mapping
  d Zweige *mpl* der thematischen Kartografierung
  f branches *fpl* de cartographie thématique
  r виды *mpl* тематического картографирования; отрасли *fpl* тематического картографирования

399 breadth-first search
  (algorithms to traverse a tree that explore the structure by enumerating each level totally before moving to a finer level)
  d breitenorientierte Suche *f*, Breitensuche *f*
  f recherche *f* en largeur [d'abord]
  r поиск *m* [преимущественно] в ширину; перебор *m* в ширину

400 break line
  d Bruchlinie *f*, Bruchkante *f*
  f ligne *f* à casser; ligne de cassure
  r линия *f* разъёма; линия отрыва; барьер *m*

401 breakpoint
  d Unterbrechungsstelle *f*, Bedarfsschaltepunkt *m*
  f point *m* d'interruption
  r точка *f* прерывания

  * **B-Rep** → 391

402 bridge
  d Brücke *f*
  f pont *m*
  r мост *m*

403 bridge
  (in mathematical topology)
  d Brücke *f*
  f pont *m*
  r замощение *n*; перешеек *m*; связывающее ребро *n*

404 brightness; luminosity
  d Helligkeit *f*, Heiterkeit *f*
  f éclairement *m*; luminosité *f*
  r освещённость *f*, светлота *f*, яркость *f*, блеск *m*

  * **browser** → 3963

405 browser window
  d Browser-Fenster *n*
  f fenêtre *f* de butineur
  r окно *n* браузера

406 browsing
  d Browsing *n*; Durchsicht *f*
  f revue *f*; survol *m*; feuilletage *m*
  r пролистывание *n*; покадровый просмотр *m*; броузинг *m*

  * **BSQ image format** → 310

  * **bubble chart** → 407

407 bubble graph; bubble chart
  d Blasendiagramm *n*
  f graphique *m* à bulles
  r пузырьковая диаграмма *f*, схема *f* [процесса], изображаемая кружочками и стрелками

  * **buffer** → 412

  * **buffer area** → 412

408 buffer calculation
  d Pufferberechnung *f*, Bufferberechnung *f*
  f calcul *m* de tampon
  r вычисление *n* буфера

409 buffer distance
  d Pufferdistanz *f*, Buffer-Distanz *f*
  f marge *f* de tampon
  r "буферное" расстояние *n*

410 buffering
  d Pufferung *f*, Puffern *n*
  f tamponnage *m*
  r буферирование *n*; буферизация *f*

411 buffer size
  d Puffergröße *f*
  f taille *f* de tampon
  r размер *m* буфера

412 buffer [zone]; buffer area
  (a zone constructed outwards from an isolated object to a specific distance)
  d Pufferzone *f*, Buffer-Zone *f*, Distanzbereich *m*
  f zone *f* tampon; couloir *m*
  r буферная зона *f*, буфер *m*

413 bug
  d Entwurfsfehler *m*
  f bogue *m*; défaut *m* de conception
  r ошибка *f* при проектировании; программная ошибка *f*, "вошь" *f*

414 building area; building zone; constructible zone
  d Grundfläche *f*, Baufläche *f*

f espace *m* à bâtir; zone *f* à bâtir; zone de
construction; zone constructible
r площадь *f* застройки; район *m* под
застройку

415 building limit
  d Baugrenze *f*
  f limite *f* à bâtir
  r граница *f* застройки

  * **building line** → 1487

  * **building zone** → 414

416 business GIS; business-mapping system
  d Business-Mapping-System *n*
  f SIG *m* commercial; système *m* de
business-mapping
  r корпоративная ГИС *f*

417 business graphics; management graphics
  d Geschäftsgrafik *f*
  f infographie *f* d'entreprise
  r деловая графика *f*

418 business graphics data
  d Geschäftsgrafikdaten *pl*
  f données *fpl* de graphique d'entreprise
  r данные *pl* деловой графики

  * **business-mapping system** → 416

  * **butt** → 997

419 bypass
  (of a street)
  d Umgehung *f*; Umgehungsstraße *f*
  f dérivation *f*; bi-passe *m*
  r обходный путь *m*; объезд *m*; перепуск *m*;
объездная дорога *f*

420 byte
  d Byte *n*
  f octet *m*
  r байт *m*

C

* **CAD** → 661, 665

* **cadaster** → 422

421 **cadastral class**
 d Katasterklasse f
 f classe f cadastrale
 r кадастровый класс m

* **cadastral map** → 2652

* **cadastral plan** → 2652

422 **cadastral register; land [survey] register; land registry; register of real estate; land records; [real estate] cadaster; cadastre**
 d Grundbuch n; Grundkataster m; Liegenschaftskataster m; Kataster m; Grundsteuerregister n; Flurbuch m
 f registre m terrien; registre hypothécaire; registre du cadastre; cadastre m [d'immeubles]; livre m foncier
 r кадастр m; поземельная книга f

* **cadastral registration** → 2090

* **cadastral survey** → 3723

423 **cadastral topographic cartographic information system**
 d amtliches topografisch-kartografisches Informationssystem n; ATKIS
 f système m d'information cartographique-topographique cadastral
 r государственная топографо-картографическая информационная система f

* **cadastre** → 422

* **CADD** → 662

424 **calibrated image map**
 d kalibrierte Imagemap f
 f image f réactive calibrée
 r калиброванное адресуемое отображение n

425 **calibrated line**
 d kalibrierte Leitung f; berichtigte Leitung
 f ligne f corrigée; ligne calibrée
 r градуированная линия f; калиброванная линия

426 **calibration**
 d Kalibrierung f; Eichung f
 f calibration f
 r калибровка f

427 **calibration analysis**
 d Kalibrierungsanalyse f
 f analyse f de calibration
 r анализ m калибровки; анализ градуировки

* **callout** → 428

428 **callout [figure]**
 (label with a pointer)
 d Referenzzahl f
 f chiffre-référence f
 r выноска f; метка-идентификатор f

* **CAM** → 660

* **CAMA** → 663

* **CAMS** → 285

* **CAP** → 664

* **cap** → 3298

* **capsulation** → 1263

429 **capture**
 d Einfang m; Fangen n; Sammlung f; Sammeln n
 f saisie f; capture f; captage m
 r захват m; улавливание n

* **CAR** → 666

430 **cardinal direction**
 d Kardinalrichtung f
 f direction f cardinale
 r кардинальное направление n

431 **cardinal point; Gauss point**
 d Kardinalpunkt m
 f point m cardinal
 r кардинальная точка f

432 **carrier-phase tracking**
 (in GPS)
 d Trägerphasen-Verfolgung f
 f poursuite m de phase [de l'onde] porteuse
 r слежение n за фазой несущей

* **carte** → 1611

433 **Cartesian coordinates; grid coordinates**
 d Kartesische Koordinaten fpl; Gitterkoordinaten fpl; Rechtwinkelkoordinaten fpl

f coordonnées *fpl* cartésiennes; coordonnées rectangulaires
r декартовы координаты *fpl*; прямоугольные координаты

434 Cartesian coordinate system
d Kartesisches Koordinatensystem *n*
f système *m* de coordonnées cartésiennes
r декартова система *f* координат

* **cartogram → 539**

435 cartographer; mapmaker
d Kartograph *m*; Kartenzeichner *m*
f cartographe *m*
r картограф *m*

* **cartographic → 436**

436 cartographic[al]
d kartografisch; kartographisch
f cartographique
r картографический

* **cartographical grid → 2274**

437 cartographic catalog
d kartografischer Katalog *m*
f catalogue *m* cartographique
r картографический каталог *m*

438 cartographic clipart
d kartografische Clipart *f*
f clipart *m* cartographique; graphique *m* prédessiné cartographique
r картографический шаблон *m* для рисунка; картографическая иллюстративная вставка *f*; картографическая аппликация *f*

439 cartographic communication; communication in cartography
d kartografische Kommunikation *f*
f communication *f* cartographique
r картографическая коммуникация *f*

440 cartographic control
d kartografische Steuerung *f*
f contrôle *m* cartographique
r картографическое управление *n*

* **cartographic data → 2263**

441 cartographic databank
d kartografische Datenbank *f*
f banque *f* de données cartographiques
r картографический банк *m* данных

442 cartographic database; CDB; [digital] map database
d Kartendatenbasis *f*

f base *f* de données cartographiques
r картографическая база *f* данных; база картографических данных

443 cartographic data visualizer; CDV
d kartografischer Datenbrowser *m*
f visionneur *m* de données cartographiques
r визуализатор *m* картографических данных

444 cartographic design; map design
d kartografisches Design *n*; Kartenprojektierung *f*
f projet *m* cartographique
r картографический дизайн *m*; [художественное] проектирование *n* карт

445 cartographic drawing; map drawing
d kartografische Zeichnung *f*; Kartenzeichnen *n*
f dessin *m* cartographique
r картографическое черчение *n*; картографический чертёж *m*

446 cartographic education; cartographic training
d kartografische Bildung *f*
f enseignement *m* cartographique
r картографическое образование *n*

447 cartographic environment
d kartografische Umwelt *f*; kartografische Umgebung *f*
f environnement *m* cartographique
r картографическая среда *f*

448 cartographic feature; [component] element of map; map feature; feature
(a group of spatial elements which together represent a real-world entity. Often used as a synonym for the term object)
d Feature *n*; Grundzug *m*; Zug *m*
f détail *m* cartographique; trait *m*; entité *f* cartographique
r составная часть *f* карты; элемент *m* карты; картографический элемент

449 cartographic feature file; CFF
d Grundzugsdatei *f*
f fichier *m* de détails cartographiques
r файл *m* [формата] CFF

450 cartographic generalization
d kartografische Generalisierung *f*; kartografische Verallgemeinerung *f*
f généralisation *f* cartographique
r картографическая генерализация *f*

* **cartographic image → 456**

451 cartographic information
d kartografische Information *f*

f information *f* cartographique
r картографическая информация *f*

452 cartographic information retrieval system
  d kartografisches Datenbediensystem *n*;
    kartografisches
    Informationsrückgewinnungssystem *n*
  f système *m* de recherche d'information
    cartographique
  r картографическая
    информационно-поисковая система *f*

453 cartographic lettering; map lettering
    inscription; lettering
  d Kartenbeschriftung *f*; Kartenschriften *n*;
    Beschriftung *f*
  f inscription *f* cartographique; écriture *f*
    cartographique; composition *f* d'écritures
  r картографическая надпись *f*; текстовое
    сопровождение *n*

454 cartographic method of research
  d kartografische Forschungsmethode *f*
  f méthode *f* de recherche cartographique
  r картографический метод *m* исследования

455 cartographic modeling; geomodeling
  d kartografisches Modellierung *f*
  f modélisation *f* cartographique
  r геомоделирование *n*

456 cartographic pattern; cartographic image
  d kartografisches Bild *n*
  f image *f* cartographique
  r картографический образ *m*

  * **cartographic printing → 2311**

  * **cartographic production → 2901**

  * **cartographic projection → 2314**

457 cartographic representation
  d kartografische Darstellung *f*
  f représentation *f* cartographique
  r картографическое представление *n*

  * **cartographic scale → 2326**

458 cartographic scanner
  d kartografischer Scanner *m*
  f scanner *m* cartographique
  r картографический сканер *m*

459 cartographic symbols; map symbols;
    conventional signs
  d Kartenzeichen *npl*; kartografische
    Symbole *npl*; Landkartensymbole *npl*;
    Signaturen *fpl*; Planzeichen *npl*
  f symboles *mpl* cartographiques

r условные картографические знаки *mpl*;
    картографические обозначения *npl*;
    символические обозначения на карте

460 cartographic technique
  d Kartentechnik *f*
  f technique *f* cartographique
  r картографическая техника *f*

461 cartographic toponymy
  d kartografische Toponymie *f*
  f toponymie *f* cartographique
  r картографическая топонимика *f*

  * **cartographic training → 446**

462 cartographic visualization
  d kartografische Visualisierung *f*
  f visualisation *f* cartographique
  r картографическое визуализирование *n*

463 cartographic work format
  d kartografisches Arbeitsformat *n*
  f format *m* de travail cartographique
  r формат *m* картографического произведения

464 cartographic workstation
  d kartografische Arbeitsstation *f*
  f station *f* de travail cartographique
  r картографическая рабочая станция *f*

465 cartography; mapping science
  d Kartografie *f*; Kartographie *f*;
    Landkartentechnik *f*
  f cartographie *f*
  r картография *f*

466 cartology
  d Kartenkunde *f*
  f cartologie *f*
  r картоведение *n*

467 cartometric index; cartometric parameter
  d kartometrischer Index *m*
  f indice *f* cartométrique
  r картометрический показатель *m*

468 cartometric operations; CO
  d kartometrische Operationen *fpl*
  f opérations *fpl* cartométriques
  r картометрические операции *fpl*

  * **cartometric parameter → 467**

469 cartometry
  d Kartometrie *f*
  f cartométrie *f*
  r картометрия *f*

470 cascaded menu; cascading menu
  d überlappendes Menü *n*

f menu *m* [en] cascade
r каскадное меню *n*; меню каскада

471 cascading
d kaskadiert; überlappend
f en cascade
r каскадный

* **cascading menu** → 470

472 cascading windows; overlaid windows
d kaskadierte Fenster *npl*
f fenêtres *fpl* en cascade
r каскадно-расположенные окна *npl*

473 case
d Fach *n*; Register *n*
f case *f*; registre *m*
r регистр *m*

* **casual error** → 16

474 categorical coverage
(an exhaustive partitioning of a
two-dimensional region into arbitrarily shaped
zones that are defined by membership in a
particular category of a classification system)
d grundsätzliche Überdeckung *f*; kategorische
Bodenbedeckung *f*
f couverture *f* catégorielle
r категорийное покрытие *n*

475 categorical data
d kategorische Daten *pl*
f données *fpl* catégoriques
r категорийные данные *pl*

476 category
d Kategorie *f*
f catégorie *f*
r категория *f*

* **catenary** → 477

477 catenary [line]; overhead contact line
d Kettenfahrleitung *f*; Kettenlinie *f*;
Kettenkurve *f*; Fahrleitung *f*;
Kettendurchhang *m*
f ligne *f* caténaire; caténaire *f*; courbe *f* de la
chaînette; chaînette *f*; ligne de contact à
suspension caténaire
r цепная линия *f*

* **CCD** → 491

* **CDB** → 442

* **CDP** → 493

* **CDV** → 443

478 celestial chart; star map; stellar chart
d Sternkarte *f*
f carte *f* céleste
r небесная карта *f*

479 celestial equator
d Himmelsäquator *m*
f équateur *m* céleste
r небесный экватор *m*

* **celestial geodesy** → 3198

480 celestial globe
d Himmelserdball *m*
f globe *m* céleste
r небесный глобус *m*

481 cell
d Zelle *f*
f cellule *f*
r ячейка *f*

482 cell height
d Zellenhöhe *f*
f hauteur *f* de cellule
r высота *f* ячейки

483 cell size
d Zellengröße *f*
f taille *f* de cellule; grandeur *f* de cellule
r размер *m* ячейки

484 cell size calibration
d Zellengrößekalibrierung *f*
f calibration *f* de taille de cellule
r калибровка *f* размера ячейки

485 cell value
d Zellwert *m*
f valeur *f* d'une cellule
r значение *n* ячейки

486 cell width
d Zellenbreite *f*
f largeur *f* de cellule
r широта *f* ячейки

487 census; enumeration
d Zensus *m*; Aufzählung *f*; Abzählung *f*;
Enumeration *f*
f dénombrement *m*; énumération *f*
r перечисление *n*; пересчёт *m*; учёт *m*
численности

* **census** → 497

* **census area** → 499

488 census block
d Volkszählungsblock *m*

f bloc *m* de recensement
r блок *m* переписи

489 census block group; block group
(a set of census blocks)
d Volkszählungsblockgruppe *f*
f groupe *m* de blocs de recensement
r группа *f* блоков переписи

490 census code
d Volkszählungscode *m*
f code *m* de recensement
r код *m* переписи

491 census county division; CCD
d Gebietseinteilung *f* für eine Erhebung;
Aufteilung *f* eines Erhebungsbezirks
f division *f* de recensement; bureau *m* local de
recensement
r переписной участок *m* округа

492 census data
d Volkszählungsdaten *pl*
f données *fpl* de recensement
r данные *pl* переписей

493 census designated place; CDP
d Arbeitsstättenzählung *f*
f place *f* [destinée] de recensement
r местоназначение *n* переписи

494 census feature class code; CFCC
d fachlicher Erhebungsbereich *m*
f code *m* spécifique de classe de recensement
r код *m* класса признаков переписи

495 census geography
d Volkszählungsgeografie *f*
f géographie *f* de recensement
r география *f* переписи

496 census map
d Volkszählungskarte *f*; Zensuskarte *f*
f carte *f* de recensement
r карта *f* переписей

**497 census of population; population census;
census**
d Bevölkerungszählung *f*; Volkszählung *f*
f recensement *m* de population
r перепись *f* [населения]; народосчисление *n*

* census region → 499

498 census subarea
d Volkszählungsunterbereich *m*
f sous-région *f* de recensement
r подобласть *f* переписи

499 census tract; CT; census region; census

area; enumeration district; enumeration
area
d Zensusgebiet *n*; Zählbezirk *m*;
Volkszählungsregion *f*;
Volkszählungsbereich *m*;
Abzählungsbezirk *m*;
Enumerationsbezirk *m*
f secteur *m* de recensement; secteur de
dépouillement; région *f* de recensement;
district *m* de recensement
r переписной район *m*; область *f* переписи;
переписной участок *m*; счётный участок

500 census tract boundary
d Zensusgebietsgrenze *f*
f limite *f* de secteur de recensement
r граница *f* переписного района

501 census tract relationship
d Zensusgebietsbeziehung *f*
f relation *f* entre secteurs de recensement
r взаимосвязь *f* между переписными
районами

**502 census tract relationship file; block
relationship file**
d Zensusgebietsbeziehung-Datei *f*
f fichier *m* de relations entre secteurs de
recensement
r файл *m* взаимосвязей между переписными
районами

503 census tract report
d Zensusgebietsbericht *m*
f rapport *m* de secteurs de recensement
r перечень *f* переписных районов с
указанием их границ

504 census tract street index; CTSI
d Zensusgebiet-Straßenindex *m*
f index *m* CTSI
r индекс *m* улиц переписных районов

505 center; centre
d Zentrum *n*
f centre *m*
r центр *m*

506 cent[e]ring
d Zentrierung *f*
f centrage *m*
r центрирование *n*

507 centerline; centreline
d zentrale Leitung *f*; Mittellinie *f*; Mittelachse *f*
f ligne *f* centrale; ligne de centre; ligne axiale;
axe *m*
r центральная линия *f*; линия центров; осевая
линия; ось *f* центров; центровая линия

508 centerline data; CLD
d Mittelachsedaten *pl*; Mittelliniedaten *pl*
f données *fpl* de ligne de centre
r данные *pl* центровой линии; данные
   центральной линии

* **center of homology → 510**

509 center of mass; mass center
d Massenmittelpunkt *m*
f centre *m* de masse
r центр *m* массы; центр инерции

**510 center of perspective; center of
    perspectivity; center of projection; center
    of homology**
d Zentrum *n* der Perspektive;
   Projektionszentrum *n*; Mittelpunkt *m* der
   Kollineation; Homologiezentrum *n*
f centre *m* de perspective; centre perspectif;
   centre de projection; centre de collinéation;
   centre d'homologie
r центр *m* перспективы; центр проекции;
   центр проектирования; центр гомологии

* **center of perspectivity → 510**

* **center of projection → 510**

511 centerpoint; centrepoint; central point
d Mittelpunkt *m*
f point *m* central
r центральная точка *f*; точка центра

* **central cartographic databank → 513**

**512 central city; central place; major centre;
    central town; urban nucleus; urban core**
d Zentralstadt *f*; Stadtkern *m*; Stadtmitte *f*
f noyau *m* urbain; ville *f* centrale;
   ville-centre *m*; localité *f* centrale; place *f*
   centrale
r главный город *m* (*района, региона*);
   город-центр *m*; [старая,] центральная
   часть *f* города

513 central[ized] cartographic databank
d zentralisierte kartografische Datenbank *f*
f banque *f* de données cartographiques
   centralisée
r централизованный картографический
   банк *m* данных

* **central meridian → 2889**

* **central place → 512**

* **central point → 511**

* **central projection → 690**

514 central tendency
d Zentraltendenz *f*
f tendance *f* centrale
r средняя тенденция *f*; центр *m*
   распределения

* **central town → 512**

* **centre → 505**

* **centreline → 507**

* **centrepoint → 511**

* **centring → 506**

* **centroid → 159**

* **centroid point → 159**

* **CEP → 544**

515 certificate of title
d Bescheinigung *f* über den Titel; Zeugnis *n*
f certificat *m* de titre [de propriété]
r сертификат *m* титула; свидетельство *n*
   правооснования

* **CFCC → 494**

* **CFF → 449**

* **CGI → 668, 670**

516 chain
   (a directed set of non-intersecting line
   segments with nodes at each end and reference
   to left and right polygons)
d Kette *f*; Folge *f*; Kettenzug *m*
f chaîne *f*
r [мерная] цепь *f*; простой список *m*;
   цепочка *f*

517 chain carrier
d Kettenübertrager *m*
f propagateur *m* de chaînes
r носитель *m* цепи; активный центр *m*
   реакции

518 chain code
d Kettencode *m*; rekurrenter Code *m*
f code *m* d'enchaînement; code en chaîne
r цепной код *m*

519 chain encoding
d Kettencodierung *f*

f codage *m* de chaîne
r цепное кодирование *n*

520 chain of triangles
d Dreieckskette *f*
f chaîne *f* de triangles
r цепь *f* треугольников; ряд *m* треугольников

521 chain service area
d Verkehrsgebiet *n*; Bedienungsgebiet *n*; Versorgungsbereich *m*
f aire *f* de service; zone *f* de service; zone de couverture
r обслуживаемая зона *f*; зона обслуживания

522 change detection
(of sequential images)
d Änderungserkennung *f*
f détection *f* de changements
r выявление *n* изменений

523 change image
(an image produced using raster algebra that shows change over time between coregistered images)
d Änderungsbild *n*
f image *f* changée
r измененное изображение *n*

524 channel geometry; bed geometry; hydraulic geometry
d Gerinnegeometrie *f*; Bettgeometrie *f*; Kanalgeometrie *f*
f géométrie *f* de lit
r геометрия *f* пласта

525 character; symbol
d Zeichen *n*; Symbol *n*
f caractère *m*; symbole *m*
r знак *m*; символ *m*

* **character attribute** → 3624

526 character font; symbol font; type font; font
d Zeichenschriftart *f*; Schrift[art] *f*; Font *n*
f police *f* [de caractères]; fonte *f*; style *m* de types
r шрифт *m*; литера *f*; комплект *m* шрифта; гарнитура *f* [шрифта]

* **character map** → 3560

527 character string; string
d String *m*; Kette *f*; Zeichenkette *f*
f chaîne *f* [de caractères]; ordre *m* symbolique
r [символьная] строка *f*; цепочка *f* символов

* **chart** → 1018, 1611

528 chart correction
d Kartenkorrektur *f*
f correction *f* d'une carte
r корректура *f* карты

529 chart datum
d Kartennull *f*
f datum *m* cartographique
r картографический репер *m*

* **check** → 532

530 check box
d Kontrollkasten *m*; Kontrollkästchen *n*
f case *f* à cocher; cocher *m*; boîte *f* d'essai; case d'option
r тестовый ящик *m*; поле *n* для галочки; графа *f* для галочки; флажок *m*; переключатель *m*; триггерная кнопка *f*; панель *f* контроля

* **check-in** → 531

531 check[ing]-in
d Abfertigung *f*; Check-in *n*
f arrivée *f*
r время *n* регистрации; отметка *f* о прибытии

532 check[out]; test; verification
d Durchprüfung *f*; Austesten *n*
f contrôle *m*; essai *m*; test *m*; vérification *f*
r контроль *m*; проверка *f*

* **checkpoint** → 739

533 chi-squared distribution
d Chi-Quadrat-Verteilung *f*
f distribution *f* de X^2; distribution chi carré
r распределение *n* хи-квадрат

534 chi-squared statistics
d Chi-Quadrat-Maßzahl *f*
f variable *f* X^2; variable *f* chi carré
r статистика *f* хи-квадрат

535 chord
d Kreissehne *f*; Sehne *f*; Bisekante *f*; Doppelsekante *f*
f corde *f*
r хорда *f*

* **chorisogram** → 539

* **chorogram** → 539

536 chorology; geographic botany; biogeography; geonemy
d Chorologie *f*
f chorologie *f*; géonémie *f*
r хорология *f* (*наука о территориальном размещении организмов*)

537 choropleth framework
(a thematic mapping technique that displays a
quantitative attribute using ordinal classes
applied as uniform symbolism over a whole
areal feature)
d Choroplethe-Rahmenwerk *n*
f infrastructure *f* de carte diagramme
r корпус *m* картодиаграммы; рама *f*
картодиаграммы

538 choropleth line
d Choroplethe[-Linie] *f*
f ligne *f* de carte diagramme
r линия *f* карто[диа]граммы

539 choropleth map; cartogram; chorogram;
chorisogram; diagram[matic] map
d Diagrammkarte *f*; Choroplethenkarte *f*;
Kartodiagramm *n*; Kartogramm *n*
f carte *f* diagramme; carton *m*
r картограмма *f*; картодиаграмма *f*

540 choropleth maps without class intervals;
continuous-tone cartogram
d Halbton-Diagrammkarte *f*; Diagrammkarte *f*
ohne Klassenbreite
f carte *f* diagramme en demi-teintes; carte
diagramme en tonalité continue
r картограмма *f* безинтервальных шкал;
непрерывная картограмма

541 choropleth zones
d Choroplethe-Zonen *fpl*
f zones *fpl* de cartes diagrammes
r зоны *fpl* картодиаграмм

* **chroma → 542**

* **chromaticity diagram → 606**

* **chromaticity value → 617**

542 chrominance; chroma
d Chrominanz *f*; Farbqualität *f*;
Farb[en]einheit *f*; Chroma *n*
f chrominance *f*; chroma *m*
r интенсивность *f* цвета; цветовая
плотность *f*

* **CIP → 729**

543 circle
d Kreis *m*
f cercle *m*
r окружность *f*; круг *m*

* **circle graph → 2753**

* **circle of probable error → 544**

* **circuit → 2234**

* **circular chart → 2753**

* **circular error of probability → 544**

544 circular error probable; circular error of
probability; CEP; circle of probable error
d Gleichwahrscheinlichkeitskreis *m*;
Streukreisradius *m*; wahrscheinlicher
kreisförmiger Fehlerbereich *m*
f cercle *m* d'erreur probable; écart *m* circulaire
probable
r вероятная круговая погрешность *f*

545 circular position error; CPE
d Streukreis-Ausrichtungsfehler *m*;
Streukreis-Lagefehler *m*
f erreur *f* circulaire de position
r круговая погрешность *f* позиции

* **circumscription → 1148**

546 circumscription
d Schlagen *n* eines Kreises; Umschrift *f*;
Umschreibung *f*
f circonscription *f*
r периферия *f*; ограничение *n*; предел *m*

* **CIR image → 613**

* **city map → 548**

547 city parcel
d Stadtparzelle *f*
f parcelle *f* de ville
r городской участок *m* [земли]

548 city plan; town plan; city map
d Stadtplan *m*
f plan *m* de ville
r план *m* города

549 class
d Klasse *f*
f classe *f*
r класс *m*

550 class center
(the set of raster values that define the mean
vector for the class in feature space)
d Klassenzentrum *n*
f centre *m* de classe
r центр *m* класса

551 class code; class identifier
d Klassenidentifikator *m*
f code *m* spécifique; identificateur *m* de classe
r код *m* класса; идентификатор *m* класса

* class identifier → 551

552 classification
  d Klassifizierung f; Klassifikation f
  f classification f; classement m
  r классификация f; распределение n;
    классирование n

553 classification accuracy
  d Klassifikationsgenauigkeit f
  f exactitude f de classification
  r точность f классификации

554 classification matrix
  d Klassifizierungsmatrix f
  f matrice f de classification
  r матрица f классификации

555 classification method
  d Klassifizierungsmethode f
  f méthode f de classification
  r метод m классификации

556 classified image
  d klassifiziertes Bild n; vergröbertes Bild
  f image f classifiée
  r классифицированное изображение n

557 classified image editor
  d klassifiziertes Bildbearbeitungsprogramm n
  f éditeur m d'images classifiées
  r редактор m классифицированных
    изображений

558 classifier
  d Klassifikator m; Klassifizierer m
  f class[ificat]eur m
  r классификатор m

559 class interval
  d Klassenbreite f; Klassenintervall n
  f intervalle m de classe
  r классовый интервал m; класс-интервал m;
    интервал группирования; интервал
    группировки

560 class list
  d Klassenliste f
  f liste f de classes
  r список m классов

561 class statistical parameter
  d statistischer Klassparameter m
  f paramètre m statistique de classe
  r статистический параметр m класса

* CLD → 508

562 clean data; valid data
  d gültige Daten pl; fehlerlose Daten

  f données fpl valables; données valides;
    données pures
  r правильные данные pl; достоверные
    данные

* clear → 954

* clearing → 954

563 clearinghouse
  d Clearinghaus n; Clearing-House n
  f centre m d'échange [et de compensation];
    chambre m de compensation
  r организация f по сбору, классификации и
    распространению информации или услуг;
    центр m обмена информацией

* clickable image → 564

564 clickable image [map]; image map; hot
    image; interactive graphic
  d klickbare Imagemap f; Bildatlas m; klickbare
    Karte f; anklickbare Map f
  f image f réactive; image cliquable; image en
    coordonnées; carte f sensible; carte-image f;
    carte imagée
  r адресуемое отображение n

565 cliff symbols; escarpment symbols
  d Steilhangssymbole npl
  f symboles mpl de falaise; symboles de berge à
    forte déclivité
  r символы mpl обрывов; символы клифов

566 climagraph; climograph; climogram
  d Klimakurve f; Klimanomogramm n;
    Klimogramm n
  f diagramme m climatique; clima[to]gramme m;
    climogramme m
  r климатическая кривая f; гидротермическая
    диаграмма f; климадиаграмма f

567 climatic change
  d Klimaveränderung f
  f échange m climatique
  r изменение n климата

568 climatic region
  d Klimazone f; klimatische Region f
  f région f climatique
  r климатическая зона f

* climogram → 566

* climograph → 566

* clinometer → 1922

569 clinometric chart; clinometric map; slope
    map
  d Neigungskarte f

f carte *f* clinométrique; carte pente
r клинометрическая карта *f*; карта склонов

* **clinometric map** → 569

* **clip** → 571

570 **clipped region**
d abgeschnittene Region *f*; geschorene Region
f région *f* coupée
r обрезанный регион *m*

571 **clip[ping]; scissoring; cutting**
d Ausschnitt *m*; Kappen *n*; Klippen *n*
f [dé]coupage *m*; écrêtage *m*; détourage *m*
r вырезание *n*; отсечение *n*;
клиппирование *n*; усечение *n*; срезание *n*;
отсекание *n*

572 **clipping window**
d Schnittfenster *n*
f fenêtre *f* de découpage
r отсекающее окно *n*

* **CLL** → 775

573 **clone window**
d Verdopplungsfenster *n*
f doublon *m* de fenêtre
r дублированное окно *n*

574 **close** *v*
d abschließen; schließen
f fermer
r замыкать; закрывать

575 **close coupling; tight coupling**
d Kurzkupplung *f*
f attelage *m* serré
r сильная связь *f*

* **closed contour** → 578

576 **closed features**
d abgeschlossene Feature *npl*
f entités *fpl* fermées
r замкнутые элементы *mpl*; замкнутые
объекты *mpl*

577 **closed figure; closed form; closed shape**
d geschlossene Form *f*
f forme *f* fermée; figure *f* fermée
r замкнутая фигура *f*; замкнутая форма *f*

* **closed form** → 577

578 **closed outline; closed contour**
d geschlossene Höhenlinie *f*
f courbe *f* de niveau fermée

r замкнутая горизонталь *f*

579 **closed [polygonal] traverse**
d geschlossener Polygonzug *m*
f cheminement *m* fermé
r замкнутый теодолитный ход *m*; замкнутый
полигон[альный ход] *m*

580 **closed polyline**
d geschlossene Polylinie *f*
f polyligne *f* fermée
r замкнутая ломаная *f*

* **closed shape** → 577

* **closed traverse** → 579

581 **close-up view**
d Nahaufnahme *f*
f vue *f* [de plan] rapprochée
r крупномасштабный вид *m*; вид крупным
планом

* **closing** → 582

* **closing error** → 1319

582 **closure; enclosure; closing**
d Abschluss *m*; Abschließung *f*; Hülle *f*;
Fermeture *n*
f fermeture *f*; clôture *f*; enceinte *f*; espace *m*
clos; enclos *m*
r замыкание *n*; закрытие *n*

* **cloud map** → 583

583 **cloud[s] map**
d Wolkenkarte *f*
f carte *f* de nuages
r карта *f* облачности; облачная карта

584 **clump**
(a set of contiguous line, node, and polygon
elements in a vector object)
d Klumpen *m*; Gruppe *f*
f lingot *m*; groupe *m*; massif *m*; bouquet *m*
r группа *f*; семейство *n*; фамилия *f*

585 **cluster; subassembly**
(group of tiles)
d Cluster *m*
f grappe *f*; cluster *m*; bouquet *m*; batterie *f* [de
dérouleurs]
r кластер *m*; блок *m*; группа *f*

586 **cluster analysis**
d Cluster-Analyse *f*
f analyse *f* cluster
r кластерный анализ *m*

587 clustering; clusterization
 d Gruppierung *f*, Clusterbildung *f*, Clusterung *f*
 f [re]groupement *m*; groupage *m*; coalescence *f*
 r группирование *n*; кластеризация *f*

* **clusterization → 587**

588 cluster labeling
 d Cluster-Etikettieren *n*
 f étiquetage *m* de grappes
 r маркирование *n* кластеров; присваивание *n* кластерам меток

589 cluster map
 d Cluster-Karte *f*
 f plan *m* de configuration
 r карта *f* кластеров

* **CLUT → 614**

* **CO → 468**

590 coastal mapping
 d Küstenkartierung *f*
 f cartographie *f* côtière
 r картографирование *n* [морского] берега

591 coastal marsh
 d Sumpfgebiet *n* an der Küste; Sumpfgebiet in Küstennähe
 f marais *m* côtier
 r прибрежный марш *m*

592 coastal zone
 d Küstengebiet *n*
 f zone *f* littorale
 r прибрежная зона *f*; береговая зона

593 coastal zone color scanner
 d Küstenzonen-Farbscanner *m*
 f scanner *m* couleur de marais côtier
 r цветовой сканер *m* прибрежной зоны

594 coastline; shoreline
 d Küstenstreifen *m*; Küstenlinie *f*, Speicheruferlinie *f*
 f bande *f* côtière; bord *m* de côte; contour *m* de côtes; littoral *m*
 r береговая линия *f*; береговая черта *f*; урез *m* воды

595 code
 d Code *m*
 f code *m*
 r код *m*

* **coding → 1264**

596 coefficient of contrast; contrast factor; contrast ratio

d Kontrastkoeffizient *m*; Kontrastverhältnis *n*
 f facteur *m* de contraste
 r коэффициент *m* контрастности

597 coextensive *adj*
 d koextensiv
 f coextensif
 r имеющий одинаковое протяжение во времени или пространстве

598 cognitive map
 d kognitive Karte *f*
 f carte *f* cognitive
 r познавательная карта *f*

* **COGO → 760**

599 Cohen's kappa
 (a measure of agreement between two classifications)
 d Kappa-Koeffizient *m*
 f cappa *m* de Cohen
 r каппа *f* Коэна

600 co-kriging
 d Co-Kriging *n*
 f co-crigeage *m*
 r ко-кригинг *m*

601 cold start
 (the process of powering up a GPS receiver initially and awaiting for it to lock onto satellites without the use of initial data)
 d Kaltstart *m*
 f amorçage *m* à froid
 r холодный старт *m*

602 collapse
 d Kollaps *m*; Zusammenbruch *m*
 f collapsus *m*; affaissement *m* [brusque]; effondrement *m*
 r свёртка *f*; коллапс *m*

* **collective use area → 2049**

* **collineatory transformation → 2918**

* **color → 619**

603 color aerial photograph
 d Farbaerofotografie *f*
 f aérophoto *f* couleur
 r цветной аэрофотоснимок *m*

604 color background
 d Farbhintergrund *m*
 f arrière-plan *m* couleur
 r цветовой фон *m*

605 color box
 d Farbkasten *m*

 f boîte *f* en couleurs
 r цветовая панель *f*

606 color chart; chromaticity diagram
 d Farbtafel *f*; Farbdiagramm *n*
 f carte *f* de couleurs; diagramme *m* de
  chromaticité
 r шкала *f* цветового охвата; цветной тест *m*;
  диаграмма *f* цветности

607 color composite
 d Farbkomposit *n*
 f composante *f* de couleur
 r составляющая *f* цвета

608 color compression
 (removing duplicate (in some cases
 near-duplicate) colors from a color map to
 make room for new colors)
 d Farbenkompression *f*
 f compression *f* de couleurs
 r сжатие *n* цвета; цветовая компрессия *f*

609 color defilation; variations in color
 d Farbenschwankungen *fpl*
 f défilement *m* de couleur
 r изменения *npl* цвета

 * **colored area** → 1416

610 colored areas
 d farbige Zonen *fpl*
 f régions *fpl* colorées
 r закрашенные области *fpl*

611 colored tile
 d farbige Musterkachel *f*
 f mosaïque *f* colorée
 r закрашенный мозаичный шаблон *m*;
  закрашенный мозаичный фрагмент *m*

612 color inflection
 d Farbinflexion *f*
 f inflexion *f* de couleur
 r изменение *n* интонации цвета

613 color infrared image; CIR image
 d Farbinfrarotbild *n*
 f image *f* infrarouge couleur
 r цветовое инфракрасное изображение *n*

 * **color lookup table** → 614

614 color map; color [lookup] table; CLUT
 d Farbkarte *f*; Farbtabelle *f*;
  Farbzuordnungstabelle *f*
 f table *f* de couleurs
 r [справочная] таблица *f* цветов; таблица
  цветности

615 color palette
 d Farbpalette *f*
 f palette *f* des couleurs
 r цветовая палитра *f*

 * **color plate** → 2259

 * **color scale** → 618

616 color separation
 d Farbseparation *f*; Farbteilung *f*; Farbauszug *m*;
  Farbentrennung *f*
 f séparation *f* des couleurs; sélection *f* des
  couleurs
 r разделение *n* цветов; цветоотделение *n*

 * **color table** → 614

617 color value; tone value; chromaticity value
 d Farbwert *m*
 f valeur *f* de couleur; valeur de chromaticité
 r значение *n* цветности; код *m* цвета;
  насыщенность *f* цвета

618 color wedge; color scale
 d Farbskala *f*
 f échelle *f* de couleurs
 r цветовая шкала *f*; шкала цветов

619 colo[u]r
 d Farbe *f*
 f couleur *f*
 r цвет *m*

620 column
 d Spalte *f*
 f colonne *f*
 r колонка *f*; столбец *m*

621 column builder
 d Spaltenbilder *m*
 f bâtisseur *m* de colonne
 r построитель *m* колонки

 * **column diagram** → 315

 * **column graph** → 315

622 column index
 d Spaltenindex *m*
 f indice *m* de colonne
 r индекс *m* колонки

623 combination
 d Kombination *f*
 f combinaison *f*
 r комбинация *f*; сочетание *n*

624 combine *v*
 d kombinieren

f combiner
r комбинировать

625 combined shading
   d kombinierte Bildabschattung *f*; gemischtes Schattieren *n*
   f nuançage *m* combiné; projection *f* d'ombre combiné; miroitement *m* combiné
   r отмывка *f* при комбинированном освещении

626 command line
   d Befehlszeile *f*
   f ligne *f* de commande
   r командная строка *f*

627 common object
   d Gemeinsamobjekt *n*
   f objet *m* commun
   r общий объект *m*

   * **common operation** → 2047

   * **common use area** → 2049

   * **communal area** → 3947

   * **communication in cartography** → 439

628 communicative reliability
   d kommunikative Zuverlässigkeit *f*
   f fiabilité *f* de communication
   r коммуникационная надёжность *f*

629 compactness
   d Kompaktheit *f*
   f compacité *f*
   r компактность *f*

630 compactness ratio
   d Kompaktheitsverhältnis *n*
   f taux *m* de compacité
   r коэффициент *m* компактности

631 comparability
   d Vergleichbarkeit *f*
   f comparabilité *f*; reproductibilité *f*
   r сравнимость *f*; сопоставимость *f*

632 compass
   d Bussole *f*; Kompass *m*
   f boussole *f*
   r буссоль *f*; компас *m*

633 compass aspect; aspect; exposure; direction of steepest slope
   d Belichtung *f*
   f exposition *f*; orientation *f*
   r экспозиция *f* [склона]; экспонирование *n*

634 compass azimuth; compass bearing; compass direction; magnetic azimuth
   d Kompassrichtung *f*
   f direction *f* de compas; direction magnétique
   r компасный азимут *m*; магнитный азимут; компасный пеленг *m*; склонение *n* компаса

   * **compass bearing** → 634

   * **compass card** → 636

635 compass declination; magnetic declination
   d Kompassmissweisung *f*; magnetische Abweichung *f*; magnetische Deklination *f*
   f déclinaison *f* magnétique
   r магнитное склонение *n*

   * **compass direction** → 634

   * **compass point** → 2810

636 compass rose; compass card
   (a circle drawn on a map which is subdivided in a clockwise direction from 0° to 360°)
   d Kompassrose *f*; Windrose *f*; Kursrose *f*
   f rose *f* des vents; rose du compas
   r девиационный круг *m* компаса; изображение *n* картушки компаса; лимб *m* картушки компаса

637 compatibility of geoimages
   d Verträglichkeit *f*; Kompatibilität *f*
   f compatibilité *f* de géo-images
   r совместимость *f* геоизображений

638 compilation
   d Kompilieren *n*; Kompilierung *f*
   f compilation *f*
   r составление *n*; компиляция *f*

   * **compilation map** → 2659

639 compilation [map] manuscript
   d Autoren[karten]manuskript *n*
   f manuscrit *m* de minute d'auteur
   r авторский оригинал *m* карты

   * **compilation manuscript** → 639

640 compilation scale
   d Kompilierungsmaßstab *m*
   f échelle *f* de compilation
   r масштаб *m* составления

   * **compilation sheet** → 2659

641 compiled map set
   d übersetzte Kartenmenge *f*
   f ensemble *m* compilé de cartes
   r скомпилированный набор *m* карт

642 complete datum
d Volldatum *n*
f datum *m* complet
r полный репер *m*

643 complete spatial randomness
d vollräumliche Wahllosigkeit *f*; vollräumliche
   Zufälligkeit *f*
f aspect *m* aléatoire spatial complet
r полная пространственная случайность *f*;
   полная пространственная
   беспорядочность *f*

644 complete survey
d Gesamtaufnahme *f*
f levé *m* d'ensemble
r суммарная съёмка *f*; обзорная съёмка

645 complex atlas
d vielseitiger Atlas *m*
f atlas *m* complexe
r комплексный атлас *m*

646 complexity
d Komplexität *f*
f complexité *f*
r сложность *f*

**647 complex map; composite map; aggregate
   map**
d komplexe Karte *f*
f carte *f* complexe
r комплексная карта *f*

648 complex mapping
d komplexe Kartierung *f*
f cartographie *f* complexe
r комплексное картографирование *n*

649 complex object
d vielseitiges Objekt *n*
f objet *m* complexe
r составной объект *m*

650 complex polygon
d Komplexpolygon *n*
f polygone *m* complexe
r составной полигон *m*

651 complex surface
d Komplexfläche *f*
f surface *f* complexe
r сложная поверхность *f*; комплексная
   поверхность

* **component element of map → 448**

652 component model
d Komponentenmodell *n*
f modèle *m* de components
r компонентная модель *f*; модель

компонентов

* **composite map → 647**

653 comprehension
d Begriffsvermögen *n*; Verständnis *n*
f compréhension *f*
r понимание *n*; выделение *n*; обозримость *f*

* **comprehensive development area map →
   2101**

654 compress *v*
d komprimieren; kompressieren; verdichten
f comprimer; compresser
r сжать; уплотнять; компрессировать

655 compressed digital terrain elevation data
d kompressierte digitale Höhenliniendaten *pl*
f données *fpl* de terrain numériques
   compressées
r уплотнённые цифровые данные *pl* для
   вертикальной наводки

656 compression
d Kompression *f*; Kompaktion *f*;
   Komprimierung *f*
f compression *f*, compaction *f*
r сжатие *n*; уплотнение *n*; компрессия *f*

657 compression method
d Kompressionsmethode *f*
f méthode *f* de compression
r метод *m* компрессирования

* **compromise map projection → 145**

**658 computational error; computing error;
   error of computation; miscalculation**
d Rechnungsfehler *m*; Rechenfehler *m*
f erreur *f* de calcul
r вычислительная ошибка *f*; ошибка в
   вычислении

659 computation of area; area calculation
d Flächenberechnung *f*
f calcul *m* d'aires
r вычисление *n* площади

**660 computer[-aided] cartography;
   computer[-aided] mapping; CAM;
   automated cartography; automated
   mapping**
d computergestützte Kartografie *f*;
   Computerkartografie *f*; automatische
   Kartografie *f*
f cartographie *f* par ordinateur; cartographie
   automatisée; cartomatique *f*
r компьютерная картография *f*;
   автоматизированная картография *f*;
   автоматическое картографирование *n*

661 computer-aided design; CAD
   d computerunterstütztes Entwerfen *n*;
     computerunterstütztes Design *n*
   f conception *f* assistée par ordinateur; CAO
   r автоматизированное проектирование *n*;
     проектирование с помощью компьютера

662 computer-aided design and drafting;
    CADD
   d computerunterstütztes Entwerfen *n* und
     Zeichnen *n*
   f conception *f* et dessin *m* assisté par ordinateur
   r автоматизированное проектирование *n* и
     изготовление *n* чертежей

   * **computer-aided drafting** → 665

   * **computer-aided mapping** → 660

   * **computer-aided mapping system** → 285

663 computer-aided mass appraisal; CAMA
   d computerunterstützte Abschätzung *f*
   f appréciation *f* de masse par ordinateur
   r автоматизированная массовая оценка *f*

664 computer-aided planning; CAP
   d computerunterstützte Planung *f*
   f planification *f* par ordinateur
   r автоматизированная планировка *f*

665 computer-assisted drafting;
    computer-aided drafting; CAD
   d computerunterstütztes Zeichnen *n*
   f dessin *m* assisté par ordinateur
   r автоматизированное изготовление *n*
     чертежей

666 computer-assisted reproduction; CAR
   d computerunterstützte Reproduktion *f*;
     computerunterstützte Wiedergabe *f*
   f reproduction *f* par ordinateur
   r автоматизированное воспроизведение *n*

667 computer atlas; electronic atlas
   d Computer-Atlas *m*
   f atlas *m* électronique
   r компьютерный атлас *m*; электронный атлас

   * **computer cartography** → 660

668 computer-generated imaging; CGI
   d computergesteuerte Wiedergabetechnik *f*;
     durch Computer generierte Abbildung *f*
   f imagination *f* générée par ordinateur;
     présentation *f* en image par ordinateur
   r изображения *npl*, генерированные
     компьютером

669 computer graphics
   d Computergrafik *f*
   f graphique *m* par ordinateur
   r компьютерная графика *f*

670 computer graphics interface; CGI
   d Computergrafik-Schnittstelle *f*
   f interface *f* d'infographie
   r интерфейс *m* машинной графики;
     стандарт *m* CGI

   * **computerized map** → 1239

   * **computer map** → 1239

   * **computer mapping** → 660

   * **computer mapping system** → 285

   * **computing error** → 658

671 concatenated key
    (in databases)
   d verketteter Schlüssel *m*; verbundener
     Schlüssel
   f clé *f* en chaîne
   r сцеплённый ключ *m*

672 concatenation
   d Konkatenation *f*
   f concaténation *f*
   r сцепление *n*; сочленение *n*; конкатенация *f*

673 concavity
   d Konkavität *f*
   f concavité *f*
   r вогнутость *f*

   * **concentrative operation** → 2393

674 concentric ring buffers
   d Puffer *mpl* mit konzentrische Kreise
   f tampons *mpl* en anneaux concentriques
   r концентрические буферы *mpl*

   * **concept map** → 678

675 conceptual accuracy
   d konzeptionelle Genauigkeit *f*
   f exactitude *f* conceptuelle
   r концептуальная точность *f*

676 conceptual description
   d Konzeptualbeschreibung *f*
   f description *f* conceptuelle
   r концептуальное описание *n*

677 conceptual framework
   d konzeptionelles Rahmenwerk *n*

f cadre *m* conceptuel; cadre théorique
r концептуальная структура *f*;
концептуальная основа *f*; концептуальные
рамки *fpl*; система *f* понятий (*в базах
данных*)

678 concept[ual] map; framework map
d konzeptionelle Karte *f*
f carte *f* conceptuelle; carte-cadre *f*
r концептуальная карта *f*

679 conceptual model
d Konzeptualmodell *n*; konzeptionelles
Modell *n*
f modèle *m* conceptuel
r концептуальная модель *f*

680 concurrency management
d Verwaltung *f* paralleler Prozesse
f management *m* de concurrence
r управление *n* параллельных процессов

681 conditional line
d bedingte Linie *f*
f ligne *f* conditionnelle
r условная линия *f*

682 conditional operator
d bedingter Operator *m*
f opérateur *m* conditionnel
r условный оператор *m*

683 conductivity map
d Leitfähigkeitskarte *f*
f carte *f* de conductivité
r карта *f* удельной проводимости

684 confidence interval
d Konfidenzintervall *n*; Vertrauensintervall *n*
f intervalle *m* de confiance
r доверительный интервал *m*

685 configuration
d Konfiguration *f*
f configuration *f*
r конфигурация *f*

* **conflation** → 877

**686 conformal projection; equal-angle
projection; anglepreserving projection;
orthomorphic projection; equiangular
projection**
d konforme Projektion *f*; winkeltreue
Abbildung *f*; orthomorphe Projektion
f projection *f* conforme; projection équiangle;
projection orthomorphique
r равноугольная проекция *f*; конформная
проекция; ортоморфная проекция

687 conformity
d Konformität *f*
f conformité *f*
r согласие *n*; соответствие *n*

688 confusion matrix
(in pattern recognition)
d Unordnungsmatrix *f*
f matrice *f* de confusion
r матрица *f* неточностей

689 conglomerate
d Konglomerat *m*
f conglomérat *m*
r конгломерат *m*; смесь *f* разнородных
элементов

**690 conic[al] projection; central projection;
perspective projection; perspective
mapping; perspective**
d Kegelprojektion *f*; Zentralprojektion *f*;
konische Abbildung *f*; perspektivische
Projektion *f*; Perspektive *f*; Perspektivität *f*
f projection *f* conique; projection *f* centrale;
projection perspective; perspective *f*
r коническая проекция *f*; центральная
проекция; перспективная проекция;
перспектива *f*

* **conic projection** → 690

691 conjoint boundary
d Gemeinsamgrenze *f*
f limite *f* conjointe; limite unie
r общая граница *f*

692 conjugate points
d weitspannende Punkte *mpl*
f points *mpl* conjugués
r сопряжённые точки *fpl*

693 connected arcs
d verbundene Bögen *mpl*; angeschlossene
Bögen
f arcs *fpl* connectées
r связанные дуги *fpl*

* **connected coverage framework** → 694

**694 connected coverages; connected coverage
frameworks**
(a general term for measurement frameworks
that involve relationships between distinct
spatial objects)
d angeschlossene Erfassungsbereiche *mpl*
f couvertures *fpl* connectées
r структуры *fpl* связанных покрытий

695 connected graph
d zusammenhängender Graph *m*; verbundener
Graph

f graphe *m* connexe
r связанный граф *m*; связной граф

696 connected line segments
d verbundene Geradensegmente *npl*
f segments *mpl* linéaires connectés
r связанные линейные сегменты *mpl*

* **connectedness → 698**

697 connected surface
d zusammenhängende Fläche *f*
f surface *f* connexe
r связная поверхность *f*

698 connectivity; connectedness
d Konnektivität *f*; Verbundenheit *f*
f connectivité *f*, connexité *f*; connectabilité *f*
r связанность *f*; связываемость *f*; связность *f*

699 connector
d Konnektor *m*
f connecteur *m*
r конъектор *m*; соединитель *m*

* **consecutive access → 3266**

700 consistency
d Konsistenz *f*; Widerspruchsfreiheit *f*
f consistance *f*
r совместимость *f*; непротиворечивость *f*

701 consolidated city
d konsolidierte Stadt *f*, zusammengelegte Innenstadt *f*
f cité *f* compactée
r консолидированный город *m*; объединённый город

702 consolidation
d Konsolidation *f*; Verfestigung *f*
f consolidation *f*
r консолидация *f*; укрупнение *n*

703 constant
d Konstante *f*; Festwert *m*
f constante *f*
r константа *f*; постоянная *f*

* **constant error → 3567**

704 constant geometric accuracy
d konstante geometrische Genauigkeit *f*
f exactitude *f* géométrique constante
r постоянная геометрическая точность *f*

705 constellation
d Konstellation *f*; Sternbild *n*
f constellation *f*
r констелляция *f*; созвездие *n*; плеяда *f*;

совокупность *f*

* **constraint → 3155**

706 constraints analysis
d Analyse *f* von Nebenbedingungen
f analyse *f* de contraintes
r анализ *m* ограничений

707 constraint satisfaction method
d Constraint-Erfüllungsmethode *f*; Methode *f* von Erfüllen von Nebenbedingungen
f méthode *f* de satisfaction de contraintes
r метод *m* поиска допустимого решения

* **constructible zone → 414**

708 construction solid geometry; constructive solid geometry; CSG
(geometry of combined solids)
d konstruktive Körpergeometrie *f*; konstruktive Vollkörpergeometrie *f*
f géométrie *f* de solides constructive
r конструктивная блочная геометрия *f*

* **constructive solid geometry → 708**

* **contact screen → 2896**

709 containment
d Eindämmung *f*; Sicherheitsbehälter *m*
f confinement *m*; contenant *m*
r сдерживание *n*; сохранение *n*

710 content standards
d Inhalt-Grundsätze *mpl*
f standards *mpl* de contenu; unités *fpl* de mesure de contenu
r нормы *fpl* содержания; стандарты *mpl* содержания

711 content standards for spatial metadata; CSSM
d Raum-Metadaten-Standards *mpl*
f standards *mpl* de contenu de métadonnées spatiaux
r нормы *fpl* содержания пространственных метаданных

* **context menu → 3289**

* **context-sensitive menu → 3289**

712 contiguity
d Kontiguität *f*; Benachbartheit *f*
f contiguïté *f*
r смежность *f*

713 contiguous data structure
d abhängige Datenstruktur *f*; fortlaufende Datenstruktur; durchgehende Datenstruktur

f structure *f* de données contiguë
r смежная структура *f* данных

714 continental map
  d Erdteilkarte *f*
  f carte *f* continentale
  r континентальная карта *f*

715 continental shelf
  d Kontinentalschelf *n*; Kontinentalsockel *m*
  f plateau *m* continental; plate-forme *f* continentale
  r континентальный шельф *m*; материковая отмель *f*

716 continuity
  d Stetigkeit *f*; Kontinuität *f*
  f continuité *f*
  r непрерывность *f*

717 continuous data; seamless data
  d kontinuierliche Daten *pl*
  f données *fpl* continues
  r бесшовные данные *pl*; непрерывные данные; сплошные данные; аналоговые данные

718 continuous distribution
  d kontinuierliche Verteilung *f*
  f distribution *f* continue
  r непрерывное распространение *n*

719 continuous feature
  d kontinuierliches Feature *n*
  f trait *m* continu
  r непрерывный характер *m* местности

720 continuous map; seamless map
  d kontinuierliche Karte *f*
  f carte *f* continue
  r бесшовная карта *f*; непрерывное отображение *n*

721 continuous phenomenon
  d kontinuierliche Erscheinung *f*
  f phénomène *m* continu
  r непрерывное явление *n*

722 continuous surface
  d kontinuierliche Fläche *f*
  f surface *f* continue
  r непрерывная поверхность *f*

723 continuous tone; contone; full tone
  d kontinuierlicher Ton *m*
  f tonalité *f* continue; ton *m* continu
  r непрерывный оттенок *m*; безрастровый тон *m*

  * **continuous-tone cartogram** → 540

  * **continuous tone photography** → 724

724 continuous tone reproduction; continuous tone photography
  d Halbtonaufnahme *f*
  f photo *f* en tonalité continue
  r безрастровый снимок *m*

725 continuous variable
  d stetige Variable *f*; kontinuierliche Variable
  f variable *f* continue
  r непрерывная переменная *f*

  * **contone** → 723

  * **contour** → 732

726 contour; outline
  d Kontur *f*; Umriss *m*
  f contour *m*
  r контур *m*; очертание *n*

  * **contour attributes** → 2671

727 contour[ed] map; line map
  d Nivellierungsplan *m*; Höhenplan *m*; Isochronenkarte *f*, Isochronenplan *m*; Höhen[schicht]linienkarte *f*; Schicht[en]linienkarte *f*
  f carte *f* en courbes de niveau; carte au trait; carte isochronique; plan *m* de nivellement
  r карта *f*, вычерченная в горизонталях; карта с горизонталями; контурная карта; карта изолиний; карта изохрон

728 contouring; outlining; bordering
  d Umrisszeichnung *f*; Berandung *f*; Einfassung *f*; Ränderung *f*; Konturen *n*
  f tracé *m* des courbes de niveau; mise *f* en contour; bordurage *m*; isolignage *m*
  r вычерчивание *n* контуров; уточнение *n* контура; оконтуривание *n*; окаймление *n*; обрамление *n*; окантовка *f*

729 contour interpolation program; CIP
  d Profilllinieninterpolationsprogramm *n*
  f programme *m* d'interpolation de lignes de niveau
  r программа *f* интерполяции горизонталей

730 contour interval; contour level; vertical interval; equidistance
  d Höhenlinienabstand *m*; Äquidistanz *f*
  f équidistance *f* [des courbes de niveau]
  r сечение *n* горизонталей на карте; высота *f* сечения рельефа; интервал *m* между горизонталями; интервал между изолиниями

  * **contour level** → 730

731 contour line
 d Konturlinie *f*
 f ligne *f* de contour
 r контурная линия *f*

**732 contour [line]; level line; level curve; height
 line; isohypse; isoheight; hypsographic[al]
 curve; structural contour; stratum**
 (on a map)
 d [strukturelle] Höhenlinie *f*;
 Höhenschichtlinie *f*; Schichtlinie *f*;
 Höhenkurve *f*; Horizontalkurve *f*; Profillinie *f*;
 Isohypse *f*
 f ligne *f* de niveau; courbe *f* de niveau; courbe
 isohypse; isohypse *f*; strate *f*
 r линия *f* уровня; уровневая линия;
 горизонталь *f*; изогипса *f*; линия равных
 высот

* contour map → 727

733 contour mapping
 d Profilliniendarstellung *f*
 f isocartographie *f*
 r картографирование *n* изолиний;
 картографирование контуров

**734 contour number; elevation number;
 contour value**
 d Höhen[schicht]linienzahl *f*; Höhenkote *f*;
 Kote *f*
 f cote *f* de courbe [de niveau]
 r цифровая отметка *f* горизонтали

* contour survey → 103

735 contour tagging
 d Höhenlinienetikettierung *f*
 f étiquetage *m* de lignes de niveau
 r тегирование *n* горизонталей; тегирование
 контуров

* contour value → 734

736 contrast correction
 d Kontrastkorrektur *f*
 f correction *f* de contraste
 r исправление *n* контраста; коррекция *f*
 контраста

* contrast factor → 596

* contrast ratio → 596

737 contrast stretching
 d Kontraststrecken *n*; Kontraststreckung *f*
 f étirement *m* de contraste
 r вытягивание *n* контраста

* control ball → 3777

738 control character
 d Steuerzeichen *n*
 f caractère *m* de contrôle
 r управляющий символ *m*

* control net → 2530

739 control point; checkpoint
 d Steuerpunkt *m*
 f point *m* de contrôle
 r контрольная точка *f*

740 control point; permanent point
 d Festpunkt *m*
 f marque *f* de canevas; point *m* de canevas;
 point permanent marqué; point pilote
 r пункт *m* плановой сети; реперная точка *f*

741 control point list
 d Steuerpunkt-Liste *f*
 f liste *f* de points de contrôle
 r список *m* контрольных точек

742 control segment
 (of GPS)
 d Kontrollsegment *n*; Steuersegment *n*
 f secteur *m* de contrôle
 r подсистема *f* наземного контроля и
 управления

743 control station
 d Bedienungspult *n*; Steuerstand *m*
 f station *f* de contrôle; station de triangulation;
 point *m* géodésique
 r управляющая станция *f*

744 conventional projection
 d konventionelle Projektion *f*
 f projection *f* conventionnelle
 r конвенциональная проекция *f*; условная
 проекция

* conventional signs → 459

745 convergence
 d Konvergenz *f*
 f convergence *f*
 r сближение *n*

* convergence angle → 746

**746 convergence of meridians; convergent
 angle; convergence angle; grid declination;
 declination of grid north; theta angle**
 d Missweisung *f*; Theta-Winkel *m*
 f convergence *f* de méridiens
 r сближение *n* меридианов

* convergent angle → 746

* conversational mode → 1979

747 conversation map
d Gespräch-Karte f
f carte f de conversations
r карта f разговоров; карта сеансов

748 conversion
d Konvertierung f; Umformung f;
Umwandlung f
f conversion f
r преобразование n; превращение n;
конверсия f

749 conversion to cartographic form
d Umsetzung f in kartografische Darstellungen
f conversion f en forme cartographique
r преобразование n в картографическом
представлении

750 converter
d Konverter m; Umwandler m; Wandler m
f convertisseur m; traducteur m
r конвертер m; преобразователь m

* convex hull → 751

751 convex hull [polygon]
d konvexe Hülle f
f polygone m en enveloppe convexe;
enveloppe f convexe
r полигон m в форме выпуклой оболочки;
выпуклая оболочка f

752 convexity
d Konvexität f
f convexité f
r выпуклость f

753 convolution
d Faltung f; Konvolution f
f convolution f
r конволюция f; свёртка f; свёртывание n

754 coordinate
d Koordinate f
f coordonnée f
r координата f

755 coordinate conversion
d Koordinatenkonvertierung f
f conversion f de coordonnées
r преобразование n координат

756 coordinate dimensioning
d koordinatenbezogene Positionierung f;
koordinatenbezogene Dimensionierung f
f dimensionnement m de coordonnées
r задание n размеров

757 coordinate display
d Koordinatendarstellung f
f affichage m de coordonnées
r изображение n координат; представление n
координат

**758 coordinated universal time; universal time
[coordinated]; UTC; Greenwich meantime;
Greenwich Civil Time; GCT**
d koordinierte Weltzeit f; mittlere
Greenwichzeit f; GMT; greenwiche
Normalzeit f
f temps m universel [coordonné]; TU; temps
moyen de Greenwich; heure f [du méridien] de
Greenwich
r универсальное [координированное]
время n [по Гринвичу]; всемирное время;
мировое время; гринвичское гражданское
время

759 coordinate file
d Koordinatendatei f
f fichier m de coordonnées
r файл m координат

* coordinate fountain → 2660

760 coordinate geometry; COGO
d Koordinatengeometrie f
f géométrie f coordonnée
r координатная геометрия f

761 coordinate graphics; line graphics
d Liniengrafik f; Grafik f von Linien und
Koordinaten
f infographie f par coordonnées; graphique m
par coordonnées et traits
r графика f координатами и линиями

762 coordinate grid; network of coordinates
d Koordinatengitter n; Koordinatennetz n
f réseau m de quadrillage de coordonnées
r координатная сетка f; сетка координат

* coordinate origin → 2660

763 coordinate pair
d Koordinatenpaar n
f paire f de coordonnées
r пара f координат

764 coordinate reader
d Koordinatenleser m
f lecteur m de coordonnées
r координатный считыватель m

* coordinates in the plane → 2775

* coordinate source → 2660

765 **coordinate system**
  d Koordinatensystem *n*
  f système *m* de coordonnées
  r координатная система *f*; система координат

766 **coordinate table**
  d Koordinatentabelle *f*
  f table *f* de coordonnées
  r таблица *f* координат

767 **coordinate transformation**
  d Koordinaten-Transformation *f*
  f transformation *f* de coordonnées
  r координатная трансформация *f*

768 **coordinatograph**
  d Koordinatenschreiber *m*
  f coordinatographe *m*
  r координатограф *m*

769 **copy; manifold**
  d Kopie *f*; Exemplar *n*
  f copie *f*; exemplaire *m*
  r копия *f*; экземпляр *m*

770 **copying**
  d Kopieren *n*; Vervielfältigen *n*
  f copiage *m*
  r копирование *n*

771 **copyright in cartography; authorship in cartography**
  d Kartenautorrecht *n*; Kartenurheberrecht *n*
  f droit *m* d'auteur en cartographie; droit d'exploitation en cartographie; copyright *m* en cartographie
  r авторское право *n* в картографии

772 **coregistration**
  d Coregistrierung *f*
  f co-enregistrement *m*
  r корегистрирование *n*

773 **corner**
  (the beginning or endpoint of any survey line)
  d Ecke *f*; Winkel *m*
  f coin *m*; corner *m*
  r угол *m*

774 **corner joins**
  (the location where three or more contiguous map sheets come together)
  d Ecken-Verbindung *f*
  f union *f* angulaire
  r угловое соединение *n*; угловое сочленение *n*

775 **corrected line length; CLL**
  d verbesserte Linienlänge *f*; berichtigte Gesamtlänge *f*; korrigierte Zeilenlänge *f*
  f longueur *f* de ligne corrigée; longueur de ligne révisée
  r исправленный формат *m* строки; исправленная ширина *f* колонки

776 **correction**
  d Korrektur *f*
  f correction *f*
  r поправка *f*

777 **correctness**
  d Richtigkeit *f*; Korrektheit *f*
  f justesse *f*
  r правильность *f*

778 **correlation**
  d Korrelation *f*
  f corrélation *f*
  r корреляция *f*; взаимная зависимость *f*

779 **correlation analysis**
  d Korrelationsanalyse *f*; Analyse *f* der Korrelation
  f analyse *f* de corrélation
  r корреляционный анализ *m*

780 **correlation matrix**
  d Korrelation-Matrix *f*
  f matrice *f* de corrélations
  r матрица *f* соотношений

781 **correlation points**
  d Korrelationspunkte *mpl*
  f points *mpl* de corrélation
  r корреляционные точки *fpl*

782 **corridor**
  d Korridor *m*; Gang *m*
  f corridor *m*; passage *m*; couloir *m*
  r узкая зона *f*; коридор *m* [подхода]; проход *m*

* **coted projection → 3720**

783 **count**
  (numerical measurement that aggregates the number of some objects within a collection unit)
  d Zählung *f*
  f comptage *m*; compte *m*
  r число *n*; счёт *m*; подсчёт *m*

784 **country; countryside; rural district; rural area; land**
  d ländlicher Bezirk *m*; Landkreis *m*
  f écart *m* de commune; district *m* communal; zone *f* rurale
  r [сельская] местность *f*; сельский район *m*

785 **country planning; country programming**
  d Landesplanung *f*

f planification *f* par pays; programmation *f* par
pays
r национальное планирование *n*

* **country programming** → 785

* **countryside** → 784

786 **county**
d Provinz *f*; Grafschaft *f*
f arrondissement *m*; département *m* (*État-Unis*);
comté *m* (*UK*)
r округ *m* (*в США*); графство *n* (*в Англии*)

787 **county boundary**
d Landkreisgrenze *f*; Grafschaftsgrenze *f*
f limite *f* de département; limite de comté
r граница *f* округа

788 **county features**
d Landkreismerkmale *npl*
f caractéristiques *fpl* de département; traits *mpl*
de département
r характеристики *fpl* округа

789 **county subdivision**
d Landkreisunterteilung *f*
f sous-division *f* de département
r подразделение *n* округа

790 **coupled boundary set**
d verkoppelter Randsatz *m*; gekoppelter
Randsatz
f ensemble *m* de frontières liées
r связное множество *n* граней

791 **course**
(line on a surface where slopes converge from
two directions)
d Bahn *f*; Kurs *m*
f cours *m*
r направление *n*; курс *m*; течение *n*

792 **covariance**
d Kovarianz *f*
f covariance *f*
r ковариантность *f*

* **coverage** → 1582, 2680

793 **coverage chart**
d Erfassungskarte *f*
f carte *f* de couverture
r карта *f* покрытия

794 **coverage contour**
d Erfassungskontur *f*
f contour *m* de couverture
r контур *m* покрытия

795 **coverage extent**
(the coordinates defining the minimum
boundary rectangle of a coverage or grid)
d Erfassungsextent *m*
f étendue *f* de couverture
r экстент *m* покрытия

796 **coverage generator**
(a software)
d Bodenbedeckungsgenerator *m*;
Erfassungsgenerator *m*
f logiciel *m* de génération de couverture
r генератор *m* покрытия

797 **coverage geometry**
d Coverage-Geometrie *f*
f géométrie *f* de recouvrement
r геометрия *f* покрытия

798 **coverage ID; cover-ID**
d Erfassungsbereichsidentifikator *m*
f identificateur *m* de couverture
r идентификатор *m* покрытия;
идентификатор оболочки

799 **coverage name**
d Erfassungsbereichsname *m*
f nom *m* de couverture
r имя *n* покрытия

800 **coverage units**
d Überdeckungseinheiten *fpl*
f unités *fpl* de couverture
r единицы *fpl* покрытия

* **cover-ID** → 798

* **covering of the plane with tiles** → 3684

* **CPE** → 545

801 **critical analysis**
d kritische Analyse *f*
f analyse *f* critique
r критический анализ *m*; критический
разбор *m*

802 **critical points**
d kritische Stellen *fpl*
f points *mpl* critiques
r критические точки *fpl*

803 **critical region**
d kritischer Bereich *m*
f région *f* critique
r критический регион *m*; критическая зона *f*

* **cropping** → 1478

804 **cropping area**
 d Beschneid[en]bereich *m*
 f zone *f* de coupage
 r область *f* отсекания; вырезанная область

 * **cross** → 3807

805 **cross-classification**
 d Mehrfacheinordnung *f*; kombinierte
   Gliederung *f*
 f classification *f* croisée
 r перекрёстная классификация *f*

806 **cross-correlation**
 d Querkorrelation *f*; Kreuzkorrelation *f*
 f corrélation *f* croisée; corrélation avec retards;
   corrélation avec décalage
 r перекрёстное соотношение *n*; взаимная
   корреляция *f*

807 **cross-correlation matrix**
 d Querkorrelation-Matrix *f*
 f matrice *f* de corrélations croisées
 r матрица *f* перекрёстных соотношений

 * **cross-cut** → 812

808 **crosshair**
 d Fadenkreuz *n*
 f réticule *m* en croix; croisée *f* de fils; fils *mpl*
   croisés; croisillon *f* de repérage
 r перекрестье *n* [нитей]; крест *m* нитей;
   · сетка *f* нитей

 * **cross-haired cursor** → 1082

 * **crosshair pointer** → 1082

809 **cross-hatching**
 d Kreuzschraffen *n*; Kreuzschraffur *f*;
   gekreuztes Schraffen *n*
 f hachure *f* croisée
 r перекрёстное штрихование *n*; перекрёстная
   штриховка *f*

810 **cross-hatch [line] pattern**
 d Schottenmuster *n*
 f mire *f* à dessin écossais
 r модель *f* перекрёстной штриховки

 * **cross-hatch pattern** → 810

811 **crossover point**
 d Übergangspunkt *m*
 f point *m* de crossover
 r точка *f* перегиба

812 **cross-section; cross-cut; profile**
 d Querschnitt *m*; Profil *n*
 f section *f* transversale; profil *m* en travers;

 coupe *f* transversale
 r поперечное сечение *n*; поперечный
   разрез *m*; профиль *m*

813 **cross-section block-diagram**
 d Querschnitt-Flussdiagramm *n*
 f schéma-bloc *m* de profil
 r профильная блок-диаграмма *f*

814 **cross shading**
 d Kreuzschattierung *f*
 f ombrage *m* croisé
 r перекрёстное оттенение *n*

815 **cross-tabulation**
 d Quertabulierung *f*
 f tabulation *f* transversale; tabulation croisée
 r перекрёстная табуляция *f*; комбинационная
   табуляция; перекрёстное табулирование *n*

816 **cross track error; XTE**
 d Fehler *m* quer zur Bahn
 f erreur *f* transversale; erreur en travers de la
   trajectoire; erreur latérale
 r боковое отклонение *n*

817 **cryptography**
 d Kryptografie *f*
 f cryptographie *f*
 r криптография *f*

 * **CSG** → 708

 * **CSSM** → 711

 * **CT** → 499

 * **CTSI** → 504

818 **cubic convolution; cubic interpolation**
   (interpolation method where the value is
   interpolated by fitting a third-order equation to
   the sixteen grid points surrounding the desired
   location)
 d kubische Konvolution *f*
 f convolution *f* cubique
 r кубическая свёртка *f*

819 **cubic environment mapping**
 d kubische Abbildung *f* der Umgebung
 f mappage *m* d'environnement cubique
 r кубическое картирование *n* окружения

 * **cubic interpolation** → 818

820 **cubic spline**
 d kubischer Spline *m*
 f spline *m* cubique
 r кубический сплайн *m*

821 cubic trend
 d kubischer Trend *m*
 f tendance *f* cubique
 r кубический тренд *m*

* **current layer → 24**

822 current target
 d aktives Ziel *n*
 f cible *f* active; destination *f* active
 r активная мишень *f*; активная цель *f*

823 cursor
 d Cursor *m*; Kursor *m*; Schreibmarke *f*;
    Lichtmarke *f*
 f curseur *m*
 r курсор *m*; метка *f* на видеоэкране

824 cursor [control] keys
 d Cursor[positionier]tasten *fpl*;
    Cursorsteuertasten *fpl*
 f touches *fpl* [de gestion] de curseur
 r клавиши *mpl* управления курсором;
    клавиши [движения] курсора

* **cursor keys → 824**

825 curvature of the Earth; Earth curvature
 d Erdkrümmung *f*
 f courbure *f* de la Terre; sphéricité *f* de la Terre;
    rotondité *f* de la Terre
 r кривизна *f* земной поверхности

826 curve; curved line
 d Kurve *f*
 f courbe *f*
 r кривая *f*

* **curve chart → 2176**

827 curve complexity
 d Komplexität *f* der Kurve
 f complexité *f* de la courbe
 r сложность *f* кривой

* **curved line → 826**

828 curved surface
 d Kurvenfläche *f*
 f surface *f* courbée
 r изогнутая поверхность *f*

829 curve-fit polyline
 d Polylinie *f* mit Form der Kurve
 f polyligne *f* lissée en courbe
 r полилиния *f*, сглаженная кривой

830 curve fitting
 d Kurvenanpassung *f*; Fitten *n* der Kurve;
    Kurvenausgleich *m*

 f lissage *m* d'une courbe; ajustement *m* d'une
    courbe
 r вычерчивание *n* кривой [по точкам];
    сглаживание *n* кривой; выравнивание *n*
    кривой

* **curve follower → 2174**

831 curve generation
 d Kurvengenerierung *f*
 f génération *f* de courbes
 r генерирование *n* кривых

* **curve graph → 2176**

* **curve of the declination → 2024**

832 curve-pattern
 d Kurvenmuster *n*
 f modèle *m* de courbe; forme *f* de courbe
 r описание *n* кривой; изображение *n* кривой

833 curvilinear coordinates
 d krummlinige Koordinaten *fpl*
 f coordonnées *fpl* curvilignes
 r криволинейные координаты *fpl*

834 curvimeter; curvometer; map-measurer
 d Curvimeter *n*; Kurvenmesser *m*
 f curvimètre *m*
 r курвиметр *m*

* **curvometer → 834**

835 custodian
    (an organization that takes responsibility to
    generate a particular kind of information for a
    defined geographic region and agrees to make
    it available to others)
 d Aufseher *m*; Wächter *m*; Hüter *m*
 f dépositaire *m* [des informations];
    conservateur *m* [de titres]; propriétaire *m* des
    informations; administrateur *m* séquestre
 r хранитель *m*; подразделение *n* охраны

836 custom color map
 d angepaßte Farbkarte *f*
 f carte *f* couleur personnalisée
 r заказная цветная таблица *f*

837 customization
 d Personalisierung *f*; kundenspezifische
    Produktgestaltung *f*
 f personnalisation *f*; production *f* à la demande
 r настройки *fpl* на требования пользователя;
    разработка *f* заказного варианта

838 custom label
 d angepaßtes Etikett *n*

   f étiquette *f* d'utilisateur
   r пользовательская метка *f*

839 custom symbol
   d kundenspezifisches Symbol *n*
   f symbole *m* personnalisé
   r пользовательский символ *m*

840 cut
   d Schnitt *m*
   f coupe *f*
   r сечение *n*; разрез *m*

841 cut-and-fill analysis; cut/fill analysis
   d Einschnitt- und Auftrag-Analyse *f*
   f analyse *f* de déblai-remblai
   r анализ *m* положительных и отрицательных
      объёмов; анализ выёмки с закладкой

   * **cut/fill analysis** → **841**

   * **cutting** → **571**

842 cyber-geography
   d Cybergeografie *f*
   f cybergéographie *f*
   r кибергеография *f*

843 cybermap
   d Cyberspace-Karte *f*
   f caste *f* de cyberespace
   r карта *f* киберпространства

844 cyberspace mapping
   d Cyberspace-Kartierung *f*
   f établissement *m* de cartes de cyberespace;
      représentation *f* de cyberespace
   r картографирование *n* киберпространства

   * **cycle** → **2234**

845 cyclic revision
   (of a map)
   d zyklische Laufenthalten *n*
   f révision *f* cyclique
   r периодическое обновление *n*

846 cylindrical equal-area projection
   d flächentreue Zylinderprojektion *f*
   f projection *f* cylindrique équivalente;
      projection isocylindrique
   r цилиндрическая равновеликая проекция *f*

847 cylindrical [map] projection
   d Zylinderprojektion *f*, Zylinderabbildung *f*
   f projection *f* cylindrique
   r цилиндрическая проекция *f*

   * **cylindrical projection** → **847**

D

* **3D** → 3670

* **DAC** → 1073

* **DAL** → 854

* **damping** → 238

848 **dangling arc**
  d hängender Bogen *m*
  f arc *m* pendant; arc non saturé
  r висячая дуга *f*

849 **dangling line**
  d hängende Linie *f*
  f ligne *f* pendante
  r висячая линия *f*

850 **dangling node; dangling vertex**
  d hängender Knoten[punkt] *m*
  f nœud *m* pendant; sommet *m* pendant
  r висячий узел *m*; висячая вершина *f*

* **dangling vertex** → 850

* **dash** → 314

* **dashed line** → 2722

851 **dash marking; stroke marking**
  d Strichmarkierung *f*
  f marquage *m* en trait
  r штриховое маркирование *n*

852 **data**
  d Daten [*n*]*pl*
  f données *fpl*
  r данные [*n*]*pl*

853 **data access**
  d Datenzugriff *m*
  f accès *m* de données
  r доступ *m* к данным

854 **data access language; DAL**
  d Datenzugriffsprache *f*
  f langage *m* d'accès de données
  r язык *m* доступа к данным

855 **data access security**
  d Datenzugriffsicherheit *f*
  f sécurité *f* d'accès de données
  r защита *f* доступа к данным

* **data acquisition** → 875

856 **data ag[e]ing**
  d Datenalterung *f*
  f vieillissement *m* de données
  r старение *n* данных

* **data aging** → 856

857 **data availability**
  d Datenverfügbarkeit *f*
  f disponibilité *f* de données
  r доступность *f* данных

858 **databank**
  d Datenbank *f*
  f banque *f* de données
  r банк *m* данных; БнД

859 **databank view**
  d Datenbanksicht *f*
  f vue *f* de banque de données
  r представление *n* банка данных

860 **database; DB**
  d Datenbasis *f*
  f base *f* de données
  r база *f* данных; БД

861 **database design**
  d Datenbasisprojektierung *f*
  f dessin *m* de base de données
  r проектирование *n* базы данных

862 **database directory; library reference
  workspace**
  d Datenbasiskatalog *m*
  f directoire *f* de base de données
  r каталог *m* базы данных

863 **database indexing**
  d Datenbasisindizierung *f*
  f indexage *m* de base de données
  r индексирование *n* базы данных

864 **database integrator; DBI**
  d Datenbasisintegrator *m*
  f intégrateur *m* de bases de données
  r интегратор *m* баз данных

865 **database integrity**
  d Datenbasisintegrität *f*
  f intégrité *f* de base de données
  r целостность *f* базы данных

866 **database lock**
  d Datenbasissperre *f*
  f verrou *m* de base de données
  r замок *m* базы данных

867 database management system; DBMS
  d Datenbasis-Managementsystem *n*
  f système *m* de gestion de bases de données; SGBD
  r система *f* управления базами данных; СУБД

868 database model
  d Datenbasis-Modell *n*
  f modèle *m* de base de données
  r модель *f* базы данных

869 database portability
  d Datenbasis-Portabilität *f*
  f portabilité *f* de base de données
  r переносимость *f* базы данных

870 database server
  d Datenbasisserver *m*
  f serveur *m* de bases de données
  r сервер *m* баз данных

871 database specification
  d Datenbasis-Spezifikation *f*
  f spécification *f* de base de données
  r спецификация *f* базы данных

872 database tools
  d Datenbasiswerkzeuge *npl*
  f instruments *mpl* de base de données
  r инструменты *mpl* базы данных

873 data bounds
  d Datengrenzen *fpl*
  f bornes *fpl* de données
  r ограничения *f* данных

874 data calibration
  d Datenkalibrierung *f*
  f calibrage *m* de données
  r калибровка *f* данных

875 data capture; data acquisition; data collection; data gathering
  d Datenerfassung *f*; Messwerterfassung *f*; Datensammeln *n*
  f acquisition *f* de données; saisie *f* de données; collection *f* de données
  r сбор *m* данных; совокупность *f* данных

876 data classification
  d Datenklassifizierung *f*
  f classification *f* de données
  r классификация *f* данных

* data collection → 875

* data collector → 1400

877 data conflation; conflation
  (in GIS)
  d Datenverbindung *f*; Verschmelzung *f*; Zusammenführung *f*
  f fusion *f* [d'information] multisource; intégration *f* de données multisource
  r объединение *n* множество вариантов [данных]; соединение *n* множество вариантов

878 data consistency
  d Datenkonsistenz *f*
  f consistance *f* de données
  r совместимость *f* данных; непротиворечивость *f* данных

879 data conversion
  d Datenkonvertierung *f*
  f conversion *f* de données
  r преобразование *n* данных

880 data coverage
  d Datenüberdeckung *f*
  f données *fpl* de couverture
  r пространственный охват *m* данных

881 data definition; data description
  d Datendefinition *f*; Datenbeschreibung *f*
  f définition *f* de données; description *f* de données
  r определение *n* данных; описание *n* данных

882 data definition language; data description language; DDL
  d Datenbeschreibungssprache *f*; Datendefinitionssprache *f*
  f langage *m* de définition de données; langage de description de données
  r язык *m* определения данных; язык описания данных

* data description → 881

* data description language → 882

883 data descriptive area; DDA
  d Datenbeschreibungszone *f*; Bereich *m* von deskriptive Daten
  f zone *f* des données descriptives
  r зона *f* описательных данных

884 data descriptive file
  d Datenbeschreibungsdatei *f*
  f fichier *m* des données descriptives
  r файл *m* описательных данных

885 data dictionary
  (detailed definitions of the codes employed for identifying objects and for attribute values)
  d Daten[bank]verzeichnis *n*
  f dictionnaire *f* de données; vocabulaire *m* de données
  r словарь *m* [базы] данных

*   **data element** → 922

886 **data entry; data input; insertion of data**
  d Dateneingabe *f*; Dateneinführung *f*
  f entrée *f* de données; introduction *f* de données;
       insertion *f* de données
  r ввод *m* данных

887 **data error**
  d Datenfehler *m*
  f erreur *f* de données
  r погрешность *f* данных

888 **data exchange; data interchange; DATEX**
  d Datenaustausch *m*; Datenvermittlung *f*
  f échange *m* de données
  r обмен *m* данными

889 **data exchange error**
  d Datenaustauschfehler *m*
  f erreur *f* d'échange de données
  r погрешность *f* обмена данными

890 **data extraction; data fetch[ing]**
  d Datenextraktion *f*; Datenauszug *m*
  f extraction *f* de données
  r извлечение *n* данных; выделение *n* данных

*   **data fetch** → 890

*   **data fetching** → 890

891 **data field; field**
       (in database)
  d Datenfeld *n*
  f champ *m* de données
  r поле *n* данных

892 **data flow**
  d Datenfluss *m*
  f flux *m* de données
  r информационный поток *m*

893 **data format**
  d Datenformat *n*
  f format *m* de données
  r формат *m* данных

*   **data gathering** → 875

894 **data history**
  d Datengeschichte *f*
  f historique *f* de données
  r хронология *f* данных

*   **data input** → 886

895 **data integrator**
       (company that gather digital map data from a

variety of public or private sources and adapt
it for a specific mapping project and target
software)
  d Datenintegrator *m*
  f intégrateur *m* de données
  r интегратор *m* данных

896 **data integrity; feature integrity**
  d Datenintegrität *f*
  f intégrité *f* de données
  r целостность *f* данных

*   **data interchange** → 888

897 **data layer; data plan**
  d Datenschicht *f*
  f plan *m* de données
  r слой *m* данных

*   **data logger** → 1400

898 **data manipulation language; DML**
  d Datenbehandlungssprache *f*
  f langage *m* de manipulation de données
  r язык *m* манипулирования данных

899 **data mask**
  d Datenmaske *f*
  f masque *f* de données
  r маска *f* данных

900 **data mining**
  d Data-Mining *n*; Datamining *n*
  f datamining *m*
  r раскопка *f* данных; интеллектуальный
       анализ *m* данных; обнаружение *n* знаний в
       базах данных

901 **data model**
  d Datenmodell *n*
  f modèle *m* de données
  r модель *f* данных

902 **data mosaic**
  d Datenmosaik *n*
  f mosaïque *f* de données
  r мозаика *f* данных

*   **data out** → 903

903 **data out[put]**
  d Datenausgabe *f*
  f sortie *f* de données; émission *f* de données
  r вывод *m* данных

904 **data packager**
       (company that repackage existing map data,
       with very little customization, for mass
       distribution)
  d Datenpackbetrieb *m*

f conditionneur *m* de données; emballeur *m* de données
r поставщик *m* данных

* **data plan → 897**

905 **data point; information point**
  d Datenpunkt *m*
  f point *m* d'information
  r один *m* из результатов [обработки данных]; частное значение *n*; экспериментальная точка *f*

906 **data quality; feature quality**
  (relative accuracy and precision of a particular GIS database)
  d Datenqualität *f*
  f qualité *f* des données
  r качество *n* данных

907 **data quality report**
  d Datenqualitätsbericht *m*
  f rapport *m* de qualité des données
  r отчёт *m* качества данных

* **data record → 3184**

908 **data reduction**
  d Reduktion *f* der Daten; Datenreduktion *f*
  f réduction *f* de données
  r сокращение *n* данных; приведение *n* данных

* **data reformation → 910**

909 **data reliability; informational reliability**
  d Datenzuverlässigkeit *f*
  f fiabilité *f* de données
  r надёжность *f* данных; информационная надёжность

910 **data reorganization; data reformation**
  d Datenumorganisierung *f*; Datenzusammenlegung *f*; Datenumstellung *f*; Datenreform *f*
  f réorganisation *f* de données
  r реорганизация *f* данных

911 **dataset**
  d Datenmenge *f*
  f ensemble *m* de données; lot *m* de données
  r множество *n* данных; массив *m* данных; набор *m* данных

912 **dataset information map**
  d Datenmengekarte *f*
  f carte *f* d'ensembles de données
  r информационная карта *f* набора данных

913 **dataset name**
  d Datenmengename *m*
  f nom *m* d'ensemble de données
  r имя *n* массива данных

914 **data sharing**
  d Datenteilung *f*; gemeinsame Datenbenutzung *f*
  f partage *m* de données
  r совместное пользование *n* данных

915 **data smoothing**
  d Datenglättung *f*
  f lissage *m* de données
  r сглаживание *n* данных

916 **data structure**
  d Datenstruktur *f*
  f structure *f* de données
  r структура *f* данных

917 **data temporality; data temporariness**
  d vorübergehender Datenzustand *m*; Dateneinstweiligkeit *f*; Datenzeitweiligkeit *f*
  f temporalité *f* de données
  r временные аспекты *mpl* данных

* **data temporariness → 917**

918 **data transfer; data transmission**
  d Datenübertragung *f*; Datenübermittlung *f*
  f transfert *m* de données; transmission *f* de données
  r передача *f* данных

* **data transmission → 918**

919 **data type**
  d Datentyp *m*
  f type *m* de données
  r тип *m* данных

920 **date**
  (a fixed point of time)
  d Datum *n*; Tagesdatum *n*
  f date *f*
  r дата *f*

921 **date line**
  d Datumsgrenze *f*
  f ligne *f* de changement de date
  r линия *f* перемены даты

* **DATEX → 888**

* **datum → 2811, 3462**

922 **datum; data element**
  (a piece of information)
  d Dateneinzelheit *f*; Datenelement *n*; Datengrundeinheit *f*; Angabe *f*; Datum *n*

f détail *m* de données; élément *m* de données; donnée *f*; datum *m*
r данное *n*; элемент *m* данных; данная величина *f*

923 datum conversion
d Datumkonversion *f*
f conversion *f* de repère; conversion de datum
r преобразование *n* репера

924 datum level
d Horizontale *f*; Bezugshorizont *m*
f niveau *m* de référence
r опорный уровень *m*; база *f*; нулевой уровень; реперная отметка *f*; условный горизонт *m*

* **datum mark** → 2811

925 datum plane
d Bezugsebene *f*; Bezugsfläche *f*; Referenzebene *f*; Vergleichsebene *f*
f surface *f* de référence; plan *m* de référence
r нулевая плоскость *f*; опорная плоскость; плоскость начала отсчёта; плоскость относимости; начальная плоскость

* **datum point** → 2811

926 datum shift
d Datumverschiebung *f*
f décalage *m* de datum
r смещение *n* начала отсчёта; смещение репера

927 day-time mapping
d Kartierung *f* am Tag
f cartographie *f* diurne
r дневное картографирование *n*

* **DB** → 860

928 3D bar chart
d 3D-Streifendiagramm *n*
f diagramme *m* à colonnes 3D
r объёмная столбчатая диаграмма *f*

* **DBI** → 864

* **DBMS** → 867

* **DC** → 1007

* **3D coordinates** → 3368

* **DCS** → 1008

* **DCW** → 1036

* **DD** → 1170

* **DDA** → 883

929 3D datum
d 3D-Datum *n*
f datum *m* 3D
r трёхмерный репер *m*; трёхмерный базис *m*; трёхмерная отметка *f* уровня

* **DDE** → 1198

* **3D digitizer** → 3383

* **3D digitizing** → 1189

930 3D digitizing techniques
d 3D-Diskretisierungstechnik *f*
f technique *f* de digitalisation 3D
r метод *m* трёхмерного дигитализирования

* **DDL** → 882

931 deactivation
d Inaktivierung *f*; Entaktivierung *f*
f déactivation *f*
r деактивация *f*

932 decay distance
d Wellenauslaufbereich *m*; Auslaufbereich *m*
f distance *f* d'amortissement; distance de dégénérescence
r дистанция *f* затухания [волн]

* **decentralized cartographic database** → 1139

* **deception of data** → 3446

933 decimal degree
d dezimaler Grad *m*
f degré *m* décimal
r десятичный градус *m*

* **decision-help information system** → 935

934 decision-making; decision support
d Entscheidungstreffen *n*; Entscheidungsfindung *f*; Entscheidungsvorbereitung *f*; Beschlussfassung *f*
f prise *f* de décision; procédé *m* de décision; découverte *f* de décision
r принятие *n* решений; выбор *m* решения

935 decision-making information system; decision-help information system; decision-support system; DSS
d Entscheidungsfindung-Informationssystem *n*; Entscheidungshilfe-System *n*

f système *m* d'information décisionnel
r информационная система *f* принятия
решений

* **decision support** → 934

936 decision-support database
d Entscheidungstreffen-Datenbasis *f*
f base *f* de données décisionnelles
r база *f* данных [для] системы принятия
решений

* **decision-support system** → 935

937 decision theory
d Entscheidungstheorie *f*
f théorie *f* de la décision
r теория *f* решения

938 decision tree
d Entscheidungsbaum *m*
f arbre *m* de décisions
r дерево *n* решений

939 declination
d Abweichung *f*, Deklination *f*
f déclinaison *f*
r склонение *n*

940 declination diagram
d Deklinationsdiagramm *n*
f diagramme *m* de déclinaison
r схема *f* сближения меридианов

* **declination of grid north** → 746

941 decoding; decryption
d Decodierung *f*; Decodieren *n*; Entschlüsseln *n*
f décodage *m*; déchiffrage *m*; décryptage *m*
r декодирование *n*; дешифрирование *n*;
расшифровка *f*

* **decompact** *v* → 943

942 decomposition
d Dekomposition *f*; Zerlegung *f*
f décomposition *f*
r декомпозиция *f*; разбиение *n*; разложение *n*

943 decompress *v*; **decompact** *v*; **uncompress** *v*;
unzip *v*
d dekomprimieren; entkomprimieren
f décompresser; décompacter; dézipper
r декомпрессировать

* **decryption** → 941

944 defect
d Defekt *m*
f défaut *m*

r дефект *m*

945 defects structure
d Defektstruktur *f*
f structure *f* des défauts
r структура *f* дефектов

946 Defense Mapping Agency; DMA
d militärgeografischer Dienst *m*;
Verteidigungskartierungsstelle *f*
f Agence *f* de cartographie de défense
r Картографическое управление *n*
Министерства обороны США

* **definiteness** → 2878

947 deflection; deflexion
d Ablenkung *f*, Ausschlag *m*
f déflexion *f*
r уклонение *n*

**948 deflection of plumb line; deviation of plumb
line; deflection of the vertical; plumb-line
deflection; plumb-line deviation; angle of
deviation of the vertical**
d Lotabweichung *f*
f déflexion *f* de la verticale; déviation *f* de la
verticale; angle *m* de déviation de la verticale
r уклонение *n* отвесной линии

* **deflection of the vertical** → 948

* **deflexion** → 947

949 degree; grade
d Grad *m*
f degré *m*; grade *m*
r градус *m*

* **degree square** → 2318

* **dejagging** → 136

950 Delaunay area
d Delaunay-Zone *f*
f zone *f* de Delaunay
r зона *f* Делоне

951 Delaunay triangulation
(a network that connects each point in a set of
points to its nearest neighbors; topological
"dual" of the Voronoi network)
d Delaunay-Triangulation *f*
f triangulation *f* de Delaunay
r триангуляция *f* Делоне

952 delay; lag[ging]; retardation
d Verzögerung *f*; Verzug *m*; Nacheilung *f*
f délai *m*; retard[ement] *m*
r запаздывание *n*; замедление *n*; задержка *f*

953 delete *v*; **erase** *v*
  d löschen
  f effacer
  r стирать

**954 deleting; deletion; clear[ing]; erasing;
      erasure; purging**
  d Löschen *n*; Löschung *f*
  f effaçage *m*; effacement *m*
  r стирание *n*; отмена *f*

  * **deletion** → 954

  * **deliberate data modification** → 3446

955 delineation
  d Ausscheidung *f*; Versatz *m*
  f désalignement *m*; délimitation *f*
  r очерчивание *n*

956 delineator
  d Leitpfosten *m*; Leitpflock *m*
  f délinéateur *m*; marqueur *m* routier
  r прибор *m*, записывающий длину и профиль
      пройдённого пути; проектировщик *m*

  * **DEM** → 1042, 1043

  * **demand** → 3140

957 demarcation; boundary marking
  d Abgrenzung *f*
  f démarcation *f*; abornement *m*; bornage *m*
  r демаркация *f*; установление *n* границ;
      разграничение *n*

  * **demarcation line** → 385

958 DEM extraction
      (creates an output elevation raster object of the
      selected model area)
  d digitale Höhenlinienmodell-Extraktion *f*
  f extraction *f* de modèle altimétrique digital
  r извлечение *n* цифровой модели высотных
      точек

959 demographic analysis
  d demografische Analyse *f*
  f analyse *f* démographique
  r демографический анализ *m*

960 demographic data
  d demografische Daten *pl*;
      Bevölkerungsdaten *pl*; Strukturwerte *mpl*
  f données *fpl* démographiques
  r демографические данные *pl*

  * **demographic development** → 2858

961 demographic forecasting

  d demografische Prognostizierung *f*
  f prédiction *f* démographique
  r демографическое прогнозирование *n*

**962 demographic projection; population
      projection**
  d Bevölkerungsvorausberechnung *f*
  f projection *f* démographique; projection de la
      population
  r предположительное исчисление *n*
      населения; демографический прогноз *m*;
      прогнозирование *n* численности населения

**963 demographic pyramid; population
      pyramid; age[-sex] pyramid**
  d Bevölkerungspyramide *f*;
      Bevölkerungsbaum *m*; Alterspyramide *f*
  f pyramide *f* de population
  r распределение *n* населения по возрастным
      группам и полу; возрастная пирамида *f*;
      возрастно-половая пирамида [населения]

  * **demographics** → 965

964 demographic transition
  d demografische Transition *f*; demografischer
      Übergang *m*
  f transition *f* démographique
  r сильное изменение *n* картины рождаемости
      и смертности; переходный период *m*
      естественного движения населения

  * **demographic trend** → 2858

965 demography; demographics
  d Demografie *f*
  f démographie *f*
  r демография *f*

966 dendrogram
  d Dendrogramm *n*
  f dendrogramme *m*
  r древовидная схема *f*

  * **denomination** → 2485

967 denormalization
  d Denormalisierung *f*; Denormierung *f*
  f dénormalisation *f*
  r денормализация *f*

968 densification
  d Verdichtung *f*; Verdichten *n*; Densifikation *f*
  f densification *f*
  r уплотнение *n*; загущение *n*; уменьшение *n*
      объёма

969 density
  d Dichte *f*; Dichtheit *f*; Dichtigkeit *f*

   f densité *f*
   r плотность *f*; густота *f*

970 density of observation
   (number of observations within an area)
   d Beobachtungsdichte *f*; Observationsdichte *f*
   f densité *f* d'observation
   r плотность *f* наблюдений

971 density of points; point density; dot density
   d Punktdichte *f*
   f densité *f* de points
   r плотность *f* точек

972 density slicing
   d Äquidensiten *n*; Density-Slicing *n*
   f isodensitométrie *f*; équidensitométrie *f*
   r расщепление *n* плотности; расслоение *n* плотности

973 department
   d Abteilung *f*; Fachabteilung *f*; Amt *n*; Ressort *n*; Gebiet *n*; Dezernat *n*
   f département *m*
   r отдел *m*; отделение *n*; ведомство *n*; департамент *m*

974 dependent variable
   d abhängige Variable *f*
   f variable *f* dépendante
   r зависимая переменная *f*

975 depreciation; amortization
   d Wertminderung *f*; Amortisation *f*; degressive Abschreibung *f*; Minderung *f* des Wertes
   f dépréciation *f*; amortissement *m*; moins-value *f*
   r обесценивание *n*; изнашивание *n*; амортизационные отчисления *npl*

976 depression
   d Depression *f*; Vertiefung *f*; Unterdrückung *f*; Niederdrückung *f*
   f dépression *f*
   r депрессия *f*

977 depression contour
   d Einsenkungskontur *f*; Einsenkungsprofil *n*
   f courbe *f* de cuvette; courbe de niveau de dépression
   r горизонталь *f* с отрицательной отметкой; замыкающая *f* понижения местности

978 depth
   d Tiefe *f*
   f profondeur *f*
   r глубина *f*

   * **depth contour** → 979

979 depth contour [line]; isobath; depth curve; submarine contour; below-sea-level contour; bottom contour
   d Isobathe *f*; Tiefenlinie *f*
   f courbe *f* isobathe; isobathe *f*; courbe de niveau bathymétrique
   r изобата *f*; линия *f* равных глубин

   * **depth curve** → 979

980 depth effect
   d Tiefenwirkung *f*
   f effet *m* de profondeur
   r эффект *m* глубины

981 depth-first search
   d tiefenorientierte Suche *f*
   f recherche *f* en profondeur [d'abord]
   r поиск *m* [преимущественно] в глубину

982 depth map
   d Tiefenkarte *f*
   f carte *f* de profondeurs
   r карта *f* дальностей; карта глубин

983 derivative map; derived map
   d abgeleitete Karte *f*; Folgekarte *f*
   f carte *f* dérivative; carte dérivée
   r производная карта *f*; извлечённая карта

984 derived column; derived field
   d abgeleitetes Feld *n*
   f champ *m* dérivé
   r выведенная колонка *f*; производное поле *n*

   * **derived field** → 984

   * **derived map** → 983

   * **derived unit** → 985

985 derived unit [of measurement]
   d kohärente Einheit *f*; abgeleitete Einheit
   f grandeur *f* dérivée; unité *f* dérivée
   r производная единица *f*

986 descending index
   d absteigender Index *m*
   f index *m* descendant
   r нисходящий индекс *m*

987 descending node
   d absteigender Knoten *m*
   f nœud *m* descendant
   r узел-потомок *m*

988 descending sort
   (of a field in a table)
   d absteigende Sortierung *f*; absteigendes Sortieren *n*

f triage *m* descendant; tri *m* décroissant
r сортировка *f* по убыванию; нисходящая
сортировка

989 description
  d Deskription *f*; Beschreibung *f*
  f description *f*
  r описание *n*; дескрипция *f*

990 descriptive data
  d deskriptive Daten *pl*; beschreibende Daten
  f données *fpl* descriptives
  r описательные данные *pl*

* **deselecting → 991**

991 deselection; deselecting; unselecting
  d Abwählung *f*
  f désélection *f*; annulation *f* de sélection
  r отмена *f* отбора; отмена выделения; отмена
  селекции

* **design → 994**

992 design; project; plan
  d Entwurf *m*; Projekt *n*; Design *n*; Plan *m*;
  Bauplan *m*
  f projet *m*; plan *m*
  r проект *m*; дизайн *m*; план *m*

993 designator
  d Bezeichner *m*
  f indicatif *m*
  r указатель *m*; [кодовое] обозначение *n*

994 design[ing]
  d Projektierung *f*
  f conception *f*; établissement *m* d'un projet
  r проектирование *n*

995 desktop GIS
  d Desktop-GIS *n*
  f SIG *m* bureautique
  r настольная ГИС *f*

996 desktop mapping
  d Tischkartierung *f*; Desktop-Kartografieren *n*
  f cartographie *f* tabulaire; cartographie sur
  ordinateur
  r настольное картографирование *n*

* **desmoothing → 3875**

997 destination; butt; goal
  d Verwendungszweck *m*; Ziel *n*; Goal *n*
  f destination *f*; but *m*; objectif *m*
  r назначение *n*; место *n* назначения; цель *f*;
  предназначение *n*

998 destination node

d Zielknoten *m*
f nœud *m* de destination
r узел *m* назначения

999 detail plate; line original
  d Strichvorlage *f*; Strichoriginal *n*
  f original *m* au trait
  r штриховой оригинал *m* [карты]

1000 detector
  d Detektor *m*
  f détecteur *m*
  r детектор *m*

1001 detector imbalance
  d Detektorabrustung *f*;
  Detektorungleichgewicht *n*
  f déséquilibre *m* de détecteur
  r неустойчивость *f* детектора; нарушение *n*
  баланса детектора

1002 developable surface
  d entwickelbare Fläche *f*
  f surface *f* développable
  r развёртывающаяся поверхность *f*

1003 developer's toolkit
  d Entwickler-Werkzeugsatz *m*
  f kit *m* d'instruments de réalisateur
  r инструментарии *mpl* разработчика

1004 development
  d Entwicklung *f*; Abwicklung *f*
  f développement *m*; élaboration *f*
  r развитие *n*; разработка *f*

* **development map → 2101**

* **development plan → 2101**

1005 deviation
  d Ablenkung *f*; Abweichung *f*; Deviation *f*
  f déviation *f*
  r девиация *f*; отклонение *n*

1006 deviation line
  d Ablenkungslinie *f*
  f ligne *f* de déviation
  r линия *f* девиации; линия отклонения

* **deviation of plumb line → 948**

1007 device coordinates; DC
  d Gerätekoordinaten *fpl*
  f coordonnées *fpl* d'appareil
  r координаты *fpl* устройства

1008 device coordinate system; DCS
  d Gerätekoordinatensystem *n*

f système *m* de coordonnées d'appareil
r координатная система *f* устройства

1009 device driver; driver; handler
  d Gerätetreiber *m*; Treiber *m*
  f driver *m* [de dispositif]; pilote *m* [de dispositif]
  r драйвер *m* [устройства]

1010 device space
  d Geräte[koordinaten]raum *m*
  f espace *m* d'appareil; espace périphérique
  r пространство *n* устройства

  * **2D feature** → 1156

  * **3D feature** → 1157

1011 3D geoimage; volumetric geoimage
  d 3D-Geobild *n*
  f géo-image *f* 3D
  r трёхмерное геоизображение *n*; объёмное геоизображение

  * **DGI** → 1144

1012 2D GIS
  d 2D-GIS *n*
  f SIG *m* 2D
  r двумерная ГИС *f*

1013 3D GIS
  d 3D-GIS *n*
  f SIG *m* 3D
  r трёхмерная ГИС *f*

  * **4D GIS** → 1467

1014 3D global geographic model
  d globales geografisches 3D-Modell *n*
  f modèle *m* géographique global 3D
  r трёхмерная глобальная географическая модель *f*

  * **DGM** → 1049

  * **DGM standard** → 1049

  * **DGN** → 1163

1015 3D hyperbolic graph
  d hyperbolischer 3D-Graph *m*
  f graphe *m* hyperbolique 3D
  r трёхмерный гиперболический граф *m*

  * **diagonal** → 1017

1016 diagonal direction
  d diagonale Richtung *f*
  f direction *f* diagonale

r диагональное направление *n*

1017 diagonal [line]
  d Diagonale *f*
  f diagonale *f*
  r диагональ *f*

1018 diagram; chart; graph; plot
  d Diagramm *n*; Schaubild *n*; Kurvenblatt *n*; Kurvenbild *n*; Grafik *f*
  f diagramme *m*; graphique *m*; courbe *f* représentative
  r диаграмма *f*; схема *f*; график *m*; лист *m* кривых

  * **diagram map** → 539

  * **diagrammatic map** → 539

1019 diagrammatic representation; diagrammatic view
  d diagrammatische Darstellung *f*; schematische Ansicht *f*; diagrammatische Ansicht
  f représentation *f* schématique; vue *f* schématique
  r схематическое представление *n*; схематическое изображение *n*

  * **diagrammatic view** → 1019

1020 diameter
  (of a network)
  d Diameter *n*; Durchmesser *m*
  f diamètre *m*
  r диаметр *m*

  * **diapason** → 2999

  * **diazo copy** → 1021

1021 diazo copy[ing]; diazo process
  (rapid method for copying documents in which the image is developed by exposure to ammonia)
  d Diazokopierverfahren *n*; Diazokopie *f*
  f diazocopie *f*; procédé *m* diazo[ïque]
  r диазотипный процесс *m*; диазотипия *f*

  * **diazo process** → 1021

1022 differential corrections
  d differentiale Korrekturen *fpl*
  f corrections *fpl* numériques
  r дифференциальные поправки *fpl*

1023 differential leveling
  d differentiale Nivellierung *f*
  f nivellement *m* différentiel
  r дифференциальное нивелирование *n*

* differential position accuracy → 1024

1024 differential position[al] accuracy
  d differentiale Positionsgenauigkeit *f*;
  differentiale Positioniergenauigkeit *f*
  f exactitude *f* de position différentielle
  r дифференциальная позиционная точность *f*

1025 differential positioning
  d differentiale Positionierung *f*
  f positionnement *m* différentiel
  r дифференциальное позиционирование *n*

1026 differential ranging
  d differentiale Entfernungsmessung *f*
  f mesure *f* de distance différentielle;
  jalonnement *m* différentiel
  r дифференциальное определение *n*
  дальности; дифференциальное определение
  дистанции; дифференциальные
  промеры *mpl*; дифференциальный замер *m*

1027 differential rectification
  d Differentialentzerrung *f*
  f redressement *m* différentiel
  r дифференциальное выпрямление *n*

1028 diffuse color
  d diffuse Farbe *f*
  f couleur *f* diffuse
  r диффузный цвет *m*

1029 diffuse reflectance
  d diffuse Reflektanz *f*
  f réflectance *f* diffuse
  r диффузная отражательная способность *f*

1030 diffusion-limited aggregation
  d Diffusion-begrenzte Aggregation *f*
  f agrégation *f* limitée par diffusion
  r агрегация *f*, ограниченная диффузией

* DIGEST → 1047

1031 digital; numeric[al]
  d digital; numerisch
  f numérique
  r цифровой; числовой; численный

* digital-analog converter → 1073

1032 digital cadastral map
  d digitale Kataster[plan]karte *f*
  f plan *m* cadastral numérique
  r цифровая кадастровая карта *f*

1033 digital camera; digitizing camera
  d Digitalkamera *f*; digitale Kammer *f*
  f caméra *f* numérique; caméra numérisante
  r цифровая камера *f*

1034 digital cartographic model
  d digitales kartografisches Modell *n*
  f modèle *m* cartographique numérique
  r цифровая картографическая модель *f*

1035 digital cartography
  d numerische Kartografie *f*
  f cartographie *f* numérique
  r цифровая картография *f*

1036 digital chart of the world; DCW
  d numerische Weltkarte *f*; digitale Weltkarte
  f carte *f* mondiale numérique
  r цифровая карта-основа *f* мира (*цифровой
  аналог карты ONC*)

1037 digital classification
  d numerische Klassifizierung *f*
  f classification *f* numérique
  r цифровая классификация *f*

1038 digital color separation
  d numerische Farbteilung *f*
  f séparation *f* des couleurs numérique
  r цифровое цветоделение *n*

1039 digital coverage
  d digitaler Erfassungsbereich *m*
  f couverture *f* numérique
  r цифровое покрытие *n*

1040 digital data; numeric data
  d numerische Daten *pl*
  f données *fpl* numériques
  r цифровые данные *pl*

1041 digital elevation data
  d digitale Höhenliniendaten *pl*; digitale
  Elevationsdaten *pl*
  f données *fpl* d'élévation numériques
  r цифровые данные *pl* высотных точек

1042 digital elevation matrix; DEM
  d digitale Höhenlinienmatrix *f*
  f matrice *f* d'élévation numérique
  r цифровая матрица *f* высотных точек

1043 digital elevation model; DEM
  d digitales Höhenlinienmodell *n*
  f modèle *m* altimétrique digital; modèle
  d'élévation numérique
  r цифровая модель *f* высотных точек

1044 digital exchange format; DXF
  d numerisches Austauschformat *n*
  f format *m* d'échange numérique
  r цифровой формат *m* обмена

1045 digital format
  d numerisches Format *n*

f format *m* numérique
r цифровой формат *m*

1046 digital geochemistry
d numerische Geochemie *f*
f géochimie *f* numérique
r цифровая геохимия *f*

1047 digital geographic information exchange standard; DIGEST
d DIGEST-Standard *m*
f standard *m* DIGEST
r стандарт *m* DIGEST

1048 digital geographic vector-based format
d numerisches geografisches vektorbasiertes Format *n*
f format *m* numérique géographique vectoriel
r цифровой векторный географический формат *m*

* **digital geospatial metadata** → 1049

1049 digital geospatial metadata [standard]; DGM [standard]
d DGM-Format *n*
f standard *m* DGM
r стандарт *m* DGM

1050 digital geotopographic information system
d digitales geotopografisches Informationssystem *n*
f système *m* d'information géotopographique numérique
r цифровая геотопографическая информационная система *f*

1051 digital image; numeric image
d digitales Bild *n*; Digitalbild *n*; digitales Abbild *n*
f image *f* numérique
r цифровое изображение *n*

1052 digital image matching
d Digitalbildanpassung *f*; digitale Bildzuordnung *f*
f adaptation *f* d'images numériques
r согласование *n* цифровых изображений

1053 digital image processing
d Digitalbildverarbeitung *f*
f traitement *m* d'image numérique
r обработка *f* цифрового изображения

1054 digit[al]ization; digitizing
d Digitalisierung *f*
f digitalisation *f*; numérisation *f*
r дигитализирование *n*; оцифровка *f*; цифрование *n*; оцифрование *n*; перевод *m* в цифровую форму; ввод *m* аналоговой информации

* **digitalization error** → 1087

1055 digit[al]ize *v*
d digitalisieren
f numériser; digitaliser
r дигитализировать; преобразовывать в цифровую форму

1056 digital landmass system
d numerisches Landmaßsystem *n*
f système *m* de masses continentales en code numérique
r система *f* отображения поверхности Земли в цифровом коде

1057 digital landscape object-structured model
d digitales objektstrukturiertes Landschaftsmodell *n*
f modèle *m* numérique de paysage structuré objet
r цифровая объектно-структурированная модель *f* местности

1058 digital line graph; DLG
(data format developed by US Geological Survey National Mapping Division that uses a topological vector model)
d digitaler Streckengraph *m*; DLG-Format *n*
f format *m* DLG
r формат *m* DLG (*обмена цифровыми картографическими данными, принятый в Геологической сумке США*)

* **digital map** → 1239

1059 digital map data
d digitale Kartendaten *pl*
f données *fpl* cartographiques numériques; données cartographiques numérisées
r цифровые картографические данные *pl*

* **digital map database** → 442

1060 digital map library
d digitale Kartenbibliothek *f*
f bibliothèque *f* de cartes numériques; cartothèque *f*
r библиотека *f* цифровых карт

* **digital orthophoto** → 1061

1061 digital orthophoto [image]
d numerische Orthofotografie *f*
f orthophoto[graphie] *f* numérique
r цифровой ортоснимок *m*

1062 digital orthophoto quadrangles standard; DOQ standard
(US standard for computer-generated image of an aerial photography)

d DOQ-Standard *m*
f standard *m* DOQ
r стандарт *m* DOQ (*стандарт Геологической съёмки США для передачи цифровых изображений высокого разрешения*)

1063 digital photogrammetry
d numerisches Messbildverfahren *n*; numerische Fotogrammetrie *f*
f photogrammétrie *f* numérique
r цифровая фотограмметрия *f*

1064 digital photography; digitized photography
d numerisches Foto *n*; digitalisierte Fotografie *f*; digitales Foto
f photo[graphie] *f* numérique; photo[graphie] numérisée
r цифровой снимок *m*; дискретизированная фотография *f*; оцифрованной снимок

1065 digital plotter
d numerischer Plotter *m*; numerischer Kurvenschreiber *m*
f traceur *m* numérique
r цифровой графопостроитель *m*; цифровой плоттер *m*

* **digital shading** → 115

1066 digital signature
d numerische Unterschrift *f*
f signature *f* numérique
r цифровая надпись *f*

1067 digital situation model; DSM
d digitales Situationsmodell *n*
f modèle *m* numérique de situation
r цифровая модель *f* ситуации

1068 digital source
d numerische Quelle *f*
f source *f* numérique
r цифровой источник *m*

1069 digital space
d digitaler Raum *m*
f espace *m* numérique
r цифровое пространство *n*

1070 digital surface
d Digitalfläche *f*
f surface *f* numérique
r цифровая поверхность *f*

1071 digital tablet; graphic tablet; tablet; digitizing board
d [numerisches] Tablett *n*; Tafelchen *n*; Abtaster *m*
f table *f* de numérisation; table numérisante; tablette *f* [à numériser]; planchette *f*

r цифровой планшет *m*; графический планшет; таблет *m*; графическое устройство *n* ввода данных; сколка *f*

* **digital taxonomy** → 2581

1072 digital terrain model; DTM
d digitales Geländemodell *n*
f modèle *m* numérique de terrain
r цифровая модель *f* рельефа; ЦМР; цифровая модель местности; ЦММ; математическая модель местности; МММ

1073 digital[-to]-analog converter; DAC
d Digital-Analog-Konverter *m*; D/A-Umwandler *m*; DAC
f convertisseur *m* numérique-analogique
r цифрово-аналоговый преобразователь *m*; ЦАП

1074 digital topographical data
d numerische topografische Daten *pl*
f données *fpl* topographiques numériques
r цифровые топографические данные *pl*

1075 digital topography
d digitale Topografie *f*
f topographie *f* numérique
r цифровая топография *f*

* **digitization** → 1054

* **digitize** *v* → 1055

1076 digitized cloud mosaic
d digitalisierte Wolkenaufnahme *f*
f mosaïque *f* numérique de nuages
r цифровой монтаж *m* снимков облачности

1077 digitized data
d digitalisierte Daten *pl*
f données *fpl* numérisées
r данные *pl*, преобразованные в цифровую форму; данные в цифровой форме; оцифрованные данные

1078 digitized device; digitizer; quantifier
d Digitalgeber *m*; Digitalisierer *m*; Digitizer *m*; digitalisierendes Gerät *n*
f dispositif *m* digitalisé; digitaliseur *m*; capteur *m* digital; numériseur *m*
r устройство *n* оцифровки; дигитайзер *m*; цифрователь *m*

1079 digitized image
d digitalisiertes Bild *n*
f image *f* numérisée
r оцифрованное изображение *n*; изображение, преобразованное в цифровую форму

1080 **digitized map**
   d bezifferte Karte *f*
   f carte *f* numérisée
   r оцифрованная карта *f*

   * **digitized photography** → 1064

   * **digitizer** → 1078

1081 **digitizer configuration**
   d Digitizer-Konfiguration *f*
   f configuration *f* de digitaliseur
   r конфигурация *f* дигитайзера

1082 **digitizer cursor; digitizer puck; puck; crosshair pointer; cross-haired cursor**
   d Fadenkreuz-Zeiger *m*
   f curseur *m* réticule; curseur réticulaire; pointeur *m* en croix; pointeur en fils croisés; viseur *m*
   r курсор *m* (*конструктивная часть цифрователя*)

1083 **digitizer driver**
   d Digitizer-Treiber *m*
   f driver *m* de digitaliseur
   r драйвер *m* дигитайзера

1084 **digitizer glass**
   d Digitalisierlupe *f*; Abtastlupe *f*
   f loupe *f* de digitaliseur
   r лупа *f* дигитайзера

1085 **digitizer menu; digitizing menu**
   d Digitalisierer-Menü *n*
   f menu *m* de digitaliseur
   r меню *n* дигитайзера

   * **digitizer puck** → 1082

1086 **digitizer template**
   d Digitalisierer-Schablone *f*
   f modèle *m* de numériseur
   r шаблон *m* дигитайзера

   * **digitizing** → 1054

   * **digitizing board** → 1071

   * **digitizing camera** → 1033

1087 **digitizing error; digitalization error**
   d Digitalisierungsfehler *m*
   f erreur *f* de digitalisation
   r погрешность *f* дигитализирования

1088 **digitizing hardware**
   d Digitalisierungshardware *f*
   f matériel *m* de digitalisation
   r аппаратное обеспечение *n*

дигитализирования

   * **digitizing menu** → 1085

1089 **digitizing module**
   d Digitalisierungsmodul *m*
   f module *m* de numérisation
   r модуль *m* дигитализирования

1090 **digitizing options**
   d Digitalisierungsoptionen *fpl*
   f options *fpl* de numérisation
   r опции *fpl* дигитализирования

1091 **digitizing session**
   d Digitalisierungssession *f*
   f session *f* de numérisation
   r сеанс *m* оцифрования; сеанс дигитализирования

1092 **digitizing software**
   d Digitalisierungssoftware *f*
   f logiciel *m* de numérisation
   r программное обеспечение *n* дигитализирования

1093 **digitizing window**
   d Digitalisierungsfenster *n*
   f fenêtre *f* de numérisation
   r окно *n* дигитализирования; окно оцифровки

1094 **Dijkstra's algorithm**
   d Dijkstra-Algorithmus *m*
   f algorithme *m* de Dijkstra
   r алгоритм *m* Дийкстра

1095 **dike**
   d Deich *m*
   f dyke *m*; filon *m* rocheux
   r сточная канава *f*; ров *m*; дамба *f*; плотина *f*; запруда *f*; перемычка *f*

1096 **dilatation filter**
   d Dilatationsfilter *m*
   f filtre *m* de dilatation
   r фильтр *m* растяжения

1097 **dilution of precision; DOP**
   d Genauigkeitsabfall *m*; Verdünnung *f* der Genauigkeit
   f dilution *f* de précision
   r показатель *m* снижения точности

1098 **2D image**
   d 2D-Bild *n*
   f image *f* 2D
   r двумерное изображение *n*; плоское изображение

1099 **2,5D image**
  d 2,5D-Bild *n*
  f image *f* 2,5D
  r 2,5-мерное изображение *n*

  * **3D image → 3969**

1100 **3D image measuring**
  d 3D-Bild-Messung *f*
  f mesure *f* d'images 3D
  r измерение *n* трёхмерных изображений

1101 **dimension**
  d Ausmaß *n*; Vermaß *n*; Dimension *f*
  f dimension *f*
  r размер *m*; оразмерение *n*

1102 **dimensionality**
  d Dimensionalität *f*
  f dimensionnalité *f*
  r размерность *f*

1103 **dimensionless representation**
  d dimensionlose Darstellung *f*
  f représentation *f* sans dimensions
  r безразмерное представление *n*

  * **DIME system → 1193**

1104 **dimmed selection**
  (about menu option or button)
  d verdunkelte Auswahl *f*; abgeblendete
     Selektion *f*
  f sélection *f* non accessible
  r недоступная выборка *f*; "серая" выборка

1105 **directed graph; oriented graph; orgraph**
  d gerichteter Graph *m*; orientierter Graph
  f graphe *m* dirigé; graphe orienté
  r направленный граф *m*; ориентированный
     граф; орграф *m*

1106 **directed link**
  d Direktverbindung *f*
  f liaison *f* orientée
  r направленная связь *f*

1107 **directed network**
  d gerichtetes Netz[werk] *n*
  f réseau *m* orienté
  r ориентированная сеть *f*; директивный
     сетевой график *m*

1108 **direction**
  d Richtung *f*
  f direction *f*
  r направление *n*

1109 **directional derivative**
  d Richtungsableitung *f*

  f dérivative *f* directionnelle
  r производная *f* по оси; производная путевой
     устойчивости; косая производная;
     производная по направлению

1110 **directional filter**
  d leitender Filter *m*
  f filtre *m* directionnel
  r направляющий фильтр *m*

  * **direction angle → 329**

1111 **direction distortion**
  d Richtungsverzerrung *f*
  f distorsion *f* de direction
  r искажение *n* направления

  * **direction of steepest slope → 633**

1112 **direction point; bearing point**
  d Richt[ungs]punkt *m*
  f point *m* visé; point de direction
  r ориентировочная точка *f*; точка
     направления наблюдения

1113 **direct signs**
  d direkte Zeichen *npl*
  f caractères *mpl* directs
  r прямые признаки *mpl*

1114 **direct volume display device; DVDD**
  d direktes 3D-Display *n*
  f afficheur *m* volumétrique à visualisation
     directe
  r объёмный дисплей *m* непосредственной
     трёхмерной визуализации

1115 **Dirichlet tessellation**
  d Dirichlet-Musterkachelung *f*
  f tessellation *f* de Dirichlet
  r тасселяция *f* Дирихле

  * **dirth → 1784**

1116 **disaggregate *v***
  d zerlegen; zerfallen; zerstreuen; zerteilen
  f désagréger
  r разделять на составные части; размельчать;
     дезагрегировать; детализировать

1117 **disaggregation**
  d Disaggregation *f*; Zerfall *m*
  f désagrégation *f*; morcellement *m* vertical
  r дезагрегирование *n*; разукрупнение *n*;
     выделение *n* составляющих

1118 **disaster mapping**
  d Unglückkartieren *n*; Katastrophe-Kartografie *f*

f mappage *m* de dépannages; cartographie *f* d'avaries
r картографирование *n* аварий

* **disconformity** → 3506

* **discount factor** → 1119

1119 discount rate; discount factor
(economic factor that deflates a future sum to make it comparable with current expenses)
d Diskontsatz *m*
f taux *m* d'escompte; taux d'actualisation
r учётный курс *m*; учётный процент *m*; льготный тариф *m*

* **discrepancy** → 1319

1120 discrete data
d diskrete Daten *pl*
f données *fpl* discrètes
r дискретные данные *pl*

1121 discrete distribution
d diskrete Verteilung *f*
f distribution *f* discrète
r дискретное распределение *n*

1122 discrete phenomena
d diskrete Erscheinungen *fpl*
f phénomènes *mpl* discrets
r дискретные явления *npl*

1123 discretization; quantization; quantizing
d Diskretisierung *f*; Quantisierung *f*; Quantelung *f*
f discrétisation *f*; quantification *f*
r дискретизация *f*; дискретизирование *n*; квантование *n*

1124 discriminant analysis; discriminatory analysis
d Diskriminanzanalyse *f*
f analyse *f* discriminante
r дискриминантный анализ *m*

* **discriminatory analysis** → 1124

1125 dispatching
d Absendung *f*; Absenden *n*; Entsendung *f*; Erledigung *f*
f répartition *f* [du travail]
r диспетчеризация *f*

* **dispersion diagram** → 3228

1126 displacement; shift[ing]; offset
d Verschiebung *f*; Schieben *n*; Schiftung *f*; Verlagerung *f*; Verdrängung *f*
f décalage *m*; déplacement *m*
r смещение *n*; перемещение *n*; сдвиг *m*

* **displacement of symbol** → 2454

* **display** → 3231, 3962

1127 displayable attribute
d Darstellungsattribut *n*
f attribut *m* affichable
r воспроизводимый атрибут *m*

1128 display frame; frame
d Einzelbild *n*
f cadre *m*
r кадр *m*

* **displaying** → 3962

1129 dissemination
d Zerstreuung *f*
f dissémination *f*
r рассеяние *n*

1130 dissolve
(of frames)
d Überblendung *f*
f fondu *m*
r наплыв *m*; плавная смена *f* изображений

1131 distance
d Abstand *m*; Entfernung *f*; Distanz *f*
f distance *f*
r расстояние *n*

* **distance address assignment** → 1132

1132 distance[-based] address assignment
d Fernadresszuordnung *f*
f adressage *m* éloignée
r удалённое присваивание *n* адреса; дистанционное присваивание *n* адреса

1133 distance distortion
d Abstandverzerrung *f*
f distorsion *f* de distance
r искажение *n* расстояний

* **distance measurement** → 3004

1134 distance measuring equipment; DME; distance meter
d Entfernungsmessgerät *n*; Entfernungsmesser *m*
f dispositif *m* de mesure de distance; appareil *m* de mesure de distance
r дальномер *m*; дальномерная аппаратура *f*; дальномерное оборудование *n*

* **distance meter** → 1134

1135 distance units
d Distanzeinheiten *fpl*

f unités *fpl* de distance
r единицы *fpl* расстояния

1136 distance vector
  d Distanzvektor *m*
  f vecteur *m* de distance
  r вектор *m* расстояния

* **distant methods** → 3123

* **distinctness** → 2878

1137 distortion; alteration
  d Distorsion *f*; Verzerrung *f*
  f distorsion *f*
  r искажение *n*; дисторсия *f*

1138 distortion isograms; lines of equal distortions; isanamorphic lines
  d Äquideformate *npl*; Isodeformate *npl*
  f lignes *fpl* d'égale altération
  r изолинии *fpl* равных искажений; изоколы *mpl*

1139 distributed cartographic database; decentralized cartographic database
  d verteilte kartografische Datenbasis *f*; dezentrale kartografische Datenbasis
  f base *f* de données cartographiques distribuée
  r распределённая картографическая база *f* данных; децентрализованная картографическая база данных

1140 distributed data
  d verteilte Daten *pl*
  f données *fpl* distribuées
  r распределённые данные *pl*

1141 distributed databank
  d verteilte Datenbank *f*
  f banque *f* de données distribuée
  r распределённый банк *m* данных

1142 distributed database
  d verteilte Datenbasis *f*
  f base *f* de données distribuée
  r распределённая база *f* данных; РБД

1143 distributed database management system
  d verteiltes Datenbasis-Managementsystem *n*
  f système *m* de gestion de base de données distribuée
  r система *f* управления распределённых баз данных; СУРБД

1144 distributed geographic information; DGI
  d verteilte geografische Information *f*
  f information *f* géographique distribuée
  r распределённая географическая информация *f*

1145 distributed information system
  d verteiltes Informationssystem *n*
  f système *m* d'information distribué
  r распределённая информационная система *f*

1146 distributed processing
  d verteilte Verarbeitung *f*
  f traitement *m* distribué
  r распределённая обработка *f*

1147 distribution density
  d Verteilungsdichte *f*
  f densité *f* de distribution
  r густота *f* распределения

1148 district; ward; circumscription
  d Bezirk *m*; Distrikt *m*; Kreis *m*; Stadtviertel *n*; Landstrich *m*; Ortsteil *m*; Stadtteil *m*
  f circonscription *f*; district *m*; secteur *m*
  r район *m*; округ *m*; участок *m*

1149 district boundary; ward boundary
  d Stadtteilgrenze *f*; Bezirkgrenze *f*
  f limite *f* de circonscription; limite de secteur
  r граница *f* [городского] административного района

* **district fill** → 1418

1150 districting
  d in Bezirke Einteilung *f*; Einteilung *f* der Wahlkreise; Districting *n*
  f découpage *m* électoral; découpage des circonscriptions; délimitation *f* des circonscriptions
  r деление *n* на округа; районирование *n*

1151 diversity
  d Diversity *n*; Unterschied *m*; Verschiedenartigkeit *f*
  f diversité *f*
  r разнообразие *n*; различие *n*; разность *f*

1152 divide
  d Scheide *f*
  f division *f*; ligne *f* de partage
  r делёж *m*; рубеж *m*; раздел *m*

1153 dividers
  d Trennwand *f*; Spitzzirkel *m*; Stechzirkel *m*
  f compas *m* à pointes sèches
  r делительный циркуль *m*; разметочный циркуль; циркуль[-измеритель] *m*

1154 division
  d Division *f*; Teilung *f*
  f division *f*; partage *m*
  r деление *n*

* **DLG** → 1058

* **DLG-E** → 1155

1155 DLG-Enhanced; DLG-E
 d Erweiterung *f* des DLG
 f DLG *m* amélioré
 r улучшенный формат *m* DLG

* **DMA** → 946

* **2D map** → 2769

* **DME** → 1134

* **DML** → 898

* **3D network** → 3367

1156 2D object; 2D feature; planimetric feature
 d 2D-Objekt *n*
 f objet *m* 2D; objet flat; objet planimétrique
 r плоский объект *m*; планиметрический
  объект

1157 3D object; 3D feature; volumetric feature
 d 3D-Objekt *n*
 f objet *m* 3D; objet volumétrique
 r трёхмерный объект *m*; 3D объект;
  объёмный объект

1158 document
 d Dokument *n*
 f document *m*
 r документ *m*

1159 domain
 d Domäne *f*; Domain *f*
 f domaine *m*
 r домен *m*; область *f*

* **domain decomposition** → 162

1160 domain name
 d Domain-Name *m*
 f nom *m* de domaine
 r имя *n* домена; название *n* домена;
  именование *n* домена

1161 domain name mapping
 d Domain-Name-Kartierung *f*
 f cartographie *f* de noms de domaines
 r картографирование *n* имён доменов

1162 domain structure
 d Domain-Struktur *f*
 f structure *f* de domaine
 r структура *f* домена

1163 domestic geographic name; DGN
 d heimischer geografischer Name *m*
 f nom *m* géographique domestique

 r местное географическое имя *n*

1164 Donald elliptic projection
 d elliptische Projektion *f* von Donald
 f projection *f* elliptique de Donald
 r эллиптическая проекция *f* Доналда

* **DOP** → 1097

1165 Doppler effect
 d Dopplereffekt *m*
 f effet *m* Doppler
 r эффект *m* Доплера

* **DOQ standard** → 1062

1166 dot
 d Punkt *m*; Punktmarke *f*
 f marque *f* ponctuelle; index *m* ponctuel;
  point *m*
 r точка *f*

* **dot chart** → 3228

* **dot density** → 971

1167 dot[-density] map
  (a thematic map depicting point occurrences
  and the distribution of these occurrences)
 d Punktdichte-Karte *f*
 f carte *f* à points; carte par maille; carte maillée
 r точечная [тематическая] карта *f*

* **dot map** → 1167

1168 dot method; absolute method
 d Punktmethode *f*
 f méthode *f* des points
 r точечный способ *m*

1169 dots per inch; dpi
 d Punkte *mpl* pro Zoll
 f points *mpl* au pouce
 r точек *fpl* на дюйм

1170 double-difference; DD
 d Doppeldifferenz *f*
 f double différence *f*
 r вторая разность *f*; сдвоённая разность

1171 double digitizing
 d Doppeldigitalisierung *f*
 f double digitalisation *f*
 r двойное оцифрование *n*; двойное
  дигитализирование *n*

**1172 double image; multiple image; echo image;
 ghost**
 d Doppelbild *n*; Mehrbild *n*; Geisterbild *n*;
  Echobild *n*

f image *f* double; image fantôme;
   écho-image *f*
r паразитное изображение *n*; повторное
   изображение; сдвоённое изображение;
   комбинированное изображение;
   многократная экспозиция *f*;
   эхо-изображение *n*; мультиплицированное
   изображение

* **double picture plotter** → 3492

1173 double-quad name
d Doppelquadname *m*
f double nom *m* de quad
r сдвоённое имя *n* квадратного участка

* **doubling** → 1196

1174 downstream
d stromabwärts [gerichtet]; stromabwärts
   gelegen; flussabwärts; mit dem Strom
f dans le sens descendant; en voie descendante;
   d'aval
r вниз по течению; по направлению трафика;
   по ходу [основного] трафика

1175 3D perspective view
d perspektivische 3D-Ansicht *f*
f vue *f* perspective 3D
r перспективный трёхмерный вид *m*

* **dpi** → 1169

1176 3D polygon
d 3D-Polygon *n*
f polygone *m* 3D
r трёхмерный многоугольник *m*; 3D
   многоугольник

1177 3D polygon mesh
d 3D-Polygonmasche *f*
f maillage *m* par polygones 3D
r трёхмерная многоугольная сеть *f*

* **drag** → 1178

1178 drag[ging]
d Ziehen *n*; Nachziehen *n*
f [en]traînement *m*; traînage *m*; glissement *m*
r таскание *n*; буксировка *f*

1179 draping
   (a perspective or panoramic rendering of
   two-dimensional features superimposed an a
   surface)
d Gewebe *n*; Drapieren *n*
f drapage *m*; plissement *m* [moulant]
r образование *n* складок; драпировка *f*

1180 draw *v* a map; trace *v* a map

d eine Karte zeichnen
f dessiner une carte
r чертить карту

1181 drawing device
d Zeichengerät *n*
f dispositif *m* de dessin
r чертёжное устройство *n*

**1182 drawing exchange format; DXF; drawing
      interchange format**
   (of AutoCAD)
d Zeichnungsaustausch-Format *n*;
   DXF-Format *n*
f format *m* DXF
r формат *m* обмена чертежей; формат DXF

1183 drawing instrument
d Zeicheninstrument *n*
f instrument *m* de dessin
r чертёжный инструмент *m*; прибор *m* для
   черчения

* **drawing interchange format** → 1182

1184 drawing order
d Zeichenordnung *f*
f ordre *m* de dessin
r порядок *m* вычерчивания

* **drawing original** → 2659

* **driver** → 1009

* **drop** → 1185

* **dropdown menu** → 2953

* **drop-off** → 238, 1365

* **dropped scan lines** → 2434

1185 drop[ping]
d Auslassen *n*; Auslassung *f*
f lâchage *m*; troncation *f*
r выпадание *n*; пропадание *n*

1186 drum plotter; belt-bed plotter
d Trommelplotter *m*; Walzenplotter *m*;
   Walzenschreiber *m*
f traceur *m* à tambour
r барабанный плоттер *m*; барабанный
   графопостроитель *m*

1187 drum scanner
d Trommelscanner *m*
f scanner *m* à tambour
r барабанный сканер *m*

1188 3D seismic survey
d seismische 3D-Aufnahme *f*

f levé *m* sismique en 3D
r трёхмерная сейсмическая съёмка *f*

* **DSM** → 1067

* **DSS** → 935

* **2D surface** → 2773

1189 **3D [surface] digitizing**
d 3D-Flächendigitalisierung *f*;
3D-Digitalisierung *f*
f numérisation *f* [de surfaces] 3D
r дигитализирование *n* 3D поверхностей; 3D
дигитализирование; трёхмерное
дигитализирование

* **DTM** → 1072

1190 **2D topology**
d 2D-Topologie *f*
f topologie *f* 2D
r двумерная топология *f*

1191 **dual-axis Fourier shape analysis**
d Zwei-Achsen-Fourier-Formanalyse *f*
f analyse *f* de formes de Fourier à deux axes
r анализ *m* двуосных фигур Фурье

1192 **dual-frequency receiver**
d Doppelfrequenz-Empfänger *m*
f récepteur *m* de double fréquence
r двухчастотный приёмник *m*; двухволновой
приёмник

1193 **dual-independent map encoding system;**
DIME system
(of US Bureau of the Census)
d DIME-System *n*
f système *m* DIME
r система *f* и формат *m* представления
данных о пространственных объектах
(*принятые в бюро переписей США до её*
замены системой TIGER в 1990г.)

1194 **duplicate data**
d duplizierte Daten *pl*
f données *fpl* dupliquées
r дублированные данные *pl*

1195 **duplicate line**
d duplizierte Linie *f*
f ligne *f* dupliquée
r дублированная линия *f*

1196 **duplicating; doubling; duplication**
d Duplikation *f*; Verdoppelung *f*; Duplizierung *f*
f duplication *f*
r удвоение *n*; дублирование *n*

* **duplication** → 1196

* **DVDD** → 1114

* **dwelling unit** → 1855

* **DXF** → 1044, 1182

1197 **dynamically-linked data view**
d dynamisch verknüpfte Datendarstellung *f*;
dynamisch gelinkte Datendarstellung
f vue *f* de données chaînées dynamiquement
r представление *n* динамически связанных
данных

1198 **dynamic data exchange; DDE**
d dynamischer Datenaustausch *m*
f échange *m* de données dynamique
r динамический обмен *m* данными

1199 **dynamic generalization**
d dynamische Generalisierung *f*
f généralisation *f* dynamique
r динамическая генерализация *f*

1200 **dynamic geoimage**
d dynamisches Geobild *n*
f géo-image *f* dynamique
r динамическое геоизображение *n*

* **dynamic mode digitizing** → 3508

1201 **dynamic segmentation**
(a method for referencing attribute
information along a network that does not
divide each segment of the network wherever
any attribute changes)
d dynamische Segmentierung *f*
f segmentation *f* dynamique
r динамическое сегментирование *n*

E

1202 Earth-centered Greenwich Cartesian coordinate system; geocentric Greenwich Cartesian coordinate system
d geozentrisches Kartesisches Greenwich-Koordinatensystem *n*
f système *m* de coordonnées géocentrique cartésienne de Greenwich
r геоцентрическая гринвичская прямоугольная система *f* координат

   * **Earth curvature** → 825

1203 Earth ellipsoid
d Erdellipsoid *n*
f ellipsoïde *m* de la Terre
r земной эллипсоид *m*

1204 Earth-fixed axis system
d erdfestes Achsensystem *n*
f trièdre *m* terrestre
r система *f* координатных осей Земли

1205 Earth observation
d Erdbeobachtung *f*
f observation *f* de la Terre
r наблюдение *n* Земли

1206 Earth-observing satellite
d Erdbeobachtung-Satellit *m*
f satellite *m* d'observation de la Terre
r спутник *m* наблюдения Земли

1207 Earth resources survey
d Überwachung *f* der Erde-Ressourcen
f observation *f* écographique; observation de ressources terrestres
r обследование *n* ресурсов Земли

   * **Earth sciences** → 1675

1208 Earth simulator
d Erdsimulator *m*
f simulateur *m* de la Terre
r симулятор *m* Земли

   * **earth slide** → 2091

1209 Earth's sphere; terrestrial globe; globe
d [terrestrischer] Erdball *m*; Globus *m*; Globuskugel *f*; Weltkugel *f*
f globe *m* [terrestre]
r земная сфера *f*; [земной] глобус *m*

   * **earth-synchronous** → 1679

1210 easting
   (a rectangular (x,y) coordinate measurement of distance east from a north-south reference line, usually a meridian used as the axis of origin within a map zone or projection)
d Rechtswert *m*
f abscisse *f* [d'un quadrillage]; coordonnée *f* rectangulaire dans la direction ouest-est
r восточное положение *n* в координатной или географической сетке; восточное склонение *n* (*магнитной стрелки*); движение *n* на восток; направление *n* на восток; отшествие *n* на восток

1211 eccentricity
d Exzentrizität *f*
f excentricité *f*
r эксцентрицитет *m*; эксцентричность *f*

1212 echo
d Echo *n*
f écho *m*
r эхо *n*

   * **echo image** → 1172

   * **ecological sensitivity index** → 1297

1213 economic geography
d Wirtschaftsgeografie *f*
f géographie *f* économique
r экономическая география *f*

1214 economics of cartographic production
d Kartenentwerfen-Ökonomik *f*
f économie *f* de production cartographique
r экономика *f* картографического производства

   * **EDA** → 1344

1215 edge
   (of a graph)
d Kante *f*
f arête *f*
r ребро *n*

1216 edge
d Rand *m*; Schranke *f*
f côte *f*; bord *m*
r граница *f* [поверхностей]; грань *f*; край *m*; кромка *f*

1217 edge-based triangle subdivision
d kantenbasierte Dreieckunterteilung *f*
f sous-division *f* de triangle à base d'arêtes
r треугольное подразделение *n*, основанное на реберном представлении

1218 edge condition; marginal condition
d Randbedingung *f*
f condition *f* limite; condition aux bornes
r краевое условие *n*; граничное условие

* **edge detect** → 1219

* **edge detect filter** → 1220

1219 edge detect[ion]; edge enhancing
d Randerkennung *f*; Flankenerkennung *f*
f détection *f* de côtes; détection de contours
r обнаружение *n* краев; выделение *n* краев

1220 edge detect[ion] filter; edge detector
d Randerkennungsfilter *m*; Flankendetektor *m*
f filtre *m* de détection de contours
r фильтр *m* обнаружения краев

* **edge detector** → 1220

* **edge enhancing** → 1219

1221 edge extraction
d Kantenextraktion *f*
f extraction *f* d'arêtes
r извлечение *n* ребер

* **edge join** → 1223

1222 edge mapping
d Kantenabbildung *f*
f mappage *m* de côtes
r отображение *n* граней; отображение ребер

* **edge match** → 1223

1223 edge match[ing]; edge join
d Kantenübereinstimmung *f*; Randanpassung *f*
f coïncidence *f* d'arêtes; raccordement *m*
r соответствие *n* ребер; совпадение *n* ребер; сводка *f*

1224 edge visibility
d Kantensichtbarkeit *f*
f visibilité *f* d'arête
r видимость *f* ребра

* **edit** → 1225

* **edit-distance** → 3329

1225 edit[ing]
d Bearbeitung *f*; Aufbereitung *f*; Editieren *n*
f mise *f* en forme; édition *f*; rédaction *f*
r редактирование *n*

1226 editing program; editor
d Bearbeitungsprogramm *n*; Editierprogramm *n*; Editor *m*

f programme *m* d'édition; éditeur *m*
r редактирующая программа *f*; редактор *m*

1227 editing tolerances
(weed, grain, edit distance, snap distance, and nodesnap distance)
d Editiertoleranzen *fpl*
f tolérances *fpl* d'édition
r допуски *mpl* редактирования

* **editor** → 1226

* **EDM** → 1238

1228 effective Earth-curvature factor
d effektiver Erdkrümmungsfaktor *m*
f facteur *m* de courbure équivalente de la Terre
r полезный коэффициент *m* кривизны земной поверхности

1229 effective instantaneous field of view
d effektiver momentaner Bildfeldwinkel *m*; effektives momentanes Bildfeld *n*
f champ *m* de vision instantané effectif
r действительный мгновенный сектор *m* обзора; действительная мгновенная зона *f* обзора

* **EIM** → 1292

* **EIS** → 1291

1230 elastic circle
d elastischer Kreis *m*
f cercle *m* élastique
r эластичный круг *m*

1231 elastic line
d elastische Linie *f*
f ligne *f* élastique
r эластичная линия *f*

1232 elastic transformation
d elastische Transformation *f*
f transformation *f* élastique
r эластичная трансформация *f*

* **electoral district** → 3972

1233 electromagnetic radiation
d elektromagnetische Strahlung *f*
f radiation *f* électromagnétique
r электромагнитное излучение *n*

1234 electromagnetic spectrum
d elektromagnetisches Spektrum *n*
f spectre *m* électromagnétique
r электромагнитный спектр *m*

1235 electromagnetic survey
d elektromagnetische Forschung *f*

f levé *m* électromagnétique
r электромагнитная разведка *f*;
электромагнитная съёмка *f*;
электромагнитные исследования *npl*

1236 electromechanical sensor
d elektromechanischer Sensor *m*
f senseur *m* électromécanique
r электромеханический сенсор *m*

1237 electronically sensible surface
d elektronisch fühlbare Fläche *f*
f surface *f* sensible électroniquement
r поверхность *f*, ощущаемая электронными
приборами

* **electronic atlas** → 667

* **electronic distance measurement** → 1238

**1238 electronic distance measuring; electronic
distance measurement; EDM**
d elektronische Entfernungsmessung *f*
f mesure *f* électronique des distances
r электронное измерение *n* расстояний

**1239 electronic map; computer[ized] map;
digital map**
d elektronische Karte *f*; digitale Karte
f carte *f* électronique; carte infographique; carte
numérique; carte informatique; carte
informatisée
r электронная карта *f*; компьютерная карта;
цифровая карта

1240 electronic planimeter
d elektronisches Planimeter *n*
f planimètre *m* électronique
r электронный планиметр *m*

1241 electrostatic plotter
d elektrostatischer Plotter *m*
f traceur *m* électrostatique
r электростатический плоттер *m*

**1242 element of area; areal element; surface
element**
d Flächenelement *n*; Oberflächenelement *n*
f élément *m* d'aire; élément de surface
r элемент *m* площади; элемент поверхности

* **element of map** → 448

**1243 element of volume; volume element; volume
pixel; voxel**
d Volumenelement *n*; Raumelement *n*; Voxel *n*
f élément *m* de volume; élément volumétrique;
pixel *m* 3D; voxel *m*
r элемент *m* трёхмерного изображения;
элемент объёма; объёмный элемент

[изображения]; воксел *m*

* **elevated guide way** → 3167

* **elevated road** → 3167

1244 elevation
d Elevation *f*
f élévation *f*
r поднимание *n*

* **elevation** → 1811

* **elevation angle** → 123

**1245 elevation benchmark; elevation datum;
figure; spotheight; spot elevation**
d Elevationsbenchmark *n*; Höhenpunkt *m* [in der
Karte]; Höhenzahl *f*
f repère *m* d'élévation; point *m* coté; cote *f*
r отметка *f* уровня; высотная отметка

1246 elevation computer
d Elevationscomputer *m*
f ordinateur *m* d'élévations
r вычислитель *m* угла возвышения

* **elevation control** → 2140

1247 elevation data
d Elevationsdaten *pl*; Höhenliniendaten *pl*
f données *fpl* d'élévation
r данные *pl* для вертикальной наводки;
угломестные данные

1248 elevation data browser
d Elevationsdaten-Browser *m*
f visionneur *m* de données d'élévation
r браузер *m* данных для вертикальной
наводки

1249 elevation data format
d Elevationsdatenformat *n*
f format *m* de données d'élévation
r формат *m* данных для вертикальной
наводки

* **elevation datum** → 1245

* **elevation difference** → 1812

1250 elevation mask
d Elevationsmaske *f*
f masque *f* d'élévation
r маска *f* высотных точек

1251 elevation model
d Höhenlinienmodell *n*
f modèle *m* altimétrique; modèle d'élévation
r модель *f* высотных точек

* **elevation number** → 734

* **elevation of a benchmark referred to a datum** → 1813

* **elevation rudder** → 1252

* **elevation scale** → 93

* **elevation tint** → 1878

* **elevation tint box** → 1879

* **elevation view** → 3940

* **elevator** → 3234

1252 elevator; elevation rudder
   d Aufzug *m*; Hebewerk *n*; Elevator *m*; Elevationsruder *m*
   f élévateur *m*; gouvernail *m* d'élévation
   r элеватор *m*; подъёмник *m*; руль *m* высоты; лифт *m*

1253 ellipse of distortion; Tissot's indicatrix
   d Verzerrungsellipse *f*; Tissot-Indikatrix *f*
   f ellipse *m* de distorsions
   r эллипс *m* искажений; индикатриса *f* Тиссо

1254 ellipsoid
   d Ellipsoid *n*
   f ellipsoïde *m*
   r эллипсоид *m*

1255 ellipsoidal coordinates
   d elliptische Koordinaten *fpl*
   f coordonnées *fpl* ellipsoïdaux
   r эллипсоидальные координаты *fpl*

* **ellipsoid height** → 1548

1256 ellipsoid of revolution; revolution ellipsoid
   d Drehungsellipsoid *n*; Rotationsellipsoid *n*
   f ellipsoïde *m* de révolution
   r эллипсоид *m* вращения

1257 elliptic projection
   d elliptische Projektion *f*
   f projection *f* elliptique
   r эллиптическая проекция *f*

1258 embedded SQL
   d eingebettete Sprache *f* der strukturierten Abfragen
   f langage *m* encastré de requêtes structurées
   r встроенный язык *m* структурированных запросов

1259 embedding; nesting
   d Einbettung *f*; Einbetten *n*; Einschachtelung *f*;
Verschachtelung *f*; Schachtelung *f*
   f encastrement *m*; imbrication *f*; emboîtage *m*; emboîtement *m*
   r вложение *n*; вложенность *f*; упаковка *f*; внедрение *n*; укладка *f*

1260 emboss[ed] effect; relief effect
   d Reliefeffekt *m*
   f effet *m* relief
   r эффект *m* рельефа

* **emboss effect** → 1260

* **emergency planning** → 1261

1261 emergency [response] planning
   d Notfallplanung *f*; Sicherheitsmaßnahme *f*
   f planification *f* des mesures d'urgence
   r планирование *n* критических ситуаций

1262 empty cell
   d leere Zelle *f*
   f cellule *f* vide
   r пустая клетка *f*

1263 encapsulation; capsulation
   d Einkapselung *f*; Kapselung *f*; Verkapselung *f*
   f [en]capsulage *m*; enrobage *m*
   r капсулирование *n*; герметизация *f*; инкапсуляция *f*

* **enclosure** → 582

1264 encoding; coding
   d Codieren *n*; Codierung *f*; Verschlüsseln *n*
   f codage *m*; codification *f*
   r кодирование *n*

1265 end coordinates
   d Endkoordinaten *fpl*
   f coordonnées *fpl* finales
   r конечные координаты *fpl*; координаты конца

* **ending point** → 1270

* **end lap** → 1465

1266 end node; endpoint node; peripheral node
   d Endpunkt-Knoten *m*; Endknoten *m*
   f nœud *m* d'extrémité; nœud périphérique
   r периферный узел *m*; конечный узел; конечная вершина *f*

1267 endnode ID
   d Endpunkt-Identifikator *m*
   f identificateur *m* de nœud périphérique
   r идентификатор *m* конечного узла

1268 end of line
 d Zeilenende *f*
 f fin *m* de ligne
 r конец *m* строки

1269 end-of-line code
 d Zeilenende-Code *m*
 f code *m* de fin de ligne
 r код *m* конца строки

  * **end overlap** → 1465

1270 endpoint; ending point; terminal point
 d Endpunkt *m*
 f point *m* final; extrémité *f*
 r конечная точка *f*

  * **endpoint node** → 1266

1271 endpoint of arc; arc endpoint
 d Endpunkt *m* des Bogens
 f point *m* d'arc final
 r конечная точка *f* дуги

  * **engineering geodesy** → 141

1272 engineering map
 d technische Karte *f*
 f plan *m* topométrique [du génie civil]
 r техническая карта *f*

1273 enlarged scale
 d verlängerter Maßstab *m*
 f échelle *f* élargie
 r увеличенный масштаб *m*

  * **enquiry** → 3140

1274 entity; graphic primitive
 d Entität *f*; [grafisches] Primitiv *n*
 f entité *f* [graphique]; primitive *f*
 r [графический] примитив *m*

1275 entity class
 d Einheitsklasse *f*; Entitätsklasse *f*
 f classe *f* d'entité
 r класс *m* примитива

1276 entity deformation
 d Einheitsdeformation *f*
 f déformation *f* d'entité
 r искажение *n* примитива

1277 entity integrity
 d Entitätenintegrität *f*
 f intégrité *f* d'entité
 r целостность *f* примитива

1278 entity name
 d Einheitsname *m*
 f nom *m* d'entité
 r имя *n* примитива

1279 entity-relationship diagram; ER-diagram
 d Einheitsbeziehungsdiagramm *n*;
   ER-Diagramm *n*
 f diagramme *m* de relations entre entités
 r диаграмма *f* взаимосвязи примитивов

1280 entity-relationship model
 d Einheitsbeziehungsmodell *n*
 f modèle *m* de relations entre entités
 r модель *f* взаимосвязи примитивов

1281 entity-relationship modeling
 d Einheitsbeziehungsmodellierung *f*
 f modelage *m* de relations entre entités
 r моделирование *n* взаимосвязи примитивов

1282 entity selection set; selected entity set
 d Einheit-Auswahlsatz *m*
 f ensemble *m* de sélection d'entités; sélection *f*
   d'entités
 r коллекция *f* отбора примитивов

1283 entity sort order
 d Ordnung *f* der Einheitssortierung
 f ordre *m* de triage d'entités
 r порядок *m* сортировки примитивов

1284 entity transformation
 d Einheitstransformation *f*
 f transformation *f* d'entité
 r преобразование *n* примитива

1285 entity type
 d Einheitstyp *m*
 f type *m* d'entité
 r тип *m* примитива

  * **enumeration** → 487

  * **enumeration area** → 499

  * **enumeration district** → 499

1286 environment; ambience
 d Umwelt *f*; Umgebung *f*; Einkreisung *f*
 f environnement *m*; ambiance *f*
 r среда *f*; окружение *n*; окрестность *f*

1287 environmental assessment
 d Umweltabschätzung *f*
 f appréciation *f* d'environnement
 r экологическая оценка *f*; экологическая
   экспертиза *f*; оценка качества окружающей
   среды

  * **environmental data** → 107

1288 environmental database
d Umweltdatenbasis *f*
f base *f* de données environnementaux
r база *f* данных об окружающей среде

1289 environmental data catalog
d Umweltdatenkatalog *m*
f catalogue *m* de données environnementaux
r каталог *m* данных об окружающей среде

1290 environmental GIS
d Umwelt-GIS *n*
f SIG *m* environnemental
r природоохранная ГИС *f*

1291 environmental impact statement; EIS
d Umweltverträglichkeitsanweisung *f*;
Umwelteinflussanweisung *f*
f dossier *m* d'impact sur l'environnement;
rapport *m* prospectif d'ambiance; déclaration *f*
relative aux incidences sur l'environnement
r формулировка *f* воздействия на природную
среду; спецификация *f* экологического
воздействия

**1292 environmental information management;
EIM**
d Umweltinformationsverwaltung *f*
f gestion *f* d'information environnementale
r управление *n* информации об окружающей
среде

1293 environmental information system
d Umweltinformationssystem *n*; UIS
f système *m* d'information environnementale
r информационная система *f* окружающей
среды

1294 environmental modeling
d Umweltmodellierung *f*
f modelage *m* d'environnement
r моделирование *n* окружающей среды

1295 environmental planning
d Umweltplanung *f*
f planification *f* environnementale
r экологическое планирование *n*;
планирование окружающей среды

**1296 environmental resources information
network**
d Umweltressourcen-Informationssystem *n*
f système *m* d'information de ressources
environnementaux
r информационная система *f* ресурсов
окружающей среды

* **environmental satellite** → **1298**

1297 environmental sensitivity index; ecological

sensitivity index; ESI
d Umweltempfindlichkeitsindex *m*
f indice *m* de sensibilité environnementale
r индекс *m* чувствительности окружающей
среды к внешнему воздействию

1298 environmental [survey] satellite; ENVISAT
d Umweltsatellit *m*
f satellite *m* pour l'étude du milieu; satellite
d'étude du milieu
r искусственный спутник *m* для
экологического мониторинга

1299 environment division
d Umweltabteilung *f*
f division *f* de l'environnement
r раздел *m* окружающей среды

1300 environment mapping
d Umgebungsabbildung *f*
f application *f* d'environnement
r отображение *n* окрестности; проекция *f*
окружения

1301 environment variable
d Umweltvariable *f*
f variable *f* d'environnement
r переменная *f* окружения

* **ENVISAT** → **1298**

* **ephemeris** → **2640**

* **ephemeris data** → **2640**

1302 epipolar
(condition of geometric coplanarity
established between a pair of stereo images to
give them the same relative orientation)
d epipolar
f épipolaire; nucléique
r эпиполярный

1303 epipolar axis
d Kernachse *f*
f axe *m* nucléique
r эпиполярная ось *f*

* **equal-angle projection** → **686**

1304 equal-area *adj*
d flächentreu
f à aire équivalente
r равновеликий

* **equal-area map projection** → **1316**

* **equal-area projection** → **1316**

1305 **equal-interval interpolation**
(a classification procedure that divides the
total range of attribute values by the number
of classes)
d Interpolation *f* in regelmässigen Intervallen
f interpolation *f* à intervalles égaux
r интерполирование *n* с равными
интервалами

1306 **Equator**
d Äquator *m*
f équateur *m*
r экватор *m*

1307 **equatorial coordinates**
d Äquatorialkoordinaten *fpl*
f coordonnées *fpl* équatoriaux
r экваториальные координаты *fpl*

* **equatorial distance** → 2568

1308 **equatorial plane**
d Äquatorebene *f*
f plan *m* équatorial; plan de l'équateur
r экваториальная плоскость *f*, плоскость
экватора

* **equatorial projection** → 3806

1309 **equatorial radius**
d äquatorialer Radius *m*
f rayon *m* équatorial
r радиус *m* Экватора

* **equiangular projection** → 686

* **equiareal projection** → 1316

* **equidistance** → 730

1310 **equidistand cylindrical projection; plate
carree projection**
d äquidistante Zylinderprojektion *f*;
Plate-Carree-Projektion *f*
f projection *f* équidistante cylindrique;
projection [des cartes] plate carrée
r цилиндрическая равнопромежуточная
проекция *f*

1311 **equidistant lines**
d abstandstreue Linien *fpl*
f lignes *fpl* équidistantes
r эквидистантные линии *fpl*;
равнопромежуточные линии

1312 **equidistant projection**
d abstandstreue Projektion *f*, äquidistante
Abbildung *f*, längentreue Projektion
f projection *f* équidistante
r равнопромежуточная проекция *f*

1313 **equidistribution**
d Gleichaufteilung *f*
f équidistribution *f*
r равнораспределение *n*

1314 **equipment; facility**
d Ausrüstung *f*, Ausstattung *f*, Apparatur *f*,
Einrichtung *f*
f équipement *m*; appareillage *m*
r оборудование *n*; аппаратура *f*,
сооружения *npl*

1315 **equipment of map; mapping equipment**
d Kartografierungsausrüstung *f*
f équipement *m* de cartographie
r оснащение *n* карты; картографическое
оборудование *n*

* **equipotential surface** → 2145

* **equipotent surface** → 2145

1316 **equivalent projection; equal-area [map]
projection; equiareal projection; authalic
projection; homolographic projection**
d flächentreue Kartenprojektion *f*, flächentreue
Abbildung *f*, natürliche Projektion *f*
f projection *f* [à aire] équivalente
r равновеликая проекция *f*

* **erase** *v* → 953

* **erasing** → 954

* **erasure** → 954

* **ER-diagram** → 1279

* **erratic error** → 16

1317 **error**
d Fehler *m*
f erreur *f*
r погрешность *f*, ошибка *f*

1318 **error analysis; error estimation**
d Fehleranalyse *f*
f analyse *f* d'erreurs
r анализ *m* ошибок

* **error estimation** → 1318

1319 **error of closure; closing error; misclosure;
discrepancy**
d Abschlussfehler *m*; Horizontabschlussfehler *m*
f erreur *f* de fermeture; écart *m* de fermeture
r невязка *f* [полигона]; незамкнутость *f*

* **error of computation** → 658

* error of observation → 2608

* ESA → 1327

* escarpment symbols → 565

* ESI → 1297

* establishment → 3269

1320 estimated value; appraised value
  d Schätz[ungs]wert *m*; geschätzter Wert *m*
  f valeur *f* d'estimation
  r оценочная стоимость *f*; оцененная
    стоимость

1321 estimation; assessment; evaluation;
     valuation; appraisal
  d Abschätzung *f*; Schätzung *f*; Bewertung *f*;
    Auswertung *f*
  f estimation *f*; appréciation *f*; évaluation *f*
  r оценка *f*; оценивание *n*

* estuarine area → 1322

1322 estuary; estuarine area
  d Ästuar *m*; Mündungsgebiet *n*
  f estuaire *m*; territoire *m* d'estuaire; zone *f*
    estuarienne
  r эстуарий *m*; дельта *f*; [широкое] устье *n*
    [реки]

1323 Euclidean distance
  d Euklidische Distanz *f*
  f distance *f* d'Euclide; distance euclidienne
  r евклидово расстояние *n*

1324 Euclidean geometry
  d Euklidische Geometrie *f*
  f géométrie *f* d'Euclide; géométrie euclidienne
  r евклидова геометрия *f*

1325 Euclidean plane
  d Euklidische Ebene *f*
  f plan *m* euclidien
  r евклидова плоскость *f*

1326 Euclidean space
  d Euklidischer Raum *m*
  f espace *m* euclidien; espace d'Euclide
  r евклидово пространство *n*

1327 European Space Agency; ESA
  d Europäische Weltraumorganisation *f*;
    Europäische Raumfahrtbehörde *f*
  f Agence *f* spatiale européenne
  r Европейское космическое агентство *n*

* evaluation → 1321

1328 evaluative map
  d Auswertungskarte *f*
  f carte *f* d'évaluation
  r оценочная карта *f*

1329 event
  d Ereignis *n*; Vorgang *m*
  f événement *m*
  r событие *n*

* event flag → 1331

1330 event location
  d Ereignislokalisierung *f*
  f localisation *f* d'événement
  r размещение *n* события; определение *n*
    местоположения явления

1331 event marker; event flag
  d Ereigniskennzeichen *n*
  f marque *f* d'événement
  r флажок *m* события

* event origin → 1332

1332 event source; event origin
  d Ereignisquelle *f*
  f source *f* d'événements
  r источник *m* явлений; источник событий

1333 event table
  d Ereignistabelle *f*
  f table *f* de localisation
  r таблица *f* явления; таблица события

1334 event theme
  d Ereignisthema *n*
  f thème *m* d'événement
  r тематический слой *m* явления; тема *f*
    события

1335 exact interpolator
  d exakter Interpolator *m*
  f interpolateur *m* exact
  r точный интерполятор *m*

1336 exaggeration
  d Übertreibung *f*; Exaggeration *f*
  f exagération *f*
  r утрирование *n* [размера или формы]

1337 exchange format; transfer format
  d Wechselformat *n*; Übertragungsformat *n*
  f format *m* d'échange
  r формат *m* обмена

1338 exclusion areas
  d Ausschlussflächen *fpl*
  f régions *fpl* d'exclusion
  r области *fpl* исключения

1339 exhaustive set
 d vollständige Menge *f*
 f ensemble *m* exhaustif
 r полное множество *n*

1340 exit; output; quit
 d Ausgang *m*; Austritt *m*
 f sortie *f*
 r выход *m*

1341 expert system
 d Expertensystem *n*
 f système *m* expert
 r экспертная система *f*

1342 explanatory inscriptions
 d explanatorische Beschriftungen *fpl*
 f inscriptions *fpl* explicatives
 r пояснительные надписи *fpl*

1343 explanatory scale
 d explanatorischer Maßstab *m*; erläuternder Maßstab
 f échelle *f* explicative
 r именованный масштаб *m*

1344 exploratory data analysis; EDA
 d explorative Datenanalyse *f*
 f analyse *f* exploratoire de données
 r анализ *m* данных об исследованиях

 * **export → 1345**

 * **exportation → 1345**

1345 export[ing]; exportation
 d Versand *m*; Export *m*; Ausfuhr *f*
 f exportation *f*
 r экспортирование *n*; износ *m*; отправка *f*; отсылка *f*

1346 exporting of vector objects
 d Vektorobjekt-Export *m*
 f exportation *f* d'objets vectoriels
 r экспортирование *n* векторных объектов

 * **exposure → 633**

1347 expression
 d Expression *f*; Ausdruck *m*
 f expression *f*
 r выражение *n*

1348 extended line
 d gestreckte Linie *f*; erweiterte Gerade *f*; abgeschlossene Gerade
 f droite *f* tendue; droite achevée
 r расширённая прямая *f*; прямая, пополненная бесконечными точками

1349 extended object
 d erweitertes Objekt *n*
 f objet *m* étendu
 r расширённый объект *m*

1350 extensive measurement
 (numerical measurement along an axis where addition provides the basic rule)
 d extensive Messung *f*
 f mesure *f* extensive; mesure étendue
 r экстенсивное измерение *n*

 * **exterior polygon → 2675**

1351 external attribute
 d äußeres Attribut *n*
 f attribut *m* externe
 r внешний объект *m*

1352 external database file
 d äußere Datenbasisdatei *f*
 f fichier *m* de base de données externe
 r файл *m* внешней базы данных

1353 external file
 d äußere Datei *f*
 f fichier *m* externe
 r внешний файл *m*

1354 externality
 d Auswirkung *f*
 f externalité *f*; effet *m* externe; avantage *m* indirect
 r внешность *f*; фактор *m* внешнего порядка; экзогенный фактор

 * **external polygon → 2675**

1355 external raster
 d externer Raster *m*
 f rastre *m* externe
 r внешний растр *m*

1356 extract *v* attribute information
 d Attributinformation schleppen
 f extraire une information d'attributs
 r извлекать атрибутную информацию

1357 extrapolation
 d Extrapolieren *n*; Extrapolierung *f*; Extrapolation *f*
 f extrapolation *f*
 r экстраполяция *f*; экстраполирование *n*

1358 extremal points
 d Extrempunkte *mpl*
 f points *mpl* extrêmes
 r экстремальные точки *fpl*

F

* face → 1488

* facet → 1360

1359 **facet normal**
d Facettennormale *f*
f normale *f* à une facette
r нормаль *f* к фацету

1360 **facet[te]**
d Facette *f*
f facette *f*
r фацет *m*; грань *f*

1361 **facet vertex**
d Facettenspitze *f*
f sommet *m* de facette
r вершина *f* фацета

* facility → 1314

1362 **facility management**
d Facility-Management *n*
f gestion *f* de sites; gestion externe du système informatique; gestion par impartition
r управление *n* средствами

* fair drafting → 1363

* fair draught → 1363

1363 **fair draught[ing]; fair drafting; fair drawing; final compilation**
(of a map)
d Reinzeichnung *f*
f rédaction *f* définitive; mise *f* au net; dessin *m* définitif; dessin au net
r издательский оригинал *m* (*карты*)

* fair drawing → 1363

1364 **fall**
(a downward slope)
d Fall *m*
f pente *f*
r пад *m*

* falloff → 238

1365 **falloff; drop-off**
d Abfall *m*
f descente *f*
r спуск *m* вниз; падение *n*

1366 **false color composite**
d Falschfarbenkomposition *f*
f [photo]composition *f* à couleurs fausses
r спектрозональный аэрофотоснимок *m*; ложноцветный снимок *m*

1367 **false easting; x-shift**
d falscher Rechtswert *m*
f abscisse *f* fausse
r ложное отшествие *n* на восток; ложная величина *f* абсциссы

1368 **false northing; y-shift**
d falscher Hochwert *m*
f ordonnée *f* [d'un quadrillage] fausse
r ложное отшествие *n* на север; ложная величина *f* ординаты [по карте]

1369 **false origin**
(of coordinate system)
d falscher Ursprung *m*
f fausse origine *f*; point *m* de référence relatif
r фальшивое начало *n*

1370 **false perspective; pseudo-perspective**
d Pseudoperspektive *f*
f pseudo-perspective *f*
r псевдо-перспектива *f*

1371 **family household; household**
d Familienhaushalt *m*; Privathaushalt *m*; Haushalt *m*
f ménage *m*
r хозяйство *n*; семейная единица *f*

1372 **fast statics**
d Schnellstatistik *f*
f statistique *f* rapide
r ускорённая статика *f*

1373 **feasibility study**
d Studie *f* über die Durchführbarkeit; Durchführbarkeitsstudie *f*; Durchführbarkeitsanalyse *f*
f étude *f* de faisabilité; étude de viabilité
r предпроектное исследование *n*

* feature → 448, 1615, 3712

1374 **feature attribute**
d Feature-Attribut *n*
f attribut *m* de caractéristique; attribut d'objet
r атрибут-характеристика *m*; атрибут *m* объекта; атрибут элемента рельефа

1375 **feature attribute table**
(a table used to store attribute information for a specific coverage feature class)
d Feature-Attributtabelle *f*
f table *f* d'attributs d'objets
r таблица *f* атрибутов объектов

1376 feature cataloguing methodology
 d Objektkatalogisierungsmethodologie *f*
 f méthodologie *f* de catalogage d'objets
 r методология *f* каталогизации объектов

1377 feature class
 d Feature-Klasse *f*, Merkmalsklasse *f*
 f classe *f* de formes; classe de caractéristiques
 r класс *m* признаков

1378 feature code
 d Feature-Code *m*
 f code *m* de caractéristique; code de forme; code de trait
 r код *m* признака; код элемента рельефа

1379 feature extraction
 (in pattern recognition)
 d Merkmalsextraktion *f*, Merkmalgewinnung *f*
 f extraction *f* de caractéristiques; extraction de formes; extraction de traits pertinents; extraction de primitives de reconnaissance; extraction d'attributs; identification *f* de structure
 r выделение *n* признаков

1380 feature extraction processor
 d Merkmalsextraktionsprozessor *m*
 f processeur *m* d'extraction de caractéristiques
 r процессор *m* выделения признаков

1381 feature extraction segment
 d Merkmalsextraktionssegment *n*
 f segment *m* d'extraction de caractéristiques
 r сегмент *m* выделения признаков

 * **feature header** → **1382**

 * **feature ID** → **1382**

 * **feature identification code** → **1382**

1382 feature identification number; feature ID; feature identification code; feature header; feature label
 d Feature-Identifikationsnummer *f*, Merkmalsidentifikator *m*
 f identificateur *m* de caractéristique; identificateur de forme; identificateur de trait
 r идентификационный номер *m* признака; идентификатор *m* формы

 * **feature integrity** → **896**

 * **feature label** → **1382**

 * **feature map** → **3114**

1383 feature map
 d Merkmalskarte *f*

 f carte *f* de caractéristiques
 r карта *f* [характерных] признаков; характеристическая карта (*изображения*)

1384 feature mapping
 d Merkmalsabbildung *f*
 f mappage *m* de caractéristiques; mappage de traits
 r отображение *n* [характеристических] признаков; отображение элементов рельефа

 * **feature quality** → **906**

1385 feature selection
 d Feature-Auswahl *f*, Merkmalsauswahl *f*
 f sélection *f* de caractéristique; sélection de forme; sélection de trait
 r выбор *m* признака; выбор формы

1386 feature separation
 d Feature-Trennung *f*, Merkmalstrennung *f*
 f séparation *f* de caractéristiques; séparation *f* de formes; séparation de traits
 r разделение *n* признаков; разделение форм

1387 feature serial number
 d Feature-Seriennummer *f*
 f nombre *m* de série de caractéristique; nombre de série de trait
 r серийный номер *m* признака

1388 feature space
 d Merkmalsraum *m*; Feature-Raum *m*
 f espace *m* de caractéristique; espace de trait
 r пространство *n* признака; пространство формы

1389 feature type
 d Relieftyp *m*
 f type *m* de relief
 r тип *m* рельефа

1390 feature type
 d Merkmaltyp *m*; Featuretyp *m*
 f type *m* de caractéristique
 r тип *m* признака

1391 feature unit
 d Merkmalseinheit *f*
 f unité *f* caractéristique
 r единица *f* рельефа

1392 feature vector
 d Relief-Vektor *m*
 f vecteur *m* de relief
 r вектор *m* рельефа

1393 federal information processing standards; FOIS (US)
 d Bundesdatenverarbeitungsstandards *mpl*

f standards *mpl* fédérales de traitement
d'information

r стандарты *mpl* на обработку информации
(*издаваемые Национальным бюро
стандартов и Институтом
информационных технологий США*)

1394 feed scanner; sheet-fed scanner
d Einzugsscanner *m*
f scanner *m* à défilement
r сканер *m* с полистовой подачей

1395 fenceline
d Zaunlinie *f*
f ligne *f* de cloison
r заграждающая линия *f*

* **fenestration** → 3999

1396 fiducial marks
d Rahmenmarken *fpl*
f repères *mpl* du cadre d'appui; repères du fond
de chambre
r координатные метки *fpl*

* **field** → 891, 1582, 2628

* **field analysis** → 1407

1397 field check; field comparison
d Feldprüf *m*; Feldkontrolle *f*; Feldvergleich *m*
f contrôle *m* d'exploitation
r выездная проверка *f*; проверка на месте;
проверка в процессе эксплуатации

* **field comparison** → 1397

1398 field data
d Felddaten *pl*; Einsatzdaten *pl*
f données *fpl* d'exploitation; données à
application
r полевые данные *pl*; эксплуатационные
данные

1399 field data collection
d Felddaten-Sammlung *f*; Felddaten-Kollektion *f*
f collection *f* de données d'exploitation
r совокупность *f* полевых данных

**1400 field data collector; data collector; data
logger**
d Felderdatenkollektor *m*
f collecteur *m* de données d'exploitation
r коллектор *m* полевых данных

* **field intensity** → 1406

1401 field length
d Feldlänge *f*
f longueur *f* de champ

r длина *f* поля

* **field mapping** → 3723

1402 field name
(in a table)
d Feldname *m*
f nom *m* de champ
r имя *n* поля

* **field names** → 1613

**1403 field of view; field of vision; FOV; visual
field; angular field [of vision]**
d Sehfeld *n*; Gesichtsfeld *n*; Blickfeld *n*;
Bildfeldwinkel *m*
f champ *m* [angulaire] visuel; champ de vision;
champ de visée
r поле *n* зрения; сектор *m* обзора; зона *f*
обзора

* **field of vision** → 1403

1404 field separator
(a sign)
d Feldtrennzeichen *n*
f séparateur *m* de champs
r разделитель *m* полей

1405 field sheet; survey sheet
d Aufnahmeblatt *n*
f feuille *f* de levé
r топографический лист *m*

1406 field strength; field intensity
d Feldstärke *f*
f force *f* de champ; intensité *f* de champ
r напряжённость *f* поля

**1407 field survey; ground survey; field analysis;
landscape analysis**
d Feldstudie *f*; direkte Befragung *f*;
Feldaufnahme *f*; Bodenaufnahme *f*;
Geländeanalyse *f*
f enquête *f* sur le terrain; étude *f* sur le terrain;
levé *m* terrestre; analyse *f* de terrain
r полевые измерения *npl*

1408 field terminator
d Feldterminator *m*
f terminateur *m* de champ
r признак *m* конца поля

1409 field unit
d Feldeinheit *f*
f unité *f* de champ
r полевой блок *m*

* **figure** → 1245, 3279

1410 figure-ground
d Figur-Grund *m*
f dessin *m* en grisé; figure *f* fond
r фоновая фигура *f*

1411 file
d Datei *f*
f fichier *m*
r файл *m*

1412 file containing surfaces and values of lots
d Besitzstands- und Schätzungsnachweisdatei *f*
f fichier *m* de surfaces et valeurs de parcelles
r файл *m* площадей и значений участков

1413 file format
d Dateiformat *n*
f format *m* de fichier
r формат *m* файла

1414 file transfer
d Dateiübertragung *f*
f transfert *m* de fichiers
r перенос *m* файлов

* **fill → 1417**

1415 fill color
d Füll[ungs]farbe *f*
f couleur *f* de remplissage
r цвет *m* заливки; цвет *m* заполнения

1416 filled area; colored area
d gefüllter Bereich *m*; farbige Zone *f*
f domaine *m* rempli; zone *f* remplie; région *f* colorée
r заполненная зона *f*; закрашенная область *f*

1417 fill[ing]
d Füllung *f*; Auffüllen *n*; Füllen *n*
f remplissage *m*; garnissage *m*
r заливка *f*; заполнение *n*; закрашивание *n*

1418 fill[ing] of a district; district fill
d Bezirkfüllung *f*
f remplissage *m* de district
r закрашивание *n* района; закрашивание округа

1419 fill name
d Füllungsname *m*
f nom *m* de remplissage
r имя *n* заливки

* **fill of a district → 1418**

1420 fill palette
d Füllungspalette *f*
f palette *f* de remplissage
r палитра *f* закраски

1421 fill pattern
d Füllungsmuster *n*
f motif *m* de remplissage
r мотив *m* закраски

* **film recorder → 2418**

1422 filter
d Filter *m*; Sieb *n*
f filtre *m*
r фильтр *m*

1423 filter criteria; filtering criteria
d Filterkriterien *npl*
f critères *mpl* de filtrage
r критерии *mpl* фильтрирования

* **filtering criteria → 1423**

* **final compilation → 1363**

1424 find *v*
d finden; suchen
f trouver; découvrir; [re]chercher
r найти; искать

1425 fine generalization
d Feingeneralisierung *f*
f généralisation *f* fine
r рафинированная генерализация *f*

1426 finite-element method
d Finite-Element-Methode *f*
f méthode *f* d'éléments finis
r метод *m* конечных элементов

1427 finite-element model
d Finite-Element-Modell *n*
f modèle *m* d'élément fini
r конечно-элементная модель *f*

1428 first normal form
d erste Normal[en]form *f*
f première forme *f* normale
r первая нормальная форма *f*

1429 first order effect
d Effekt *m* erster Ordnung
f effet *m* de premier ordre
r эффект *m* первого порядка

* **first projection → 1851**

1430 fix *v*
d befestigen; fixieren
f fixer
r фиксировать; закреплять

* **fixed error → 3567**

1431 fixed spacing
d fester Abstand *m*
f espacement *m* fixe
r фиксированное расстояние *n*

1432 flatbed plotter
d Flachbettplotter *m*
f table *f* traçante à plat; traceur *m* à plat
r планшетный графопостроитель *m*; плоский графопостроитель

1433 flatbed scanner
d Flachbettscanner *m*
f scanner *m* à plat
r планшетный сканер *m*; плоский сканер

1434 flat drop-off
d flacher Abfall *m*; platter Abfall
f descente *f* plate
r плоский наклон *m*

1435 flat Earth model
d flaches Erdmodell *n*
f modèle *m* plat de la Terre
r плоская модель *f* Земли

* **flat file → 1436**

1436 flat [image] file
d flache Datei *f*; nichthierarchische Datei; lineare Datei
f fichier *m* plat; fichier non hiérarchique
r плоский файл *m*; неиерархический файл

1437 flight simulator database
d Flugsimulator-Datenbasis *f*
f base *f* de données de simulateur de vol
r база *f* данных авиасимулятора

1438 flight simulator systems
d Flugsimulatorsysteme *npl*
f systèmes *mpl* de simulateurs de vol
r авиасимуляционные системы *fpl*

* **floating menu → 2859**

1439 flood
d Fluttung *f*
f inondation *f*
r заливание *n*; затопление *n*

1440 flood control map
d Hochwasserschutzkarte *f*
f carte *f* de zones d'inondation
r карта *f* зон затопления

* **floodplain → 1441**

1441 floodplain [district]
d Hochwasserüberschwemmungsgebiet *n*;
Überschwemmungsgebiet *n*;
Überflutungsgebiet *n*;
Überschwemmungsfläche *f*
f champ *m* d'inondation; zone *f* d'inondation; plaine *f* inondable; plaine alluviale
r аллювиальная равнина *f*; намывная равнина; заливная терраса *f*; долина *f* разлива; заливной район *m*

1442 flow accumulation
(the total number of cells, including non-neighboring cells, that drain into a selected cell)
d Flussakkumulation *f*; Abfluss *m*
f accumulation *f* en coulée
r поточное накопление *n*

1443 flowchart; flow diagram; block-diagram; block scheme; box plot
d Flussdiagramm *n*; Ablaufdiagramm *n*; Ablaufschema *n*; Blockdiagramm *n*; Block[schalt]bild *n*; Blockschema *n*; Boxplot *n*
f diagramme *m* de flux; diagramme de blocs; schéma-bloc *m*; organigramme *m*
r блок-схема *f*; схема *f* последовательности; диаграмма *f* последовательности; блок-диаграмма *f*

* **flow diagram → 1443**

1444 flow direction
d Flussrichtung *f*; Ablaufrichtung *f*; Strömungsrichtung *f*
f direction *f* d'écoulement; sens *m* d'écoulement; sens du courant
r направление *n* потока

1445 flow line; stream line; line of stream
d Flusslinie *f*; Stromlinie *f*; Ablauflinie *f*
f ligne *f* de flux
r линия *f* потока

1446 flow path
(the drainage path through a watershed that begins at any selected point and runs to one of the outlets of the study site)
d Flussweg *m*
f voie *f* de passage; canal *m* de passage; passage *m*
r проток *m*

1447 flow path raster
d Flussweg-Raster *m*
f rastre *m* de passage
r растр *m* протока

1448 flow simulation
d Simulation *f* von Strömungen; Verkehrssimulation *f*

f simulation *f* de flux
r симуляция *f* потока; симуляция трафика

1449 fluctuation
 d Fluktuation *f*; Schwankung *f*
 f fluctuation *f*
 r флуктуация *f*; колебание *n*

1450 flying carpet
 (a graphic depiction of a one or more layers as
 an isometric view)
 d fliegender Teppich *m*
 f tapis *m* volant
 r ковёр-самолёт *m*

1451 fog
 d Schleier *m*; Nebel *m*
 f voile *m*; brouillard *m*
 r вуаль *f*; туман *m*

 * **FOIS → 1393**

1452 follower
 d Folge[r]stufe *f*
 f suiveur *m*
 r повторитель *m*; следящий элемент *m*

 * **font → 526**

1453 font palette
 d Schriftpalette *f*
 f palette *f* de polices
 r шрифтовая палитра *f*

1454 footnote; bottom note
 d Fußnote *f*; Anmerkung *f*
 f note *f* en bas de page
 r подстрочное примечание *n*; сноска *f*

1455 forecasting; prediction; prognosis
 d Vorhersage *f*; Prognostizierung *f*; Prädiktion *f*
 f prédiction *f*; prévision *f*
 r прогнозирование *n*; предсказание *n*

 * **forecast map → 2905**

1456 foreground
 d Vordergrund *m*
 f premier plan *m*; avant plan
 r передний план *m*

1457 foreign key
 (item in a relational table that contains a value
 identifying rows in another table)
 d Fremdschlüssel *m*; fremder Kennbegriff *m*
 f clé *f* étrangère
 r внешний ключ *m*

 * **foreshortening → 3061**

1458 forest district
 d Forstrevier *n*
 f district *m* forestier
 r лесной район *m*

1459 forestry map
 d Forstwirtschaftskarte *f*
 f carte *f* de foresterie
 r карта *f* лесного хозяйства; карта леса

1460 form
 d Formblatt *n*
 f formulaire *m*
 r формуляр *m*; бланк *m*

 * **form → 3279**

1461 format conversion
 d Formatkonvertierung *f*
 f conversion *f* de formats
 r преобразование *n* форматов;
   конвертирование *n* форматов

1462 format error
 d Formatfehler *m*
 f erreur *f* de format
 r погрешность *f* формата

1463 formline
 d Formlinie *f*
 f courbe *f* figurative; courbe à l'effet
 r гипотетическая горизонталь *f*

1464 forms interface
 d Formblatt-Schnittstelle *f*; Formblatt-Interface *f*
 f interface *f* de formes
 r интерфейс *m* формуляров

 * **forward lap → 1465**

**1465 forward [over]lap; end [over]lap;
   longitudinal overlap**
 d Längsüberdeckung *f*
 f recouvrement *m* longitudinal
 r продольное перекрытие *n*

**1466 four-color problem; problem of coloring
   maps in four colors**
 d Vierfarbenproblem *n*
 f problème des quatre couleurs
 r задача *f* о четырёх красках

 * **four-corners area → 2956**

1467 four-dimensional GIS; 4D GIS
 d vierdimensionales GIS *n*; 4D-GIS *n*
 f SIG *m* 4D
 r четырёхмерная ГИС *f*

1468 Fourier polygons
 d Fourier-Polygone *npl*

f polygones *mpl* de Fourier
r многоугольники *mpl* Фурье

1469 Fourier transformation
    d Fourier-Transformation *f*
    f transformation *f* de Fourier
    r преобразование *n* Фурье

    * **FOV** → **1403**

1470 fractal
    d Fractal *n*
    f fractale *f*
    r фрактал *m*

1471 fractal image; fractal pattern
    d fractales Bild *n*
    f image *f* fractale
    r фрактальное изображение *n*

    * **fractal pattern** → **1471**

1472 fractal surface
    d fractale Fläche *f*; fractale Oberfläche *f*
    f surface *f* fractale
    r фрактальная поверхность *f*

1473 fractional dimension
    d teilweise Dimension *f*; schrittweise Dimension
    f dimension *f* fractionnaire
    r секционированное оразмерение *n*;
        фракционный размер *m*

    * **frame** → **374, 1128, 1476, 2530**

1474 frame identification
    d Bezugsrahmenidentifikation *f*
    f identification *f* de repère
    r обозначение *n* репера

    * **frame of a map** → **376**

1475 frame of reference; reference frame[work]
    d Bezugsrahmen *m*; Basisrahmen *m*
    f repère *m*; position *f* de référence
    r репер *m*; точка *f* отсчёта

    * **frame scan** → **3024**

1476 framework; frame; skeleton
    d Rahmenwerk *n*; Fachwerk *n*; Gestell *n*;
        Gerüst *n*
    f châssis *m*; armature *f*; ossature *f*;
        cadre-porteur *m*
    r корпус *m*; каркас *m*; остов *m*

    * **framework map** → **678**

    * **framework of control points** → **2530**

1477 framework of control points for stereoplotting
    d Stereoplotten-Festpunktnetz *n*
    f canevas *m* de traçage stéréoscopique
    r геодезическая сеть *f* стереоскопического
        вычерчивания

1478 framing; cropping
    d Framing *n*; Umrahmung *f*; Einrahmung *f*;
        Rahmung *f*
    f tramage *m*; cadrage *m*
    r формирование *n* кадра; кадрирование *n*;
        кадрировка *f*

1479 framing system
    d Rahmungssystem *n*
    f système *m* de tramage
    r система *f* кадрирования

1480 free node
    d freier Knoten *m*
    f nœud *m* libre
    r свободный узел *m*

1481 free spatial demand curve
    d freie spatiale Bedarfskurve *f*
    f courbe *f* de demande spatiale libre
    r свободная пространственная кривая *f*
        спроса

1482 freezing of coordinates
    d Koordinatengefrierung *f*
    f blocage *m* de coordonnées; congélation *f* de
        coordonnées
    r блокировка *f* координат; замораживание *n*
        координат

1483 freezing of layer
    d Schichtgefrierung *f*
    f congélation *f* de plan
    r замораживание *n* слоя

1484 frequency dissemination
    d Frequenzzerstreuung *f*
    f dissémination *f* de fréquence
    r рассеяние *n* частоты

1485 frequency domain filter
    d Frequenzbereichsfilter *m*
    f filtre *m* de domaine fréquentiel
    r фильтр *m* частотного интервала

    * **freshen** *v* **up** → **3078**

1486 from-node
    (of an arc's two endpoints, the one first
        digitized)
    d vom Knoten
    f de nœud
    r от вершины

1487 frontage; building line; alignment
   d Angrenzer *m*; Anlieger *m*; Baulinie *f*
   f devanture *f*; alignement *m*
   r участок *m* между зданием и дорогой; граница *f* земельного участка (*по дороге, реке*); протяжённость *f* в определённом направлении

1488 front face; face
   d Frontseite *f*; Vorderseite *f*; Vorderfläche *f*
   f face *f* [frontale]
   r лицевая сторона *f*; лицевая поверхность *f*; передняя грань *f*; фасад *m*; лицо *n*

   * **frontier point** → 390

   * **full tone** → 723

1489 fully-analytical aerotriangulation
   d voll analytische Aerotriangulation *f*
   f aérotriangulation *f* entièrement analytique
   r вполне аналитическая аэротриангуляция *f*

1490 functional region; nodal region
   d funktionale Region *f*
   f région *f* fonctionnelle
   r центральный регион *m*; узловой регион

1491 functional relationship
   d funktionale Beziehung *f*
   f relation *f* fonctionnelle
   r функциональная связь *f*

1492 functional surface
   d Funktionsfläche *f*
   f surface *f* fonctionnelle
   r функциональная поверхность *f*

1493 fundamental plane of spherical coordinates
   d Fundamentalebene *f* der sphärische Koordinaten; Grundkreis *m*
   f plan *m* fondamental de coordonnées sphériques
   r основа *f* сферических координат

1494 fuzziness
   d Unbestimmtheit *f*; Flau *m*
   f flou *m*
   r нечёткость *f*; нерезкость *f*; размытость *f*; расплывчатость *f*

1495 fuzzy; hazy; blurred; non sharp; unsharp
   d unscharf; unbestimmt; Fuzzy-
   f flou; non distinct
   r размытый; нечёткий; нерезкий; неясный; расплывчатый

1496 fuzzy boundary
   d unscharfe Grenze *f*
   f limite *f* floue

   r нечёткая граница *f*

1497 fuzzy classification
   d unscharfe Klassifizierung *f*
   f classification *f* floue
   r нечёткая классификация *f*; размытая классификация

1498 fuzzy clustering
   d Fuzzy-Clusterung *f*
   f groupage *m* flou
   r нечёткая кластеризация *f*

1499 fuzzy graph
   d unbestimmter Graph *m*
   f graphe *m* flou
   r нечёткий граф *m*

1500 fuzzy logic
   d Fuzzylogik *f*; unscharfe Logik *f*
   f logique *f* floue
   r нечёткая логика *f*; непрерывная логика; размытая логика

1501 fuzzy logic model
   d Fuzzylogik-Modell *n*
   f modèle *m* de logique floue
   r модель *f* нечёткой логики

1502 fuzzy membership
   d unscharfe Mitgliedschaft *f*
   f appartenance *f* floue
   r нечёткое число *n* членов

1503 fuzzy node
   d unbestimmter Knoten *m*
   f nœud *m* flou
   r нечёткий узел *m*

1504 fuzzy object
   d unbestimmtes Objekt *n*
   f objet *m* flou
   r нерезкий объект *m*; нечёткий объект; размытый объект

1505 fuzzy set
   d unscharfe Menge *f*; Fuzzy-Menge *f*
   f ensemble *m* flou
   r нечёткое множество *n*; размытое множество

1506 fuzzy set theory
   (an extension to set theory that permits an object to have a degree of membership)
   d Fuzzy-Menge-Theorie *f*
   f théorie *f* des ensembles flous
   r теория *f* нечётких множеств; теория размытых множеств

1507 fuzzy tolerance
(a distance within which intersections and
points will be treated as coincident)
d unbestimmte Toleranz *f*
f tolérance *f* floue
r нечёткий допуск *m*

G

1508 gadget
d Gadget *n*
f gadget *m*; pontil *m* à griffes
r графический объект *m*, неограниченный
    окном

1509 Gall-Peters projection
d Gall-Peters-Projektion *f*
f projection *f* de Gall-Peters
r проекция *f* Гала-Питерса

1510 gap
d Abstand *m*; Lücke *f*; Spalte *m*
f lacune *f*; interstice *f*; fente *f*; bâillement *m*;
    écart *m*; vide *m*
r промежуток *m*; зазор *m*; щель *f*

* **Gauss distribution** → 2556

1511 Gaussian algorithm
d Gaussscher Algorithmus *m*
f algorithme *m* de Gauss
r гауссовский алгоритм *m*

1512 Gaussian approximation
d Gausssche Approximation *f*
f approximation *f* de Gauss
r гауссова аппроксимация *f*

1513 Gaussian coordinates
d Gausssche Koordinaten *fpl*
f coordonnées *fpl* de Gauss
r гауссовские координаты *fpl*

* **Gaussian distribution** → 2556

1514 Gauss-Krüger projection
d Gauss-Krüger-Projektion *f*
f projection *f* de Gauss-Krüger
r проекция *f* Гаусса-Крюгера

* **Gauss-Laplace distribution** → 2556

* **Gauss point** → 431

**1515 gazetteer; index of place names; list of place
    names**
d geografisches Lexikon *n*
f nomenclature *f* toponymique; catalogue *m* des
    noms de lieux
r указатель *m* географических названий;
    газеттир *m*

* **GBF** → 1576

* **GCM** → 1518

* **GCT** → 758

* **GCTP** → 1519

* **GDF** → 1586

* **GDOP** → 1657

* **GEBCO** → 1517

* **GEMS** → 1704

1516 general atlas
d allgemeiner Atlas *m*
f atlas *m* général
r общегеографический атлас *m*

**1517 general bathymetric chart of the oceans;
    GEBCO**
d allgemeine bathymetrische Meereskarte *f*
f carte *f* générale bathymétrique des océans
r общая батиметрическая карта *f* океанов

1518 general circulation model; GCM
d allgemeines Zirkulationsmodell *n*; allgemeiner
    Kreislauf *m*
f modèle *m* de circulation générale
r модель *f* общей циркуляции [атмосферы];
    МОЦ; модель глобальной циркуляции

**1519 general coordinate transformation
    package; geographic coordinate
    transformation package; GCTP**
    (a software)
d Software *f* für geografische
    Koordinatentransformation
f logiciel *m* de transformation de coordonnées
    géographiques
r программный пакет *m* трансформации
    географических координат

1520 generalization
    (conversion of a geographic representation to
    one with less resolution and less information
    content)
d Generalisierung *f*; Generalisation *f*;
    Verallgemeinerung *f*
f généralisation *f*
r генерализация *f*

1521 generalization data group
d Generationsdatengruppe *f*
f groupe *m* de données de généralisation
r группа *f* данных о генерализации

1522 **generalization degree**
   d Generalisierungsgrad *m*
   f degré *m* de généralisation
   r степень *f* генерализации

1523 **generalization operators**
   d Generalisierungsoperatoren *mpl*
   f opérateurs *mpl* de généralisation
   r генерализационные операторы *mpl*;
        операторы генерализации

1524 **generalization scale**
   d Generalisierungsmaßstab *m*
   f échelle *f* de généralisation
   r масштаб *m* генерализации; масштаб
        обобщения

1525 **generalized**
   d generalisiert
   f généralisé
   r обобщённый

1526 **generalized coordinates**
   d generalisierte Koordinaten *fpl*;
        verallgemeinerte Koordinaten
   f coordonnées *fpl* généralisées
   r обобщённые координаты *fpl*

   * **general map** → 3551

1527 **generic mapping tools**
   d generische Kartenwerkzeuge *npl*
   f instruments *mpl* de mappage génériques
   r типичные инструменты *mpl*
        картографирования

1528 **genetic algorithms**
   d genetische Algorithmen *mpl*
   f algorithmes *mpl* génétiques
   r генетические алгоритмы *mpl*

1529 **geocentric coordinates**
   d geozentrische Koordinaten *fpl*
   f coordonnées *fpl* géocentriques
   r геоцентрические координаты *fpl*

1530 **geocentric datum**
   d geozentrisches Datum *n*
   f datum *m* géocentrique
   r геоцентрический репер *m*

   * **geocentric Greenwich Cartesian coordinate
      system** → 1202

1531 **geocentric latitude**
   d geozentrische Breite *f*
   f latitude *f* géocentrique
   r геоцентрическая широта *f*

1532 **geocentric longitude**

   d geozentrische Länge *f*
   f longitude *f* géocentrique
   r геоцентрическая долгота *f*

1533 **geocentric meridian**
   d geozentrischer Meridian *m*
   f méridien *m* géocentrique
   r геоцентрический меридиан *m*

1534 **geocentric parallel**
   d geozentrischer Breitengrad *m*
   f parallèle *m* géocentrique
   r геоцентрическая параллель *f*

   * **geocode** → 1601

1535 **geocoded census**
   d geocodierte Zählung *f*; geocodierter Zensus *m*
   f recensement *m* géocodé
   r геокодированная перепись *f* [населения]

   * **geocoding** → 1592

1536 **geocoding error**
   d Geocodierungsfehler *m*
   f erreur *f* de géocodage
   r погрешность *f* геокодирования

   * **geodata** → 1584ʹ

1537 **geodata network**
   d Geodatennetz *n*
   f réseau *m* de données géographiques
   r сеть *f* географических данных

1538 **geodata server**
   d Geodatenserver *m*
   f serveur *m* de données géographiques
   r сервер *m* географических данных

1539 **geodemographics**
   d Geodemografie *f*
   f géodémographie *f*
   r геодемография *f*

   * **geodesic** → 1541

1540 **geodesic; geodetic** *adj*
   d geodätisch
   f géodésique
   r геодезический

   * **geodesic circle** → 1544

1541 **geodesic [line]; geodetic line; geodetic
        length**
        (of a surface)
   d Geodätische *f*; geodätische Linie *f*; Geodäte *f*
   f ligne *f* géodésique; géodésique *f*
   r геодезическая линия *f*

* **geodesic triangle** → 1559

1542 **geodesy**
   d Geodäsie *f*
   f géodésie *f*
   r геодезия *f*

* **geodesy on the ellipsoid** → 3440

* **geodetic** *adj* → 1540

1543 **geodetic azimuth; surveying azimuth**
   d geodätischer Azimut *m*
   f azimut *m* géodésique
   r геодезический азимут *m*

1544 **geodetic circle; geodesic circle**
   d geodätischer Kreis *m*
   f cercle *m* géodésique
   r геодезическая окружность *f*

* **geodetic control** → 2530

* **geodetic control network** → 2530

1545 **geodetic coordinates**
   d geodätische Koordinaten *fpl*
   f coordonnées *fpl* géodésiques
   r геодезические координаты *fpl*

1546 **geodetic curvature; tangential curvature**
   d Tangentialkrümmung *f*
   f courbure *f* géodésique; courbure tangentielle
   r геодезическая кривизна *f*

1547 **geodetic data**
   d geodätische Daten *pl*
   f données *fpl* géodésiques
   r геодезические данные *pl*

* **geodetic datum** → 3462

1548 **geodetic height; ellipsoid height**
   d geodätische Höhe *f*
   f hauteur *f* géodésique
   r геодезическая высота *f*

1549 **geodetic instrument**
   d geodätisches Instrument *n*
   f instrument *m* géodésique
   r геодезический прибор *m*

* **geodetic latitude** → 205

* **geodetic length** → 1541

* **geodetic line** → 1541

1550 **geodetic longitude**
   d geodätische Länge *f*

   f longitude *f* géodésique
   r геодезическая долгота *f*

1551 **geodetic mapping; Beltrami['s] mapping**
   d geodätische Abbildung *f*; Beltramische Abbildung; bahntreue Abbildung
   f application *f* géodésique; application de Beltrami
   r геодезическое отображение *n*; отображение Бельтрами; отображение, сохраняющее траектории

1552 **geodetic measurements**
   d geodätische Messungen *fpl*
   f mesurages *mpl* géodésiques
   r геодезические измерения *npl*

1553 **geodetic meridian**
   d geodätischer Meridian *m*
   f méridien *m* géodésique
   r геодезический меридиан *m*

* **geodetic net** → 2530

* **geodetic network** → 2530

1554 **geodetic parallel**
   d geodätischer Breitengrad *m*
   f parallèle *m* géodésique
   r геодезическая параллель *f*

1555 **geodetic reference system; GRS**
   d geodätisches Referenzsystem *n*
   f système *m* de référence géodésique
   r геодезическая референцная система *f*

1556 **geodetic survey; survey**
   d [geodätische] Aufnahme *f*
   f levé *m* [géodésique]; service *m* de géodésie
   r [геодезическая] съёмка *f*

1557 **geodetic surveying**
   d geodätische Überwachung *f*
   f surveillance *f* géodésique
   r геодезическое обследование *n*

1558 **geodetic torsion**
   d geodätische Torsion *f*; geodätische Windung *f*
   f torsion *f* géodésique
   r геодезическое кручение *n*

1559 **geodetic triangle; geodesic triangle**
   d geodätisches Dreieck *n*
   f triangle *m* géodésique
   r геодезический треугольник *m*

1560 **geodetic zenith distance**
   d geodätische Zenitdistanz *f*
   f distance *f* zénithale géodésique
   r геодезическое зенитное расстояние *n*

1561 geodimensions
  d geografische Dimensionen *fpl*
  f géodimensions *fpl*
  r географические измерения *npl*

1562 geodynamics
  d Geodynamik *f*
  f géodynamique *f*
  r геодинамика *f*

1563 geofacilities database
  d Geomöglichkeiten-Datenbasis *f*
  f base *f* de données de moyens géographiques
  r база *f* данных географических средств

  * **geofacility** → **1619**

1564 geo-field
  d geografischer Bereich *m*
  f champ *m* géographique
  r географическая область *f*; географическая сфера *f* деятельности

1565 geographer
  d Geograph *m*
  f géographe *m*
  r географ *m*

  * **geographic** → **1569**

1566 geographic access
  d geografischer Zugriff *m*
  f accès *m* géographique
  r географический доступ *m*

1567 geographic access rights
  d geografische Zugriffsrechte *npl*
  f droits *mpl* d'accès géographique
  r права *npl* географического действия; полномочия *npl* географического доступа

1568 geographic address
  d geografische Adresse *f*
  f adresse *f* géographique
  r географический адрес *m*

1569 geographic[al]
  d geografisch; geographisch
  f géographique
  r географический

  * **geographical longitude** → **206**

  * **geographically dispersed** → **1570**

1570 geographically distributed; geographically dispersed
  d geografisch verteilt
  f géographiquement dispersé;

  géographiquement distribué
  r территориально распределённый; территориально рассредоточенный

1571 geographically distributed system
  d geografisch verteiltes System *n*
  f système *m* géographiquement distribué
  r территориально распределённая система *f*

1572 geographically referenced; geographically referred; georeferenced; georeferred
  d geografisch codiert; georeferenziert
  f géoréférencé; géographiquement spécifique; localisé
  r географически упоминаемый; географически относимый

1573 geographically referenced data; georeferenced data
  d georeferenzierte Daten *pl*
  f données *fpl* géoréférencées; données à référence géospatiale; données localisées
  r географически относимые данные *pl*

  * **geographically referred** → **1572**

1574 geographic application
  d geografische Anwendung *f*
  f application *f* géographique
  r географическое применение *n*

1575 geographic atlas; atlas
  d [geografischer] Atlas *m*; Kartenwerk *n*
  f atlas *m* [géographique]
  r [географический] атлас *m*

1576 geographic base files; GBF
  d geografischer Grunddatensatz *m*
  f fichiers *mpl* de base géographique
  r массив *m* базы географических данных

  * **geographic botany** → **536**

  * **geographic calibration** → **1619**

1577 geographic cartography
  d geografische Kartografie *f*
  f cartographie *f* géographique
  r географическая картография *f*

1578 geographic catalog of political and statistical areas
  d geografischer Katalog *m* der politischen und statistischen Bereiche
  f catalogue *m* géographique de régions politiques et statistiques
  r географический каталог *m* политических и статистических регионов

1579 **geographic category**
  d geografische Kategorie *f*
  f catégorie *f* géographique
  r географическая категория *f*

1580 **geographic center**
  d geografisches Zentrum *n*
  f centre *m* géographique
  r географический центр *m*

  * **geographic code → 1601**

1581 **geographic coordinates; graticule reference**
  d geografische Koordinaten *fpl*
  f coordonnées *fpl* géographiques
  r географические координаты *fpl*

  * **geographic coordinate transformation package → 1519**

1582 **geographic coverage; coverage; field**
  d geografischer Erfassungsbereich *m*; erfasster Bereich *m*; Geländeausschnitt *m*
  f portée *f* géographique; zone *f* géographique; champ *m*
  r географическая зона *f* [наблюдения]; [географическая] область *f* покрытия; покрытие *n*

1583 **geographic cycle**
  d geografischer Kreis *m*
  f cycle *m* géographique
  r географический цикл *m*

1584 **geographic data; geographic information; geodata; geospatial data; locational data**
  (spatial data that is referenced to a location on the Earth's surface)
  d geografische Daten *pl*; Geodaten *pl*; Geofachdaten *pl*; geografische Information *f*
  f données *fpl* géographiques; information *f* géographique
  r географические данные *pl*; геопространственные данные; географическая информация *f*

1585 **geographic database**
  d geografische Datenbasis *f*; Geodatenbasis *f*
  f base *f* de données géographiques; géobase *f*
  r географическая база *f* данных

1586 **geographic data file; GDF**
  d Geodaten-Datei *f*
  f fichier *m* de données géographiques
  r файл *m* цифровых картографических данных (*файловый формат, предложенный для цифровой электронной карты Европы в рамках проекта DEMETER*)

1587 **geographic data management**
  d Geodaten-Management *n*
  f gestion *f* de données géographiques
  r управление *n* географических данных

1588 **geographic dataset**
  d geografische Datenmenge *f*
  f ensemble *m* de données géographiques
  r совокупность *f* географических данных

1589 **geographic data technology**
  d Geodaten-Technologie *f*
  f technologie *f* de données géographiques
  r технология *f* географических данных

1590 **geographic display**
  (mapping geographic location of remote server)
  d geografisches Display *n*
  f affichage *m* géographique
  r географический дисплей *m*

1591 **geographic division**
  d geografische Teilung *f*; geografischer Abschnitt *m*
  f division *f* géographique; sectionnement *m* géographique
  r географическое разделение *n*

1592 **geographic encoding; geocoding**
  d Geocodierung *f*
  f géocodage *m*
  r геокодирование *n*

1593 **geographic [encoding and]referencing; map georeferencing; georeferencing; map registration**
  d Georeferenzierung *f*
  f géoréférencement *m*; calage *m* d'une carte
  r стандартизация *f* представления географических данных; географическое кодирование *n* и стандартизация *f*

1594 **geographic engine**
  d geografische Engine *f*
  f moteur *m* géographique
  r географический процессор *m*

1595 **geographic entity**
  d geografische Entität *f*
  f entité *f* géographique
  r географический примитив *m*

1596 **geographic entry system**
  d geografische Dateneingabesystem *n*
  f système *m* d'entrée de données géographiques
  r система *f* ввода географических данных

1597 **geographic equivalent**
  d geografisches Äquivalent *n*

f équivalent *m* géographique
r географический эквивалент *m*

* **geographic facility** → 1619

* **geographic feature** → 1615

* **geographic graticule** → 2274

* **geographic grid** → 2274

1598 geographic grid system; grid system
   d geografisches Koordinatensystem *n*;
    vermaschtes Netz *n*
   f système *m* de coordonnées géographiques;
    réseau *m* maillé
   r система *f* географических координат;
    сеточная система

1599 geographic hierarchy
   d geografische Hierarchie *f*
   f hiérarchie *f* géographique
   r географическая иерархия *f*

**1600 geographic identification code scheme;
GICS**
   d Geocodierungsschema *n*
   f schème *m* de codage géographique
   r схема *f* географического кодирования

**1601 geographic identifier; geographic code;
geocode**
   d geografischer Identifikator *m*; Geocode *m*
   f code *m* géographique
   r географический код *m*; геокод *m*

1602 geographic identity
   d geografische Identität *f*
   f identité *f* géographique
   r географическая идентичность *f*

1603 geographic index
   d geografisches Register *n*
   f index *m* géographique; index de
    bibliographies géographiques
   r географический регистр *m*

1604 geographic index database
   d Datenbasis *f* des geografischen Registers
   f base *f* de données d'indices géographiques
   r база *f* данных географических регистров

* **geographic information** → 1584

1605 geographic information development
   d Geodatenentwicklung *f*
   f développement *m* d'information géographique
   r развитие *n* географической информации

* **geographic information infrastructure** →
1677

**1606 Geographic Information Retrieval and
Analysis System; GIRAS** (US)
   d System *n* der Geodatenwiedergewinnung- und
    Analyse
   f Système *m* de recherche et analyse
    d'information géographique
   r Картографическая система *f*
    географической съёмки

**1607 geographic information science;
geoinformatics; geomatics**
   d Geoinformatik *f*
   f géo-informatique *f*; géomatique *f*
   r геоинформатика *f*; геоматика *f*

1608 geographic information system; GIS
   d Geoinformationssystem *n*; geografisches
    Informationssystem *n*; GIS
   f système *m* d'information géographique; SIG
   r географическая информационная система *f*;
    геоинформационная система; ГИС

1609 geographic inquiry
   d geografische Abfrage *f*
   f interrogation *f* géographique
   r географический запрос *m*

1610 geographic knowledge system; GKS
   d geografisches wissenbasiertes System *n*
   f système *m* de connaissance géographique
   r система *f* географических знаний

* **geographic latitude** → 205

1611 geographic map; map; chart; carte
   d Landkarte *f*; Karte *f*
   f carte *f* [géographique]
   r [географическая] карта *f*

1612 geographic metaphor
   d geografische Metapher *f*
   f métaphore *f* géographique
   r географическая метафора *f*

**1613 geographic names; place names; field
names**
   d geografische Namen *mpl*; Flurnamen *mpl*
   f noms *mpl* géographiques
   r географические наименования *npl*;
    топонимы *mpl*

**1614 geographic names information system;
GNIS**
   d Informationssystem *n* der geografischen
    Namen

f système *m* d'information de noms
géographiques
r информационная система *f* географических
названий

**1615 geographic object; geo-object; [geographic]
feature**
d Geoobjekt *n*
f objet *m* géographique; accident *m*
géographique
r географический объект *m*

1616 geographic operators
(for spatial analysis, intersection, point in
polygon, area in area etc.)
d geografische Operatoren *mpl*
f opérateurs *mpl* géographiques
r географические операторы *mpl*

1617 geographic pattern
d Raummuster *n*
f profil *m* géographique
r географическое распределение *n*

1618 geographic pole; pole
d geografischer Pol *m*
f pôle *m* géographique
r [географический] полюс *m*

* **geographic presentation** → 1622

**1619 geographic reference; georeference;
geographic calibration; geographic facility;
geofacility**
d geografische Referenz *f*; Georeferenz *f*
f lieu *m* géographique; géoréférence *f*
r географическая ссылка *f*

1620 geographic reference identification number
d Identifikationsnummer *f* für eine geografische
Referenz
f numéro *m* d'identification du lieu
géographique
r идентификатор *m* системы географических
координат

* **geographic referencing** → 1593

1621 geographic relationship
d geografische Beziehung *f*
f relation *f* géographique
r географическая [взаимо]связь *f*

**1622 geographic representation; geographic
presentation; geographic visualization;
georepresentation**
d geografische Darstellung *f*
f présentation *f* géographique
r географическое представление *n*

1623 geographic resource; geo-resource;

geosource
d geografische Ressource *f*; Georessource *f*
f géo-ressource *f*
r географический ресурс *m*

1624 geographic sciences
d geografische Wissenschaften *fpl*
f sciences *fpl* géographiques
r географические науки *fpl*

1625 geographic search
d geografische Suche *f*
f recherche *f* géographique
r географический поиск *m*

1626 geographic security
d geografische Sicherheit *f*
f sécurité *f* géographique
r географическая защита *f*

1627 geographic site; site
(a reference to the physical attributes of a
location)
d Lage *f*; Ortslage *f*; Standort *m*
f site *m* géographique
r место[положение] *n*; узел *m*

**1628 geographic statistics; georeferenced
statistics; geostatistics**
d Geostatistik *f*; georeferenzierte Statistik *f*
f statistique *f* géographique; statistique
géoréférencée
r географическая статистика *f*;
геостатистика *f*; статистика географических
ссылок

1629 geographic statistics program
d Geostatistiksprogramm *n*
f programme *m* de statistique géographique
r программа *f* географической статистики

1630 geographic theme; theme
d [geografisches] Thema *n*
f thème *m* [géographique]
r [географическая] тема *f*

1631 geographic traceroute
d geografische Traceroute *f*
f routage *m* géographique
r трассировка *f* топологии географического
размещения; маршрутизация *f*
географических объектов

* **geographic traceroute application** → 1632

**1632 geographic traceroute program; geographic
traceroute application**
d geografisches Traceroute-Programm *n*
f application *f* de routage géographique
r программа *f* трассировки топологии
географического расположения

1633 geographic unit
 d geografische Einheit *f*
 f unité *f* géographique
 r географическая единица *f*

1634 geographic view
 d geografische Ansicht *f*
 f vue *f* géographique
 r географический вид *m*

* **geographic visualization** → 1622

1635 geography markup language; GML
 d Geografie-Markierungssprache *f*
 f langage *m* de balisage de géographie
 r язык *m* разметки для географии

1636 geography of Internet address space
 d Geografie *f* des Internet-Adressraums
 f géographie *f* d'espace des adresses Internet
 r география *f* адресного пространства Интернет

1637 geoid
 (three-dimensional shape of the Earth defined by the surface where gravity has the value associated with mean sea level)
 d Geoid *n*
 f géoïde *m*
 r геоид *m*

1638 geoid[al] height; orthometric height
 d orthometrische Höhe *f*
 f cote *f* orthométrique; hauteur *f* orthométrique; HO
 r ортометрическая высота *f*

* **geoid height** → 1638

1639 geoimage
 d Geobild *n*
 f géo-image *f*
 r геоизображение *n*

* **geoinformatic education** → 1640

* **geoinformatic mapping** → 1642

* **geoinformatics** → 1607

1640 geoinformatic training; geoinformatic education
 d Geoinformatik-Bildung *f*
 f enseignement *m* géo-informatique
 r геоинформационное образование *n*

1641 geoinformational conception
 d Geoinformatik-Konzeption *f*
 f conception *f* géo-informatique; géoconception *f*

 r геоинформационная концепция *f*

1642 geoinformational mapping; geoinformatic mapping
 d Geoinformatik-Kartierung *f*
 f cartographie *f* géo-informatique
 r геоинформационное картографирование *n*

* **GEOLAN** → 1643

* **geological local area network** → 1643

1643 geological [survey] local area network; GEOLAN
 d lokales Netzwerk *n* von geologischer Aufnahme
 f réseau *m* local géologique
 r геологическая местная сеть *f*

1644 geologic data
 d geologische Daten *pl*
 f données *fpl* géologiques
 r геологические данные *pl*

1645 geologic formation
 d geologische Formation *f*; geologische Bildung *f*
 f formation *f* géologique
 r геологическое образование *n*

1646 geologic map
 d geologische Karte *f*
 f carte *f* géologique
 r геологическая карта *f*

1647 geologic mapping
 d geologische Kartografie *f*
 f cartographie *f* géologique
 r геологическое картирование *n*

1648 geomagnetic field
 d Erdmagnetfeld *n*; Magnetfeld *n* der Erde
 f champ *m* géomagnétique
 r геомагнитное поле *n*

1649 geomagnetic survey
 d geomagnetische Überwachung *f*
 f surveillance *f* géomagnétique
 r геомагнитная съёмка *f*

1650 geomarketing
 d geografische Vermarktung *f*; geografisches Marketing *n*; Geomarketing *n*
 f geomarketing *m*
 r географический маркетинг *m*; геомаркетинг *m*

* **geomatics** → 1607

1651 geometric accuracy
   d geometrische Genauigkeit *f*
   f exactitude *f* géométrique
   r геометрическая точность *f*

1652 geometric complexity
   d geometrische Komplexität *f*
   f complexité *f* géométrique
   r геометрическая сложность *f*

1653 geometric correction; geometric rectification
   (removing geometric distortion from a raster or a vector object)
   d geometrische Korrektur *f*
   f correction *f* géométrique
   r геометрическая поправка *f*; геометрическая коррекция *f*

1654 geometric coverage limits
   d geometrische Erfassungsbereichsgrenzen *fpl*
   f limites *fpl* de couverture géométriques
   r геометрические зоны *fpl* действия

1655 geometric data
   d geometrische Daten *pl*; Geometriedaten *pl*
   f données *fpl* géométriques
   r геометрические данные *pl*

1656 geometric database
   d geometrische Datenbasis *f*
   f base *f* de données géométriques
   r база *f* геометрических данных

1657 geometric dilution of precision; GDOP
   d geometrischer Genauigkeitsabfall *m*
   f dilution *f* géométrique de précision
   r геометрический фактор *m* точности; показатель *m* снижения точности, обусловленный геометрическими факторами

1658 geometric distortion
   d geometrische Verzerrung *f*
   f distorsion *f* géométrique
   r геометрическое искажение *n*

1659 geometric distortions correction
   d Korrektur *f* der geometrischen Verzerrungen
   f correction *f* de distorsions géométriques
   r исправление *n* геометрических искажений

1660 geometric graphics
   d geometrische Grafik *f*
   f graphique *f* géométrique
   r геометрическая графика *f*; графика для построения геометрических фигур

1661 geometric model
   d geometrisches Modell *n*
   f modèle *m* géométrique
   r геометрическая модель *f*

1662 geometric operation
   d geometrische Operation *f*
   f opération *f* géométrique
   r геометрическая операция *f*

1663 geometric primitive
   d geometrisches Primitiv *n*; geometrisches Grundelement *n*
   f primitive *f* géométrique
   r геометрический примитив *m*

* **geometric rectification** → 1653

1664 geometric surface
   d geometrische Fläche *f*
   f surface *f* géométrique
   r геометрическая поверхность *f*

1665 geometric transformation
   d geometrische Transformation *f*
   f transformation *f* géométrique
   r геометрическая трансформация *f*; геометрическое преобразование *n*

* **geomodeling** → 455

1666 geomorphometry
   d Geomorphometrie *f*
   f géomorphométrie *f*
   r геоморфометрия *f*; количественная геоморфология *f*

* **geonemy** → 536

* **geo-object** → 1615

1667 geophysical data center
   d geophysikalisches Datenzentrum *n*
   f centre *m* de données géophysiques
   r центр *m* геофизических данных

1668 geophysical events
   d geophysikalische Ereignisse *npl*
   f événements *mpl* géophysiques
   r геофизические события *npl*

1669 geoportal
   d Geoportal *n*
   f géoportail *m*
   r геопортал *m*

1670 geopotential number
   d geopotentieller Wert *m*; Potentialwert *m*
   f nombre *m* géopotentiel
   r геопотенциальное число *n*

1671 geoprocessing
  d Verarbeitung *f* von geografischen Daten
  f traitement *m* de données géographiques;
  géoprocessing *m*
  r обработка *f* данных, связанных с науками о
  Земле

  * **GEOREF** → 4011

  * **georeference** → 1619

  * **georeferenced** → 1572

  * **georeferenced data** → 1573

1672 georeferenced image map
  d geografisch codierte Imagemap *f*
  f image *f* réactive géoréférencée
  r географически относимое адресуемое
  отображение *n*

  * **georeferenced statistics** → 1628

  * **georeferencing** → 1593

  * **georeferencing point** → 2323

  * **georeferred** → 1572

1673 georelational conception
  d georelationale Konzeption *f*
  f conception *f* géorelationnelle
  r геореляционная концепция *f*

1674 georelational [data] model
  d georelationales Datenmodell *n*
  f modèle *m* de données géorelationnel
  r геореляционная модель *f* данных

  * **georelational model** → 1674

  * **georepresentation** → 1622

  * **geo-resource** → 1623

1675 geosciences; Earth sciences
  d Geowissenschaften *fpl*
  f sciences *fpl* de la Terre
  r науки *fpl* о Земле

  * **geosource** → 1623

1676 geosource gateway
  d Georessourcen-Gateway *n*
  f passerelle *f* de ressources géographiques
  r межсетевой интерфейс *m* географических
  ресурсов; шлюз *m* географических
  ресурсов

  * **geospatial data** → 1584

**1677 geospatial data infrastructure; geographic
  information infrastructure; global
  information infrastructure; GII**
  d Geodateninfrastruktur *f*
  f infrastructure *f* de données géospatiaux;
  infrastructure d'information géographique
  r инфраструктура *f* геопространственных
  данных; инфраструктура географической
  информации

1678 geospatial data model
  d Geodatenmodell *n*
  f modèle *m* de données géospatiaux
  r модель *f* геопространственных данных

**1679 geostationary; geosynchronous;
  earth-synchronous**
  d geostationär; geosynchron
  f géostationnaire; géosynchrone
  r геостационарный

1680 geostationary navigation overlay system
  d geostationäres Navigationssystem *n*
  f système *m* de navigation par recouvrement
  géostationnaire
  r геостационарная оверлейная
  навигационная система *f*

1681 geostationary orbit
  d geostationärer Orbit *m*
  f orbite *f* géostationnaire
  r геостационарная орбита *f*

  * **geostatistics** → 1628

  * **geosynchronous** → 1679

1682 geosynchronous orbit
  d geosynchroner Orbit *m*
  f orbite *f* géosynchrone
  r геосинхронная орбита *f*

  * **ghost** → 1172

  * **GICS** → 1600

1683 GIF format
  d GIF-Format *n*
  f format *m* GIF
  r формат *m* GIF

  * **GII** → 1677

  * **GIRAS** → 1606

  * **GIS** → 1608, 1707

1684 GIS application
d GIS-Anwendung *f*
f application *f* SIG
r ГИС-приложение *n*

1685 GIS architecture
d GIS-Systemarchitektur *f*
f architecture *f* de SIG
r архитектура *f* ГИС

1686 GIS-assisted civil information system
d GIS-gestütztes Bürgerinformationssystem *n*
f système *m* d'information civil assisté par SIG
r гражданская информационная система *f* на основе ГИС

1687 GIS-assisted management system
d GIS-gestütztes Managementsystem *n*
f système *m* de gestion assisté par SIG
r система *f* управления с помощью ГИС

1688 GIS-based analysis
d GIS-basierte Analyse *f*
f analyse *f* à base de SIG
r геоинформационный анализ *m*

1689 GIS base layer
d GIS-Basisschicht *f*
f couche *f* de base SIG
r базовый слой *m* ГИС

1690 GIS benchmark
d GIS-Benchmark *n*
f étalonnage *m* de SIG
r эталонное тестирование *n* ГИС

1691 GIS database
d GIS-Datenbasis *f*
f base *f* de données SIG
r база *f* данных ГИС

1692 GIS database technology
d Datenbasistechnologie *f* für GIS
f technologie *f* de base de données géospatiaux; technologie de base de données SIG
r технология *f* базы геопространственных данных

1693 GIS functionality; GIS functions
d GIS-Funktionalität *f*
f fonctionnalité *f* de SIG
r функциональные возможности *fpl* ГИС

* **GIS functions → 1693**

1694 GIS[-functions] server
d GIS[-Funktions]-Server *m*
f serveur *m* de SIG
r сервер *m* географических информационных систем

1695 GIS in Internet
d GIS-Internetlösungen *fpl*; GIS *n* im Internet
f Internet-SIG *m*
r ГИС *f* в Интернет

1696 GIS market
d GIS-Markt *m*
f marché *m* de SIG
r рынок *m* географических информационных систем

1697 GIS product
d GIS-Produkt *n*
f produit *m* de SIG
r ГИС-продукт *m*

1698 GIS project
d GIS-Projekt *n*
f projet *m* SIG
r ГИС-проект *m*; геоинформационный проект

* **GIS server → 1694**

1699 GIS shareware package
d GIS-Shareware *f*
f paquet *m* contributif SIG
r некоммерсиальный ГИС-пакет *m*

1700 GIS software
d GIS-Software *f*
f logiciel *m* SIG
r программное обеспечение *n* ГИС

1701 GIS technology
d GIS-Technologie *f*
f technologie *f* de SIG
r геоинформационная технология *f*; ГИС-технология *f*

1702 GIS user
d GIS-Benutzer *m*
f utilisateur *m* de SIG
r пользователь *m* ГИС

* **GKS → 1610**

1703 global coordinate space
d globaler Koordinatenraum *m*
f espace *m* de coordonnées global
r глобальное пространство *n* координат

* **global coordinate system → 4008**

1704 global environmental monitoring system; GEMS
d globale Umweltmonitoringsystem *n*
f système *m* global de monitorage de l'environnement
r глобальная система *f* мониторинга окружающей среды; ГСМОС

1705 global environmental system
d globales Umweltsystem *n*
f système *m* global de l'environnement
r глобальная система *f* окружающей среды

1706 global GIS
d globales GIS *n*
f SIG *m* global
r глобальная ГИС *f*; планетарная ГИС

1707 global indexing system; GIS
d globales Indizierungssystem *n*
f système *m* d'indexage global
r глобальная система *f* индексирования

* **global information infrastructure → 1677**

1708 globalization
d Globalisierung *f*
f globalisation *f*
r глобализация *f*

1709 globally unique identifier
d global einheitlicher Identifikator *m*
f identificateur *m* unique globalement
r глобально-однозначный идентификатор *m*;
уникальный идентификатор

1710 global positioning system; GPS
d globales Positioniersystem *n*; GPS
f système *m* de positionnement par satellites;
système mondial de [radio]repérage; système
de positionnement à capacité globale; GPS
r глобальная позиционирующая система *f*;
ГПС; спутниковая система
позиционирования; глобальная
локационная система; глобальная система
рекогносцировки

**1711 global resource information database;
GRID**
d globale Ressourcendatenbasis *f*
f base *f* de données de ressources globale
r глобальная природно-ресурсная база *f*
данных; ГРИД (*информационная система
и международная программа, выполняемая
в рамках ГСМОС (GEMS) при ЮНЕП*)

1712 global spatial data system
d globales Raumdatensystem *n*
f système *m* global de données spatiaux
r глобальная система *f* пространственных
данных

1713 global topology
d globale Topologie *f*
f topologie *f* globale
r общая топология *f*

* **globe → 1209**

1714 glyph; graphic symbol
d Glyphe *n*; Bildzeichen *n*
f glyphe *m*; symbole *m* graphique; signe *m*
graphique
r стрелочный глиф *m*

* **GML → 1635**

* **GNIS → 1614**

1715 gnomonic [map] projection
d gnomonische Kartenprojektion *f*
f projection *f* gnomonique
r гномоническая [картографическая]
проекция *f*

* **gnomonic projection → 1715**

* **gnosiological conception → 2440**

* **goal → 997**

* **gore → 1716**

1716 gore [lot]
(a thin triangular piece of land, the boundaries
of which are defined by surveys of adjacent
properties)
d Keil *m*; Zwickel *m*
f fuseau *m*; terrain *m* enclavé; godet *m*
r участок *m* земли клином

1717 governmental unit; GU
d staatliche Einheit *f*
f unité *f* gouvernementale
r государственная единица *f*

* **GPS → 1710**

1718 GPS-based navigation system
d GPS-basiertes Navigationssystem *n*
f système *m* de navigation à base de GPS
r навигационная система *f* с помощью ГПС

1719 GPS constellation
d GPS-Konstellation *f*
f constellation *f* GPS
r ГПС-констелляция *f*

1720 GPS control monitor
d GPS-Kontrollmonitor *m*
f moniteur *m* de contrôle GPS
r контрольное устройство *n* ГПС

1721 GPS navigation
d GPS-Navigation *f*
f navigation *f* par GPS
r навигация *f* с помощи ГПС

1722 GPS receiver
d GPS-Empfänger *m*

f récepteur *m* GPS
r приёмник *m* позиционирования

1723 GPS software
d GPS-Software *f*
f logiciel *m* de GPS
r программное обеспечение *n* ГПС

1724 grabber button
d Grabber-Schaltfläche *f*
f bouton *m* accrocheur
r бутон *m* захвата кадра; бутон захвата
  изображения; "ладошка" *f*

1725 gradation
d Gradation *f*
f gradation *f*
r градация *f*

  * **grade** → 949

  * **grademeter** → 1922

1726 gradient
d Gradient *m*
f gradient *m*
r градиент *m*

1727 gradient analysis
d Gradientenanalyse *f*
f analyse *f* de gradient
r градиентный анализ *m*

  * **gradient angle** → 125

1728 gradient density
d Gradient[en]dichte *f*
f densité *f* graduelle
r градиентная плотность *f*

1729 gradient estimation
d Gradient[en]abschätzung *f*
f évaluation *f* de gradient
r оценка *f* крутости; оценка градиента

1730 gradient filtering
d Gradientfilterung *f*
f filtrage *m* de gradient
r фильтрирование *n* градиента;
  фильтрирование крутости

**1731 gradient filters; sharpening filters; Sobel
    filters**
d Schärfefilter *mpl*
f filtres *mpl* distincts
r контрастные фильтры *mpl*

  * **gradient of slope** → 126

1732 graduated point symbols

d graduierte Punktsymbole *npl*
f signes *mpl* symboliques dégradés
r градуированные шкалы *fpl* значков;
  градуированные условные знаки *mpl*

  * **graduation** → 3211

  * **graininess** → 1734

1733 grain tolerance
d Grain-Toleranz *f* (*minimaler Abstand zweier
  Vertices in Kurven*)
f tolérance *f* de grain
r допуск *m* зерна

1734 granularity; graininess
d Körnigkeit *f*
f granularité *f*
r зернистость *f*; гранулярность *f*;
  дискретность *f*

  * **graph** → 1018

1735 graph
d Graph *m*
f graphe *m*
r граф *m*

  * **graphic** → 1736

1736 graphic[al]
d grafisch
f graphique
r графический

  * **graphical and analytical technique methods**
    → 1737

**1737 graphical and analytical techniques;
    graphical and analytical technique methods**
d grafische und analytische Methoden *fpl*
f méthodes *fpl* graphiques et analytiques
r графоаналитические приёмы *mpl*

1738 graphic elements
d grafische Elemente *npl*
f éléments *mpl* graphiques
r графические элементы *mpl*

  * **graphic factor** → 1749

1739 graphic image
d grafisches Bild *n*
f image *f* graphique
r графическое изображение *n*; графический
  образ *m*

1740 graphic input device
d grafisches Eingabegerät *n*

f dispositif *m* d'entrée graphique
r устройство *n* графического ввода

1741 graphic output device
   d grafisches Ausgabegerät *n*
   f dispositif *m* de sortie graphique
   r устройство *n* графического вывода

1742 graphic overlay
   d grafisches Overlay *n*
   f calque *m* graphique
   r графический оверлей *m*; графическая
     накладка *f*

   * **graphic primitive** → 1274

1743 graphic representation
   d grafische Darstellung *f*
   f représentation *f* graphique
   r графическое представление *n*

1744 graphic scale; linear scale; bar scale
   d grafischer Maßstab *m*
   f échelle *f* graphique
   r графический масштаб *m*; линейный
     масштаб

1745 graphics data structure
   d Grafik-Datenstruktur *f*
   f structure *f* de données graphiques
   r структура *f* графических данных

   * **graphics film recorder** → 2418

1746 graphics page
   d grafische Seite *f*
   f page *f* graphique
   r графическая страница *f*

1747 graphic super[im]position
   d grafische Superponierung *f*; grafische
     Überlagerung *f*
   f sur[im]position *f* graphique
   r графическая суперпозиция *f*; графическое
     наложение *n*

   * **graphic superposition** → 1747

   * **graphic symbol** → 1714

   * **graphic tablet** → 1071

1748 graphic user interface; GUI
   d grafische Benutzerschnittstelle *f*
   f interface *f* graphique utilisateur
   r графический интерфейс *m* пользователя;
     графический пользовательский интерфейс

**1749 graphic variable; graphic factor;
     semiological factor**

d grafische Variable *f*
f variable *f* graphique
r графическая переменная *f*

1750 graph paper; plotting paper
   d Millimeterpapier *n*
   f papier *m* millimétré
   r миллиметровая бумага *f*

1751 graph visualization
   d Graph-Darstellung *f*
   f visualisation *f* par graphique
   r изображение *n* в виде диаграммы

1752 graph window
   d Diagramm-Fenster *n*
   f fenêtre *f* de graphique
   r окно *n* диаграммы

   * **graticule** → 2274

   * **graticule intersection** → 1770

   * **graticule reference** → 1581

1753 graticule ticks
   d Teilstriche *mpl*
   f amorces *fpl* de réseau géographique; amorces
     de canevas
   r отметки *fpl* [географической] сетки

   * **grating** → 3923

   * **gravimetry** → 1757

1754 gravity coverage
   d Schwere-Abdeckung *f*
   f couverture *f* de levé gravimétrique
   r покрытие *n* гравиметрической съёмкой;
     охват *m* гравиметрической съёмкой

1755 gravity net
   d Schwerenetz *n*
   f réseau *m* gravimétrique
   r гравиметрическая сеть *f*

1756 gravity surface
   d Schwereoberfläche *f*
   f surface *f* de gravité
   r поверхность *f* гравитации

1757 gravity survey; gravimetry
   d Gravimetrie *f*
   f levé *m* gravimétrique; gravimétrie *f*
   r гравиметрические исследования *npl*;
     гравиразведочные работы *fpl*;
     гравиметрическая съёмка *f*; гравиметрия *f*

**1758 gray level; grey level; shade of gray; gray
     shade**
   d Graustufe *f*

f niveau *m* de gris
r уровень *m* серого; уровень яркости

1759 **grayscale; greyscale; shading scale**
   d Grau[stufen]skala *f*; Graustufung *f*
   f escalier *m* de demi-teintes gris; échelle *f* de
     gris
   r шкала *f* яркостей; шкала полутонов;
     полутоновая шкала; шкала оттенков серого
     цвета

   * **gray shade** → 1758

1760 **graytone[d] image; graytoned raster**
   d Grauwertbild *n*
   f image *f* en demi-teintes de gris
   r изображение *n* оттенками серого цвета

   * **graytoned raster** → 1760

   * **graytone image** → 1760

1761 **great circle**
   d Großkreis *m*
   f grand cercle *m*
   r окружность *f* большого круга; большой
     круг *m*

   * **great circle line** → 2661

   * **Greenwich Civil Time** → 758

   * **Greenwich meantime** → 758

   * **Greenwich meridian** → 2889

   * **grey level** → 1758

   * **greyscale** → 1759

1762 **greyscale map**
   d Grau[stufen]karte *f*
   f carte *f* de demi-teintes gris
   r полутоновая карта *f*

   * **GRID** → 1711

   * **grid** → 3005

   * **grid azimuth** → 329

   * **grid bearing** → 329

1763 **grid boundaries**
   d Gittergrenzen *fpl*
   f limites *fpl* de grille
   r границы *fpl* сетки

   * **grid box** → 1764

1764 **grid cell; grid box; grid square; resolution
     cell**
   d Rasterelement *n*; Rasterzelle *f*; Gitterzelle *f*;
     Auflösungsraumelement *n*
   f cellule *f* de grille; point *m* de trame; point de
     résolution
   r элемент *m* растра; клетка *f* растра

   * **grid coordinates** → 433

   * **grid declination** → 746

1765 **gridded data set**
   d Grid-Datenmenge *f*
   f ensemble *m* de données carroyés
   r множество *n* матричных и растровых
     данных

   * **gridding** → 1766, 3923

1766 **gridding [of manuscript]; grid tracing**
   d Gitterverfolgung *f*
   f traçage *m* de réseau de coordonnées
   r трассировка *f* координатной сетки

1767 **grid extent**
   d Gitter-Extent *m*
   f étendue *f* de grille
   r экстент *m* сетки

1768 **grid figures**
   d Gitternummern *fpl*
   f numérotation *f* du quadrillage
   r цифры *fpl* сетки

1769 **grid generation**
   d Gittergenerierung *f*
   f génération *f* d'un quadrillage
   r генерирование *n* сетки

1770 **grid intersection; graticule intersection**
   d Gitterkreuz *n*
   f intersection *f* du quadrillage; croisillon *f*
   r пересечение *n* в сетке

1771 **grid interval**
   d Gitterintervall *n*
   f espacement *m* des lignes du quadrillage
   r интервал *m* сетки

   * **gridline** → 3018

1772 **grid map; raster map**
     (a map in which the information is carried in
     the form of grid cells)
   d Rasterkarte *f*
   f carte *f* rastre
   r растровая карта *f*

1773 grid mode
(of digitizing)
d Gittermodus *m*
f mode-grille *m*
r сеточный режим *m*

1774 grid north
d Gitternorden *m*
f nord *m* d'un carroyage; nord de la grille; nord du quadrillage
r условное направление *n* северного меридиана; север *m* картографической сетки; сеточный север

1775 grid point data
d Rasterpunktdaten *pl*; Gitterpunktdaten *pl*
f données *fpl* de points de grille
r данные *pl* в узлах координатной сетки

1776 grid references
d Gitterreferenzen *fpl*
f nœuds *mpl* de grille
r узлы *mpl* координатной сетки

1777 grid reference system
d Gitterreferenzsystem *n*
f carroyage *m* de référence; système *m* de référence de carroyage
r система *f* прямоугольных координат

1778 grid scale factor
d Gitter-Maßstabsfaktor *m*; Gitter-Ähnlichkeitsfaktor *m*
f facteur *m* d'échelle de grille
r масштабный коэффициент *m* сетки; масштаб *m* сетки

1779 grid south
d Gittersüden *m*
f sud *m* de la grille
r условное направление *n* южного меридиана; юг *m* картографической сетки; сеточный юг

* **grid square** → 1764

1780 grid street pattern
d Gitterstraßenmuster *n*
f modèle *m* quadrillé de rues
r сеточная модель *f* улиц

1781 grid surface
d Gitterfläche *f*
f surface *f* de grille
r поверхность *f* сетки; сеточная поверхность

* **grid system** → 1598

* **grid thematic map** → 1782

1782 grid theme; grid thematic map
d Raster-Themakarte *f*
f carte *f* thématique rastre
r растровая тематическая карта *f*

* **grid tracing** → 1766

1783 grid zone
d Gitterzone *f*
f zone *f* de grille
r зона *f* координатной сетки; зона решётки

1784 ground; soil; dirth
d Erde *f*; Grund *m*
f terre *f*; sol *m*
r земля *f*; почва *f*; грунт *m*

1785 ground control
d Bodenkontrolle *f*; Erdsteuerung *f*
f contrôle *m* au sol
r наземный контроль *m*

1786 ground control point
d Erdsteuerpunkt *m*; Boden[kontroll]station *f*; Passpunkt *m*
f point *m* de contrôle terrestre; point d'appui terrestre
r наземная контрольная точка *f*; наземный контрольный пункт *m*

1787 ground elevation
d Geländehöhe *f*
f élévation *f* du sol
r нулевая высотная отметка *f*

* **ground features** → 3605

1788 ground impedance
d Erdimpedanz *f*
f impédance *f* de terre
r земной импеданс *m*

* **ground line** → 292

* **ground map** → 3338

1789 ground patch area
d Grund-Patch-Area *n*
f tache *f* de prise de vue; tache élémentaire; tache d'analyse; tachèle *f*
r фрагмент *m* земли

1790 ground resolution
d Erdauflösung *f*
f résolution *f* terrestre
r разрешение *n* на местности; разрешение деталей земной поверхности

* **ground survey** → 1407

1791 ground tied structures
 d verbundene Bodenstrukturen *fpl*
 f structures *fpl* terrestres liées
 r связанные земные структуры *fpl*

1792 ground track
 d Kurs *m* über Grund
 f trace *f*; trajectoire *f* au sol
 r трасса *f* орбиты

 * **ground truth** → **1793**

1793 ground truth[ing]
 d Grundtest *m*; Ground-Truth *n*
 f réalité *f* du terrain; réalité sur le terrain;
   vérité-terrain *f*
 r наземная проверка *f*; наземный контроль *m*
   данных

1794 ground truth point; ground truth site
 d Ground-Truth-Punkt *m*
 f site *f* témoin
 r пункт *m* наземного контроля

 * **ground truth site** → **1794**

1795 ground wave
 d Bodenwelle *f*
 f onde *f* de sol; onde directe
 r земная волна *f*; земная радиоволна *f*

 * **group function** → **1796**

1796 group[ing] function
 d Gruppierungsfunktion *f*
 f fonction *f* de groupement
 r функция *f* группирования

 * **GRS** → **1555**

 * **GU** → **1717**

 * **GUI** → **1748**

H

1797 habitat; biotop
d Fundort *m*; Lebensraum *m*
f habitat *m*; biotope *m*
r биотоп *m*; место *n* обитания; ареал *m*
обитания; среда обитания

1798 habitat model
d Habitatmodell *n*
f modèle *m* d'habitat
r модель *f* места обитания

* **hachure** → 3271

1799 halftone
d Halbton *m*
f demi-teinte *f*; demi-ton *m*; simili *m*
r полутон *m*; полукраска *f*

1800 Hamiltonian circuit
d Hamilton'sche Kontur *f*
f circuit *m* de Hamilton
r гамильтонова цепь *f*; гамильтонов контур *m*

1801 hand-drafted map; hand[-drawn] map
d mit der Hand gefertigtes kartografisches
Erzeugnis *n*; Basiswetterkarte *f*
f carte *f* réalisée à la main
r выполненная вручную карта *f*

* **hand-drawn map** → 1801

* **handheld** → 2689

* **handheld computer** → 2689

* **handler** → 1009

* **hand map** → 1801

* **harbor chart** → 1802

1802 harbo[u]r chart; port plan
d Hafenplan *m*
f plan *m* de port
r морской план *m*; портовый план

1803 hard classification; strong classification
d harte Klassifizierung *f*; kräftige
Klassifikation *f*
f classification *f* dure; classification rigide
r жёсткая классификация *f*

* **hard-copy map** → 2697

1804 hardware
d Hardware *f*; Gerätetechnik *f*
f matériel *m*; hardware *m*
r аппаратное обеспечение *n*; аппаратные
средства *npl*; технические средства

1805 hardware platform
d Hardwareplattform *f*
f plate-forme *f* [en] matériel
r аппаратная платформа *f*

* **HARN** → 1829

* **hatch** → 3271

1806 hatched background
d schraffierter Hintergrund *m*
f arrière-plan *m* hachuré
r штриховой фон *m*

1807 hatch frequency
d Schraffurfrequenz *f*
f fréquence *f* de hachure
r частота *f* штриховки

* **hatching** → 3275

1808 hatch lines
d Schraffurlinien *fpl*
f lignes *fpl* de hachure
r линии *fpl* штриховки

1809 hazardous zone
d gefährliche Zone *f*
f zone *f* dangereuse
r опасная зона *f*

* **hazy** → 1495

* **HDM** → 1860

**1810 head-mounted display; HMD; head-up
display; HUD; head-mounted screen;
videocasque**
d Kopfbildschirm *m*; Videohelm *m*
f écran *m* monté sur la tête; visiocasque *f*;
casque *f* de visualisation; viseur *m* tête-haute;
visière *f* stéréoscopique; casque de vision 3D
r шлем-дисплей *m*; головной экран *m*;
видеокаска *f*

* **head-mounted screen** → 1810

* **"heads-up" digitizing** → 2627

* **head-up display** → 1810

* **height** → 1811

1811 height [above sea level]; elevation; altitude
d Höhe *f* über Normal-Null; Höhe über NN;
  Höhe [über Meer]; Meereshöhe *f*, Aufriss *m*
f hauteur *f*; altitude *f*; élévation *f*
r высота *f*

**1812 height difference; altitude difference;
elevation difference**
d Höhenunterschied *m*
f dénivelée *f*; dénivellation *f*; faille-pli *f*; écart *m*
  de hauteur
r денивелирование *n*

* **height line → 732**

**1813 height of a benchmark referred to a datum;
elevation of a benchmark referred to a
datum**
d Höhenlage *f* eines Fixpunktes in Bezug auf
  eine Horizontale
f hauteur *f* d'un repère par rapport à un plan
r высота *f* репера, отнесена к началу отсчёта

* **height plane → 2144**

1814 heliographic longitude
d heliografische Länge *f*
f longitude *f* héliographique
r гелиографическая долгота *f*; солнечная
  долгота

1815 hemisphere
d Halbkugel *f*
f hémisphère *f*
r полусфера *f*; полушар *m*

1816 heterogeneity
d Heterogenität *f*
f hétérogénéité *f*
r гетерогенность *f*; разнородность *f*

1817 heuristic
d heuristisch
f [h]euristique
r эвристический

1818 heuristics
d Heuristik *f*
f [h]euristique *f*
r эвристика *f*

1819 hextree
d Hexagonalbaum *m*
f arbre *m* à six branches
r гексотомическое дерево *n*

1820 hidden line; invisible line
d verdeckte Linie *f*; verdeckte Bildkante *f*
f ligne *f* cachée
r невидимая линия *f*; скрытая линия

* **hidden-line elimination → 1821**

**1821 hidden-line removal; hidden-line
elimination**
d Entfernung *f* der verdeckten Linien
f élimination *f* de lignes cachées
r устранение *n* скрытых линий; удаление *n*
  невидимых линий

1822 hide *v*
d verstecken; verdecken; ausblenden
f masquer; cacher
r скрыть

1823 hierarchical cartographic database
d hierarchische kartografische Datenbasis *f*
f base *f* de données cartographiques
  hiérarchique
r иерархическая картографическая база *f*
  данных

1824 hierarchical classification
d hierarchische Klassifizierung *f*
f classification *f* hiérarchique
r иерархическая классификация *f*

1825 hierarchical database
d hierarchische Datenbasis *f*
f base *f* de données hiérarchique
r иерархическая база *f* данных

1826 hierarchical [data] model
d hierarchisches Datenmodell *n*
f modèle *m* [de données] hiérarchique
r иерархическая модель *f* [данных]

1827 hierarchical geographic presentation
d hierarchische geografische Darstellung *f*
f présentation *f* géographique hiérarchique
r иерархическое географическое
  представление *n*

* **hierarchical model → 1826**

1828 hierarchy
d Hierarchie *f*
f hiérarchie *f*
r иерархия *f*

1829 high-accuracy reference networks; HARN
d hochpräzise Referenznetzwerke *npl*
f réseaux *mpl* référentiels à haute exactitude
r эталонные сети *fpl* высокой точностью

1830 higher geodesy; higher survey[ing]
d höhere Geodäsie *f*
f haute géodésie *f*
r высшая геодезия *f*

* **higher survey → 1830**

* **higher surveying** → 1830

* **highland** → 3878

1831 high-pass filter
d Hochpassfilter *m*
f filtre *m* passe-haut
r фильтр *m* верхних частот;
высокочастотный фильтр

1832 high-precision geodetic network
d hochpräzises geodätisches Netz *n*
f réseau *m* géodésique à haute précision
r геодезическая сеть *f* высокой точностью

1833 high-resolution sea-floor mapping system
d hochauflösendes
Meeresboden-Kartografierungssystem *n*
f système *m* à haute résolution de cartographie
des fonds marins
r система *f* картографирования ложе моря с
высокой разрешающей способностью

1834 high water
d Hochwasser *n*
f hautes eaux *fpl*
r паводок *m*; высокий уровень *m* воды;
высокая вода *f*; половодье *n*

1835 high water line
d Hochwasserlinie *f*; Flutgrenze *f*
f laisse *f* des hautes eaux
r линия *f* паводка

1836 hillshade image
d Schummerungsbild *n*
f image *f* d'estompage
r изображение *n* отмывки

1837 hill shading
d Schummerung *f*
f estompage *m*
r полутоновое изображение *n* рельефа

1838 hinterland; back country; back land
d Hinterland *n*; Umland *n*; Rückland *n*
f arrière-pays *m*
r удалённый от промышленного центра
район *m*; район вглубь от прибрежной
полосы или границы; внутренний район

* **HIS** → 1870

* **histogram** → 315

1839 historical view
d historische Ansicht *f*
f vue *f* historique
r ретроспективный взгляд *m*

* **HMD** → 1810

* **homogeneity** → 3854

* **homogeneous regions** → 3855

1840 homolographic map
d Karte *f* mit flächentreue Projektion
f carte *f* à projection équivalente
r карта *f* в равновеликой проекции

* **homolographic projection** → 1316

1841 horizon
d Horizont *m*
f horizon *m*
r горизонт *m*

1842 horizontal accuracy
d horizontale Genauigkeit *f*
f exactitude *f* horizontale
r точность *f* определения местоположения;
точность плановой привязки

1843 horizontal angle
d horizontaler Winkel *m*; Horizontalwinkel *m*
f angle *m* horizontal
r горизонтальный угол *m*

1844 horizontal circle level; alidade level
d Horizontierlibelle *f*; Alhidadenlibelle *f*;
Stehachsenlibelle *f*
f nivelle *f* de l'alidade; niveau *m* de l'alidade;
niveau de verticalité
r уровень *m* алидады

* **horizontal control** → 2530

1845 horizontal coordinates
d horizontale Koordinaten *fpl*;
Grundrisskoordinaten *fpl*
f coordonnées *fpl* horizontaux; coordonnées
planimétriques
r горизонтальные координаты *fpl*; плоские
координаты; плановые координаты

1846 horizontal datum
d Horizontaldatum *n*
f datum *m* horizontal
r горизонтальная линия *f* отсчёта;
горизонтальная линия приведения;
горизонтальная ось *f* координат

1847 horizontal dilution of precision
d horizontaler Genauigkeitsabfall *m*
f dilution *f* horizontale de précision
r снижение *n* точности определения
положения в горизонтальной плоскости;
уменьшение *n* точности положения в
горизонтальной плоскости

1848 **horizontal interpolator**
  d horizontaler Interpolator *m*
  f interpolateur *m* horizontal
  r горизонтальный интерполятор *m*

* **horizontal net → 2530**

1849 **horizontal plane**
  d horizontale Ebene *f*; waagerechte Ebene
  f plan *m* horizontal
  r горизонтальная плоскость *f*

1850 **horizontal position**
  d horizontale Position *f*
  f position *f* horizontale
  r горизонтальная позиция *f*

1851 **horizontal projection; first projection**
       (in two-plane projection)
  d Horizontalprojektion *f*; erste Projektion *f*;
       Grundriss *m*
  f projection *f* horizontale; coupe *f* horizontale
  r горизонтальная проекция *f*;
       горизонтальный разрез *m*

1852 **horizontal resolution**
  d horizontale Auflösung *f*
  f résolution *f* horizontale
  r горизонтальная разрешающая
       способность *f*

* **hot image → 564**

1853 **hot views**
  d dynamische Ansichten *fpl*
  f vues *fpl* dynamiques
  r динамические представления *npl*

* **house address → 2242**

* **household → 1371**

1854 **household survey**
  d Haushaltserhebung *f*
  f enquête *f* sur les ménages; enquête auprès des
       ménages
  r обследование *n* домашних хозяйств

1855 **housing unit; HU; dwelling unit**
  d Wohneinheit *f*; stehendes Haus *n*
  f unité *f* de logement; unité d'habitation
  r жилищная единица *f*; жилая единица

* **HU → 1855**

1856 **hub**
       (in a network)
  d Kern *m*; Nabe *f*; Hub *m*
  f noyau *m*; concentrateur *m*
  r концентратор *m*; ядро *n*

* **HUC → 1871**

* **HUD → 1810**

1857 **human map**
       (a map showing information about how people
       use the land)
  d anthropogene Karte *f*
  f carte *f* humaine
  r гуманитарная карта *f*

1858 **hybrid editor; raster-vector editor**
  d Hybrideditor *m*; Raster-Vektor-Editor *m*
  f éditeur *m* hybride
  r гибридный редактор *m*;
       растрово-векторный редактор

1859 **hybrid GIS; raster-vector GIS**
  d hybrides GIS *n*
  f SIG *m* hybrid
  r гибридная ГИС *f*

* **hydraulic geometry → 524**

1860 **hydroclimate data network; HDM**
  d hydroklimatisches Datennetz *n*
  f réseau *m* de données d'hydroclimate; réseau de
       données hydroclimatiques
  r сеть *f* гидроклиматических данных

* **hydrographic → 1861**

1861 **hydrographic[al]**
  d hydrografisch
  f hydrographique
  r гидрографический

1862 **hydrographic chart; nautical chart**
  d hydrografische Karte *f*; Seekarte *f*
  f carte *f* hydrographique
  r гидрографическая карта *f*

1863 **hydrographic names**
  d hydrografische Namen *mpl*
  f noms *mpl* hydrographiques
  r гидронимы *mpl*

1864 **hydrographic survey**
  d Seevermessung *f*
  f levé *m* hydrologique
  r гидрографическая съёмка *f*; промер *m*

1865 **hydrography**
  d Hydrografie *f*
  f hydrographie *f*
  r гидрография *f*

1866 **hydro-isopleth map**
  d Hydroisoplethenkarte *f*

f carte *f* des isoplèthes de la nappe phréatique
r карта *f* гидро-изоплет

1867 hydrologic atlas
d hydrologischer Atlas *m*
f atlas *m* hydrologique
r гидрологический атлас *m*

1868 hydrologic benchmark
d hydrologisches Benchmark *n*
f repère *m* hydrologique
r гидрологический репер *m*

1869 hydrologic benchmark network
d hydrologisches Benchmark-Netz *n*
f réseau *m* de repères hydrologiques
r сеть *f* гидрологических реперов

1870 hydrologic information system; HIS
d hydrologisches Informationssystem *n*
f système *m* d'information hydrologique
r гидрологическая информационная
система *f*

1871 hydrologic unit code; HUC
d Code *m* von hydrologischer Einheit
f code *m* d'unité hydrologique
r код *m* гидрологической единицы

1872 hyperspectral image analysis
d hyperspektrale Bildanalyse *f*
f analyse *f* d'images hyperspectraux
r анализ *m* гиперспектральных изображений

1873 hyperspectral image sensor
d hyperspektraler Bildsensor *m*
f capteur *m* d'imagerie hyperspectral
r гиперспектральный датчик *m* изображения;
гиперспектральный формирователь *m*
изображений

1874 hypertext map
d Hypertext-Karte *f*
f carte *f* hypertexte
r гипертекстовая карта *f*

* **hypsographical curve → 732**

* **hypsographic curve → 732**

1875 hypsographic map
d hypsografische Karte *f*
f carte *f* hypsographique
r гипсографическая карта *f*

1876 hypsometric map
d hypsometrische Karte *f*
f carte *f* hypsométrique
r гипсометрическая карта *f*

1877 hypsometric method
d hypsometrische Methode *f*
f méthode *f* hypsométrique
r гипсометрический способ *m*

**1878 hypsometric tint; altitude tint; elevation
tint**
d hypsometrische Farbe *f*;
Höhenschichtenfarbe *f*; Höhenstufenfarbe *f*
f teinte *f* hypsométrique; coloriage *m*
hypsométrique
r гипсометрическая окраска *f*

**1879 hypsometric tint scale; layer box; elevation
tint box**
d hypsometrische Farbenskala *f*
f échelle *f* de coloriage hypsométrique
r шкала *f* гипсометрической окраски;
гипсометрическая шкала

1880 hypsometry
d Hypsometrie *f*
f hypsométrie *f*
r гипсометрия *f*

I

* **I/0 devices** → 1960

* **IAC** → 1981

1881 icon; pictogram
  d Ikone *f*; Piktogramm *n*
  f icône *f*; pictogramme *m*
  r икона *f*; значок *m*; пиктограмма *f*

* **ID** → 1883

1882 identification number
  d Identifikationsnummer *f*; Kennummer *f*
  f numéro *m* d'identification; identifiant *m*
    numérique
  r идентификационный номер *m*; условное
    цифровое обозначение *n*

1883 identifier; ID
  d Identifikator *m*; Bezeichner *m*
  f identificateur *m*
  r идентификатор *m*

1884 identity
  d Identität *f*
  f identité *f*
  r тождественность *f*; идентичность *f*

**1885 identity function; identity map[ping];
    identity permutation**
  d Identitätsabbildung *f*
  f application *f* identique; permutation *f*
    identique; transformation *f* identique
  r тождественное отображение *n*

* **identity map** → 1885

* **identity mapping** → 1885

* **identity permutation** → 1885

* **IDW** → 2007

* **IGDS** → 1977

* **IGES** → 1951

* **IGIS** → 1969, 1974

1886 illumination
  d Beleuchtung *f*; Illuminierung *f*

  f illumination *f*
  r иллюминация *f*; свечение *n*

1887 illumination geometry
  d Beleuchtungsgeometrie *f*
  f géométrie *f* d'illumination
  r геометрия *f* иллюминации

1888 image; pattern; picture
  d Bild *n*; Abbild *n*
  f image *f*
  r образ *m*; изображение *n*

1889 image addition; image summation
  d Bildaddition *f*; Bildsummierung *f*
  f addition *f* d'images; sommation *f* d'images
  r суммирование *n* изображений

1890 image algebra
  d Bildalgebra *f*
  f algèbre *f* d'images
  r алгебра *f* изображений

1891 image analysis; picture analysis
  d Bildanalyse *f*
  f analyse *f* d'images
  r анализ *m* изображений

1892 image arithmetic
  d Bildarithmetik *f*
  f arithmétique *f* d'images
  r арифметика *f* изображений

1893 image-based information system
  d Bild-basiertes Informationssystem *n*
  f système *m* d'information à base d'images
  r информационная система *f*, основанная на
    анализе изображений

1894 image catalog
  d Bildkatalog *m*
  f catalogue *m* d'images
  r каталог *m* изображений

1895 image classification
  d Bildklassifizierung *f*
  f classification *f* d'images
  r классификация *f* изображений

1896 image comparison
  d Bildvergleich *m*
  f comparaison *f* d'images
  r сравнение *n* изображений

1897 image composition
  d Bildsynthese *f*
  f composition *f* d'image; synthèse *f* d'image
  r синтез *m* изображения

1898 image data
d Bilddaten *pl*
f données *fpl* [d']image
r данные *pl* об изображении

1899 image databank; picture databank; visual databank
d Bilderdatenbank *f*
f banque *f* d'images
r банк *m* данных изображений

1900 image differencing
d Bildunterscheiden *n*
f différenciation *f* d'images
r дифференцирование *n* изображений; различение *n* изображений

1901 image digitizing
d Bilddigitalisierung *f*
f digitalisation *f* d'image
r дигитализирование *n* изображения

1902 image division
d Bildteilung *f*
f division *f* d'images
r разделение *n* изображений

* **image dot → 2758**

1903 image enhancements
d Bildanreicherung *f*
f amélioration *f* d'image
r улучшение *n* [качества] изображения

1904 image integrator
d Bildintegrator *m*
f intégrateur *m* d'images
r интегратор *m* изображений

* **image map → 564**

1905 image measuring
d Bildmessung *f*
f mesure *f* d'images
r инструментальное дешифрирование *n* изображений; измерительное дешифрирование изображений

1906 image pair
d Bildpaar *n*
f paire *f* d'images
r пара *f* изображений

1907 image processing
d Bildverarbeitung *f*
f traitement *m* d'images
r обработка *f* изображений

1908 image processing system; IPS
d Bildverarbeitungsystem *n*
f système *m* de traitement d'images
r система *f* обработки изображений

1909 image reclassification
d Bildneuzuordnung *f*
f reclassement *m* d'images
r переклассификация *f* изображений

* **image recognition → 2717**

1910 image registration
d Bildregistrierung *f*
f enregistrement *m* d'images
r регистрация *f* изображений

1911 image segmentation
d Bildsegmentierung *f*
f segmentation *f* d'image
r сегментация *f* изображения

1912 image subtraction
d Bildsubtraktion *f*
f soustraction *f* d'images
r вычитание *n* изображений

* **image summation → 1889**

1913 impassable street; impasse
d unbefahrbarene Straße *f*; Engpass *m*
f rue *f* impassible; impasse *f*
r непроходимая улица *f*; непроезжая улица; тупик *m*

* **impasse → 1913**

1914 impedance
d Impedanz *f*
f impédance *f*
r импеданс *m*

1915 implementation
d Implementierung *f*; Durchführung *f*
f implémentation *f*; réalisation *f*
r внедрение *n*; реализация *f*; ввод *m* в действие

* **import → 1916**

1916 import[ing]
d Importieren *n*
f importation *f*
r импортирование *n*; внесение *n*

1917 imprecision; inaccuracy
d Ungenauigkeit *f*; Unrichtigkeit *f*
f non-précision *f*; imprécision *f*; inexactitude *f*
r неточность *f*

1918 imprint
d Presseindruck *m*; Eindruck *m*; Impressum *n*

f empreinte *f*; impression *f*
r штамп *m*; след *m*; отпечаток *m*; выходные данные *pl*

* **inaccuracy** → 1917

1919 incidence
d Inzidenz *f*
f incidence *f*
r инцидентность *f*; падение *n*

* **incidence** → 1921

* **incidence angle** → 126

1920 incident map
d Inzidentkarte *f*
f carte *f* d'incidents
r карта *f* инцидентов

1921 inclination; incidence; slant; slope; tilt; pitch; talus
d Neigung *f*; Steigung *f*; Böschung *f*; Schräge *f*; Hang *m*
f inclinaison *f*; déclivité *f*; pente *f*; talus *m*
r наклонение *n*; наклон *m*; скос *m*; уклон *m*

* **inclination angle** → 126

* **inclined projection** → 2603

1922 inclinometer; clinometer; tilt meter; grademeter
d Inklinometer *n*; Klinometer *n*; Neigungsmesser *m*; Steigungsmesser *m*; Kränungsmesser *m*
f inclinomètre *m*; clinomètre *m*; indicateur *m* de pente; couronne *f* de gisement
r инклинометр *m*; инклинатор *m*; компас *m* наклонения; наклонная буссоль *f*; прибор *m* для измерения угла наклона

1923 inclusion
d Einschluss *m*; Inklusion *f*
f inclusion *f*
r включение *n*

* **incorporated municipality** → 1924

1924 incorporated place; incorporated municipality
d amtliche Stelle *f*
f communauté *f* urbaine
r населённый пункт *m*, включённый в список городов

* **increasing** → 188

1925 increment
d Zuwachs *m*; Inkrement *n*

f incrément *m*; accroissement *m*
r приращение *n*; прирост *m*; инкремент *m*

1926 independent variable
d unabhängige Variable *f*; unabhängige Veränderliche *f*
f variable *f* indépendante
r независимая переменная *f*

1927 index
d Index *m*
f indice *m*
r индекс *m*

1928 index *v* a contour line
d Höhenlinie indizieren
f indexer une ligne de niveau
r оцифровывать горизонталь

* **index contour** → 1929

1929 index contour [line]
d Haupthöhenlinie *f*; Zähllinie *f*
f courbe *f* [de niveau] maîtresse; ligne *f* de comptage
r основная [утолщенная] горизонталь *f*; опорный контур *m*

1930 index coverage; index overlay
d Index-Überdeckung *f*; Index-Überlagerung *f*
f couverture *f* d'indice
r охват *m* индекса; индексное перекрытые *n*

1931 indexing
d Indizieren *n*; Indizierung *f*
f indexage *m*; indexation *f*
r индексация *f*; индексирование *n*

* **index key** → 3103

* **index map** → 1933

* **index of place names** → 1515

* **index overlay** → 1930

1932 index overlay model
d Index-Überlagerungsmodell *n*
f modèle *m* de couverture d'indice
r модель *f* индексного перекрытия

* **index plan** → 1933

* **index sheet** → 1933

1933 index to adjoining sheets; inter-chart relationship diagram; index map; key map; index sheet; index plan
d Index *m* von angrenzenden Blättern; Indexkarte *f*; Hinweiskarte *f*; Übersichtsbild *n*

f carte *f* répertoire; schéma *m* d'assemblage de
cartes; tableau *m* d'assemblage; feuille *f*
d'assemblage; plan *m* d'assemblage
r схема *f* расположения соседних листов
карт; сборная таблица *f*; сборный лист *m*
[карты]; схема расположения профилей

1934 indication
d Indikation *f*; Melden *n*; Anzeige *f*
f indication *f*
r индикация *f*

* **indicator** → 3295

1935 indicator kriging
d Indikator-Kriging *n*
f krigeage *m* d'indicateur
r индикаторный кригинг *m*

1936 indicator point
d Indikatorpunkt *m*
f point *m* d'indicateur
r точка *f* признака; точка указателя

1937 indirect measurement
d Indirektmessung *f*; indirekte Messung *f*
f mesurage *m* indirect
r косвенное измерение *n*

1938 indirect signs
d indirekte Zeichen *npl*
f caractères *mpl* indirects
r косвенные признаки *mpl*; индикационные
признаки

1939 inflection
d Inflexion *f*; Beugung *f*
f inflexion *f*
r сгибание *n*; изгиб *m*

* **influence zone** → 4032

* **informational reliability** → 909

* **information map** → 1943

1940 information mapping
d Informationskartierung *f*
f cartographie *f* d'information
r картографирование *n* информации

1941 information mapping technology
d Informationskartierung-Technologie *f*
f technologie *f* de cartographie d'information
r технология *f* картографирования
информации

* **information point** → 905

1942 information retrieval

d Informationswiedergewinnung *f*;
Informationswiederauffindung *f*
f récupération *f* d'information; recherche *f*
d'information
r поиск *m* информации

1943 information [space] map
d Informationskarte *f*
f carte *f* d'espace informationnel
r информационная карта *f*

1944 information support
d Datenunterstützung *f*
f soutien *m* d'information
r информационное обеспечение *n*;
информационная поддержка *f*

1945 information system
d Informationssystem *n*
f système *m* d'information
r информационная система *f*; ИС

1946 infrared scanner
d Infrarotscanner *m*
f scanner *m* [en] infrarouge
r инфракрасный сканер *m*

1947 infrared thermal mapper; thermal mapper
d [thermischer] Infrarotmapper *m*
f dispositif *m* de cartographie thermique
r устройство *n* термического
картографирования

1948 infrastructure
d Infrastruktur *f*
f infrastructure *f*
r инфраструктура *f*

1949 infrastructure planning
d Infrastruktur-Planung *f*
f planification *f* d'infrastructure
r планировка *f* инфраструктуры

1950 inheritance
d Vererbung *f*
f inhéritance *f*
r [у]наследование *n* [свойств]; наследство *n*

**1951 initial graphics exchange specification;
IGES**
d IGES-Vorschrift *f*
f standard *m* de communication graphique;
standard IGES
r формат *m* для передачи инженерной
графики; стандарт *m* IGES

1952 initialization
d Initialisierung *f*; Initialisieren *n*; Einleiten *n*
f initialisation *f*
r инициализация *f*; задание *n* начальных
условий

1953 ink-jet plotter
  d Tintenstrahlplotter *m*
  f traceur *m* à jet d'encre; traceur à bulle d'encre
  r струйный плоттер *m*

1954 innermost contour
  d innerste Höhenlinie *f*; innerste Isohypse *f*
  f courbe *f* de niveau la plus interne
  r самая внутренняя горизонталь *f*

1955 inner node
  d innerer Knoten *m*
  f nœud *m* interne
  r внутренний узел *m*

  * **inner point** → **1990**

1956 inner polygon; inside polygon
  d inneres Polygon *n*
  f polygone *m* interne
  r внутренний полигон *m*

1957 inner texture
  d innere Textur *f*
  f texture *f* interne
  r внутренняя текстура *f*

1958 input data
  d Eingabedaten *pl*
  f données *fpl* d'entrée
  r входные данные *pl*

1959 input device
  d Eingabegerät *n*
  f dispositif *m* d'entrée
  r устройство *n* ввода

1960 input/output devices; I/0 devices
  d Eingabe/Ausgabe-Geräte *npl*
  f dispositifs *mpl* d'entrée/sortie
  r периферийные устройства *npl* ввода и
      вывода

  * **inquiry** → **3140**

1961 insert *v*; inset *v*; paste *v*
  d einfügen; einschieben
  f insérer; imbriquer
  r вставлять; включать; вкладывать

  * **insertion of data** → **886**

  * **inset *v*** → **1961**

1962 inset map; map inset
  d Teilkarte *f*; Nebenkarte *f*; Beikarte *f*
  f papillon *m*; cartouche *f*; carton *m* annexe d'une
      carte
  r [карта-]врезка *f*

* **inside polygon** → **1956**

1963 install *v*
  d installieren
  f installer
  r инсталлировать

1964 instance
  d Instanz *f*
  f instance *f*
  r отдельный факт *m*; экземпляр *m*

1965 instantaneous field of view
  d unmittelbares Sehfeld *n*
  f champ *m* de vision instantané
  r мгновенный сектор *m* обзора; мгновенная
      зона *f* обзора

1966 instantiation
      (in VR)
  d Instanziierung *f*
  f instanciation *f*
  r конкретизация *f*; реализация *f*;
      подтверждение *n*

1967 instrument
  d Instrument *n*
  f instrument *m*
  r инструмент *m*

1968 instrument error
  d Instrumentfehler *m*
  f erreur *f* d'instrument
  r погрешность *f* инструмента

1969 integrated GIS; IGIS
  d integriertes GIS *n*
  f SIG *m* intégré
  r интегрированная ГИС *f*

1970 integrated regional database
  d integrierte regionale Datenbasis *f*
  f base *f* de données régionale intégrée
  r интегрированная региональная база *f*
      данных

1971 integrated survey
  d integrierte Aufnahme *f*
  f levé *m* intégré
  r интегрированная съёмка *f*

1972 integrated terrain unit mapping; ITUM
  d Festlegen *n* von integrierten Grenzen von
      Geländeeinheiten
  f cartographie *f* intégrée d'unités terrestres
  r интегрированное радиолокационное
      картографирование *n* поверхности Земли

  * **integrating instrument** → **2783**

1973 integration
 d Integration *f*; Integrierung *f*
 f intégration *f*
 r интеграция *f*; интегрирование *n*;
 встраивание *n*

1974 intelligent GIS; IGIS
 d intelligentes GIS *n*
 f SIG *m* intelligent
 r интеллигентная ГИС *f*

1975 interactive digitizing
 d interaktive Digitalisierung *f*
 f numérisation *f* interactive
 r диалоговое дигитализирование *n*

 * **interactive following of a contour** → 1978

1976 interactive geocoding
 d interaktive Geocodierung *f*
 f géocodage *m* interactif
 r интерактивное геокодирование *n*;
 диалоговое геокодирование

 * **interactive graphic** → 564

1977 interactive graphics design software; IGDS
 (of Intergraph)
 d interaktive grafische Gestaltung-Software *f*
 f logiciel *m* de dessin graphique interactif
 r программное обеспечение *n* диалогового
 графического проектирования

**1978 interactive line-following; interactive
 following of a contour**
 d interaktiver Linienverlauf *m*; interaktive
 Linienverfolgung *f*
 f filage *m* interactif d'une courbe [de niveau];
 poursuite *f* interactive de ligne
 r диалоговое слежение *n* горизонтали

 * **interactive mapping** → 2625

1979 interactive mode; conversational mode
 d Dialogbetrieb *m*; interaktiver Betrieb *m*;
 Dialogverkehr *m*
 f mode *m* de dialogue; mode interactif
 r диалоговый режим *m*; интерактивный
 режим

1980 interactive processing
 d interaktive Verarbeitung *f*;
 Dialogverarbeitung *f*
 f traitement *m* interactif
 r интерактивная обработка *f*; диалоговая
 обработка

1981 interapplication communication; IAC
 d applikationsübergreifende Kommunikation *f*
 f communication *f* entre applications

 r связь *f* между приложениями

 * **inter-chart relationship diagram** → 1933

1982 interface
 d Schnittstelle *f*
 f interface *f*
 r интерфейс *m*

1983 interior area; interior domain
 d innerer Bereich *m*; Innengebiet *n*
 f zone *f* intérieure; domaine *m* intérieur
 r внутренняя область *f*; внутренняя зона *f*

 * **interior domain** → 1983

1984 interior orientation
 d innere Orientierung *f*
 f orientation *f* interne
 r внутреннее ориентирование *n*; внутренняя
 ориентация *f*

 * **interior point** → 1990

 * **intermediate contour** → 1985

1985 intermediate contour [line]
 d Zwischenkonturlinie *f*
 f ligne *f* de contour intermédiaire; trait *m* de
 contour intermédiaire
 r промежуточная горизонталь *f*

1986 intermediate node
 d Zwischenknoten *m*
 f nœud *m* intermédiaire
 r промежуточный узел *m*

1987 intermediate scale
 d Zwischenmaßstab *m*
 f échelle *f* intermédiaire
 r средний масштаб *m*

1988 internal conversion
 d innere Konversion *f*
 f conversion *f* interne
 r внутреннее преобразование *n*

1989 internal number
 d interne Nummer *f*
 f nombre *m* interne
 r внутренний номер *m*

1990 internal point; interior point; inner point
 d innerer Punkt *m*; interner Punkt; Innenpunkt *m*
 f point *m* interne; point intérieur
 r внутренняя точка *f*

1991 internal pointer
 d interner Zeiger *m*

f pointeur *m* interne
r внутренний указатель *m*

1992 international map
d internationale Karte *f*
f carte *f* internationale
r международная карта *f*

1993 Internet mapping
d Internet-Kartierung *f*
f cartographie *f* d'Internet
r картографирование *n* Интернет-связей

1994 Internet map server; map server
d [Internet-]Kartenserver *m*
f serveur *m* de cartes [d'Internet]
r картографический сервер *m*

1995 Internet topology
d Internet-Topologie *f*
f topologie *f* d'Internet
r топология *f* Интернета

1996 interoperability
d Interoperabilität *f*; übergreifende
   Funktionsfähigkeit *f*
f interopérabilité *f*; fonctionnement *m* en
   réciprocité
r интероперабельность *f*; взаимодействие *n*;
   взаимозаменяемость *f*; оперативная
   совместимость *f*

1997 interpenetration
d gegenseitige Durchdringung *f*
f interpénétration *f*
r взаимное проникание *n*; взаимное
   проникновение *n*

1998 interpolation
d Interpolation *f*
f interpolation *f*
r интерполяция *f*; интерполирование *n*

1999 interpolation method
d Interpolationsmethode *f*
f méthode *f* d'interpolation
r метод *m* интерполяции

2000 interpretation
d Übersetzung *f*; Interpretation *f*
f interprétation *f*
r интерпретация *f*

* **interrogation** → 3140

* **intersect** → 2001

2001 intersect[ion]
   (the topological integration of two spatial data
   sets that preserves features that fall within the

   area common to both input data sets)
d Schnittmenge *f*
f intersection *f*
r пересечение *n*

2002 interval
d Zwischenraum *m*; Intervall *n*
f intervalle *m*
r интервал *m*

2003 interval arithmetics
d Intervallarithmetik *f*
f arithmétique *f* d'intervalle
r интервальная арифметика *f*; интервальные
   вычисления *npl*

* **intervisibility** → 2814

2004 invariance
   (properties that remain unchanged despite
   transformations of the numbers used to
   represent a measurement)
d Beständigkeit *f*; Invariante *f*
f invariance *f*
r инвариантность *f*; неизменность *f*

2005 inventory map
d Lager[bestands]karte *f*
f carte *f* d'inventaire
r инвентаризационная карта *f*

* **inventory of maps** → 3495

2006 inverse distance weighted interpolation
d inverse Distanzgewichtung-Interpolation *f*
f interpolation *f* pondérée inversement par les
   distances
r интерполяция *f* обратных взвешенных
   расстояний

2007 inverse distance weighting; IDW
d inverse Distanzgewichtung *f*
f pondération *f* de distances inverses
r взвешивание *n* обратных расстояний

2008 inverse Fourier transformation
d inverse Fourier-Transformation *f*
f transformation *f* de Fourier inverse
r обратная трансформация *f* Фурье

2009 inverse mapping
d inverse Abbildung *f*
f application *f* inverse
r обратное отображение *n*

2010 inversion error
d inverser Fehler *m*
f erreur *f* d'inversion
r ошибка *f* инверсии

2011 **invisible layer**
   d unsichtbare Schicht *f*
   f couche *f* non visible
   r невидимый слой *m*

  * **invisible line** → 1820

2012 **ionospheric delay**
   (of signals)
   d ionosphärische Verzögerung *f*
   f délai *m* ionosphérique
   r задержка *f* [сигналов] в ионосфере

2013 **ionospheric errors**
   d ionosphärische Fehler *mpl*
   f erreurs *fpl* ionosphériques
   r ионосферные погрешности *fpl*

  * **IPS** → 1908

  * **irregular error** → 16

2014 **irregularity**
   d Unregelmäßigkeit *f*; Regelwidrigkeit *f*;
     Verstoß *m*
   f irrégularité *f*
   r нарушение *n* порядка; неравномерность *f*;
     иррегулярность *f*

2015 **irregular line**
   d ungleiche Linie *f*
   f ligne *f* non régulière
   r неровная линия *f*

2016 **irregularly distributed points**
   d ungleich verteilte Punkte *mpl*; ungleichmäßig
     verteilte Punkte
   f points *mpl* distribués irrégulièrement
   r неравномерно распределённые точки *fpl*

2017 **irregularly spaced data**
   d Daten *pl* in unregelmäßigen Abständen
   f données *fpl* espacées irrégulièrement
   r неравномерно распределённые данные *pl*

  * **irregular polygon** → 2018

2018 **irregular[-shaped] polygon**
   d unregelmäßiges Polygon *n*
   f polygone *m* irrégulier
   r неправильный многоугольник *m*

  * **isanamorphic lines** → 1138

2019 **isarhythm map**
   d Isarhythmenkarte *f*
   f carte *f* d'isarhythmes
   r изаритмическая карта *f*

  * **island** → 2020

2020 **island [polygon]; isolated region**
   d Inselfläche *f*; Insel *f*
   f île *f*; polygone *m* isolé
   r островок *m*; изолированный участок *m*;
     изолированный многоугольник *m*

2021 **isobar**
   d Isobar *n*
   f isobare *f*
   r изобара *f*; изобар *m*

2022 **isobaric form**
   d isobare Form *f*
   f forme *f* isobare
   r изобарическая форма *f*

  * **isobath** → 979

2023 **isodemographic map**
   d isodemografische Karte *f*
   f carte *f* isodémographique
   r изодемографическая карта *f*

  * **isogonal chart** → 2025

2024 **isogonal line; isogonic line; isogone; curve
of the declination**
   d Linie *f* gleicher magnetischer Mißwertung;
     Isogon *n*
   f ligne *f* isogone
   r изогона *f*; равноугольник *m*

  * **isogone** → 2024

2025 **isogonic chart; isogonal chart**
   d winkeltreue Karte *f*
   f carte *f* isogone
   r карта *f* с равными склонениями;
     равноугольная карта; изогональная карта

  * **isogonic line** → 2024

  * **isogram** → 2031

  * **isogram block-diagram** → 2029

  * **isogram method** → 2410

  * **isoheight** → 732

  * **isohypse** → 732

2026 **isolated location**
   d isolierte Stelle *f*; isolierte Lage *f*
   f localisation *f* isolée
   r изолированное расположение *n*

  * **isolated region** → 2020

2027 isolated system
 d isoliertes System *n*
 f système *m* isolé
 r изолированная система *f*

2028 isolation
 d Isolierung *f*; Isolation *f*
 f isolation *f*
 r изоляция *f*

 * isoline → 2031

2029 isoline block-diagram; isogram block-diagram
 d Isoplethenblockdiagramm *n*
 f diagramme *m* bloc d'isolignes
 r изолинейная блок-диаграмма *f*

2030 isoline framework; isometric framework
 (a measurement framework that establishes control by a systematic set of slices through an attribute to obtain lines that represent the surface)
 d isometrisches Netz *n*
 f réseau *m* d'isolignes
 r изометрическая сеть *f*

 * isometric framework → 2030

2031 isometric line; isoline; isogram; isontic line; isopleth
 d Isolinie *f*; Wertefeldlinie *f*; Wertlinie *f*; Wertgleiche *f*; Intensitäts[wert]linie *f*; Isoplethe *f*
 f isoligne *f*; isoplèthe *f*; courbe *f* d'isovaleurs
 r изолиния *f*; изоплета *f*

2032 isometric projection
 d isometrische Projektion *f*
 f projection *f* isométrique
 r изометрическая проекция *f*

2033 isometric view
 d isometrische Ansicht *f*
 f vue *f* isométrique
 r изометрическое представление *n*

 * isontic line → 2031

 * isopleth → 2031

2034 isopleth map
 d Isoplethenkarte *f*; Isolinienkarte *f*
 f carte *f* d'isoplèthes
 r карта *f* изоплет

 * isopleth method → 2410

 * isotherm → 2036

2035 isothermal layer
 d isotherme Schicht *f*
 f couche *f* isotherme
 r изотермический слой *m*

2036 isothermal line; isotherm
 d Isotherme *f*
 f isotherme *f*
 r изотерма *f*

2037 isotherm chart
 d Isothermkarte *f*
 f carte *f* d'isothermes
 r изотермическая карта *f*

2038 isotropic curve
 d isotropische Kurve *f*
 f courbe *f* isotrope
 r изотропная кривая *f*

2039 isotropic layer
 d isotropische Schicht *f*
 f couche *f* isotrope
 r изотропный слой *m*

2040 isotropy
 d Isotropie *f*
 f isotropie *f*
 r изотропия *f*; изотропность *f*

2041 item
 d Einzelheit *f*; Artikel *m*; Einheit *f*; Element *n*
 f unité *f*; détail *m*; élément *m*
 r отдельный предмет *m*; элемент *m*; единица *f*

2042 item indexing
 d Artikelindizierung *f*
 f indexage *m* d'article
 r индексирование *n* элемента [данных]

 * ITUM → 1972

J

2043 jack-knifing
(an iterative process for estimating the errors associated with spline interpolation)
d Jackknifing *n*
f écrasement *m* de boisage; mise *f* en ciseaux
r складывание *n* вперёд с захватом

* **jaggies** → 92

* **join** → 2044

2044 join[ing]; union
d Verbund *m*; Verbindung *f*; Vereinigung *f*; Anschluss *m*
f jointure *f*; joint *m*; union *f*
r соединение *n*; объединение *n*; смыкание *n*; сшивание *n*; джойн *m*

2045 joining lines
d vereinigte Linien *fpl*
f lignes *fpl* jointes
r соединённые линии *fpl*

2046 joining nodes
d vereinigte Knoten *mpl*
f nœuds *mpl* joints
r соединённые узлы *mpl*

2047 joint operation; common operation
d Betriebsgemeinschaft *f*; Zusammenarbeit *f*
f exploitation *f* conjointe; coexploitation *f*
r совместная работа *f*

2048 joint photographic expert group format; JP[E]G format
d JP[E]G-Format *n*
f format *m* JP[E]G
r формат *m* JP[E]G

2049 joint use area; collective use area; common use area
d Bereich *m* gemeinschaftlicher Nutzung
f zone *f* d'usage conjoint; zone d'utilisation en commun
r зона *f* совместного пользования

2050 joystick
d Handsteuergeber *m*; Joystick *m*; Steuerhebel *m*; Steuerknüppel *m*
f manche *m* à balai; levier *m*; manette *f*; poignée *f*
r джойстик *m*; рычажный указатель *m*; координатная ручка *f*

* **JPEG format** → 2048

* **JPG format** → 2048

* **junction** → 2533

2051 junction
d Verknüpfung *f*; Anschluss *m*
f jonction *f*; carrefour *m*
r соединение *n*; слияние *n* [рек или пластов]

* **junction point** → 3677

2052 jurisdiction boundaries
d Gerichtsbarkeitsgrenzen *fpl*
f limites *fpl* de juridiction
r границы *fpl* подведомственной области; границы сферы полномочий

2053 jurisdiction code
d Straßenverkehrsgesetz *n*
f code *m* de juridiction
r код *m* юрисдикции

K

2054 Kelsh plotter
  d Kelsh-Plotter *m*
  f traceur *m* de Kelsh
  r плоттер *m* Келша

2055 key
  d Schlüssel *m*
  f clé *m*
  r ключ *m*

  * **key map** → 1933

2056 kinematic GPS
  d kinematisches GPS *n*
  f GPS *m* cinématique
  r кинематическая ГПС *f*

2057 kinematics
  d Kinematik *f*
  f cinématique *f*
  r кинематика *f*

2058 knowledge base
  d Wissensbasis *f*
  f base *f* de connaissance
  r база *f* знаний; БЗ

2059 kriging
      (a geostatistical technique for interpolation
      that uses information about the spatial
      autocorrelation in the vicinity of each point to
      provide "optimal" interpolation)
  d Kriging *n*
  f krigeage *m*
  r кригинг *m*; криджинг *m*

2060 kriging variance
  d Kriging-Varianz *f*
  f variance *f* de crigeage
  r вариантность *f* кригинга

L

2061 label; tag
  d Etikett *n*; Kennsatz *m*; Marke *f*
  f étiquette *f*; label *m*
  r метка *f*; этикет *m*

2062 labeled polygon
  d markiertes Polygon *n*; etikettiertes Polygon
  f polygone *m* étiqueté
  r помеченный полигон *m*

2063 labeled region
  d markierte Region *f*
  f région *f* étiquetée
  r помеченный регион *m*

2064 labeling procedure
  d Etikettierungsprozedur *f*
  f procédure *f* d'étiquetage
  r процедура *f* присваивания меток

  * **labeled** → 2065

  * **labeling** → 2066

2065 label[l]ed; tagged; marked; annotated
  d gekennzeichnet; markiert
  f étiqueté; libellé; marqué; signalé
  r помеченный; отмеченный

2066 label[l]ing; tagging
  d Kennzeichnung *f*; Auszeichnen *n*;
     Etikettieren *n*; Etikettierung *f*
  f étiquetage *m*; marquage *m*
  r присваивание *n* [объектам] меток; запись *f*
     меток; тегирование *n*; сопровождение *n*
     данных признаками

2067 label number
  d Kennsatznummer *f*; Etikettennummer *f*
  f numéro *m* d'étiquette
  r номер *m* метки

2068 label point
  d Kennsatzpunkt *m*
  f point *m* d'étiquette; localisant *m*
  r внутренняя точка *f* полигона

  * **lag** → 952

  * **lagging** → 952

2069 lake and marsh map
  d See- und Sumpfkarte *f*

  f carte *f* des lacs et marais
  r карта *f* озёр и болот

2070 Lambert azimuthal equal-area projection
  d flächentreue azimutale Lambert-Projektion *f*;
     natürliche Kartenprojektion *f*
  f projection *f* azimutale équivalente de Lambert
  r равновеликая азимутальная проекция *f*
     Ламберта

2071 Lambert [conformal conic] projection
  d Kegelprojektion *f* nach Lambert
  f projection *f* conique conforme de Lambert
  r проекция *f* Ламберта; нормальная
     коническая проекция

  * **Lambert projection** → 2071

  * **LAN** → 2206

  * **land** → 784

2072 land
  d Land *n*; Gelände *n*
  f terre *f*
  r земля *f*

2073 land
  d Lötauge *n*; Festland *n*
  f bien *m* immeuble
  r недвижимое имущество *n*

  * **land acquisition** → 20

  * **land capability map** → 2076

2074 land classification
  d Landklassifizierung *f*
  f classification *f* de terres; classification de
     biens immeubles
  r классификация *f* земель; паспортизация *f*
     земель

2075 land classification map
  d Landklassifizierungskarte *f*
  f carte *f* de classification de biens immeubles
  r карта *f* классификации земель

2076 land condition map; land capability map
  d Landbedingungskarte *f*;
     Landtauglichkeitskarte *f*
  f carte *f* d'entraînement à sec; carte des
     possibilités d'exploitation des terres
  r карта *f* состояния земель; карта
     пригодности земли

2077 land cover
  d Bodenbedeckung *f*; Landüberdeckung *f*
  f couverture *f* du sol
  r растительный покров *m*

2078 land cover database
  d Bodenbedeckungsdatenbasis *f*
  f base *f* de données de couverture du sol
  r база *f* данных растительного покрова

2079 land cover map
  d Bodenbedeckungskarte *f*
  f carte *f* de couverture du sol
  r карта *f* растительного покрова

2080 land cover statistics
  d Bodenbedeckungsstatistik *f*
  f statistique *f* de couverture du sol
  r статистика *f* растительного покрова

2081 land database
  d Geländedatenbasis *f*; Grunddatenbasis *f*
  f base *f* de données de la terre
  r поземельная база *f* данных

  *   **landfall → 2091**

  *   **landform → 3605**

  *   **landform map → 2082**

2082 landform[-type] map; terrain-type map
  d Geländeformenkarte *f*; morphografische
     Karte *f*
  f carte *f* de géomorphologie
  r геоморфологическая карта *f*

2083 land information system; LIS
  d Landesinformationssystem *n*;
     Informationssystem *n* für statistische Daten
     eines Landes
  f système *m* d'information sur le territoire;
     base *f* de données localisées
  r земельная информационная система *f*; ЗИС

2084 land line
  d Bodenleitung *f*; Geländelinie *f*
  f ligne *f* terrestre
  r наземная линия *f*; наземный профиль *m*;
     граница *f* полосы отвода

2085 land-line adjustment
  d Bodenleitungeinstellung *f*
  f ajustement *m* de ligne terrestre
  r уравнивание *n* наземной линии

2086 land management
  d Bodenbewirtschaftung *f*; Bodenbearbeitung *f*;
     Raumordnung *f*
  f aménagement *m* du territoire; gestion *f* des
     terres
  r организация *f* землепользования;
     землеустройство *n*

2087 landmark

  d Landmarke *f*; Grenzstein *m*; Wahrzeichen *n*
  f repère *m* terrestre; amer *m* terrestre
  r ориентир *m* на местности; маркировочный
     знак *m*; межевой знак; наземный ориентир;
     веха *f*

2088 landmark feature identification number
  d Landmarke-Identifikationsnummer *f*
  f identificateur *m* de repère terrestre; nombre *m*
     d'identification de repère terrestre
  r идентификатор *m* маркировочного знака;
     идентификатор ориентира на местности

  *   **land parcel → 2089**

2089 land plot; plot; soil lot; lot; allotment; land
      parcel; parcel
      (a measured piece of land)
  d Grundstück *m*; Teilstück *m*; Parzelle *f*
  f part *m* d'un terrain; parcelle *f*; lotissement *m*
  r надел *m*; [небольшой] участок *m* земли;
     доля *f*; делянка *f*; парцелла *f*

  *   **land records → 422**

  *   **land register → 422**

  *   **land register map → 2652**

2090 land registration; cadastral registration
  d Katastrierung *f*
  f cadastrage *m*; cadastration *f*
  r регистрация *f* земельных участков; учёт *m*
     земель; земельная регистрация

  *   **land registry → 422**

  *   **landscape analysis → 1407**

2091 landslide; earth slide; landfall
  d Bergrutsch *m*; Erdrutsch *m*; Erdfließen *n*
  f éboulement *m* de terrain; glissement *m* de
     terrain; fontis *m*
  r оползень *m*; обвал *m* [земли]

2092 landslide hazard model
  d Bergrutsch-Risiko-Modell *n*
  f modèle *m* de risque d'éboulements
  r модель *f* риска к обвалу

2093 landslide susceptibility
  d Bergrutsch-Suszeptibilität *f*
  f susceptibilité *f* d'éboulements
  r чувствительность *f* к обвалу

  *   **land survey → 3723**

2094 land-surveyor; surveyor
  d Landmesser *m*

f arpenteur *m*; géodésien *m*; compléteur *m*;
géomètre *m*; topographe *m*
r землемер *m*

* **land survey register → 422**

2095 land system
d Landsystem *n*
f système *m* terrestre
r земляная система *f*; ландшафтное
районирование *n* территории; система
землепользования

2096 land unit
d Landeinheit *f*
f unité *f* de terrain
r земельная единица *f*

2097 landuse
d Bodennutzung *f*; Flächennutzung *f*;
Landnutzung *f*
f utilisation *f* du sol; utilisation des terres;
exploitation *f* du terre; occupation *f* du sol
r землепользование *n*; использование *n*
земель; использование угодий

2098 landuse boundary
d Landnutzungsgrenze *f*
f limite *f* d'utilisation du sol
r граница *f* землепользования

2099 landuse classification system
d Landnutzung-Klassifizierungssystem *n*
f système *m* de classification selon l'utilisation
du sol
r система *f* классификации земель по
характеру их использования

2100 landuse management system; LUMS
d Landnutzung-Managementsystem *n*
f système *m* de gestion d'utilisation du sol
r система *f* управления землепользования

**2101 landuse map; landuse plan; comprehensive
development area map; development plan;
development map**
d Landnutzungskarte *f*; Flächennutzungsplan *m*;
Bodennutzungsplan *m*; Bebauungsplan *m*
f carte *f* d'utilisation du sol; plan *m* d'occupation
du sol; carte d'aménagement
r карта *f* землепользования

* **landuse plan → 2101**

2102 landuse planning
d Planung *f* der Bodennutzung;
Landnutzungsplanung *f*
f planification *f* d'exploitation du terre
r планирование *n* землепользования

2103 landuse survey
d Flächennutzungserhebung *f*
f enquête *f* sur l'utilisation du sol
r съёмка *f* землепользования

2104 landuse view
d Landnutzungsaussicht *f*
f vue *f* d'utilisation du sol
r представление *n* землепользования

2105 landuse zone
d Landnutzungszone *f*
f zone *f* d'utilisation du sol
r зона *f* землепользования

* **lap → 2679**

2106 large-area triangulation; LAT
d Großflächentriangulation *f*
f triangulation *f* de grandes surfaces
r триангуляция *f* крупного масштаба

2107 large-format atlas
d Großformat-Atlas *m*
f atlas *m* de gros format
r большой атлас *m*

2108 large-scale map
d Großmaßstab-Karte *f*; großmaßstäbige Karte *f*
f carte *f* à grande échelle
r крупномасштабная карта *f*

2109 laser plotter
d Laserplotter *m*
f traceur *m* laser
r лазерный плоттер *m*

* **LAT → 2106**

2110 lateral connection
d Seitenverbindung *f*
f connexion *f* latérale
r боковая связь *f*

2111 lateral lap; side lap
d Seitenüberlappung *f*; Querüberdeckung *f*
f recouvrement *m* latéral
r боковое перекрытие *n*

2112 lath
d Dachlatte *f*; Latte *f*
f latte *f*; liteau *m*
r штукатурка *f*; дрань *f*; рейка *f*

* **latitude → 205**

2113 latitude-longitude
d Breite-Länge *f*
f latitude-longitude *f*
r широта-долгота *f*

2114 latitude of origin; origin latitude
  d Ursprungsbreite *f*
  f latitude *f* d'origine
  r исходная широта *f*

2115 lattice
  d Gitter *n*
  f lattis *m*; treillis *m*; semis *m*; réseau *m*
  r решётка *f*; сетка *f*

2116 layer; overlay
  d Schicht *f*
  f couche *f*
  r слой *m*

2117 layer-based GIS
  d Schicht-basierendes GIS *n*; auf Schichten beruhendes GIS
  f SIG *m* à base de couches
  r послойно-организованная ГИС *f*

  *   layer box → 1879

2118 layer control
  d Schichtsteuerung *f*
  f contrôle *m* de couches
  r управление *n* слоями

2119 layered representation; multilayered representation
  d Mehrschichtdarstellung *f*
  f représentation *f* en couches; représentation stratifiée
  r послойное представление *n*; слоистое представление; многослойное представление

2120 layer index
  d Schichtindex *m*
  f indice *m* de couche
  r индекс *m* слоя

2121 layering; slicing
  d Schichtenteilung *f*; Schichtung *f*; Schneiden *n*
  f division *f* en couches; tranchage *m*
  r разбиение *n* на слои; расслоение *n*; многоуровневое представление *n*

2122 laying out; marking out; pegging out; staking
  d Absteckung *f*; Vermarkung *f*
  f marquage *m* [par piquets]
  r маркирование *n*; наложение *n*

  *   layout → 2123

2123 layout [chart]
  d topografische Anordnung *f*, Aufstellungsweise *f*; Layout *n*; Aufstellungsplan *m*;

Aufmachung *f*
  f disposition *f* [topologique]; layout *m*; mise *f* en page; carte *f* de disposition; carte topologique
  r планировка *f*; схема *f* расположения; топологическая схема

2124 layout window
  d Layoutfenster *n*
  f fenêtre *f* du layout
  r окно *n* макета

2125 layover
  d Wende *f*; Aufenthalt *m*
  f déversement *m* radar; basculement *m* (*inversion des pentes sur une image radar*)
  r остановка *f*; задержка *f* [в пути]; "дорожки" *fpl*

2126 leading edge
  d Vorderkante *f*
  f bord *m* avant
  r ведущий край *m*

2127 lead line
  d Führungsanteil *m*
  f ligne *f* de guidage; conduite *m* d'écoulement; ligne de sonde
  r ведущая линия *f*

2128 least-cost path
  d kostengünstiger Leitweg *m*; kostenoptimierter Leitweg
  f chemin *m* du moindre coût; chemin au plus bas coût
  r маршрут *m* движения с минимальными издержками

2129 least-cost path problem
  d Problem *n* des kostengünstigen Leitweges
  f problème *m* de chemin au plus bas coût
  r расчёт *m* маршрута движения с минимальными издержками

2130 least-squares adjustment; least-squares fitting
  (of curves)
  d Ausgleich *m* durch die Methode der kleinsten Quadrate
  f ajustage *m* par méthode des moindres carrés
  r подбор *m* методом наименьших квадратов

  *   least-squares fitting → 2130

2131 least-squares method
  d Methode *f* der kleinsten Quadrate
  f méthode *f* des moindres carrés
  r метод *m* наименьших квадратов

  *   LED plotter → 2147

2132 **left-right topology**
d seitenverkehrte Topologie *f*
f topologie *f* de gauche à droite
r топология *f* связей левых и правых сторон
полигонов

* **legend** → 2278

2133 **legend box; legend frame**
d Legendrahmen *m*
f boîte *f* de légende
r коробка *f* условных обозначений; панель *f*
легенды

* **legend frame** → 2133

2134 **legend type**
d Legendtyp *m*
f type *m* de légende
r тип *m* легенды; тип условных обозначений

2135 **legend window**
d Legendfenster *n*
f fenêtre *f* de légende
r окно *n* легенды

2136 **length of arc; arc length**
d Bogenlänge *f*
f longueur *f* d'arc
r длина *f* дуги

* **lettering** → 453

2137 **levee**
d Damm *m*; Morgenempfang *m*
f levée *f*; digue *f*
r насып *m*; приподнятый берег *m* реки;
намывной вал *m* реки

2138 **level**
d Niveau *n*; Pegel *m*
f niveau *m*
r уровен *m*

2139 **level**
(an instrument)
d Nivellierinstrument *n*
f instrument *m* de mesure de niveau
r нивелир *m*

* **level control** → 2140

2140 **level control [network]; leveling network;**
elevation control; vertical control; vertical
net; network of heights
d Höhenfixpunkte *mpl*; Höhennetz *n*;
Standregelung *f*
f canevas *m* altimétrique
r нивелирная геодезическая сеть *f*

* **level curve** → 732

* **leveling** → 2141

* **leveling error of closure** → 3934

* **leveling network** → 2140

* **level line** → 732

2141 **level[l]ing**
d Nivellierung *f*; Nivellement *n*
f nivellement *m*; calage *m*
r нивелирование *n*

2142 **level map**
d Niveauliniendarstellung *f*
f carte *f* à courbes de niveau
r таблица *f* выходных уровней

2143 **level of mapping**
d Abbildungsniveau *n*
f niveau *m* d'application
r уровень *m* отображения

2144 **level plane; height plane**
d Höhenebene *f*
f plan *m* de niveau
r плоскость *f* уровня

2145 **level surface; equipotent[ial] surface;**
potential surface
(surface which at every point is perpendicular
to the plumbline or the direction in which
gravity acts)
d Äquipotentialfläche *f*; Potentialfläche *f*;
Niveaufläche *f*
f surface *f* équipotentielle; surface de niveau
r эквипотенциальная поверхность *f*;
равномощная поверхность

* **level theodolite** → 3578

2146 **library**
d Bibliothek *f*
f bibliothèque *f*
r библиотека *f*

* **library reference workspace** → 862

2147 **light-emitting diode plotter; LED plotter**
d LED-Ploter *m*
f traceur *m* à diodes électroluminescentes
r светодиодный плоттер *m*

* **light path** → 3032

2148 **light source; luminous source**
d Lichtquelle *f*

f source *f* de lumière; source lumineuse; source
d'éclairage
r источник *m* света; световой источник

2149 **limit of accuracy of chronology; time
resolution; temporal resolution**
d zeitliche Auflösung *f*; Zeitauflösung *f*
f finesse *f* de la chronologie; résolution *f* de
temps; résolution temporaire
r разрешающая способность *f* по времени;
разрешение *n* по времени

2150 **line**
d Linie *f*; Leitung *f*
f ligne *f*
r линия *f*

2151 **lineage**
d Abstammung *f*
f généalogie *f* (*des données*)
r генеалогия *f* (*данных*)

2152 **linear-angular**
d Längen- und Winkel-
f linéaire-angulaire
r линейно-угловой

2153 **linear-angular network**
d Längen- und Winkelnetz *n*
f réseau *m* linéaire-angulaire
r линейно-угловая сеть *f*

2154 **linear error**
d linearer Fehler *m*
f erreur *f* linéaire
r линейная погрешность *f*

2155 **linear event**
d Linienereignis *n*
f événement *m* linéaire
r линейная ось *f* синфазности

* **linear feature** → 2162

2156 **linear filter**
d linearer Filter *m*
f filtre *m* linéaire
r линейный фильтр *m*

2157 **linear interpolation**
d lineare Interpolation *f*
f interpolation *f* linéaire
r линейная интерполяция *f*

2158 **linear interpolator**
d linearer Interpolator *m*
f interpolateur *m* linéaire
r линейный интерполятор *m*

* **linearization** → 3028

2159 **linearly connected structure**
d geradlinig verbundene Struktur *f*
f structure *f* liée rectilignement
r линейно-связанная структура *f*

2160 **linear map**
d Linienkarte *f*
f carte *f* linéaire
r линейная карта *f*

2161 **linear network**
d Liniennetz *n*
f réseau *m* linéaire
r линейная сеть *f*

2162 **linear object; linear feature; line feature**
d lineares Objekt *n*
f objet *m* linéaire; entité *f* linéaire
r линейный объект *m*; одномерный объект

2163 **linear optimization**
d lineare Optimierung *f*
f optimisation *f* linéaire
r линейная оптимизация *f*

* **linear projective mapping** → 2918

2164 **linear referencing method**
d lineare Referenzmethode *f*
f méthode *f* de référencement linéaire
r метод *m* линейного эталонирования

2165 **linear referencing system; LRS**
d lineares Referenzsystem *n*
f système *m* de référencement linéaire
r система *f* линейного эталонирования

* **linear regression** → 3303

* **linear scale** → 1744

2166 **linear scaling factor**
d linearer Maßstabsfaktor *m*
f facteur *m* d'échelle linéaire
r линейный масштабный коэффициент *m*

2167 **linear stretching**
d lineare Streckung *f*
f étirage *m* linéaire
r линейное вытягивание *n*

2168 **linear structure**
d lineare Struktur *f*
f structure *f* linéaire
r линейная структура *f*

2169 **linear trend**
d linearer Trend *m*
f tendance *f* linéaire
r линейный тренд *m*

2170 **linear triangulation**
  d lineare Triangulation *f*
  f triangulation *f* linéaire
  r линейная триангуляция *f*

* **line chart** → 2176

2171 **line copy**
  d zeiliger Schablonentisch *m*
  f reproduction *f* d'un original au trait; copie *f* au trait
  r копия *f* штрихового оригинала

2172 **line detection**
  d Linienerkennung *f*
  f détection *f* de lignes
  r выделение *n* линейных элементов изображения

2173 **line extraction**
  d Linienextraktion *f*
  f extraction *f* de traits géométriques
  r извлечение *n* линий

* **line feature** → 2162

2174 **line follower; curve follower; tracer**
  (a semi-automatic device in which a laser beam is used to trace out lines from a source map and convert them to digital form)
  d Kurvenleser *m*
  f lecteur *m* de courbes; suiveur *m* de courbes
  r повторитель *m* кривых

2175 **line-following algorithm**
  d Linienverlauf-Algorithmus *m*; Linienverfolgungsalgorithmus *m*
  f algorithme *m* de filage de courbe; algorithme de poursuite de ligne
  r алгоритм *m* прослеживания горизонтали

2176 **line graph; curve chart; line chart; curve graph**
  d Kantengraph *m*
  f graphique *m* linéaire; graphique en segments
  r линейная диаграмма *f*

* **line graphics** → 761

2177 **line indicator**
  d lineare Anzeige *f*, linearer Indikator *m*
  f indicateur *m* linéaire
  r линейный признак *m*

2178 **line-in-polygon operation**
  d Linie-im-Polygon-Operation *f*
  f opération *f* de définition de ligne à polygone
  r операция *f* определения принадлежности линии полигону

2179 **line length**
  d Zeilenlänge *f*
  f longueur *f* de ligne
  r формат *m* строки; ширина *f* колонки

2180 **line length code**
  d Zeilenlänge-Code *m*
  f code *m* de longueur de ligne
  r код *m* длины линии

* **line map** → 727

2181 **line mapping**
  d Linienkartierung *f*
  f mappage *m* de lignes
  r проведение *n* линию на карте

2182 **line negative**
  d Schraffennegativ *n*
  f négatif *m* de trait
  r штриховой негатив *m*

2183 **line number; row number**
  d Zeilennummer *f*
  f numéro *m* de ligne
  r номер *m* строки

2184 **line of latitude**
  d Breite-Linie *f*
  f ligne *f* de latitude
  r линия *f* широты

2185 **line of longitude**
  d Länge-Linie *f*
  f ligne *f* de longitude
  r линия *f* долготы

2186 **line of nodes; node line**
  d Knotenlinie *f*
  f ligne *f* de nœuds
  r линия *f* узлов

2187 **line of projection**
  d Projektionslinie *f*
  f ligne *f* de projection
  r проектирующая линия *f*

* **line of stream** → 1445

2188 **line omissing**
  d Liniendurchlaß *m*
  f omission *f* de ligne
  r пропуск *m* линии; прерывание *n* линии

* **line original** → 999

* **line section** → 2189

2189 **line segment; line section**
  d Linienabschnitt *m*

f segment *m* de ligne; portion *f* de ligne; tronçon *m*
r линейный сегмент *m*

2190 line smoothing
d Linienglättung *f*
f lissage *m* de ligne
r сглаживание *n* линии

* **lines of equal distortions; → 1138**

2191 line spacing
d Linienabstand *m*
f espacement *m* entre lignes
r межмаршрутное расстояние *n*

2192 line style
d Linienstil *m*
f style *m* de ligne
r стиль *m* линии

2193 line symbols
d Zeichen *npl* von Linien; Liniensymbole *npl*
f lignes *fpl*
r линейные [условные] знаки *mpl*

2194 line thinning
d Linienverdünnung *f*; Linienausdünnung *f*
f amincissement *m* de ligne; atténuation *f* de ligne
r утончение *n* линий

2195 line tracing
d Linienverfolgung *f*
f traçage *m* de lignes
r трассировка *f* линий

2196 linetype
d Linientyp *m*
f type *m* de ligne
r тип *m* линии

2197 line weeding
d Linienjäten *n*
f triage *m* de ligne; élagage *m* de ligne; émondage *m* de ligne
r устранение *n* избыточных промежуточных точек в цифровой записи линий

* **link → 2199, 3248**

* **linkage → 2199**

* **link and node structure → 153**

2198 linked tables
d verknüpfte Tabellen *fpl*
f tables *fpl* liées
r связанные таблицы *fpl*

2199 link[ing]; bind[ing]; linkage
d Verknüpfung *f*; Verbinden *n*; Binden *n*; Link *n*; Verflechtung *f*; Seilzug *m*
f liaison *f*; lien *m*; ligature *f*; embase *f*; reliure *f*
r связь *f*; связывание *n*; связка *f*

* **LIS → 2083**

2200 list box
d Listenfeld *n*
f boîte *f* de liste; zone *f* de liste déroulante
r поле *n* списка; списковое поле; панель *f* перечня

* **list generator → 3132**

* **list of place names → 1515**

2201 literal
d Literal *n*; buchstäbliche Konstante *f*
f littéral *m*; libellé *m*; constante *f* littérale
r литерал *m*; литеральная константа *f*

2202 live access layer
d direkte Zugriffsschicht *f*
f couche *f* à accès en direct
r слой *m* оперативным доступом

2203 live access table
d Tabelle *f* mit direktem Zugriff
f table *f* d'accès vive
r таблица *f* оперативным доступом

2204 live remote DBMS access
d direkter DBMS-Fernzugang *m*
f accès *m* à distance au SGBD
r оперативный дистанционный доступ *m* к СУБД

2205 loadable driver
d ablauffähiger Treiber *m*; ladbarer Treiber
f pilote *m* chargeable
r загружаемый драйвер *m*; нерезидентный драйвер

2206 local area network; LAN
d lokales Netz[werk] *n*
f réseau *m* local
r локальная вычислительная сеть *f*; ЛВС; местная сеть

2207 local area rubbersheeting
d lokales Einpassen *n* mit Gummibandfunktion; lokaler Gummiband *m*
f étirement *m* par fil élastique local
r локальное эластичное соединение *n*

2208 local consistency
d lokale Stetigkeit *f*; örtliche Konsistenz *f*

f consistance f locale
r локальная совместимость f

2209 local coordinate system
d lokales Koordinatensystem n; regionales Koordinatensystem
f système m de coordonnées local
r местная система f координат

2210 local databank
d lokale Datenbank f
f banque f de données locale
r локальный банк m данных

2211 local drain direction
d lokale Entwässerungsrichtung f
f direction f locale cours d'eau
r локальное направление n дренажа

2212 local GIS
d lokales GIS n
f SIG m local
r локальная ГИС f; местная ГИС

2213 locality
d Örtlichkeit f; örtliche Lage f
f localité f
r локальность f; окрестности f pl

* **localization → 2215**

2214 locating grid
d Lokalisierungsgitter n
f grille f de localisation
r указательная сетка f; сетка-указательница f

2215 location; localization; placement
d Stellung f, Lokalisierung f; Ortung f; Plazierung f; Aufstellung f
f location f; localisation f; [em]placement m
r расположение n; размещение n; локализация f

* **locational analysis → 3311**

* **locational data → 1584**

2216 location-allocation procedure
d Location-Allocation-Prozedur f
f procédure f de localisation-allocation
r процедура f расположения/размещения

2217 locational reference
d Lagereferenz f; lagebezogener Verweis m
f référence f de position
r позиционный эталон m

2218 locational referencing system
d Lagereferenzsystem n
f système m référentiel de position

r позиционная эталонная система f

2219 locational symbol
d Lagesymbol n
f symbole m de position
r условный знак m местоположения; позиционный символ m

* **location analysis → 3311**

2220 location quotient
d Lokationsquotient m
f indice m de position
r коэффициент m местоположения; фактор m местоположения

2221 locator
d Lokalisierer m
f relévateur m [de coordonnées]; localisateur m; dispositif m de localisation
r устройство n ввода позиции

* **lock → 2222**

* **lock-in → 2222**

2222 lock[ing]; lockout; blocking; lock-in
d Verriegelung f; Sperrung f; Blockierung f
f verrouillage m; blocage m
r запирание n; блокирование n; блокировка f

* **lockout → 2222**

2223 lofting
d Schnürbodenverfahren n
f génération f de volume
r лофтинг m; крупномасштабная вертикальная конвекция f; крупномасштабные восходящие течения n pl; крупномасштабный вертикальный перенос m

2224 logarithmic scale
d logarithmische Skala f
f échelle f logarithmique
r логарифмическая шкала f

2225 log file
d Sicherstellungsdatei f
f fichier m d'enregistrement
r регистрационный файл m

2226 logical accuracy
d logische Genauigkeit f
f exactitude f logique
r логическая точность f

2227 logical connector
d logischer Verbinder m; logischer Anschluss m

f connecteur m logique
r логический соединитель m

* **logical expression** → 369

* **logically continuous database** → 3239

* **logical operator** → 372

2228 logical query
d logische Abfrage f
f interrogation f logique
r логический запрос m

2229 logical relationship
d logische Beziehung f
f relation f logique
r логическая [взаимо]связь f

2230 logical selection
d logische Auswahl f
f sélection f logique
r логический выбор m

* **long-haul network** → 3996

* **longitude** → 206

2231 longitudinal check
d längslaufende Überprüfung f
f essai m longitudinal
r продольный контроль m; контроль вдоль
дорожек

* **longitudinal overlap** → 1465

2232 long[-lived] transaction
d Langläufer m; langweiliger Abschluss m;
langsame Transaktion f
f transaction f longue
r продолжительная транзакция f

* **long transaction** → 2232

2233 lookup table
d Lookup-Tabelle f; Nachschlagtabelle f
f table f de référence
r таблица f просмотра; просмотровая
таблица; справочная таблица

2234 loop; circuit; cycle
(of a network)
d Schleife f; Zyklus m
f boucle f; lacet m
r петля f

2235 loose coupling
d lose Kopplung f
f couplage m lâche
r связь f меньше критической; слабая связь;

нежёсткое соединение n

* **lot** → 2089

2236 low-pass filter
d Tiefpassfilter m
f filtre m passe-bas
r фильтр m низких частот; низкочастотный
фильтр

2237 low water
(minimum height reached by a falling tide)
d niedriger Wasserspiegel m
f basses eaux fpl; étiage m; reflux m; marée f
basse
r отлив m; межень m; низкая вода f

2238 low water line
d Niedrigwasserlinie f
f laisse f de basse mer
r линия f наибольшего отлива

* **loxodrome** → 3160

* **loxodromic curve** → 3160

* **loxodromic line** → 3160

* **loxodromic spiral** → 3160

* **LRS** → 2165

* **luminosity** → 404

* **luminous source** → 2148

* **LUMS** → 2100

M

* MA → 2416

* macro → 2239

* macrocode → 2239

* macrocommand → 2239

2239 macroinstruction; macrocommand;
macrocode; macro[s]
d Makrobefehl *m*; Makrokommando *n*;
Makroinstruktion *f*; Makros *n*
f macro-instruction *f*; macrocommande *f*;
macro *m*
r макрокоманда *f*; макроинструкция *f*;
макро[с] *m*

* macros → 2239

* magnetic azimuth → 634

* magnetic declination → 635

2240 magnetic north
d magnetische Nordrichtung *f*
f nord *m* magnétique
r магнитный север *m*; северный магнитный
полюс *m*; магнитное направление *n*
северного меридиана

2241 Mahalanobis distance
(the distance between region mean and class
distribution center)
d Mahalanobis-Distanz *f*
f distance *f* de Mahalanobis
r обобщённое расстояние *n*; расстояние
Махаланобиса

2242 mailing address; house address; address;
structure number
d [Mail-]Adresse *f*
f adresse *f* d'envoi
r [почтовый] адрес *m*

* major centre → 512

* management graphics → 417

* manifold → 769

2243 man-made feature
d vom Menschen verursachtes Objekt *n*;
künstlich erzeugtes Objekt; anthropogenes
Objekt
f objet *m* dû à l'homme; objet d'origine
artificielle; objet d'origine humaine; détail *m*
culturel
r искусственный объект *m*; антропогенный
объект

2244 manner of cartographic representation;
mode of cartographic representation
d kartografischer Darstellungsmodus *m*
f mode *m* de représentation cartographique
r способ *m* картографического изображения

2245 manual digitizing
d manuelle Digitalisierung *f*
f numérisation *f* manuelle
r ручное дигитализирование *n*

* many-to-many relation → 2246

2246 many-to-many relation[ship]
d m:n-Beziehung *f*; Viel-zu-Viel-Beziehung *f*
f association *f* de type n:m
r соотношение *n* многие-ко-многим

* many-to-one relation → 2247

2247 many-to-one relation[ship]
d m:1-Beziehung *f*; Viel-zu-Eins-Beziehung *f*
f association *f* de type m:1; relation *f*
plusieurs-à-un
r соотношение *n* один-ко-многим

2248 map *v*
d abbilden; aufzeichnen; mappen
f correspondre
r отображать; устанавливать соответствие

* map → 1611

2249 map *v*; survey *v*
d kartografieren; umsetzen auf; aufnehmen;
planen
f cartographier; lever
r картографировать; составлять карту;
производить съёмку местности

* map accuracy → 2298

2250 map adjustment; map reconciliation
d Karteneinstellung *f*; Kartenversöhnung *f*;
Kartenabstimmung *f*
f ajustage *m* de cartes
r согласование *n* карт

2251 map ag[e]ing
d Kartenalterung *f*
f vieillissement *m* de carte
r старение *n* карты

* **map aging** → 2251

2252 map algebra
 d Map-Algebra *f*; kartografische Algebra *f*
 f algèbre *f* cartographique
 r картографическая алгебра *f*

2253 map amendment
 d Kartenänderung *f*; Kartenberichtigung *f*;
 Kartenverbesserung *f*
 f amendement *m* dans une carte
 r поправка *f* в карте

2254 map archiving
 d Kartenarchivierung *f*
 f archivage *m* en cartographie
 r архивирование *n* карт

* **map assembly** → 2285

2255 map bibliography
 d kartografische Bibliographie *f*
 f bibliographie *f* cartographique
 r картографическая библиография *f*;
 картобиблиография *f*

* **map border** → 376

* **map browser** → 2348

* **map capacity** → 2276

2256 map clipping object
 d kartografisches Schnittobjekt *n*
 f objet *m* de découpage cartographique
 r картографический отсекающий объект *m*

2257 map collection; map gallery
 d Kartensammlung *f*
 f cartothèque *f* cartographique
 r коллекция *f* карт; галерея *f* карт

2258 map-coloring problem
 d Färbungsproblem *n* der Karten
 f problème *m* de coloriage des cartes
 r задача *f* о раскрашивании карт

**2259 map color plate; color plate; map
 color-separated copy; map separates; map
 separation plate; map individual image**
 d Kartenfarboriginal *n*; Kartenfarbvorlage *f*
 f original *m* couleur [de carte]
 r цветоотдельный оригинал *m* карты

* **map color-separated copy** → 2259

* **map compilation** → 2297

2260 map components
 d Kartenkomponenten *fpl*

 f composantes *fpl* d'une carte
 r компоненты *fpl* карты

2261 map coordinate system
 d kartografisches Koordinatensystem *n*;
 Kartenkoordinatensystem *n*
 f système *m* de coordonnées de carte
 r координатная система *f* карты

2262 map coverage
 d Kartenabdeckung *f*
 f surface *f* couverte
 r картографическая изученность *f*

2263 map data; cartographic data
 d Kartendaten *pl*; kartografische Daten *pl*
 f données *fpl* cartographiques
 r картографические данные *pl*

* **map database** → 442

2264 map depot; map library
 d Kartenbibliothek *f*
 f bibliothèque *f* de cartes
 r картохранилище *n*; библиотека *f* карт

* **map design** → 444

2265 map development
 d Kartenentwicklung *f*; kartografische
 Projektierung *f*
 f développement *m* de carte
 r построение *n* карты

2266 map digitizing
 d Digitalisierung *f* von Karten
 f numérisation *f* de carte; numérisation
 cartographique
 r оцифрование *n* карты

2267 map display
 d Kartendarstellung *f*
 f affichage *m* cartographique
 r картографическое представление *n*

* **map division** → 2300

* **map drawing** → 445

2268 map editing function
 d Kartenbearbeitungsfunktion *f*
 f fonction *f* d'édition cartographique
 r картографическая функция *f*
 редактирования

* **map edition** → 2317

* **map estimation** → 2269

2269 map evaluation; map estimation
d Kartenbewertung *f*
f évaluation *f* de carte
r оценка *f* карты

2270 map extent
d Abbildungsextent *m*
f étendue *f* d'application
r экстент *m* отображения; расширение *n* отображения

* **map extract** → 2328

* **map feature** → 448

2271 map feature coding
d Grundzugscodierung *f*
f codage *m* d'éléments cartographiques
r кодирование *n* картографических элементов

2272 map file
d Kartendatei *f*
f fichier *m* carte
r картный файл *m*; файл карты; листинговый файл

2273 map formatting
d Kartenformatierung *f*
f formatage *m* de carte
r форматирование *n* карты

* **map frame** → 376

* **map gallery** → 2257

* **map georeferencing** → 1593

2274 map graticule; geographic graticule; graticule; map grid; cartographical grid; geographic grid
d Kartennetz *n*; Kartengitter *n*; kartografisches Netz *n*; Gitternetz *n*; Gradnetz *n*
f grille *f* [cartographique]; réseau *m* cartographique; graticule *f* cartographique; carroyage *m* de méridiens et parallèles; quadrillage *m* géographique; réseau géographique
r сетка *f* на карте; картографическая сетка; географическая сетка

* **map grid** → 2274

2275 map [image] record
d Kartenbildsatz *m*
f enregistrement *m* cartographique
r картографическая запись *f*

* **map individual image** → 2259

2276 map informativity; map capacity
d Karteninformativität *f*
f informativité *f* de carte
r информативность *f* карты

* **map inset** → 1962

* **map interpretation** → 2321

* **map join** → 2286

2277 map language
d Kartensprache *f*
f langage *m* de carte
r язык *m* карты

* **map layer** → 2291

2278 map legend; legend; sheet memory
d Legende *f*; Zeichenerklärung *f*
f légende *f* [cartographique]
r легенда *f* [карты]; экспликация *f*; условное обозначение *n*

* **map lettering inscription** → 453

2279 map lettering language
d Kartenbeschriftungssprache *f*
f système *m* d'écriture cartographique
r язык *m* картографических надписей

* **map library** → 2264

2280 map limits
d Kartengrenzen *fpl*
f limites *fpl* de carte
r границы *fpl* карты

2281 map linking
d Kartenverknüpfung *f*
f liaison *f* des cartes
r связывание *n* карт

2282 map locator
d Kartenzeiger *m*
f localisateur *m* de carte
r определитель *m* листов карты

* **mapmaker** → 435

* **mapmaking** → 2901

* **map margin** → 376

2283 map matching
d Kartenvergleich *m*
f coïncidence *f* des cartes
r сопоставление *n* карт

* **map-measurer** → 834

2284 map measuring accuracy
d Genauigkeit *f* der Kartenmessungen
f précision *f* de mesure cartographique;
exactitude *f* de mesure cartographique
r точность *f* измерений по картам

2285 map montage; map assembly
d Kartenmontage *f*
f assemblage *m* de carte
r компоновка *f* карты

2286 map mosaicing; mosaicing; map join
d [kartografische] Mosaiking *n*; Mosaiken *n*
f mosaïquage *m* [cartographique]
r сшивка *f* карт

2287 map name
d Name *m* der Karte
f nom *m* de carte
r имя *n* карты

* **map numbering** → 3287

2288 map of the world; world map
d Weltkarte *f*; Erdkarte *f*
f carte *f* mondiale; mappemonde *f*
r карта *f* мира; карта земных полушарий

2289 map orientation
d Kartenorientierung *f*
f orientation *f* de carte
r карта *f* [локальной] ориентации

2290 map *v* out
d kartieren
f mapper
r картировать

2291 map overlay; map layer
d kartografische Schicht *f*; Map-Overlay *n*
f transparent *m* cartographique
r картографический слой *m*

2292 map overlay and statistical system format; MOSS format
d MOSS-Format *n*
f format *m* MOSS
r формат *m* MOSS

2293 mappable table
d abgebildete Tabelle *f*
f table *f* mappé
r таблица *f*, представленная с помощью карты; закартированная таблица

2294 map pair
d Kartenpaar *n*
f paire *f* de cartes
r пара *f* карт

2295 mapped phenomena
d abgebildete Erscheinungen *fpl*; gemappte Erscheinungen
f phénomènes *mpl* mappés
r закартированные явления *npl*

2296 mapped variable
d abbildete Variable *f*
f variable *f* mappée
r отображённая переменная *f*

2297 mapping; map compilation
d Kartierung *f*; Kartenentwerfen *n*
f compilation *f* cartographique
r картографирование *n*; картирование *n*; составление *n* карт; картосоставление *n*

* **mapping** → 2908

2298 map[ping] accuracy; map precision
d Kartierungsgenauigkeit *f*; Kartenpräzision *f*
f exactitude *f* cartographique; précision *f* cartographique
r точность *f* картографирования; [геометрическая] точность карты

2299 mapping agency
d Kartierungsstelle *f*
f agence *f* de cartographie
r картографическое агентство *n*

2300 map[ping] division
d kartografische Teilung *f*
f division *f* cartographique
r картографическое деление *n*

* **mapping equipment** → 1315

2301 mapping function
d Abbildungsfunktion *f*
f fonction *f* d'application
r функция *f* отображения

2302 mapping instrument
d Kartiergerät *n*; Kartierinstrument *n*
f instrument *m* de cartographie
r картографический инструмент *m*

2303 map[ping] resolution
d kartografische Auflösung *f*
f résolution *f* cartographique; résolution d'une carte
r картографическая разрешающая способность *f*

* **mapping science** → 465

2304 mapping system
d Kartierungssystem *n*; Kartografierungssystem *n*

f système *m* cartographique
r картографическая система *f*

2305 mapping toolkit
    d Kartenwerkzeugsatz *m*
    f kit *m* d'instruments cartographiques
    r инструментарии *mpl* картографирования

2306 map[ping] unit
    (a set of areas drawn on a map to represent a
    well-defined feature or set of features)
    d Abbildungseinheit *f*; kartografische Einheit *f*;
    Kartiereinheit *f*
    f unité *f* [de mesure] cartographique
    r картографическая единица *f*

2307 map plane
    d Kartenebene *f*
    f plan *m* de carte
    r плоскость *f* карты

2308 map plotting
    d Kartenplotten *n*
    f traçage *m* de carte; tracé *m* de carte
    r вычерчивание *n* карты

2309 map pragmatics
    d kartografische Pragmatik *f*
    f pragmatique *f* cartographique
    r картографическая прагматика *f*

    * **map precision** → **2298**

2310 map preparation
    d Kartenvorbereitung *f*; Kartenerstellung *f*
    f préparation *f* de carte
    r изготовление *n* карты

2311 map printing; cartographic printing
    d Kartendruck *m*; kartografische
    Druckerzeugnisse *npl*
    f impression *f* cartographique
    r картографическое печатание *n*

**2312 map produced from the digitized
    information**
    d aus den digital verschlüsselten Informationen
    konstruierte Karte *f*
    f carte *f* produite à partir d'information
    digitalisée
    r карта *f*, созданная на базе
    дигитализированной информации

    * **map production** → **2901**

2313 map program
    d Abbildungsprogramm *n*
    f programme *m* d'application
    r программа *f* отображения

2314 map projection; cartographic projection
    d kartografische Abbildung *f*;
    Kartenprojektion *f*; kartografische Projektion *f*
    f projection *f* cartographique
    r картографическая проекция *f*

2315 map projection distortion
    d Kartenprojektion-Verzerrung *f*
    f distorsion *f* de projection cartographique
    r искажение *n* картографической проекции

2316 map projector
    d Kartenprojektor *m*
    f projecteur *m* cartographique
    r картографический проектор *m*

2317 map publication; map edition
    d Kartenausgabe *f*
    f publication *f* de cartes
    r издание *n* карт

    * **map quad** → **2318**

**2318 map quad[rangle]; topographic
    quadrangle; quad[rangle] map; degree
    square**
    d Gradabteilungskarte *f*; Gradabteilungsblatt *n*;
    topografisches Quadrangel *n*; Planquadrat *n*
    f carte *f* limitée par des méridiens et des
    parallèles; degré *m* carré
    r стандартный лист *m* топографической
    карты; топографический
    четырёхугольник *m*; картографическая
    трапеция *f*

2319 map quality
    d Kartenqualität *f*
    f qualité *f* de carte
    r качество *n* карт

2320 map query
    d Kartenabfrage *f*
    f interrogation *f* cartographique
    r картографический запрос *m*

2321 map reading; map interpretation
    d Kartenübersetzung *f*
    f lecture *f* de carte; interprétation *f* de carte
    r чтение *n* карты; распознавание *n* карты

2322 map reading system
    d Kartenübersetzungssystem *n*
    f système *m* de lecture de cartes
    r система *f* распознавания карт

    * **map reconciliation** → **2250**

    * **map record** → **2275**

    * **map region** → **3087**

* **map registration** → 1593

2323 map registration point; georeferencing point
d Georeferenzierungspunkt *m*
f point *m* de géoréférencement
r точка *f* географического кодирования

2324 map reliability
d Kartenzuverlässigkeit *f*; Zuverlässigkeit *f* einer Karte
f fiabilité *f* d'une carte
r надёжность *f* карты

* **map resolution** → 2303

2325 map revision
d Laufendhalten *n* einer Karte; Fortführen *n* einer Karte
f révision *f* d'une carte
r обновление *n* карты

2326 map scale; cartographic scale
d Kartenmaßstab *m*; kartografischer Maßstab *m*
f échelle *f* de carte; échelle cartographique
r масштаб *m* карты; картографический масштаб

2327 map script
d kartografischer Skript *m*
f script *m* cartographique
r картографический скрипт *m*

2328 map section; map segment; map extract
d Kartensegment *n*; Kartenausschnitt *m*
f découpage *m* cartographique
r картографический сегмент *m*; картографическое извлечение *n*

* **map segment** → 2328

2329 map semantics
d kartografische Semantik *f*
f sémantique *f* cartographique
r картографическая семантика *f*

2330 map semiotics
d kartografische Semiotik *f*
f sémiotique *f* cartographique
r картографическая семиотика *f*

* **map separates** → 2259

* **map separation plate** → 2259

2331 map series
d Kartenserie *f*
f série *f* de cartes
r серия *f* карт

* **map server** → 1994

2332 map set
d Kartensatz *m*
f suite *f* de cartes
r набор *m* карт

2333 map sheet; sheet
d Kartenblatt *n*
f feuille *f* d'une carte; coupure *f*
r лист *m* карты

2334 map sheet boundaries
d Grenzen *fpl* des Kartenblatts
f limites *fpl* de feuille de carte
r границы *fpl* листа карты

2335 maps of nature and society interaction
d Karten *fpl* der Natur- und Sozialwechselwirkung
f cartes *fpl* d'interaction naturelle et sociale
r карты *fpl* взаимодействия природы и общества

2336 map squaring; quadrate mapping
d Gitternetzanlegung *f*
f carroyage *m* cartographique
r разбиение *n* карты на квадраты

2337 map stylistics
d kartografische Stilistik *f*
f stylistique *f* cartographique
r картографическая стилистика *f*

2338 map summary
d kartografisches Resümee *n*
f résumé *m* cartographique; sommaire *m* cartographique
r картографическая сводка *f*; картографическое резюме *n*

2339 map surface
d Kartenfeld *n*; Karteninhaltsfläche *f*
f surface *f* cartographiée
r поверхность *f* карты

2340 map symbology
d Landkartensymbolsätze *mpl*
f symbologie *f* cartographique
r картографическая символика *f*

* **map symbols** → 459

2341 map syntactics; map syntax
d kartografische Syntax *f*
f syntactique *f* cartographique; syntaxe *m* cartographique
r картографическая синтактика *f*

* **map syntax** → 2341

2342 map techniques
d kartografische Methoden *fpl*
f techniques *fpl* cartographiques
r приёмы *mpl* анализа карт

2343 map theme
d Kartenthema *n*
f thème *m* cartographique
r картографическая тема *f*

2344 map title
d Kartentitel *m*
f titre *m* de carte
r заголовок *m* карты; название *n* карты

2345 map transformation
d Kartentransformation *f*
f transformation *f* de cartes
r преобразование *n* карт

* **map unit** → 2306

2346 map use
d Kartennutzung *f*
f utilisation *f* de carte
r использование *n* карт

2347 map view
d kartografische Ansicht *f*
f vue *f* cartographique
r картографический вид *m*

2348 map viewer; map browser
d kartografischer Browser *m*
f logiciel *m* de visualisation cartographique;
visionneur *m* cartographique; butineur *m*
cartographique
r картографический визуализатор *m*;
картографический браузер *m*

* **marginal condition** → 1218

* **marginal data** → 2350

2349 marginal [data] representation
d Grenzdatendarstellung *f*
f représentation *f* de données marginaux
r зарамочное оформление *n* [карты]

2350 marginal information; marginal data
d Kartenrandangaben *fpl*; Grenzdaten *pl*;
marginale Daten *pl*
f données *fpl* marginaux; renseignements *mpl*
marginaux
r зарамочные данные *pl* [карты]

* **marginal representation** → 2349

* **margin of a map** → 376

2351 marine mapping
d Meereskartografie *f*
f cartographie *f* maritime
r морское картографирование *n*

* **marked** → 2065

2352 marker
d Marker *m*; Markierer *m*; Richtpunkt *m*
f marqueur *m*
r маркер *m*

2353 marker palette
d Marker-Palette *f*
f palette *f* de marqueur
r палитра *f* маркера

2354 marker shape
d Markerform *f*
f forme *f* de marqueur
r форма *f* маркера

2355 marker size
d Markergröße *f*
f taille *f* de marqueur
r размер *m* маркера

2356 marker symbol
d Markerzeichen *n*
f symbole *m* de marqueur
r символ *m* маркера

2357 marking features
d Markierungsfeature *npl*
f entités *fpl* de marquage
r элементы *mpl* маркировки

2358 marking of a survey point; monumentation
d Festlegen *n* eines Vermessungspunkts
f matérialisation *f* d'un repère
r определение *n* граничного геодезического
знака

* **marking out** → 2122

* **marquee** → 3258

* **marsh** → 368

* **MAS** → 2422

2359 mask
d Maske *f*
f masque *f*
r маска *f*

2360 masking
d Maskieren *n*; Maskierung *f*; Ausblenden *n*
f masquage *m*
r маскирование *n*

2361 masking tape
 d Abdeckband *n*; Abdeckklebeband *n*;
   Kantenklebeband *n*
 f ruban *m* à masquer; ruban de papier-cache;
   ruban-cache *m*; bande *f* adhésive
 r изоляционная лента *f*; маскирующая лента

2362 mass balance
 d Massenbilanz *f*; Materialbilanz *f*; Stoffbilanz *f*
 f bilan *m* massique; bilan matière
 r баланс *m* массы; равновесие *n* материалов;
   противовес *m*

 * **mass center** → 509

2363 mass computation
 d Massenberechnung *f*
 f calcul *m* de masse
 r вычисление *n* массы

2364 mass point
 d Massenpunkt *m*
 f point *m* de masse
 r материальная точка *f*

2365 matching
 d Abgleich *m*; Anpassung *f*;
   Paarigkeitsvergleich *m*
 f appariement *m*; coïncidence *f*
 r согласование *n* [признаками];
   совмещение *n*; совпадение *n*;
   сопоставление *n*

2366 match tolerance
 d Abgleichstoleranz *f*
 f tolérance *f* de coïncidence
 r допуск *m* совпадения

2367 mathematical base of maps
 d mathematische Kartenbasis *f*
 f base *f* mathématique de cartes
 r математическая основа *f* карт

2368 mathematical cartography
 d mathematische Kartografie *f*
 f cartographie *f* mathématique
 r математическая картография *f*

2369 mathematical function
 d mathematische Funktion *f*
 f fonction *f* mathématique
 r математическая функция *f*

2370 mathematical modeling
 d mathematische Modellierung *f*
 f modelage *m* mathématique
 r математическое моделирование *n*

2371 matrix
 d Matrix *f*

 f matrice *f*
 r матрица *f*

 * **matrix image** → 3015

 * **matrix processor** → 184

2372 maximized window
 d maximiertes Fenster *n*
 f fenêtre *f* maximisée
 r максимизированное окно *n*

2373 maximum likelihood
   (a method embodying probability theory for
   fitting a mathematical model to a set of data)
 d maximale Mutmaßlichkeit *f*; maximale
   Plausibilität *f*
 f vraisemblance *f* maximale
 r максимальное правдоподобие *n*

2374 maximum likelihood classification
 d Maximum-Likelihood-Klassifizierung *f*;
   Klassifizierung *f* der größten
   Wahrscheinlichkeit
 f classification *f* selon le maximum de
   vraisemblance; classification du maximum de
   vraisemblance
 r максимально правдоподобная
   классификация *f*

 * **mean** → 2379

2375 mean area
 d Mittelfläche *f*; Skelettfläche *f*
 f surface *f* moyenne; surface médiane
 r средняя площадь *f*

2376 meander line
 d Mäanderlinie *f*
 f ligne *f* de méandre
 r меандрообразная линия *f*; меандровая
   линия

2377 meander survey
 d Mäandervermessung *f*
 f levé *m* par cheminement
 r меандрообразная съёмка *f*

 * **mean filter** → 291

2378 mean sea level; MSL
 d mittlere Meereshöhe *f*; mittlere Seehöhe *f*
 f niveau *m* moyen de la mer
 r средний уровень *m* моря; СУМ

2379 mean [value]
 d Mittelwert *m*; Mittel *n*
 f valeur *f* moyenne; moyenne *f*
 r средняя величина *f*; среднее значение *n*;
   среднее *n*

2380 **measurement framework**
(a scheme that establishes rules for control of
other components of a phenomenon that
permit the measurement of one component)
d Messung-Rahmenwerk *n*
f équipement *m* de mesure
r инструментарий *m* измерения

* **measurement of altitudes** → 104

* **measurement of heights** → 104

2381 **measure of an angle; angle measure;**
angular measure
d Winkelmaß *n*
f mesure *f* d'angle; mesure angulaire
r угловая мера *f*

2382 **measuring accuracy; accuracy of**
measurement
d Genauigkeit *f* der Messung
f précision *f* de mesure; exactitude *f* de mesure
r точность *f* измерений

2383 **measuring grid**
d Messunggitter *n*
f grille *f* de mesure
r мерная решётка *f*; палетка *f*

* **measuring in** → 3549

2384 **measuring point**
d Messpunkt *m*; Gleissicherungspunkt *m*;
Messstelle *f*
f point *m* de mesure
r точка *f* измерения; точка замера; замерный
пункт *m*

2385 **medial-axis transformation**
d Transformation *f* der Mittellinie
f transformation *f* d'axe médian
r преобразование *n* срединной оси (*в*
технике сжатия изображений)

2386 **median**
d Mediane *f*; Seitenhalbierende *f*
f médiane *f*
r медиана *f*

2387 **median filter**
d Median-Filter *m*
f filtre *m* médian
r медианный фильтр *m*

2388 **medium scale map**
d Karte *f* mittlerer Größe; mittelgroße Karte
f carte *f* à moyenne échelle; carte à échelle
sous-synoptique
r среднемасштабная карта *f*

* **memory** → 3500

2389 **mensuration**
d Vermessung *f*; Messkunde *f*
f mensuration *f*
r определение *n* размеров; измерение *n*

2390 **mental attribute**
d dem Geist zugeordnete Eigenschaft *f*
f attribut *m* mental
r ментальный атрибут *m*

2391 **menu**
d Menü *n*
f menu *m*
r меню *n*

2392 **menu item; menu point**
d Menüelement *n*; Menüpunkt *m*
f élément *m* de menu
r элемент *m* меню

* **menu point** → 2392

* **Mercator projection** → 3867

* **Mercator's projection** → 3867

2393 **merger; merging; concentrative operation**
d Fusion *f*; Unternehmenszusammenschluss *m*
f fusion *f*; fusionnement *m*
r слияние *n*

* **merging** → 2393

2394 **merging object**
d Verschmelzungsobjekt *n*
f objet *m* fusionné
r сходящийся объект *m*

2395 **meridian**
d Meridian *m*; Längenkreis *m*
f méridien *m*
r меридиан *m*

* **meridional projection** → 3806

2396 **mesh**
d Masche *f*
f filet *f*; maille *f*; réseau *m* de facettes;
maillage *m*
r [простая планарная] сеть *f*; сетка *f*

2397 **mesh adjacency**
d Maschen-Adjazenz *f*
f adjacence *f* de mailles
r смежность *f* сетей

2398 **mesh-edge incidence**
d Masche-Knoten-Inzidenz *f*

f incidence *f* maille-arête
r инцидентность *f* сети-ребра

* **mesh facet → 3545**

2399 **meshing**
 d Mischen *n*; Zusammenmischen *n*;
 Verschmelzen *n*
 f interclassement *m*
 r смешивание *n*

* **mesh point → 2529**

2400 **metacartography**
 d Metakartografie *f*
 f métacartographie *f*
 r метакартография *f*

2401 **metadata; metainformation**
 d Metadaten *pl*; Metainformation *f*
 f métadonnées *fpl*
 r метаданные *pl*

2402 **metadata base**
 d Metadatenbasis *f*
 f base *f* de métadonnées
 r база *f* метаданных

2403 **metafile**
 d Metadatei *f*
 f métafichier *m*
 r метафайл *m*

* **metainformation → 2401**

2404 **metainformation system**
 d Metainformationssystem *n*
 f système de méta-information
 r метаинформационная система *f*

2405 **metamorphism**
 d Metamorphismus *m*
 f métamorphisme *m*
 r метаморфизм *m*

* **meteorological map → 3982**

2406 **meteorological model**
 d meteorologisches Modell *n*
 f modèle *m* météorologique
 r метеорологическая модель *f*

2407 **metes and bounds method**
 d Abmarkungsmethode *f*
 f méthode *f* de la projection horizontale;
 méthode de description par tenants et
 aboutissants
 r метод *m* описания пределов земельного
 участка в терминах протяженности и
 векторов границ

* **method of area → 2408**

2408 **method of area [symbols]**
 d Flächensymbole-Methode *f*
 f méthode *f* [de présentation de symboles] de
 régions
 r способ *m* картографического изображения
 ареалов

2409 **method of [cartographic] symbols**
 d Methode *f* der kartografischen Symbole
 f méthode *f* [de présentation] de symboles
 cartographiques
 r способ *m* [картографического
 изображения] значков

2410 **method of isolines; isogram method;**
 isopleth method
 d Isolinien-Methode *f*
 f méthode *f* d'isolignes
 r способ *m* [картографического
 изображения] изолиний

2411 **method of line symbols**
 d Liniensymbol-Methode *f*
 f méthode *f* [de présentation] de lignes
 r способ *m* линейных знаков

2412 **method of motion symbols; method of**
 vectors
 d Methode *f* der Bewegungssymbole
 f méthode *f* [de présentation] de symboles de
 mouvement
 r способ *m* картографического изображения
 знаков движения

2413 **method of qualitative background**
 d Methode *f* des qualitativen Hintergrunds
 f méthode *f* d'arrière-plan qualitatif
 r способ *m* качественного фона

2414 **method of quantitative background**
 d Methode *f* des quantitativen Hintergrunds
 f méthode *f* d'arrière-plan quantitatif
 r способ *m* количественного фона

* **method of symbols → 2409**

* **method of vectors → 2412**

2415 **metric system**
 d metrisches System *n*
 f système *m* métrique
 r метрическая система *f*

2416 **metropolitan area; MA; metropolitan**
 region
 d Großstadtbereich *m*; Stadtgebiet *n*

f aire *f* métropolitaine; région *f* métropolitaine
r метропольный ареал *m*; площадь *f* города с пригородами; город *m* с пригородами

2417 metropolitan boundary
d Stadtgrenze *f*
f limite *f* métropolitaine
r граница *f* метропольного района

* **metropolitan region** → 2416

* **MGRS** → 2421

2418 microfilm-plotter; photographic film recorder; photoplotter; graphics film recorder; film recorder; video hard copy unit
d Fotoplotter *m*
f phototraceur *m*; système *m* de phototraçage
r микрофильм-плоттер *m*; фотоплоттер *m*

2419 microwave carrier signal
d Mikrowellenträgersignal *n*
f signal *m* convoyeur de micro-ondes; porteuse *f* du signal de micro-ondes
r микроволновый несущий сигнал *m*

2420 military atlas
d militärer [geografischer] Atlas *m*
f atlas *m* militaire
r военный атлас *m*

2421 military grid reference system; MGRS
(an extension of the UTM system)
d Militärgitterreferenzsystem *n*
f carroyage *m* de référence militaire
r военная система *f* прямоугольных координат

2422 millimeterwave atmospheric sounder; MAS
d atmosphärische Millimeterwellensonde *f*
f sondeur *m* atmosphérique d'onde millimétrique
r атмосферный зонд *m* миллиметрового диапазона

* **mine-survey** → 2427

* **mine surveying science** → 2427

2423 minimum bounding rectangle
d minimales begrenztes Rechteck *n*; kleinstes achsenparalleles umschließendes Rechteck
f boîte *f* délimitée minimale
r минимальный ограничительный прямоугольник *m*

2424 minimum distance
d minimale Distanz *f*

f distance *f* minimale
r минимальное расстояние *n*

2425 minimum distance classifier
d Klassifikator *m* mit minimaler Distanz
f classificateur *m* à distance minimale
r классификатор *m*, построенный по критерию минимального расстояния

2426 minimum distance mapping
d Minimal-Distanz-Abbildung *f*
f mappage *m* de distance minimale
r отображение *n* минимального расстояния

2427 mining geodesy; mine surveying science; mine-survey
d Markscheidekunde *f*
f géodésie *f* de mines; topographie *f* minière
r маркшейдерское дело *n*

2428 minute
d Minute *f*
f minute *f*
r минута *f*

* **mire** → 368

* **mirroring** → 3433

* **mirror reflection** → 3433

* **miscalculation** → 658

2429 misclassification
d Missklassifikation *f*; Klassifikationsfehler *m*
f classification *f* erronée; erreur *f* de classification
r ошибочная классификация *f*; неправильная классификация; ошибка *f* классификации

2430 misclassification matrix
d Missklassifikationsmatrix *f*
f matrice *f* de classifications erronées
r матрица *f* ошибок классификации

* **misclosure** → 1319

2431 misclosure distance
d Offenlegungsdistanz *f*; Offenbarungsdistanz *f*; Enthüllungsdistanz *f*
f distance *f* d'écart de fermeture
r расстояние *n* незамкнутости; расстояние невязки

* **mislabeled** → 3869

2432 mislabeled polygon
d nichtetikettiertes Polygon *n*; ungekennzeichnetes Polygon

f polygone *m* non étiqueté
r непомеченный полигон *m*

* **mislabelled → 3869**

2433 missing node
d fehlender Knoten *m*
f nœud *m* manquant
r отсутствующий узел *m*

2434 missing scan lines; dropped scan lines
d fehlende Bildzeilen *fpl*; herausgefallene
  Abtastzeilen *fpl*
f lignes *fpl* de balayage manquantes
r отсутствующие полосы *fpl* сканирования;
  потерянные строки *fpl* развёртки

2435 mixed element; mixed pixel; mixel
d vermischtes Pixel *n*
f élément *m* mixte; pixel *m* mixte
r миксел *m*

* **mixed pixel → 2435**

2436 mixed pixel problem
d Problem *n* der vermischten Pixel
f problème *m* des pixels mixtes
r задача *f* о микселах

* **mixel → 2435**

2437 mobile GIS
d mobiles GIS *n*
f SIG *m* mobil
r мобильная ГИС *f*

* **modal noise → 3425**

* **modal split → 2438**

2438 modal split[ting]
d Verkehrsmittelwahl *f*; Verkehrsteilung *f*;
  Modentrennung *f*
f répartition *f* entre les modes [de transport];
  fractionnement *m* modal; séparation *f* de
  modes
r расщепление *n* мод; селекция *f* мод

* **mode filter → 3978**

2439 model area
  (the area of overlap in a pair of stereo images
  that is selected for creation of a DEM)
d Modellraum *m*
f surface *f* de modèle
r модельная площадь *f*; модельное
  пространство *n*

**2440 modeling and cognitive conception;
  gnosiological conception**

d gnosiologische Konzeption *f*
f conception *f* gnoséologique
r модельно-познавательная концепция *f*

2441 modeling of volumes
d Volumenmodellierung *f*
f modelage *m* de volumes
r моделирование *n* объёмов

* **mode of cartographic representation →
  2244**

2442 modifiable areal unit problem
d Problem *n* der Gebietseinheit
f problème *m* de l'unité territoriale modifiable
r задача *f* об изменяемой единице площади

* **modified cylindrical projection → 2939**

2443 modified projection
d modifizierte Projektion *f*
f projection *f* modifiée
r модифицированная проекция *f*

2444 Mollweide projection
d Mollweide-Projektion *f*
f projection *f* de Mollweide
r проекция *f* Молвайда

2445 monitor station
d Monitorstation *f*
f poste *m* de contrôle
r контрольный пост *m*

* **monochrome aerial photograph → 359**

* **montage → 2453**

* **monument → 386**

* **monumentation → 2358**

2446 morphological filter
d morphologischer Filter *m*
f filtre *m* morphologique
r морфологический фильтр *m*

2447 morphology
d Morphologie *f*
f morphologie *f*
r морфология *f*

**2448 morphometric index; morphometric
  parameter**
d morphometrischer Parameter *m*
f indice *m* morphométrique
r морфометрический показатель *m*

* **morphometric parameter → 2448**

2449 morphometry
  d Morphometrie *f*
  f morphométrie *f*
  r морфометрия *f*

2450 Morton matrix
  d Morton-Matrix *f*
  f matrice *f* de Morton
  r матрица *f* Мортона

2451 Morton order
  d Morton-Ordnung *f*
  f ordre *m* de Morton
  r порядок *m* Мортона

  * **mosaic** → **62**

  * **mosaicing** → **2286**

  * **MOSS format** → **2292**

  * **motif** → **2716**

  * **mount** → **2453**

2452 mountain[eous] region
  d bergige Region *f*; gebirgige Region
  f région *f* montagneuse
  r горная область *f*

  * **mountain region** → **2452**

2453 mount[ing]; montage
  d Montage *f*
  f montage *m*
  r монтаж *m*; компоновка *f*

2454 moving of symbol; displacement of symbol
  d Symbolverschiebung *f*
  f déplacement *m* de symbole
  r перемещение *n* знака

  * **MSL** → **2378**

  * **MSS** → **2477**

  * **multiband** *adj* → **2455**

2455 multichannel; multiline; multiband *adj*
  d Mehrkanal-; mehrkanalig
  f multicanal
  r многозональный; многоканальный

2456 multichannel receiver
  d Mehrkanalempfänger *m*
  f récepteur *m* multicanal
  r многозональный приёмник *m*;
      многоканальный приёмник

2457 multi-criteria analysis

  d Mehrkriterienanalyse *f*
  f analyse *f* multicritère
  r анализ *m* по множеству критериев

2458 multi-criteria decision-making
  d Mehrkriterienentscheidungsfindung *f*
  f prise *f* de décision multicritère
  r принятие *n* решений по множеству
      критериев

2459 multidimensional scaling
  d mehrdimensionale Skalierung *f*
  f mise *f* à échelle multidimensionnelle
  r многомерное масштабирование *n*

2460 multidimensional space
  d mehrdimensionaler Raum *m*
  f espace *m* multidimensionnel
  r многомерное пространство *n*

  * **multilayered representation** → **2119**

  * **multiline** → **2455, 2464**

  * **multilook image** → **2978**

2461 multimedia GIS
  d Multimedia-GIS *n*
  f SIG *m* multimédia
  r мультимедийная ГИС *f*

2462 multipath
  d Mehrstrecke *f*; Mehrweg *m*
  f trajet *m* multiple
  r многолучевость *f*; многопутность *f*

2463 multipath error
  d Mehrstreckenfehler *m*
  f erreur *f* de trajets multiples
  r погрешность *f* многолучевости

  * **multiple image** → **1172**

2464 multiple line; multiline
  d Mehrleitung *f*
  f multiligne *f*
  r мультилиния *f*; линия *f* из нескольких
      параллельных линий; многопроводная
      линия

2465 multiple regression
  d mehrfache Regression *f*; Mehrfachregression *f*
  f régression *f* multiple
  r множественная регрессия *f*

  * **multiple representation** → **2470**

2466 multiplexing receiver
  d Multiplexempfänger *m*

f récepteur *m* de multiplexage
r мультиплексный приёмник *m*

2467 multipurpose cadaster
  d Kataster *m* mit verschiedenen Verwendungen;
     Vielzweckkataster *m*;
     Multifunktionskataster *m*
  f cadastre *m* à objectifs multiples; cadastre
     multifonction
  r многоцелевой кадастр *m*;
     многофункциональный кадастр

2468 multipurpose geographic data system
  d Geodatensystem *n* mit verschiedenen
     Verwendungen
  f système *m* de données géographiques à
     objectifs multiples
  r многоцелевая система *f* географических
     данных

2469 multiscale GIS
  d Mehrskalen-GIS *n*; Multiskalen-GIS *n*
  f SIG *m* multi-échelle
  r полимасштабная ГИС *f*;
     масштабно-независимая ГИС

**2470 multiscale representation; multiple
     representation**
  d Mehrskalendarstellung *f*
  f représentation *f* multi-échelle
  r полимасштабное представление *n*;
     множественное представление

2471 multisegment path
  d Multisegment-Weg *m*
  f chemin *m* multisegment
  r многосегментный пут *m*; многосегментная
     траектория *f*

2472 multisensor image
  d Multisensorbild *n*
  f image *f* multicapteur
  r многосенсорное изображение *n*;
     изображение нескольких датчиков

2473 multispectral data
  d multispektrale Daten *pl*
  f données *fpl* multispectraux
  r многоспектральные данные *pl*

2474 multispectral electrooptical imaging sensor
  d multispektraler elektrooptischer Bildsensor *m*
  f senseur *m* électrooptique multispectral
  r многоспектральный электрооптический
     датчик *m*

2475 multispectral photography
  d multispektrale Fotografie *f*
  f photographie *f* multispectrale
  r многоспектральный снимок *m*

2476 multispectral pixel-to-pixel classification
  d multispektrale Pixel-zu-Pixel-Klassifizierung *f*
  f classification *f* multispectrale pixel à pixel
  r многоспектральная пиксельная
     классификация *f*

* multispectral resource sampler → 2477

* multispectral scanner → 2477

**2477 multispectral scanning system; MSS;
     multispectral scanner; multispectral
     resource sampler**
  d Multispektralscanner *m*; multispektraler
     Abtaster *m*
  f scanner *m* multispectral; explorateur *m*
     multispectral
  r многоспектральное сканирующее
     устройство *n*; МСУ; многоспектральное
     опробирующее устройство

2478 multi-tile
  d Mehrmusterkachel *f*
  f multi-mosaïque *f*
  r множественный элемент *m* мозаичного
     изображения

2479 multi-valued surface
  d mehrwertige Fläche *f*
  f surface *f* à plusieurs valeurs; surface
     multi-valuée
  r многозначная поверхность *f*

2480 multi-variable thematic map
  d multivariate Themakarte *f*
  f carte *f* thématique à variables multiples
  r тематическая карта *f* на базе множества
     переменных

2481 mutual azimuths
  d gegenseitige Azimute *mpl*
  f azimuts *mpl* mutuels
  r взаимные азимуты *mpl*

N

2482 nadir
   d Fußpunkt *m*; Nadir *m*
   f nadir *m*
   r надир *m*; самый низкий уровень *m*

2483 name
   d Name *m*
   f nom *m*
   r имя *n*

2484 names overlay; names plate
   d Namenoriginal *n*
   f planche *f* d'écritures
   r оригинал *m* надписей

   * **names plate** → 2484

2485 naming; denomination
   d Namengebung *f*; Benennung *f*
   f dénomination *f*; désignation *f*; nommage *m*
   r наименование *n*; именование *n*; присваивание *n* имён

2486 naming convention
   d Benennung-Konvention *f*
   f convention *f* de nommage
   r соглашение *n* о наименовании

   * **NAT** → 2516, 2535

2487 national atlas
   d Nationalatlas *m*; Landesatlas *m*; landeskundlicher Atlas *m*
   f atlas *m* national
   r национальный атлас *m*

2488 national cadastral information system
   d amtliches Liegenschaftskataster-Informationssystem *n*; ALKIS
   f système *m* d'information national cadastral
   r государственная кадастральная информационная система *f*

2489 national control point information system
   d amtliches Festpunktinformationssystem *n*; AFIS
   f système *m* d'information national de points d'appui
   r государственная информационная система *f* пунктов плановой сети

2490 national geodetic vertical datum
   d Landeshorizont *m*
   f datum *m* vertical géodésique national
   r национальный исходный горизонт *m*; национальное начало *n* отсчёта высоты

2491 national geospatial database
   d nationale Geodatenbasis *f*
   f base *f* de données géospatiaux nationale
   r национальная база *f* геопространственных данных

   * **national grid** → 2492

2492 national grid [system]
   d nationales Verbundnetz *n*
   f réseau *m* maillé national
   r национальная система *f* координат

2493 national map accuracy standards; NMAS
   d nationale Kartenpräzisionstandards *mpl*
   f standards *mpl* de précision cartographique
   r национальные стандарты *mpl* точности картографирования

2494 National Oceanic and Atmospheric Administration; NOAA (US)
   d NOAA
   f Administration *f* nationale des océans et de l'atmosphère
   r Национальное управление *n* по освоению океана и атмосферы США

2495 national transfer format (UK)
   d nationales Übertragungsformat *n*; NTF-Format *n*
   f format *m* de transfert national; format NTF
   r национальный британский формат *m* обмена [векторными данными]; формат NTF

2496 native mode
   d nativer Modus *m*
   f mode *m* naturel
   r режим *m* работы в собственной системе команд

   * **natural break** → 2497

2497 natural break [algorithm]
   d Naturbruchalgorithmus *m*
   f algorithme *m* de pause naturelle
   r алгоритм *m* естественного перерыва

2498 natural environment
   d natürliche Umwelt *f*; Naturlandschaft *f*
   f environnement *m* naturel
   r естественное окружение *n*; естественная среда *f*

2499 natural networks
   d natürliche Netzwerke *npl*

f réseaux *mpl* naturels
r естественные сети *fpl*

2500 natural phenomena modeling
    d Modellierung *f* der Naturerscheinungen
    f modelage *m* de phénomènes naturels
    r моделирование *n* естественных явлений

    * **natural resource inventory** → **2501**

2501 natural resource[s] inventory
    d Inventarisierung *f* der natürlichen Ressourcen
    f inventaire des ressources naturelles
    r кадастр *m* природных ресурсов

2502 natural resources management
    d Verwaltung *f* der natürlichen Ressourcen;
    natürliches Ressourcenmanagement *n*
    f gestion *f* de ressources naturelles
    r управление *n* природных ресурсов

    * **natural scale** → **3139**

    * **nautical chart** → **1862**

2503 navigation
    d Navigation *f*
    f navigation *f*
    r навигация *f*; передвижение *n*

2504 navigation system
    d Navigationssystem *n*
    f système *m* de navigation
    r навигационная система *f*

    * **navigator** → **3963**

2505 navigator
    (a handheld GPS unit, capable of tracking
    GPS satellites)
    d Navigator *m*
    f navigateur *m*
    r навигатор *m*

2506 nearest neighbor
    d nächster Nachbar *m*
    f voisin *m* le plus proche
    r ближайшая соседняя запись *f* (*в базе*
    данных); ближайший сосед *m*

2507 nearest neighbor algorithm
    d Nächster-Nachbar-Algorithmus *m*
    f algorithme *m* du plus proche voisin
    r алгоритм *m* ближайшего соседа

2508 nearest neighbor analysis
    d nächste Nachbarschaftsanalyse *f*
    f analyse *f* de voisinage le plus proche
    r поиск *m* ближайшего соседа

    * **nearest neighbor interpolation** → **2509**

2509 nearest neighbor sampling; nearest
    neighbor interpolation
    d Nächster-Nachbar-Abtastung *f*
    f échantillonnage *m* du plus proche voisin
    r сканирование *n* ближайшего соседа

2510 nearest node
    d nächster Knoten *m*
    f nœud *m* le plus proche
    r ближайший узел *m*

    * **nearness** → **2929**

    * **neat border** → **2511**

2511 neatline; neat border
    (a border line around a map)
    d unvermischte Kante *f*
    f bordure *f* pure; bordure de finition; bord *m* de
    dessin
    r чистая кайма *f*; чистое обрамление *n*;
    чистый бордюр *m*; чистая окантовка *f*

2512 neighborhood analysis; proximity analysis;
    proximal analysis
    d Nachbarschaftsanalyse *f*; Näherungsanalyse *f*
    f analyse *f* de voisinage
    r анализ *m* близости; анализ соседства;
    анализ окрестностей

2513 neighborhood operator
    d Näherungsoperator *m*
    f opérateur *m* de voisinage
    r оператор *m* формирования окрестностей
    [элемента изображения]

2514 neighboring points; adjacent points;
    surrounding points
    d Nachbarpunkte *mpl*
    f points *mpl* voisins
    r соседние точки *fpl*; близлежащие точки

    * **neighborhood** → **2929**

    * **neighbourhood** → **2929**

    * **nesting** → **1259**

2515 network address
    d Netz[werk]adresse *f*
    f adresse *f* de réseau
    r сетевой адрес *m*

2516 network address translation; NAT
    d Netzadresse-Translation *f*
    f translation *f* d'adresse de réseau
    r трансляция *f* сетевого адреса

2517 network analysis
 d Netzwerkanalyse *f*
 f analyse *f* de réseaux
 r сетевой анализ *m*; анализ сетей

2518 network calculation
 d Netzberechnung *f*
 f calcul *m* de réseau
 r расчёт *m* сети; вычисление *n* сети

2519 network cartographic database
 d kartografische Netzdatenbasis *f*
 f base *f* de données de réseau cartographique
 r сетевая картографическая база *f* данных

2520 network coverage
 d Netzerfassung *f*
 f couverture *m* de réseau
 r охват *m* сети

2521 network data model
 d Netzwerkdatenmodell *n*
 f modèle *m* de données de type réseau
 r модель *f* сетевых данных

2522 network densification
 d Netzverdichtung *f*
 f densification *f* de réseau
 r уплотнение *n* сети

2523 network element
 d Netzwerkelement *n*
 f élément *m* de réseau
 r элемент *m* сети; сетевой элемент

2524 network file system
 d Netzwerkdateisystem *n*
 f système *m* de fichiers de réseau
 r сетевая файловая система *f*

2525 network framework; network structure
 d Netzstruktur *f*
 f structure *f* de réseau
 r [инфра]структура *f* сети; каркас *m* сети

2526 network information system
 d Netzwerkinformationssystem *n*
 f système *m* d'information réseau
 r сетевая информационная система *f*

2527 network link
 d Netzverbindung *f*
 f liaison *f* de réseau
 r сетевая связь *f*

2528 network map
 d Netz[werk]karte *f*
 f carte *f* de réseau
 r сетевая карта *f*; карта сети

2529 network node; network point; mesh point
 d Netz[werk]knoten *m*
 f nœud *m* de réseau; point *m* de réseau
 r узел *m* сети; точка *f* сетки

2530 network of control stations; control net; geodetic [control] net[work]; geodetic control; horizontal control; horizontal net; plane control; frame; framework of control points
 d geodätisches Grundnetz *n*; geodätisches Netz *n*; trigonometrisches Festpunktfeld *n*; Festpunktnetz *n*; geodätische Grundlage *f*
 f réseau *m* [géodésique]; réseau de points d'appui; canevas *m* [géodésique]; système *m* géodésique
 r [плановая] геодезическая сеть *f*; опорная геодезическая сеть; сеть опорных точек

* **network of coordinates** → 762

* **network of heights** → 2140

* **network point** → 2529

* **network structure** → 2525

* **network topology** → 3764

2531 network tracing
 d Netztrassieren *n*; Netzverfolgung *f*
 f traçage *m* de réseau; tracement *m* de réseau
 r трассировка *f* сети

2532 neutral point; vanishing point; VP
 d Verschwindungspunkt *m*; Fluchtpunkt *m*
 f point *m* neutre; point de fuite; PF
 r нейтральная точка *f*

* **NMAS** → 2493

* **NOAA** → 2494

* **nodal degree** → 3896

* **nodal region** → 1490

2533 node; junction
 (of a graph)
 d Knoten *m*
 f nœud *m*
 r узел *m*

2534 node adjacency
 d Knoten-Adjazenz *f*
 f adjacence *f* de nœuds
 r смежность *f* узлов

2535 node attribute table; NAT
 d Knotenattributtabelle *f*

f table f d'attributs de nœuds
r таблица f атрибутов узлов

2536 node-edge incidence
d Knoten-Kanten-Inzidenz f
f incidence f nœud-arête
r инцидентность f узла-ребра

* **nodeid → 2537**

2537 node ID; nodeid
d Knotenidentifikator m
f code m d'identification de nœud;
   identificateur m de nœud
r идентификатор m узла

* **node line → 2186**

2538 node match
d Knotenanpassung f
f appariement m de nœuds
r совпадение n узлов; совмещение n узлов

2539 node match tolerance
d Knotenanpassungstoleranz f
f tolérance f d'appariement de nœuds
r допуск m совпадения узлов; допуск
   совмещения узлов

2540 node neighborhood
d Knoten-Nachbarschaft f; Knoten-Näherung f
f voisinage m de nœud
r окрестность f узла

* **node valency → 3896**

2541 noise contour
d Lärmkontur f
f contour m isopsophique; carte f de bruit
r контур m распространения шума

**2542 noise elimination; noise suppression; noise
      removal**
d Entfernung f des Rauschens
f élimination f de bruit
r устранение n искажения

* **noise removal → 2542**

* **noise suppression → 2542**

2543 nomenclature
d Nomenklatur f; Namensregister n;
   Fachbezeichnung f
f nomenclature f
r номенклатура f

2544 nominal
   (a level of measurement based assigning
   objects into discrete categories)

d Nennwert m
f nominal m
r номинал m

2545 nominal orbit plane
d nominelle Orbitalebene f
f plan m d'orbite nominal
r номинальная плоскость f орбиты

* **nominal scale → 2892**

2546 nomogram; alignment chart
d Nomogramm n
f nomogramme m
r номограмма f

**2547 non-deterministic polynomial time
      complete problem; NP-complete problem**
d NP-vollständiges Problem n
f problème m non déterministe polynomial
   complet; problème NP complet
r НП-полная задача f (о классе
   комбинаторных задач с нелинейной
   полиномиальной оценкой числа вариантов)

2548 non-Earth coordinate system
d Non-Earth-Koordinatensystem n
f système m de coordonnées non terrestres
r система f негеографических координат

2549 non-exact interpolator
d nichtexakter Interpolator m
f interpolateur m non exact
r неточный интерполятор m

* **nonformatted data → 3852**

2550 nonisometric line
d nichtisometrische Linie f
f ligne f non isométrique
r неизометрическая линия f

* **nonisotropic → 132**

2551 nonlinear kriging
d nichtlineares Kriging n
f krigeage m non linéaire
r нелинейный кригинг m

2552 nonmetropolitan area
d nichtgroßstädtisches Gebiet n
f aire f non métropolitaine; région f non
   métropolitaine
r неметропольный ареал m; пригородный
   ареал

2553 nonplanar intersection
d nichtplanarer Schnitt m; nichtplanare
   Kreuzung f; nichtplanarer Knoten m

f intersection *f* non planaire
r неплоское пересечение *n*

2554 nonplanar network
   d nichtplanares Netzwerk *n*
   f réseau *m* non planaire
   r непланарная сеть *f*, непланарная цепь *f*,
    непланарная схема *f*

   * **nonpoint emission source** → 165

   * **nonpoint source** → 165

   * **non sharp** → 1495

   * **non-spatial** → 190

   * **non-spatial data** → 192

   * **non-spatial object** → 193

2555 non-stationarity
   d Nichtstationarität *f*
   f non-stationnarité *f*
   r нестационарность *f*

   * **normal** → 2560

   * **normal aspect of a map projection** → 2562

2556 normal distribution; Gauss[-Laplace] distribution; Gaussian distribution
   d Normalverteilung *f*; Gaussverteilung *f*;
    Gausssche Verteilung *f*
   f distribution *f* normale; distribution de
    [Laplace-]Gauss; distribution gaussienne
   r нормальное распределение *n*;
    распределение Гаусса; гауссово
    распределение

2557 normal form
   d Normal[en]form *f*
   f forme *f* normale
   r нормальная форма *f*; нормальный вид *m*

2558 normal height
   d normale Höhe *f*
   f hauteur *f* normale
   r нормальная высота *f*

2559 normalization
   d Norm[alis]ierung *f*; Normung *f*
   f normalisation *f*
   r нормализация *f*; норм[ализ]ирование *n*

2560 normal [line]
   d Normale *f*
   f normale *f*
   r нормаль *f*

2561 normal polarity
   d normale Polarität *f*
   f polarité *f* normale
   r прямая полярность *f*

2562 normal projection; normal aspect of a map projection
   d Normalprojektion *f*
   f projection *f* normale
   r нормальная проекция *f*

2563 normal slicing
   d normales Schneiden *n*
   f tranchage *m* normal
   r нормальное разделение *n* на слои

2564 north
   d Norden *m*
   f nord *m*
   r север *m*

2565 north arrow
   d Nordpfeil *m*
   f flèche *f* du Nord; flèche d'orientation
   r стрелка-указатель *f* "север-юг"

2566 northeast
   d Nordosten *m*
   f nord-est *m*
   r северо-восток *m*

2567 northern
   d nördlich
   f du nord; septentrional; boréal
   r северный

2568 northing; equatorial distance
   d Hochwert *m*
   f ordonnée *f* [d'un quadrillage]; coordonnée *f*
    rectangulaire dans la direction S-N; valeur *f* de
    l'ordonnée dans le système Gauss-Krüger
   r нордовая разность *f* широт; величина *f*
    ординаты [по карте]

2569 North pole
   d Nordpol *m*
   f pôle *m* Nord; pôle boréal
   r Северный полюс *m*

2570 northwest
   d Nordwest *m*
   f nord-ouest *m*
   r северо-запад *m*

   * **NP-complete problem** → 2547

2571 nucleus of population; population nucleus
   d Siedlungskern *m*
   f noyau *m* de peuplement
   r ядро *n* популяции

2572 **null cell**
    (in topology)
  d Nullzelle *f*
  f cellule *f* nulle
  r нулевая клетка *f*

2573 **null hypothesis**
  d Nullhypothese *f*
  f hypothèse *f* nulle
  r нулевая гипотеза *f*

2574 **null value**
  d Nullwert *m*
  f valeur *f* zéro
  r нулевое значение *n*

2575 **numbering**
  d Numerierung *f*
  f numérotage *m*
  r нумерование *n*; нумерация *f*; разграфка *f*

2576 **numbering area**
  d Numerierungsbereich *m*
  f zone *f* de numérotage
  r зона *f* нумерации

2577 **numbering area code; numbering zone code**
  d NA-Code *m*; Vorwahl-Code *m*;
    Vorwahlnummer *f*
  f indicatif *m* de zone de numérotage
  r код *m* зоны нумерации

  * **numbering zone code** → 2577

2578 **number of control points**
  d Steuerpunktzahl *f*; Zahl *f* der Steuerpunkte
  f nombre *m* de points de contrôle
  r число *n* контрольных точек

2579 **number of regions**
  d Regionenzahl *f*; Zahl *f* der Regionen
  f nombre *m* de régions
  r число *n* регионов

2580 **number of rows; row number; row count**
    (of a rectangular array or of a table)
  d Zeilenzahl *f*; Zeilenanzahl *f*; Zahl *f* der Zeilen
  f nombre *m* de lignes; nombre de rangées
  r число *n* строк

  * **numeric** → 1031

  * **numerical** → 1031

2581 **numeric attribute**
  d Zahlenattribut *n*
  f attribut *m* numérique
  r числовой атрибут *m*

2582 **numeric axes**

  d numerische Achsen *fpl*
  f axes *mpl* numériques
  r цифровые оси *fpl*

  * **numeric data** → 1040

  * **numeric image** → 1051

2583 **numeric taxonomy; digital taxonomy**
  d numerische Taxonomie *f*
  f taxonomie *f* numérique
  r цифровая таксономия *f*

O

2584 object
d Objekt *n*
f objet *m*
r объект *m*

2585 object class
d Objektklasse *f*
f classe *f* d'objets
r класс *m* объектов

2586 object code
d Objekt-Code *m*
f code d'objet
r код *m* объекта

* object combination → 2587

2587 object combining; object combination
d Objektkombination *f*
f combinaison *f* d'objets
r комбинирование *n* объектов

2588 object framework
d Objektgestell *n*
f structure *f* d'objets
r структура *f* объектов

2589 object geometry
d Objektgeometrie *f*
f géométrie *f* d'objet
r геометрия *f* объекта

2590 object identifier
d Objektschlüssel *m*
f identificateur *m* d'objet
r идентификатор *m* объекта

2591 objective function
d Zielfunktion *f*
f fonction *f* d'objectif; fonction objective
r целевая функция *f*; функция цели

2592 object key catalog; OSKA
d Objektschlüsselkatalog *m*
f catalogue *m* de clés d'objets
r каталог *m* объектных ключей

2593 object management group; OMG
d Objektmanagement-Gruppe *f*
f groupe *m* de gestion des objets
r группа *f* по управлению объектами

2594 object model

d Objektmodell *n*
f modèle *m* d'objet
r модель *f* объекта

2595 object orientation
d Objektorientierung *f*; Orientierung *f* des Objekts
f orientation *f* d'objet
r ориентация *f* объекта

2596 object-oriented database
d objektorientierte Datenbasis *f*
f base *f* de données orientée objet
r объектно-ориентированная база *f* данных

2597 object-oriented data modeling
d objektorientierte Datenmodellierung *f*
f modelage *m* de données orienté objet
r объектно-ориентированное моделирование *n* данных

2598 object-oriented geoprocessing
d objektorientierte Verarbeitung *f* von geografischen Daten
f traitement *m* de données géographiques orienté objet
r объектно-ориентированная обработка *f* данных, связанных с науками о Земле

2599 object representation
d Objektdarstellung *f*
f représentation *f* d'objet
r представление *n* объекта

2600 object's nodes
d Objektknoten *mpl*
f nœuds *m* d'objet
r узлы *mpl* объекта

2601 oblique aerial photograph; perspective aerial photograph
d perspektivische Aerofotografie *f*
f aérophoto *f* perspective
r перспективный аэрофотоснимок *m*

2602 oblique hill shading
d Schrägschummerung *f*
f estompage *m* d'ombre oblique
r косая отмывка *f* возвышений

2603 oblique projection; skew projection; inclined projection
d schiefwinklige Projektion *f*; schiefe Projektion; Schrägprojektion *f*; gewöhnliche Abbildung *f*
f projection *f* oblique
r косая проекция *f*; косоугольная проекция; скошенная проекция; наклонная проекция

2604 oblique shading
d Schrägschattierung *f*

f ombrage *m* oblique
r отмывка *f* при боковом освещении

2605 oblique view
d schiefe Ansicht *f*
f vue *f* oblique
r косой вид *m*

2606 obliquity; skewness
d Verirrung *f*; Schrägheit *f*
f obliquité *f*
r косое направление *n*; конусность *f*;
наклонение *n*

2607 observation
d Beobachtung *f*; Observation *f*
f observation *f*
r наблюдение *n*

2608 observational error; error of observation
d Beobachtungsfehler *m*
f erreur *f* d'observation
r ошибка *f* наблюдения; погрешность *f* при
наблюдении

* **occupancy → 2609**

2609 occupation; occupancy; seizure; seizing
d Besetzung *f*; Belegung *f*; Ausnutzung *f*;
Nutzung *f*
f occupation *f*; encombrement *m*;
engagement *m*; prise *f*
r занятие *n*; занятость *f*

2610 oceanographic atlas
d Meeresatlas *m*
f atlas *m* océanographique
r океанографический атлас *m*

2611 oceanographic model
d ozeanografisches Modell *n*
f modèle *m* océanographique
r океанографическая модель *f*

2612 oceanographic survey
d Hochseevermessung *f*; ozeanografische
Erhebung *f*
f levé *m* océanographique; relevé *m*
océanographique
r океанская съёмка *f*

* **octatree → 2613**

2613 octtree; octatree
d Oktanenbaum *m*; Oktagonbaum *m*; Oktbaum *n*
f arbre *m* d'octants
r октотомическое дерево *n*; октарное дерево;
октодерево *n*; дерево октантов

* **ODBC → 2629**

2614 official map
(the legal adopted map of streets and
highways)
d Dienstkarte *f*; amtliche Karte *f*
f carte *f* officielle
r служебная карта *f*

2615 offline; autonomous
d abgetrennt; getrennt; autonom; Offline-
f hors ligne; autonome; offline
r автономный; независимый; офлайн

* **offset → 1126**

2616 offshore
d ablandig; Hochsee-
f au large [des côtes]; en mer; off-shore
r за пределами морской границы; за
границей; заокеанский; оффшорный

* **OGC → 2631**

* **OGIS → 2630**

* **OMG → 2593**

2617 OMG standard
d OMG-Standard *m*
f standard *m* OMG
r стандарт *m* OMG

2618 one-dimensional
d eindimensional
f à une dimension
r одноразмерный

2619 one-dimensional datum
d eindimensionales Datum *n*
f datum *m* à une dimension
r одноразмерное начало *n* отсчёта;
одноразмерный репер *m*; одномерный
датум *m*

2620 one-look image; single-look image
d Einbild *n*
f image *f* simple
r однопоисковое изображение *n*; простое
изображение

2621 one-to-one relationship
d 1:1-Beziehung *f*
f association *f* de type 1:1
r соотношение *n* один-к-одному

2622 one-variable thematic map
d einvariable Themakarte *f*
f carte *f* thématique à variable unique
r тематическая карта *f* одной переменной

2623 online access
d Online-Zugriff *m*

f accès *m* en ligne
r диалоговый доступ *m*; сетевой доступ

2624 online catalog
d Online-Katalog *m*
f catalogue *m* en ligne
r диалоговый каталог *m*; сетевой каталог

2625 online mapping; interactive mapping
d Online-Kartografie *f*; interaktive Kartierung *f*
f cartographie *f* interactive
r диалоговое картографирование *n*

2626 online query
d Online-Abfrage *f*
f interrogation *f* interactive
r диалоговый запрос *m*

2627 on-screen digitizing; "heads-up" digitizing
    (a digitizing station that provides a graphical
    user interface on the screen of a workstation)
d Bildschirm-Digitalisierung *f*
f digitalisation *f* d'écran
r экранное оцифрование *n*

2628 open area; field
d offene Fläche *f*
f espace *m* découvert; lieu *m* découvert; zone *f* découverte
r открытая площадь *f*; открытый участок *m*

2629 open database connectivity; ODBC
d offene Datenbankschnittstelle *f*
f connexion *f* aux bases de données ouverte
r открытая связь *f* с базами данных; интерфейс *m* ODBC

2630 open geodata interoperability specification; OGIS
d Spezifikation *f* für offene Geodaten-Interoperabilität
f spécification *f* d'interopérabilité ouverte de géodonnées; norme *f* d'interopérabilité de systèmes d'information géographique
r спецификация *f* открытой интероперабельности географических данных

2631 Open GIS Consortium; OGC
d Open-GIS-Konsortium *n*
f Consortium *m* de SIG ouverts
r открытый ГИС Консорциум *m*

2632 open system
d offenes System *n*
f système *m* ouverte
r открытая система *f*

* **open traverse** → 3843

2633 operating system; OS
d Operationssystem *n*; Betriebssystem *n*
f système *m* d'exploitation
r операционная система *f*

2634 operational constellation
d Betriebskonstellation *f*
f constellation *f* opérationnelle
r операционная совокупность *f*

* **operational research** → 2635

2635 operations research; operational research
d Operationsforschung *f*; Betriebsforschung *f*
f recherche *f* opérationnelle; recherche d'opérations
r исследование *n* операций

* **opposite angles** → 3930

* **optical generalization** → 3122

2636 optimal path; optimum path
d optimaler Weg *m*; optimale Route *f*
f chemin *m* optimal
r оптимальный пут *m*; оптимальный маршрут *m*

* **optimal path selection** → 3259

2637 optimal routing dispatch system
d Dispachierungssystem *n* für optimale Führung
f système *m* du type dispatching de routage optimal
r система *f* диспетчерского контроля оптимальной маршрутизации

* **optimum path** → 2636

2638 options button
d Optionenknopf *m*
f bouton *m* d'options
r кнопка *f* настройки

2639 orbit
d Orbit *m*; Umlaufbahn *f*
f orbite *f*
r орбита *f*

2640 orbital data; ephemeris [data]
d Ephemeriden *fpl*
f données *fpl* orbitaires; éphémérides *fpl*
r орбитальные данные *pl*; эфемериды *fpl*; астрономические таблицы *fpl*

2641 orbital remote sensor
d Kreisfernaufnehmer *m*
f télédétecteur *m* orbitaire; télédétecteur orbital
r дистанционный орбитальный зонд *m*

2642 orbit plane
  d Orbitalebene *f*
  f plan *m* orbitaire; plan orbital
  r плоскость *f* орбиты

  *   **orbit space** → 3787

2643 order
  d Ordnung *f*
  f ordre *m*
  r порядок *m*

2644 ordered list
  d geordnete Liste *f*
  f liste *f* ordonnée
  r упорядоченный список *m*

2645 ordering
  d Ordnung *f*; Anordnung *f*; Ordnen *n*;
     Anordnen *n*
  f ordonnancement *m*
  r упорядочение *n*

2646 order statistics
  d Anordnungswerte *mpl*; Ranggröße *f*
  f statistique *f* d'ordre; fonction *f* de l'ordre de
     l'observation
  r порядковая статистика *f*

  *   **ordinal** → 2648

2647 ordinal class
  d Ordinalklasse *f*
  f classe *f* ordinale
  r ординальный класс *m*

2648 ordinal [number]
  d Ordinalzahl *f*; Ordinale *f*
  f nombre *m* ordinal; ordinal *m*
  r ординал *m*; порядковое число *n*;
     ординальное число; порядковое
     числительное *n*

2649 ordinary kriging
  d gewöhnliches Kriging *n*
  f krigeage *m* ordinaire
  r обычный кригинг *m*

2650 ordinate
  d Ordinate *f*
  f ordonnée *f*
  r ордината *f*

  *   **ordnance datum** → 3237

  *   **ordnance map** → 2652

2651 Ordnance Survey; OS (UK)
  d Amtliche Landvermessung *f*; Amtliche
     Kartenbehörde *f*

  f Service *m* de la cartographie d'artillerie
  r Артиллерийская съёмка *f* (*государственная
     топографо-геодезическая и
     картографическая служба
     Великобритании*)

**2652 ordnance[-survey] map; cadastral map;
     cadastral plan** (UK); **land register map;
     registry map; property map** (US); **plate**
  d Kataster[plan]karte *f*; Liegenschaftskarte *f*;
     amtliche Landvermessungskarte *f*; Flurkarte *f*;
     Flurstücksnummer *f*
  f carte *f* de plan cadastral; plan *m* cadastral
  r кадастровый план *m*; кадастровая карта *f*;
     карта [по]земельного кадастра; регистровая
     карта

2653 ordnance-survey transfer format
  d Katasterkarten-Übertragungsformat *n*
  f format *m* de transfert de plan cadastral
  r формат *m* переноса кадастральных планов

2654 organizational reliability
  d organisatorische Zuverlässigkeit *f*
  f fiabilité *f* organisationnelle
  r организационная надёжность *f*

  *   **orgraph** → 1105

2655 orientation
  d Orientierung *f*
  f orientation *f*
  r ориентация *f*; ориентировка *f*

2656 orientation point
  d Orientierungspunkt *m*
  f point *m* d'orientation
  r ориентир *m*

  *   **oriented graph** → 1105

2657 origin
  d Ursprung *m*; Nullpunkt *m*
  f origine *f*
  r источник *m*

2658 original map
  d Originalkarte *f*
  f carte *f* originale
  r оригинал *m* карты

**2659 original plot; drawing original; compilation
     map; compilation sheet; base sheet**
  d Autorenkarte *f*
  f minute *f* d'auteur; feuille *f* de compilation
  r составительский оригинал *m* карты

  *   **original table** → 3347

  *   **origin latitude** → 2114

* origin node → 3468

2660 origin of coordinates; origin of coordinate
     system; coordinate origin; coordinate
     source; coordinate fountain
     d Koordinatenursprung *f*; Nullpunkt *m* des
       Koordinatensystems;
       Koordinatenanfangspunkt *m*
     f origine *f* de coordonnées; origine de système
       de coordonnées
     r начало *n* координат; начало координатной
       системы

* origin of coordinate system → 2660

2661 orthodrome; orthodromic line; great circle
     line
     d orthodromische Linie *f*
     f ligne *f* orthodromique; orthodromie *f*
     r ортодрома *f*; ортодромия *f*

* orthodromic line → 2661

2662 orthogonal meridian
     d orthogonaler Meridian *m*
     f méridien *m* orthogonal
     r ортогональный меридиан *m*

2663 orthogonal projection; perpendicular
     projection; orthographic projection
     d orthogonale Projektion *f*;
       Orthogonalprojektion *f*; senkrechte
       Parallelprojektion *f*
     f projection *f* orthogonale; projection
       orthographique
     r ортогональная проекция *f*; прямоугольная
       проекция; ортогональное проектирование *n*

2664 orthogonal space
     d Orthogonalraum *m*
     f espace *m* orthogonal
     r ортогональное пространство *n*

* orthographic projection → 2663

2665 orthography of geographical names
     d geografische Namenorthografie *f*;
       geografische Namenschreibweise *f*
     f orthographie *f* des noms géographiques
     r правила *npl* написания объектов на картах

* orthometric height → 1638

* orthomorphic projection → 686

* orthophoto → 2666

2666 orthophoto[graph]
     (photograph having the properties of an
     orthographic projection)

d Orthofotografie *f*; Orthofoto *n*
f orthophoto[graphie] *f*
r ортоснимок *m*

2667 orthophotographic map; orthophotomap;
     orthophotoplan
     (map produced by assembling
     orthophotographs at a specified uniform scale
     in a map format)
     d Orthofotokarte *f*; Orthofotoplan *m*
     f orthophotocarte *f*; orthophotoplan *m*; carte *f*
       orthophotographique
     r ортофотокарта *f*; ортофотоплан *m*

* orthophotomap → 2667

* orthophotoplan → 2667

2668 orthophotoscope; orthoprojector
     (photomechanical device used in conjunction
     with a double-projection stereoplotter for
     producing orthophotograph)
     d Orthofotoskop *m*; Orthoprojektor *m*
     f orthophotoscope *m*; orthoprojecteur *m*
     r ортогональный оператор *m* проектирования

* orthophototransformation → 2669

* orthoprojector → 2668

2669 orthorectification;
     ortho[photo]transformation
     d Orthorektifikation *f*
     f orthoredressement *m*
     r орторектификация *f*;
       ортотрансформирование *n*

* orthotransformation → 2669

* OS → 2633, 2651

* OSKA → 2592

2670 outer layer
     d äußere Schicht *f*
     f couche *f* externe
     r внешний слой *m*

* outline → 726

2671 outline attributes; contour attributes
     d Umrissattribute *npl*; Konturattribute *npl*
     f attributs *mpl* de contour
     r атрибуты *mpl* контура

* outline map → 3551

* outlining → 728

2672 outlying area; outlying district; outlying region; periphery; outskirts
d Außenbereich *m*; Reservefläche *f*; Peripherie *f*; Stadtrandgebiet *n*
f zone *f* périphérique; région *f* périphérique; zone non affectée; zone réservée
r край *m*; отроги *fpl*

* **outlying district** → 2672

* **outlying region** → 2672

* **output** → 1340

2673 output data
d Ausgabedaten *pl*
f données *fpl* sortantes
r исходные данные *pl*

2674 outside dimension
d äußere Dimension *f*
f dimension *f* hors-tout
r габарит *m*

2675 outside polygon; exterior polygon; external polygon
d äußeres Polygon *n*; Außenpolygon *n*
f polygone *m* extérieur; polygone externe
r внешний полигон *m*

* **outskirts** → 2672

2676 oval
d Eilinie *f*; Oval *n*
f ovale *m*
r овал *m*

2677 over- and underpass information
(information about the relative vertical position of features such as roads, railways and water)
d Straßenüberführungs- und Unterführungsinformation *f*; Überführungs- und Unterführungsinformation *f*
f information *f* de passages supérieurs et inférieurs
r информация *f* о проездах и эстакадах

2678 overedge
(any portion of a map lying outside the nominal map border)
d Stichbreite *f*
f surjet *m*
r часть *f* карты вне рамки

* **overhanging road** → 3167

* **overhead contact line** → 477

* **overlaid windows** → 472

2679 overlap; lap
d Überlappung *f*
f chevauchement *m*; recouvrement *m*; empiètement *m*
r перекрытие *n*; совмещение *n*

* **overlay** → 2116, 3534

2680 overlay; coverage
d Overlay *n*; Überzug *m*; Auflage *f*
f calque *m*; décalque *m*; segment *m* de recouvrement; overlay *m*
r оверлей *m*; перекрытие *n*

* **overlaying** → 3534

* **overprint** → 2681

2681 overprint[ing]
d Überdruckung *f*
f surimpression *f*
r надпечатание *n*

* **overshoot** → 2683

2682 overshoot inaccuracy
d Überschwingen-Ungenauigkeit *f*
f inexactitude *f* de dépassement
r неточность *f* из-за перерегулирования

2683 overshoot[ing]
d Überschwingweite *f*; Überschwingen *n*; Überschreitung *f*; Überregelung *f*; Überoszillation *f*; Übersteuerung *f*; Hinausschießen *n*
f [sur]dépassement *m*; surréglage *m*; surrégulation *f*; dépassement [de niveau] de réglage; suroscillation *f*
r перерегулирование *n*; отклонение *n* от заданного значения; выброс *m* сигнала

2684 overview
d Überblick *m*; Übersicht *f*
f vue *f* synoptique; aperçu *m*
r обзор *m*

P

2685 packing
d Packung f; Verpackung f
f empilement m
r упаковка f

2686 page extent
d Seitenextent m
f étendue f de page
r экстент m страницы

2687 palette
d Palette f
f palette f
r палитра f; цветовая гамма f

2688 palette file
d Palettedatei f
f fichier m de palette
r файл m палитры

* **palm-size computer** → 2689

* **palmtop** → 2689

2689 palmtop [computer]; palm-size computer; pocket computer; handheld [computer]
d Palmtop m; Kleinstcomputer m; Handheld-Computer m; Handmikrorechner m
f ordinateur m de poche; ordinateur qui tient dans la main
r карманный компьютер m

2690 pampas
d Pampa f
f pampa f
r пампасы mpl

* **pan** → 2694

2691 panchromatic
d farbempfindlich
f panchromatique
r панхроматический

2692 panchromatic band
d panchromatischer Bereich m
f spectre m panchromatique
r панхроматическая лента f

* **paneling** → 2693

2693 panel[l]ing
d Verkleidung f; Wandbekleidung f

f lambrissement m; lambrissage m
r обшивка f; отделка f филёнками

2694 pan[ning]
d Verschieben n; Schwenken n; Panning n
f panoramique m; mouvement m panoramique; pan m
r панорамирование n

* **panoramic rendering** → 2734

2695 pantograph
d Pantograph m
f pantographe m
r пантограф m

2696 paperless
d papierlos
f sans papier
r безбумажный

2697 paper map; hard-copy map
d Papierkarte f; Hardcopy-Karte f
f carte f de papier; carte sur papier; carte imprimée
r бумажная карта f; печатная карта

2698 parallax
d Parallaxe f
f parallaxe m
r параллакс m

2699 parallel
d Breitengrad m
f parallèle m
r параллель f

2700 parallelepiped classifier
d Parallelflächner-Klassifizierung f
f classification f parallélépipède
r классификация f методом параллелепипеда

2701 parallel of latitude
d Breitenkreis m; Parallelkreis m
f parallèle m de latitude; cercle m de latitude
r географическая параллель f

2702 parallel projection
d Parallelprojektion f
f projection f parallèle
r параллельная проекция f

2703 parameterized image
d parametrisiertes Bild n
f image f paramétrée
r параметризированное изображение n

2704 parametric surface
d parametrische Fläche f

f surface *f* paramétrique
r параметрическая поверхность *f*

2705 parametric tests
d parametrische Teste *mpl*
f essais *mpl* paramétriques
r параметрические тесты *mpl*;
параметрические критерии *mpl*

* **parcel** → **2089**

2706 parsing; syntax analysis
d Parsing *n*; syntaktische Analyse *f*;
Syntaxanalyse *f*
f analyse *f* lexicale; analyse syntaxique
r параметрический анализ *m*;
синтаксический анализ; синтаксический
разбор *m*

* **partial sum** → **3528**

* **partial total** → **3528**

2707 particular scale
d partikulärer Maßstab *m*
f échelle *f* particulière
r частный масштаб *m* [карты]

2708 passe
d Gebirgssattel *m*
f sauf-conduit *m*
r седловина *f*

* **paste** *v* → **1961**

* **PAT** → **2832**

* **patch** → **3545**

* **path** → **3177**

2709 path analysis
d Pfadanalyse *f*
f analyse *f* de chemins
r анализ *m* путей; анализ траекторией

2710 pathfinding
d Pfadsuchwahl *f*; Pfadbestimmung *f*;
Pfadbeurteilung *f*
f recherche *f* de chemins
r отыскание *n* путей

2711 path length
d Weglänge *f*; Streckenlänge *f*
f longueur *f* de trajectoire; longueur de chemin
r длина *f* пути; протяжённость *f* трассы;
длина траектории

2712 pathname
d Pfadname *m*

f nom *m* de chemin
r имя *n* пути; имя маршрута

2713 path optimization
d Routenoptimierung *f*
f optimisation *f* de route
r оптимизация *f* маршрута

2714 path segment
d Wegstrecke *f*
f segment *m* de route
r сегмент *m* пути; сегмент маршрута;
сегмент дороги

2715 path selection
d Wahl *f* des Pfades; Pfadselektion *f*
f sélection *f* de chemin; sélection de voie
r выбор *m* маршрута

* **pattern** → **1888, 3595**

2716 pattern; motif
d Motiv *n*
f motif *m*; forme *f* [géométrique régulière]
r мотив *m*; узор *m*; рисунок *m*

**2717 pattern recognition; PR; image
recognition**
d Bildmustererkennung *f*; Bilderkennung *f*;
Mustererkennung *f*; Gestalterkennung *f*
f reconnaissance *f* d'images; reconnaissance de
modèles
r распознавание *n* изображений;
распознавание образов

* **paving** → **3684**

* **PC** → **2730**

* **P-code** → **2879**

* **PDOP** → **2866**

2718 peak
d Spitze *f*; Scheitel *m*
f crête *f*; pic *m*; top *m*
r вершина *f*; верх *m*; пик *m*

2719 Peano-Hilton ordering
d Peano-Hilton-Ordnung *f*
f ordre *m* de Peano-Hilton
r упорядочение *n* Пиано-Гилтона

* **Peano keys** → **2720**

2720 Peano numbers; Peano keys
d Peanosche Nummern *fpl*; Peano-Nummern *fpl*
f nombres *mpl* de Peano
r числа *npl* Пиано

2721 Peano's curve
d Peanosche Kurve *f*; Peano-Kurve *f*
f courbe *f* de Peano
r кривая *f* Пиано

2722 pecked line; dashed line
d Strichlinie *f*
f ligne *f* brisée; ligne en traits; trait *m* tireté; tireté *m*
r пунктирная линия *f*; штриховая линия; штрих-линия *f*; штрихпунктирная линия; пунктир *m*

* **pegging out** → 2122

* **pel** → 2758

2723 pen palette
d Griffel-Palette *f*
f palette *f* de plumes
r перьевая палитра *f*

2724 pen plotter
d Stiftplotter *m*
f traceur *m* de plumes
r перьевой плоттер *m*

2725 perihelion
d Sonnennähe *f*
f périhélie *m*
r перигелий *m*

2726 perimeter
d Umfang; Perimeter *m*
f périmètre *m*
r периметр *m*

* **peripheral devices** → 2727

* **peripheral equipment** → 2727

* **peripheral node** → 1266

2727 peripherals; peripheral devices; peripheral equipment
d periphere Einheiten *fpl*; Peripheriegeräte *npl*; Anschlussgeräte *npl*
f dispositifs *mpl* périphériques; unités *fpl* périphériques; périphériques *mpl*
r периферийные устройства *npl*; внешние устройства; периферийное оборудование *n*; периферия *f*

* **periphery** → 2672

* **permanent point** → 740

2728 permissible address range
d zulässiger Adressenbereich *m*
f diapason *m* d'adresses permissif

r допустимый диапазон *m* адресов

* **perpendicular projection** → 2663

2729 persistent lock
d persistente Sperre *f*
f perte *f* persistante
r устойчивые потери *fpl*; постоянные потери

2730 personal computer; PC
d persönlicher Rechner *m*; Personalcomputer *m*; PC
f ordinateur *m* personnel
r персональный компьютер *m*; ПК

2731 personal mapping
d persönliche Kartografierung *f*
f cartographie *f* personnelle
r индивидуальное картографирование *n*

* **perspective** → 690

* **perspective aerial photograph** → 2601

2732 perspective drawing instrument
d perspektivisches Zeicheninstrument *n*; perspektivisches Zeichengerät *n*
f instrument *m* de dessin perspectif
r перспектограф *m*

* **perspective mapping** → 690

2733 perspective plot
d perspektivisches Plotten *n*
f traçage *m* perspective
r перспективное вычерчивание *n*

* **perspective projection** → 690

2734 perspective rendering; panoramic rendering
d perspektivisches Rendering *n*
f rendu *m* perspectif
r перспективный рендеринг *m*

2735 phase lock
d Phasenrastung *f*
f verrouillage *m* de phase
r фазовая автоподстройка *f* частоты; ФАПЧ; фазовая синхронизация *f*

2736 phase measurement method
d Phasenmessungsmethode *f*
f méthode *f* de mesure de phase
r фазовый метод *m* [измерения дальностей]

2737 phenomenon
d Erscheinung *f*
f phénomène *m*
r явление *n*; феномен *m*

* photo → 2742

2738 photo base
d Bildbasis f
f base f photographique; base à l'échelle des
photographies
r база f данных из съёмок

2739 photocopier
d Fotokopierer m
f photocopieur m; photocopieuse f
r фоторепродукционная камера f;
фотокопирующее устройство n

2740 photogrammetrically digitized
d fotogrammetrisch digitalisiert
f numérisé photogrammétriquement
r фотограмметрически оцифрованный

2741 photogrammetry
d Messbildverfahren n; Fotogrammetrie f
f photogrammétrie f
r фотограмметрия f

2742 photo[graph]
d Fotografie f; Foto n; Aufnahme f
f photo[graphie] f
r [фото]снимок m; фотосъёмка f

2743 photographic film
d fotografischer Film m
f pellicule m photographique
r фотоплёнок m

* photographic film recorder → 2418

2744 photographic hill shading
d fotomechanische Schummerung f;
Relieffotografie f
f estompage m photographique
r фоторельеф m

2745 photographic map; photomap
d Fotokarte f; Luftbildkarte f
f photocarte f
r фотокарта f

2746 photographic plotter; stereograph
d stereografischer Plotter m
f stéréographe m
r стереограф m

2747 photo index key
d Fotoindexschlüssel m
f carte-index f de photographies
r индексная карта f фотографий

* photomap → 2745

2748 photoplan

d Fotoplan m
f photoplan m
r фотоплан m

* photoplotter → 2418

2749 photosphere
d Fotosphäre f
f photosphère m
r фотосфера f

2750 photo-theodolite survey
d Fototheodolitaufnahme f
f levé m photothéodolite
r фототеодолитная съёмка f

* phototopography → 64

2751 physiographic map
d orografische Karte f
f carte f orographique
r физико-географическая карта f

* pick → 2752

2752 pick[ing]
d Identifizierung f
f désignation f
r указание n; отбор m; подбор m

* pictogram → 1881

* picture → 1888

* picture analysis → 1891

* picture cell → 2758

* picture databank → 1899

* picture dot → 2758

* picture element → 2758

* pie chart → 2753

2753 pie graph; pie chart; circular chart; circle
graph; sector diagram
d Kreisdiagramm n; Sektordiagramm n;
Tortengrafik f
f diagramme m circulaire; diagramme à
secteurs; graphique m circulaire; graphique
sectoriel; camembert m; graphique en tarte
r круговая диаграмма f; секторная диаграмма

2754 pilot-project
d Pilotprojekt m; Modellversuch m;
Pilotvorhaben n

f projet *m* pilote
r пилот-проект *m*

2755 pin map
d Pin-Karte *f*
f dessin *m* de la configuration des broches
r карта *f* значков

2756 pipeline
d Pipeline *f*
f pipeline *m*
r трубопровод *m*; магистраль *f*;
нефтепровод *m*; канал *m* информации;
конвейер *m*

2757 pit
d Grübchen *n*; Piquerloch *n*
f piqûre *f*; trou *m*
r впадина *f*

* **pitch → 1921**

**2758 pixel; picture element; pel; picture cell;
image dot; picture dot**
d Pixel *n*; Bildpunkt *m*; Bildelement *n*
f pixel *m*; élément *m* pictural; élément d'image
r пиксел *m*; элемент *m* изображения; точка *f*
изображения

2759 pixel classification
d Pixel-Klassifizierung *f*
f classification *f* de pixels
r классификация *f* пикселов

2760 pixel interpolator
d Pixel-Interpolator *m*
f interpolateur *m* de pixels
r интерполятор *m* пикселов

2761 pixel-to-pixel classifier
d Subpixel-Klassifizierer *m*;
Pixel-zu-Pixel-Klassifizierer *m*
f class[ificat]eur *m* [spectral avancé permettant
de détecter] de thèmes composant seulement
des fractions de pixels
r классификатор *m* [мультиспектральных
изображений] с разделением смешанных
пикселов

2762 place
d Platz *m*; Stelle *f*; Ort *m*
f place *f*; endroit *m*
r пункт *m*; место *n*

2763 place boundary
d Ortsgrenze *f*
f limite *f* de place
r граница *f* места

2764 place code

d Ortscode *m*; Platzcode *m*
f code *m* de place
r код *m* места

2765 place identification
d Ortidentifizierung *f*
f identification *f* de place
r идентификация *f* местоположения;
определение *n* местности

* **placement → 2215**

* **placement grid → 3183**

* **place names → 1613**

2766 place utility
d Ortleitung *f*
f avantage *m* du lieu
r польза *f* в месте спроса; польза при
широком распределении; удобство *n* места

* **plan → 992, 3714**

2767 planar graph; plane graph
(an arrangement of nodes and lines such that
the lines intersect only at nodes, thus
remaining embedded in a plane)
d Planargraph *m*
f graphe *m* plan; graphe plat
r планарный граф *m*; плоский граф

2768 planar intersection
d flacher Durchschnitt *m*
f intersection *f* plane
r плоское пересечение *n*

2769 planar map; 2D map
d flache Karte *f*
f carte *f* planaire
r планарная карта *f*

2770 planar map projections
d flache Kartenprojektionen *fpl*
f projections *fpl* cartographiques planes
r планарные картографические проекции *fpl*

2771 planar network
d Planarnetz[werk] *n*
f réseau *m* plane
r планарная сеть *f*

2772 planar polygon
d Planarpolygon *n*
f polygone *m* plan
r планарный многоугольник *m*; плоский
многоугольник

2773 planar surface; plane surface; 2D surface
d planare Fläche *f*; 2D-Fläche *f*

f surface *f* plane; surface 2D
r плоская поверхность *f*; двумерная
поверхность; 2D поверхность

2774 plane
   d Ebene *f*
   f plan *m*
   r плоскость *f*

   *   **plane control** → **2530**

2775 plane coordinates; coordinates in the plane
   d ebene Koordinaten *fpl*;
   Ebenenkoordinaten *fpl*; Koordinaten in der
   Ebene
   f coordonnées *fpl* planes; coordonnées dans le
   plan; coordonnées planaires
   r координаты *fpl* на плоскости; плоскостные
   координаты

2776 plane coordinate system
   d Ebenenkoordinatensystem *n*
   f système *m* de coordonnées plane
   r плоская координатная система *f*

   *   **plane geometry** → **2791**

   *   **plane graph** → **2767**

2777 plane of stratification; stratification plane; bedding plane
   d Schichtungsebene *f*; Stratifikationsfläche *f*
   f plan *m* de stratification; plan de couches
   r поверхность *f* напластования; плоскость *f*
   напластования

   *   **plane surface** → **2773**

2778 plane surveying
   d ebene Überwachung *f*
   f surveillance *f* plane
   r горизонтальная съёмка *f*

2779 planet
   d Planet *m*
   f planète *f*
   r планета *f*

   *   **plane-table** → **3550**

   *   **plane-table sheet** → **3550**

   *   **plane-table survey** → **2780**

2780 plane-table [topographic] survey
   d [topografische] Messtischaufnahme *f*
   f levé *m* topographique à la planchette
   r мензульная топографическая съёмка *f*

2781 planetary globe

2782 planetary mapping
   d planetarischer Erdball *m*; planetarische
   Weltkugel *f*
   f globe *m* planétaire
   r планетный глобус *m*

2782 planetary mapping
   d planetarische Kartierung *f*
   f cartographie *f* planétaire
   r планетное картографирование *n*

2783 planimeter; integrating instrument
   d Planimeter *n*
   f planimètre *m*
   r планиметр *m*

2784 planimetric base
   (a map showing in a two dimensional fashion,
   anything that would be visible from an aerial
   view photo)
   d planimetrische Basis *f*
   f base *f* planimétrique
   r планиметрическая основа *f*

2785 planimetric base mapping
   d Kartierung *f* der planimetrischen Basis
   f cartographie *f* planimétrique de base
   r картографирование *n* планиметрической
   основы

2786 planimetric data
   d planimetrische Daten *pl*
   f données *fpl* planimétriques
   r планиметрические данные *pl*

   *   **planimetric error** → **2864**

   *   **planimetric feature** → **1156**

2787 planimetric image
   d planimetrisches Bild *n*
   f image *f* planimétrique
   r планиметрическое изображение *n*

2788 planimetric map
   d planimetrische Karte *f*
   f carte *f* planimétrique
   r планиметрическая карта *f*

   *   **planimetric representation** → **2790**

2789 planimetric survey
   d Grundrissaufnahme *f*
   f levé *m* planimétrique
   r планиметрическая съёмка *f*

2790 planimetric view; planimetric representation
   d Grundrissdarstellung *f*; Situationsdarstellung *f*

f vue *f* planimétrique; représentation *f* planimétrique
r планиметрический вид *m*; планиметрическое представление *n*

2791 planimetry; plane geometry
  d Planimetrie *f*
  f planimétrie *f*; géométrie *f* plane
  r планиметрия *f*

2792 planning information system
  d Planung-Informationssystem *n*
  f système *m* d'information de planification
  r информационная система *f* по планированию

2793 plant life map
  d Pflanzenweltkarte *f*
  f carte *f* de végétaux
  r карта *f* растительной жизни; карта флоры

  * **plate** → 2652

  * **plate carree projection** → 1310

  * **plot** → 1018, 2089

  * **plot area** → 2796

2794 plotted output; plotting output
  d Plottausgabe *f*
  f sortie *f* de traçage
  r вывод *m* на графопостроитель

2795 plotter
  d Plotter *m*; Kurvenschreiber *m*
  f traceur *m* [de courbes]; traceur graphique
  r графопостроитель *m*; плоттер *m*; автоматический координатограф *m*

2796 plotter area; plot[ting] area
  d Plottbereich *m*
  f aire *f* de traçage; zone *f* de traçage
  r размер *m* рабочего поля плоттера; область *f* вычерчивания

  * **plotting area** → 2796

  * **plotting output** → 2794

  * **plotting paper** → 1750

2797 plotting speed
  d Plottengeschwindigkeit *f*
  f vitesse *f* de tracement; vitesse de tracé
  r скорость *f* прорисовки

  * **plumb-line deflection** → 948

  * **plumb-line deviation** → 948

  * **pocket computer** → 2689

2798 point
  d Punkt *m*
  f point *m*
  r точка *f*

2799 point address
  d Punktadresse *f*
  f adresse *f* de point
  r адрес *m* точки

2800 point aggregation
  d Punkt-Aggregation *f*
  f agrégation *f* de points
  r совокупность *f* точек

  * **point density** → 971

2801 pointer
  d Zeiger *m*; Hinweis *m*; Pointer *m*
  f pointeur *m*
  r указатель *m*

  * **point feature** → 2809

2802 point-feature label placement
  d Punktobjekt-Etikettenstellung *f*, Punktbeschriftung *f*
  f placement *m* d'étiquettes d'objets ponctuels
  r размещение *n* меток точечных объектов

2803 point file
  d Punktdatei *f*
  f fichier *m* de points
  r файл *m* точек

2804 pointing device
  d Positionsanzeiger *m*; Steuervorrichtung *f*
  f dispositif *m* de pointage; dispositif de désignation
  r указательное устройство *n*; указующее устройство; устройство управления позицией

2805 point-in-polygon operation
  (procedure to determine which points fall into which polygons)
  d Punkt-im-Polygon-Operation *f*
  f opération *f* de définition de point à polygone
  r операция *f* определения принадлежности точки полигону

  * **point map** → 2806

2806 point map[ping]
  d Punktabbildung *f*
  f application *f* ponctuelle
  r точечное отображение *n*

2807 point mode digitizing; variable interval digitizing
 d Punkt-Modus *m* des Digitalisierens
 f numérisation *f* par points
 r поточечное оцифрование *n*

2808 point neighborhood
 d Punkt-Nachbarschaft *f*; Punkt-Näherung *f*
 f voisinage *m* de point
 r окрестность *f* точки

2809 point object; point feature
 d Punktobjekt *n*
 f objet *m* point; objet ponctuel; entité *f* ponctuelle
 r точечный объект *m*

2810 point of compass; compass point; rhumb
 d Himmelsrichtung *f*
 f quart *m*; rhumb *m*; aire-de-vent *m*
 r компасный румб *m*

2811 point of reference; reference point; datum point; datum mark; datum
 d Bezugspunkt *m*; Referenzpunkt *m*; Stützpunkt *m*; Datum *n*
 f point *m* de référence; point d'appui; point de départ; point de repère; datum *m*
 r опорная точка *f*; исходная точка; относимая точка; исходный пункт *m*; начало *n* отсчёта; точка приведения; точка привязки; датум *m*

 * point of view → 3957

2812 point symbols
 d Punktsymbole *npl*
 f signes *mpl* symboliques
 r условные внемасштабные знаки *mpl*

2813 point-to-area conversion
 d Punkt-zu-Zone-Konvertierung *f*
 f conversion *f* point à zone
 r преобразование *n* точки в регион

2814 point-to-point visibility; intervisibility
 d Punkt-zu-Punkt-Sichtbarkeit *f*
 f visibilité *f* mutuelle de deux points
 r взаимная видимость *f* двух точек

2815 point transfer
 d Punktübertragung *f*
 f transfert *m* de points
 r передача *f* точек

2816 polar coordinates
 d Polarkoordinaten *fpl*
 f coordonnées *fpl* polaires
 r полярные координаты *fpl*

 * polar coordinates in the space → 3437

2817 polar coordinate system
 d Polarkoordinatensystem *n*
 f système *m* de coordonnées polaires
 r полярная координатная система *f*

2818 polar distance
 d Polardistanz *f*
 f distance *f* polaire
 r полярное расстояние *n*

2819 polarity
 d Polarität *f*
 f polarité *f*
 r полярность *f*

2820 polar projection
 d Polarprojektion *f*
 f projection *f* polaire
 r полярная проекция *f*

 * pole → 1618

2821 political boundaries
 d politische Grenzen *fpl*
 f limites *fpl* politiques
 r политические границы *fpl*

2822 polyconic projection
 d polykonische Projektion *f*
 f projection *f* polyconique
 r поликоническая проекция *f*

2823 polygon
 d Vieleck *n*; Polygon *n*
 f polygone *m*
 r полигон *m*; многоугольник *m*

2824 polygon adjacency
 d Polygon-Adjazenz *f*
 f adjacence *f* de polgones
 r смежность *f* полигонов

2825 polygonal approximation
 d Polygonapproximation *f*
 f approximation *f* polygonale
 r кусочно-линейная аппроксимация *f*

2826 polygonal arc
 d Polygonbogen *m*
 f arc *m* polygonal
 r ломаная дуга *f*

2827 polygonal clipping; polygon clipping
 d Polygonkappen *n*; Polygonschnitt *m*
 f découpage *m* polygonal
 r полигональное отсекание *n*

 * polygonal curve → 2852

2828 **polygonal error**
d polygonaler Fehler *m*
f erreur *f* polygonale
r полигональная погрешность *f*

* **polygonal line** → 2852

2829 **polygonal network; traverse network**
d Polygonnetz *n*
f réseau *m* de polygones
r сетка *f* полигонов; полигональная сеть *f*

* **polygonal object** → 166

* **polygonal path** → 2852

2830 **polygonal traverse; traverse**
d Polygonzug *m*
f cheminement *m* polygonal
r полигональный ход *m*; полигонный ход

2831 **polygon-arc topology**
d Polygon-Bogen-Topologie *f*
f topologie *f* polygone-arc
r топология *f* многоугольников и дуг

2832 **polygon attribute table; PAT**
d Polygonattributtabelle *f*
f table *f* d'attributs de polygones
r таблица *f* атрибутов полигонов

2833 **polygon-based modeling**
d Polygon-basierte Modellierung *f*
f modélisation *f* orientée à polygones;
modélisation polygonale
r моделирование *n* [поверхности объекта] с
помощью многоугольников;
полигональное моделирование

2834 **polygon boundary**
d Polygongrenze *f*
f limite *f* d'un polygone
r граница *f* полигона

* **polygon clipping** → 2827

2835 **polygon connection**
d Polygonschaltung *f*
f connexion *f* en polygone
r полигональное соединение *n*

2836 **polygon-converter**
(data model of the US Bureau of the Census)
d Polygon-Converter *m*; POLYVRT
f convertisseur *m* polygone
r полигональный конвертор *m*

2837 **polygon coverage**
d Polygonabdeckung *f*
f couverture *f* de polygone

r покрытие *n* полигона

* **polygon dissolving** → 2838

2838 **polygon dissolving[/merging]**
d Polygonauflösung *f* [mit Vereinigung f]
f dissolution *f* [avec fusion] de polygones;
agrégation *f* [avec fusion] de polygones
r объединение *n* смежных полигонов [с
уничтожением границ между ними]

* **polygon edge** → 164

2839 **polygon filling**
d Polygonfüllung *f*
f remplissage *m* de polygones
r закрашивание *n* многоугольников;
заполнение *n* полигонов

2840 **polygon-filling unit**
d Polygonfüll[ungs]block *m*;
Polygonfüll[ungs]einheit *f*
f bloc *m* de remplissage de polygones
r блок *m* [сплошного] закрашивания
многоугольников; блок заполнения
полигонов

2841 **polygon generation**
d Polygongenerierung *f*
f génération *f* de polygones
r генерирование *n* многоугольников;
генерирование полигонов

* **polygon ID** → 2842

2842 **polygon identification number; polygon
identifier; polygon ID**
d Polygon-Identifikator *m*
f nombre *m* d'identification de polygone
r идентификатор *m* полигона

* **polygon identifier** → 2842

2843 **polygon-in-polygon operation**
d Polygon-im-Polygon-Operation *f*
f opération *f* de définition d'un polygone à
polygone
r наложение *n* двух полигональных слоев

2844 **polygonization**
d Polygonzerlegung *f*; Polygonbildung *f*;
Polygonierung *f*
f polygonisation *f*
r разделение *n* на многоугольники;
разделение на полигоны

2845 **polygonize** *v*
d polygonizieren
f polygoniser
r разделять на многоугольники

2846 polygon mesh
 d Polygonmasche *f*
 f maillage *m* par polygones
 r многоугольная сеть *f*

2847 polygon overlay
 d Polygonüberlagerung *f*
 f recouvrement *m* de polygone
 r полигональное перекрытие *n*

2848 polygon processing
 d Polygonverarbeitung *f*
 f traitement *m* de polygones
 r обработка *f* полигонов

2849 polygon topology
 d Polygontopologie *f*
 f topologie *f* de polygone
 r топология *f* полигона

2850 polygon triangulation
 d Polygon-Triangulation *f*
 f triangulation *f* de polygones
 r триангуляция *f* полигонов

2851 polyhedral projection
 d Polyederabbildung *f*; polyedrische
    Abbildung *f*
 f projection *f* polyédrique
 r многогранная проекция *f*; полиэдрическая
    проекция

**2852 polyline; polygonal line; polygonal curve;
    polygonal path**
 d Polylinie *f*
 f polyligne *f*
 r полилиния *f*; ломаная [линия] *f*

2853 polymorphism
 d Polymorphismus *m*
 f polymorphisme *m*
 r полиморфизм *m*

2854 population
 d Bevölkerung *f*; Population *f*; Gesamtheit *f*
 f population *f*
 r популяция *f*; совокупность *f*

 * population census → 497

2855 population characteristics
 d Bevölkerungscharakteristik *f*
 f caractéristiques *fpl* de la population
 r выборочная характеристика *f*;
    характеристика совокупности;
    характеристика народонаселения

2856 population density map
 d Bevölkerungsdichte-Karte *f*;
    Populationsdichte-Karte *f*;

    Belegungsdichte-Karte *f*;
    Besetzungsdichte-Karte *f*
 f carte *f* de densité de la population; carte de
    densité démographique
 r карта *f* густоты населения

 * population nucleus → 2571

 * population projection → 962

 * population pyramid → 963

2857 population redistribution
 d Bevölkerungsumverteilung *f*
 f redistribution *f* de la population
 r перераспределение *n* населения

**2858 population trend; demographic trend;
    demographic development**
 d demografische Tendenz *f*;
    Bevölkerungsentwicklung *f*
 f tendance *f* démographique; évolution *f*
    démographique
 r тенденция *f* изменения структуры и
    численности населения; демографический
    тренд *m*

2859 popup menu; floating menu
 d Popup-Menü *n*; Auftauch-Menü *n*
 f menu *m* relevant; menu à liste directe; menu
    superposable; menu en incrustation
 r раскрывающееся меню *n*; всплывающее
    меню

2860 popup window
 d Aufspringfenster *n*
 f fenêtre *f* à liste directe; fenêtre flash; fenêtre
    en incrustation
 r всплывающее окно *n*

2861 portability
 d Portabilität *f*; Tragbarkeit *f*
 f portabilité *f*
 r переносимость *f*; перемещаемость *f*

 * port plan → 1802

2862 position
 d Position *f*; Lage *f*
 f position *f*
 r позиция *f*

 * position accuracy → 2863

2863 position[al] accuracy; position[al] precision
 d Lagegenauigkeit *f*; Positionsgenauigkeit *f*;
    Positioniergenauigkeit *f*
 f précision *f* de position[nement]; exactitude *f*
    de position
 r позиционная точность *f*

2864 **position[al] error; planimetric error**
d Lagefehler *m*; Positionsfehler *m*
f erreur *f* de position[nement]
r позиционная погрешность *f*; погрешность позиционирования

* **positional precision** → 2863

2865 **positional tolerance**
d Positionstoleranz *f*; Positioniertoleranz *f*
f tolérance *f* de position; tolérance de localisation
r позиционный допуск *m*

2866 **position dilution of precision; PDOP**
d Positionsgenauigkeitsabfall *m*
f dilution *f* de précision de positionnement
r показатель *m* снижения точности определения положения в пространстве

2867 **position display**
d Positionsdarstellung *f*
f affichage *m* de position
r изображение *n* местоположения

* **position error** → 2864

2868 **position fixing**
d Positionsbestimmung *f*; Ortsbestimmung *f*; Standortfeststellung *f*
f détermination *f* de position; relèvement *m* de positions
r фиксирование *n* позиции

2869 **positioning**
d Positionierung *f*; Positionieren *n*
f positionnement *m*
r позиционирование *n*

* **position precision** → 2863

2870 **position undershoot; undershoot**
d Unterschwingen *n*
f sous-dépassement *m*; sous-correction *f*
r недоход *m* до заданной координаты

2871 **postal code**
d Postleitzahl *f*; PLZ
f code *m* postal; numéro *m* [d'acheminement] postal
r почтовый индекс *m*

2872 **post-classification**
d Nachklassifizierung *f*
f post-classification *f*
r последующая классификация *f*

2873 **post-digitizing tasks**
d Nachdigitalisierungsprobleme *npl*
f taches *fpl* après numérisation

r задачи *fpl* после дигитализирования

2874 **post-processing**
d Nachbearbeitung *f*
f post-traitement *m*
r последующая обработка *f*

* **potential surface** → 2145

2875 **potential surface calculation**
d Potentialflächeberechnung *f*
f calcul *m* de surface équipotentielle; calcul de surface de niveau
r исчисление *n* эквипотенциальной поверхности

* **PPS** → 2877

* **PR** → 2717

2876 **precedence**
d Vorrang *m*; Vorhergehen *n*; Vorrecht *n*; Präzedenz *f*
f précédence *f*
r предшествование *n*

2877 **precise positioning service; PPS**
d Präzisionortungsdienst *m*; GPS-Service *m* zur Positionsbestimmung für autorisierte Nutzer
f service *m* de positionnement précise
r служба *f* точного определения местоположения; точный позиционный сервис *m*

2878 **precision; definiteness; distinctness**
(level of measurement and exactness of description in a GIS database)
d Präzision *f*
f précision *f*
r чёткость *f*; отчётливость *f*; определённость *f*

* **precision** → 17

2879 **precision code; P-code**
(precise or protected code which is modulated on both L1 and L2 carrier frequencies)
d Code *m* mit hoher Genauigkeit für autorisierte Nutzer; P-Code *m*
f code *m* P
r П-код *m*

2880 **precision digitizing**
d Präzisionsdigitalisierung *f*
f numérisation *f* précise
r точное дигитализирование *n*

2881 **precision digitizing tablet**
d Präzisionstablett *n*

f table *f* de numérisation de précision
r точный цифровой планшет *m*

* **prediction** → 1455

2882 **pre-digitizing tasks**
  d Vordigitalisierungsprobleme *npl*
  f taches *fpl* avant numérisation
  r задачи *fpl* до цифрования

2883 **preliminary map**
  d vorläufige Karte *f*
  f carte *f* préliminaire
  r черновая карта *f*

2884 **preliminary topology**
  d vorläufige Topologie *f*
  f topologie *f* préliminaire
  r черновая топология *f*

2885 **preservation of area**
  d Flächentreue *f*
  f conservation *f* de l'aire
  r сохранение *n* площади

2886 **pressure**
  d Druck *m*; Pressung *f*
  f pression *f*
  r давление *n*

* **primary data** → 3030

2887 **primary digital surface model**
  d primäres digitales Oberflächenmodell *n*
  f modèle *m* primaire numérique de surface
  r первичная цифровая модель *f* поверхности

2888 **primary key**
  (in a database)
  d Primärschlüssel *m*; Erst-Schlüssel *m*
  f clé *m* primaire
  r первичный ключ *m*

2889 **prime meridian; principal meridian;**
  central meridian; zero [meridian];
  Greenwich meridian
  d Hauptmeridian *m*; Meridian *m* von Greenwich
  f méridien *m* d'origine
  r начальный меридиан *m*

2890 **primitive**
  d Primitiv *n*; Grundelement *n*
  f primitive *f*
  r примитив *m*

2891 **principal component analysis**
  d Hauptkomponenten-Analyse *f*
  f analyse *f* en composantes principales
  r анализ *m* основных компонентов; анализ
     главных компонентов

* **principal meridian** → 2889

2892 **principal scale; nominal scale**
  (of a map)
  d Hauptmaßstab *m*
  f échelle *f* principale; échelle nominale
  r главный масштаб *m*

2893 **principle of least effort**
  d Prinzip *n* des geringsten Aufwands
  f principe *m* du moindre effort
  r принцип *m* наименьшего сопротивления

2894 **printer**
  d Drucker *m*
  f imprimante *f*
  r принтер *m*; печатающее устройство *n*

2895 **printing**
  d Drucken *n*
  f impression *f*
  r печатание *n*

2896 **printing frame; contact screen**
  d Druckrahmen *m*; Kopierrahmen *m*
  f châssis *m* de copie; châssis-presse *m*
  r копировальная рама *f*

2897 **privacy transposition**
  d persönliche Datentransposition *f*
  f transposition *f* d'information privée
  r преобразование *n* секретных данных

2898 **probability**
  d Wahrscheinlichkeit *f*
  f probabilité *f*
  r вероятность *f*

2899 **probability distribution function**
  d Wahrscheinlichkeitsverteilungsfunktion *f*
  f fonction *f* de distribution de probabilité
  r функция *f* распределения вероятности

* **problem of coloring maps in four colors** →
  1466

2900 **processing tolerances**
  (fuzzy, tic match, dangle length)
  d Verarbeitungstoleranzen *fpl*
  f tolérances *fpl* de traitement
  r допуски *mpl* обработки

2901 **production of maps; map production;**
  mapmaking; cartographic production
  d Kartenreproduktion *f*
  f production *f* cartographique
  r картографическое производство *n*

* **profile** → 812

2902 profile recorder
d Profilregistriergerät *n*
f enregistreur *m* de profils
r профилограф *m*; профилометр *m*

2903 profiling
d Profilaufzeichnung *f*; Bestimmung *f* des Profils
f établissement *m* de profil; copiage *m*
r анализ *m* профиля; контурная обработка *f*; обработка на копировальном станке; обработка по копиру

* **prognosis** → 1455

* **prognostic chart** → 2905

2904 prognostic contour chart; prontour chart
d Isohypsenvorhersagekarte *f*
f carte *f* d'isohypses prévue
r прогнозная контурная карта *f*; прогнозная карта абсолютной топографии

2905 prognostic map; prognostic chart; forecast map
d Vorhersagekarte *f*
f carte *f* prévue
r прогнозная карта *f*

* **project** → 992

2906 projected tolerance
d projektierte Toleranz *f*
f tolérance *f* projective
r проективный допуск *m*

* **projecting** → 2908

2907 projection
(as an image)
d Projektion *f*
f projection *f*
r проекция *f*

2908 projection; projecting; mapping
(as an operation)
d Projizieren *n*; Abbildung *f*; Mapping *n*
f projection *f*; application *f*; mappage *m*; mapping *m*
r проектирование *n*; проецирование *n*; операция *f* проецирования; отображение *n*

* **projection axis** → 292

* **projection change** → 2909

2909 projection conversion; projection change; projection transformation
d Projektionskonvertierung *f*
f conversion *f* de projections

r трансформация *f* проекций

2910 projection coordinates
d Projektionskoordinaten *fpl*
f coordonnées *fpl* de projection
r координаты *fpl* проекции

2911 projection device; projector
d Projektor *m*
f projecteur *m*
r проектор *m*

2912 projection equation
d Abbildungsgleichung *f*
f équation *f* de projection
r уравнение *n* проекции

2913 projection file
d Projektionsdatei *f*
f fichier *m* de projection
r файл *m* проекции

2914 projection parameters
d Projektionsparameter *mpl*
f paramètres *mpl* de projection
r параметры *mpl* проекции

* **projection scale** → 3217

2915 projection surface
d Projektionsfläche *f*
f surface *f* de projection
r поверхность *f* проекции

* **projection transformation** → 2909

* **projection with heights** → 3720

2916 projective algebra
d projektive Algebra *f*
f algèbre *f* projective
r проективная алгебра *f*

2917 projective geometry
d projektive Geometrie *f*
f géométrie *f* projective
r проективная геометрия *f*

2918 projective transformation; linear projective mapping; collineatory transformation
d projektive Transformation *f*; lineare projektive Abbildung *f*; kollineare Abbildung
f transformation *f* projective; application *f* projective linéaire; transformation colinéaire
r проективное преобразование *n*; линейное проективное отображение *n*; коллинеарное преобразование

* **projector** → 2911

* **prontour chart** → 2904

2919 **proof plot**
   d Probenzeichnung *f*
   f dessin *m* d'épreuve
   r пробный чертёж *m*

2920 **propagation**
   d Fortpflanzung *f*; Propagierung *f*;
      Verbreitung *f*
   f propagation *f*
   r распространение *n*

* **property map** → 2652

2921 **proportionality**
   d Proportionalität *f*
   f proportionnalité *f*
   r пропорциональность *f*

2922 **proportionally object resizing**
   d proportionale Objekt-Größenänderung *f*
   f redimensionnement *m* proportionnellement
      d'objet
   r пропорциональное переоразмерение *n*
      объекта

2923 **proportional symbols**
      (a thematic mapping technique that displays a
      quantitative attribute by varying the size of a
      symbol)
   d Proportionalsymbole *npl*
   f symboles *mpl* proportionnels
   r пропорциональные символы *mpl*

2924 **proportioning**
   d Proportionieren *n*
   f proportionnément *m*
   r проектирование *n* размера; подбор *m*
      состава

2925 **protocol**
   d Protokoll *n*
   f protocole *m*
   r протокол *m*

2926 **prototype**
   d Vorbild *n*; Urbild *n*; Prototyp *m*
   f prototype *m*
   r прототип *m*; [опытный] образец *m*

2927 **provisional coordinates**
   d vorläufige Koordinaten *fpl*
   f coordonnées *fpl* provisoires
   r неуточнённые координаты *fpl*;
      приближённые координаты

* **proximal analysis** → 2512

* **proximal polygons** → 3665

2928 **proximal tolerance**
   d proximale Toleranz *f*
   f tolérance *f* proximale
   r близкий допуск *m*; проксимальный допуск

2929 **proximity; neighbo[u]rhood; nearness**
   d Näherung *f*; Nachbarschaft *f*; Nähe *f*
   f proximité *f*; ambiance *f*; voisinage *m*
   r близость *f*; соседство *n*

* **proximity analysis** → 2512

2930 **proximity map**
   d Näherungskarte *f*
   f carte *f* de proximité
   r карта *f* окрестностей

* **proximity polygons** → 3665

2931 **proximity search**
   d Annäherungssuche *f*
   f recherche *f* [par opérateurs] de proximité
   r поиск *m* близости

2932 **proximity zone**
   d Näherungszone *f*
   f zone *f* de proximité
   r зона *f* близости

2933 **proxy data**
   d Proxydaten *pl*
   f données *fpl* proximaux
   r предварительные данные *pl*

2934 **pseudocolor**
   d Falschfarbe *f*
   f pseudocouleur *f*
   r псевдоцвет *m*

2935 **pseudocolor enhancement**
   d Pseudofarb[en]anreicherung *f*;
      Pseudofarb[en]erhöhung *f*
   f amélioration *f* de pseudocouleur
   r улучшение *n* псевдоцвета; расширение *n*
      псевдоцвета; псевдоцветное добавление *n*

2936 **pseudocolor image**
   d Falschfarb[en]bild *n*; Pseudofarbbild *n*
   f image *f* pseudocouleur
   r псевдоцветное изображение *n*

2937 **pseudocolor transform**
   d Pseudofarb[en]änderung *f*
   f transformation *f* pseudocouleur
   r трансформация *f* псевдоцвета

2938 **pseudoconical projection**
   d Pseudokegelprojektion *f*; pseudokonische
      Abbildung *f*

f projection *f* pseudo-conique
r псевдоконическая проекция *f*

2939 pseudocylindrical projection; modified cylindrical projection
d Pseudozylinderprojektion *f*
f projection *f* pseudo-cylindrique
r псевдоцилиндрическая проекция *f*

2940 pseudonode
d Pseudoknoten *m*
f pseudonœud *m*
r псевдоузел *m*

* **pseudo-perspective** → 1370

* **pseudo-random code** → 2941

2941 pseudo-random [noise] code
d Pseudozufallsstörungscode *m*;
pseudozufälliger Rauschencode *m*
f code *m* de bruit pseudo-aléatoire
r псевдослучайный код *m*; псевдослучайный
шум *m*; псевдослучайная
последовательность *f*

2942 pseudorange
d Pseudostrecke *f*
f pseudo-distance *f*
r псевдодальность *f*

2943 pseudorange measurement
d Pseudostreckenmessung *f*
f mesure *f* de pseudo-distance
r измерение *n* псевдодальности

2944 pseudorange navigation
d Pseudostreckennavigation *f*
f navigation *f* à pseudo-distance
r навигация *f* по псевдодальности

2945 pseudoscopic view
d pseudoskopische Betrachtung *f*
f vue *f* pseudoscopique
r псевдоскопический вид *m*

2946 pseudostatics
d Pseudostatik *f*
f pseudostatique *f*
r псевдостатика *f*

2947 pseudo-voting districts
d Pseudo-Wahlkreis *m*
f circonscription *f* pseudo-électorale
r псевдо-избирательный округ *m*

2948 psychomotor error
(of manual digitizing)
d psychomotorischer Fehler *m*
f erreur *f* psychomotrice

r психомоторная погрешность *f*

2949 public information
d öffentliche Information *f*
f information *f* publique
r общедоступная информация *f*

2950 public land survey system
d öffentliches Landvermessungssystem *n*
f système *m* d'arpentage public
r система *f* государственной
триангуляционной съёмки; система
общедоступного топографического
измерения; система открытой съёмки
местности

2951 public land system
d öffentliches terrestrisches System *n*
f système *m* public terrestre
r система *f* государственной земли

2952 public record
d allgemeiner Satz *m*
f enregistrement *m* public
r общедоступная запись *f*

* **puck** → 1082

2953 pulldown menu; dropdown menu; tear-off menu
d Pulldown-Menü *n*; Dropdown-Menü *n*;
Rollmenü *n*
f menu *m* déroulant
r развёрнутое меню *n*; падающее меню;
ниспадающее меню; выдвижное меню;
спускающееся меню; опускающееся меню

* **purging** → 954

2954 pyramidal chart
d pyramidales Diagramm *n*
f diagramme *m* pyramidal
r пирамидальная диаграмма *f*

2955 pyramid layers
d pyramidale Schichten *fpl*
f couches *fpl* pyramidaux
r пирамидные слои *mpl*

Q

* QBE → 2973

* Q-tree → 2965

* quad → 2959, 2960

2956 quad-corner area; four-corners area
  d Vier-Ecken-Gebiet *n*
  f région *f* de "quatre coins"
  r четырёхугольный участок *m*

2957 quad file
  d Quadratdatei *f*
  f fichier *m* de quads
  r файл *m* элементарных квадратных
    участков; файл четырёхугольных участков

  *quad map → 2318

2958 quad name
  d Quadratname *m*
  f nom *m* de quad
  r имя *n* квадратного участка

2959 quadrangle; quad; tetragon
  d Viereck *n*; Quadrangel *n*
  f quadrilatère *m*; quadrangle *m*; quad *m*
  r четырёхвершинник *m*; четырёхугольник *m*

* quadrangle map → 2318

2960 quadrant; quarter; quad
  d Quadrant *m*
  f quadrant *m*
  r квадрант *m*; квадратный участок *m*;
    квадратный блок *m*

2961 quadrant snap
  d Quadrantfang *m*
  f accrochage *m* au quadrant
  r привязка *f* к квадранту

* quadrat → 3455

* quadrate mapping → 2336

2962 quadratic polynomial
  d quadratisches Polynom *n*
  f polynôme *m* quadratique
  r квадратический полином *m*

2963 quadratic spline
  d quadratischer Spline *m*

  f spline *m* quadratique
  r квадратический сплайн *m*

2964 quadratic trend
  d quadratischer Trend *m*
  f tendance *f* quadratique
  r квадратический тренд *m*

2965 quadtree; Q-tree
  d Quadtree *n*; Viererbaum *m*
  f arbre *m* quaternaire; quad-tree *m*; tétrarbre *m*
  r квадратомическое дерево *n*;
    квадродерево *n*; дерево квадрантов; дерево
    квадратов; Q-дерево *n*; 4-дерево *n*

2966 quadtree areas
  d Quadtree-Gebiete *npl*
  f zones *fpl* d'arbres quaternaires
  r области *fpl* квадродеревьев

2967 quadtree structure
  d Quadtree-Struktur *f*
  f structure *f* d'arbre quaternaire
  r структура *f* квадродерева

* **quantifier** → 1078

2968 quantitative terrain analysis
  d quantitative Geländeanalyse *f*;
    Mengenanalyse *f* von Gelände
  f analyse *f* quantitative de terrain
  r количественный анализ *m* рельефа

* **quantization** → 1123

2969 quantize *v*
  d digital darstellen
  f quantifier
  r квантовать; разбивать на подгруппы

* **quantizing** → 1123

* **quarter** → 2960, 3150

2970 quarter scan line
  d Quadrant-Bildzeile *f*
  f ligne *f* de balayage de quadrant
  r полоса *f* сканирования квадранта

2971 quasi-geocentric coordinates
  d quasi-geozentrische Koordinaten *fpl*
  f coordonnées *fpl* quasi-géocentriques
  r квазигеоцентрические координаты *fpl*

2972 quasi-geoid
  d Quasigeoid *n*
  f quasi-géoïde *m*
  r квазигеоид *m*

* **query** → 3140

2973 query-by-example; QBE
  d Abfrage *f* durch Beispiel;
    Query-by-Example *n*
  f requête *f* définie par un exemple
  r запрос *m* по шаблону

2974 query language
  d Abfrage-Sprache *f*
  f langage *m* d'interrogation
  r язык *m* запросов

2975 query table
  d Quellentabelle *f*
  f table *f* d'interrogations
  r запросная таблица *f*

2976 quick digitalization
  d Schnelldigitalisierung *f*
  f digitalisation *f* rapide
  r быстрое оцифрование *n*

2977 quick[-look] plot
  d Schnellplotten *n*
  f traçage *m* rapide; traçage immédiat; traçage
    obtenu en première lecture
  r оперативное нанесение *n* на карту

  * **quick plot** → 2977

  * **quit** → 1340

R

2978 radar multiple-look image; multilook image
d Multi-Look-Radarbild *n*
f image *f* plurielle [par radar]
r многопоисковое [радарное] изображение *n*

2979 radar remote sensing
d Radarfernerkundung *f*; Radarfernabtastung *f*
f télédétection *f* radar
r радарное дистанционное зондирование *n*

2980 radial distortion
d Radialverzerrung *f*
f distorsion *f* radiale
r радиальное искажение *n*

2981 radial error
d Radialfehler *m*
f erreur *f* radiale
r радиальная погрешность *f*

2982 radial line
d Radialleitung *f*
f ligne *f* radiale
r радиальная линия *f*; линия положения,
определяемая азимутом

2983 radiation
d Strahlung *f*
f radiation *f*, rayonnement *m*
r излучение *n*

2984 radio button
d Radioschaltfläche *f*; rundes Optionsfeld *n*
f bouton *m* radio; bouton d'option; cercle *m*
d'option
r радиокнопка *f*; зависимая кнопка *f*

2985 radiometer
d Radiometer *m*
f radiomètre *m*
r радиометр *m*

2986 radiometric calibration
d radiometrische Kalibrierung *f*
f calibration *f* radiométrique
r радиометрическая калибровка *f*

2987 radiometric correction; spectral correction
d radiometrische Korrektur *f*
f correction *f* radiométrique
r радиометрическая коррекция *f*;
спектральная коррекция

2988 radiometric distortion
d radiometrische Verzerrung *f*
f distorsion *f* radiométrique
r радиометрическое искажение *n*

2989 radiometric resolution
d radiometrische Auflösung *f*
f résolution *f* radiométrique
r радиометрическое разрешение *n*

2990 radiometric sensitivity
d radiometrische Empfindlichkeit *f*
f sensitivité *f* radiométrique
r радиометрическая чувствительность *f*

2991 radiometry
d Radiometrie *f*
f radiométrie *f*
r радиометрия *f*

2992 radiopositioning
d Radiopositionierung *f*
f radiopositionnement *m*
r радиопозиционирование *n*

2993 radio signal
d Funksignal *n*; Funkzeichen *n*
f radiosignal *m*
r радиосигнал *m*

2994 radius
d Radius *m*; Halbmesser *m*
f rayon *m*
r радиус *m*

2995 raingauge
d Regenmesser *m*
f pluviomètre *m*
r дождемер *m*; осадкомер *m*; плювиограф *m*

2996 rain map
d Regenkarte *f*
f carte *f* des pluies
r карта *f* осадков

2997 random data
d zufällige Daten *pl*
f données *fpl* aléatoires
r случайные данные *pl*

*   **random error → 16**

2998 range
d Entfernung *f*
f portée *f*
r дальность *f*

*   **range → 3360**

2999 range; diapason
  d Umfang *m*; Reichweite *f*; Wertebereich *m*;
    Diapason *m*
  f rangée *f*; plage *f*; gamme *f*
  r охват *m*; область *f* значений; диапазон *m*

  * **range chart → 3000**

3000 ranged map; range chart
  (a type of thematic map)
  d Reichweiten-Karte *f*; eingeordnete Karte *f*
  f carte *f* en rangées; carte de contrôle de
    l'étendue
  r ранжированная карта *f*

3001 range error
  d Fehler *m* nach der Länge
  f erreur *f* de distance; étendue *f* d'une erreur
  r ошибка *f* [по] дальности; отклонение *n* по
    дальности

3002 range extraction
  d Wertebereichsextraktion *f*;
    Diapasonextraktion *f*
  f extraction *f* de rangées
  r выделение *n* диапазонов

  * **range finding → 3004**

  * **range measurement → 3004**

**3003 range of sight; range of visibility; range of
vision**
  d Sehweite *f*
  f portée *f* de la vue; portée de visibilité
  r дальность *f* видимости; предел *m*
    видимости; расстояние *n* видимости

  * **range of visibility → 3003**

  * **range of vision → 3003**

**3004 ranging; range finding; range
measurement; distance measurement**
  d Entfernungsmessung *f*; Streckenmessung *f*;
    Bereichsauswahl *f*
  f mesure *f* de distance; sélection *f* de gamme;
    repérage *m*
  r определение *n* дальности; промеры *mpl*;
    замер *m* дальности; определение дистанции

  * **ranking → 176**

  * **raster → 3015**

3005 raster; scan pattern; grid
  (of a screen)
  d Raster *m*; Grid *n*
  f trame *f*; rastre *m*
  r растр *m*

3006 raster algebra
  d Raster-Algebra *f*
  f algèbre *f* rastre
  r алгебра *f* [обработки] растровых
    изображений

3007 raster background
  d Rasterhintergrund *m*
  f fond *m* rastre
  r растровый фон *m*

3008 raster data
  d Rasterdaten *pl*
  f données *fpl* de trame; données rastrées
  r растровые данные *pl*

3009 raster [data] model; bitmap model
  d Modell *n* der Rasterdaten
  f modèle *m* rastre; modèle de données de trame
  r растровая модель *f*; модель данных растра

3010 raster density
  d Rasterdichte *f*
  f densité *f* de rastre; densité de trame
  r плотность *f* растра

3011 raster editing
  d Rasterbearbeitung *f*
  f édition *f* de rastre
  r редактирование *n* значений пикселов

3012 raster file
  d Rasterdatei *f*
  f fichier *m* rastre; fichier trame
  r растровый файл *m*

3013 raster format
  d Rasterformat *n*
  f format *m* rastre
  r растровый формат *m*

3014 raster handler
  d Raster-Steuerungsprogramm *n*
  f programme *m* traiteur de rastre
  r программа *f* управления растра

**3015 raster [image]; matrix image; bitmap
[image]**
  d Rasterbild *n*; Bitmap *n*; Bitmap-Bild *n*;
    Pixelraster *m*
  f image *f* rastre; image matricielle; image
    pixélisée; image à points; bitmap *m*;
    topogramme *m* binaire
  r растровое изображение *n*; побитовое
    отображение *n*; поразрядное отображение

3016 raster image processing
  d Rasterbildverarbeitung *f*
  f traitement *m* d'images rastres
  r обработка *f* растровых изображений

3017 **raster image processor**
  d Rasterbildprozessor *m*; Raster-Prozessor *m*
  f processeur *m* d'images rastres
  r процессор *m* растровых изображений

  * **rasterization** → 3923

  * **rasterizator** → 3924

3018 **raster[ized] line; gridline**
  d Rasterlinie *f*
  f ligne *f* de rastre; ligne de trame; ligne de grille
  r линия *f* растра; растровая линия; линия сетки

  * **raster line** → 3018

  * **raster map** → 1772

  * **raster model** → 3009

3019 **raster object; bitmap object**
  d Rasterobjekt *n*; Bitmap-Objekt *n*
  f objet *m* rastre; objet bitmap
  r растровый объект *m*

3020 **raster plotter; scan converter**
  d Rasterplotter *m*
  f traceur *m* par ligne; traceur à trame
  r растровый плоттер *m*; растровый графопостроитель *m*

3021 **raster preprocessing**
  d Rastervorverarbeitung *f*
  f prétraitement *m* de rastre
  r предварительная обработка *f* растра

3022 **raster refresh**
  d Rasterauffrischung *f*
  f rafraîchissement *m* de rastre
  r регенерация *f* растра

3023 **raster representation; raster view**
  d Rasterdarstellung *f*
  f représentation *f* trame; vue *f* matricielle
  r растровое представление *n*

  * **raster scan** → 3024

3024 **raster scan[ning]; frame scan**
  d Rasterscannen *n*; Rasterabtastung *f*; Rasterpunktabfühlung *f*
  f balayage *m* [de] trame; balayage [récurrent]; balayage ligne par ligne
  r растровое сканирование *n*

3025 **raster snap**
  d Rasterfang *m*
  f accrochage *m* au rastre
  r привязка *f* к растру

3026 **raster space**
  d Rasterraum *m*
  f espace *m* rastre
  r растровое пространство *n*

3027 **raster structure**
  d Rasterstruktur *f*
  f structure *f* rastre
  r растровая структура *f*

3028 **raster[-to]-vector conversion; vectorization; linearization**
  d Raster-Vektor-Konvertierung *f*
  f conversion *f* de rastre en vecteur; vectorisation *f*
  r растрово-векторное преобразование *n*; векторизация *f*

3029 **raster-to-vector processor; vectorizer**
  d Raster-Vektor-Konvertor *m*
  f convertisseur *m* de rastre en vecteur
  r растрово-векторный процессор *m*; [программа-]векторизатор *m*

  * **raster-vector conversion** → 3028

  * **raster-vector editor** → 1858

  * **raster-vector GIS** → 1859

  * **raster view** → 3023

3030 **raw data; primary data**
  d Rohdaten *pl*; Urdaten *pl*; unbearbeitete Daten *pl*; Ursprungsdaten *pl*; Originaldaten *pl*
  f données *fpl* brutes; données crues
  r необработанные данные *pl*

3031 **ray-casting**
  (computational technique used to simulate a visual scene with optical effects, variations in light sources and other effects)
  d Ray-Casting *n*
  f transtypage *m* de rayons
  r отслеживание *n* лучей

3032 **ray path; ray trajectory; light path**
  d Strahlengang *m*; Strahlenbahn *f*; Strahlenverlauf *m*; Strahlweg *m*
  f trajectoire *f* de rayons [optiques]; trajet *m* de lumière; trajet de rayons; marche *f* de rayons
  r ход *m* лучей

3033 **ray tracing**
  d Ray-Tracing *n*; Strahlverfolgung *f*
  f raytracing *m*; lancé *m* de rayons
  r трассировка *f* лучей

  * **ray trajectory** → 3032

* **RDBMS** → 3100

3034 real estate analysis
d Grundbesitz-Analyse *f*;
Grundeigentum-Analyse *f*
f analyse *f* de propriété foncière
r анализ *m* недвижимости

* **real estate cadaster** → 422

3035 realism
d Realismus *m*
f réalisme *m*
r реализм *m*

* **reality** → 26

3036 reality visualization
d Realitätsvisualisierung *f*
f visualisation *f* de réalité
r визуализация *f* реальности

3037 realtime corrections
d Echtzeit-Korrekturen *fpl*
f corrections *fpl* en temps réel
r исправления *npl* в реальном времени

3038 realtime geographic visualization
d geografische Visualisierung *f* in Realzeit
f visualisation *f* géographique en temps réel
r географическая визуализация *f* в реальном времени

3039 realtime kinematics; TRK
d Realzeit-Kinematik *f*
f cinématique *f* en temps réel
r кинематика *f* в реальном времени

3040 realtime positioning
d Realzeit-Positionieren *n*;
Echtzeit-Positionierung *f*;
Realzeit-Ortsbestimmung *f*
f positionnement *m* en temps réel
r позиционирование *n* в реальном времени

3041 real-world phenomena
d Erscheinungen *fpl* der realen Welt
f phénomènes *mpl* de monde réel
r явления *npl* реального мира

* **rearrangement** → 3129

3042 reassignment
d Neuzuweisung *f*
f réattribution *f*
r переназначение *n*

3043 receiver; recipient
d Empfänger *m*
f récepteur *m*; destinataire *m*
r приёмник *m*; получатель *m*; реципиент *m*; адресат *m*

3044 receiver independent exchange format; RINEX format
(GPS data format)
d empfängerunabhängiges Austauschformat *n*; RINEX-Format *n*
f format *m* RINEX
r формат *m* RINEX

3045 receiver noise
d Empfänger-Rauschen *n*
f bruit *m* de récepteur
r помехи *fpl* приёмника

* **recipient** → 3043

3046 reclassification
d Reklassifikation *f*, Umgliederung *f*;
Umbuchung *f*; Neuzuordnung *f*;
Neueinteilung *f*
f reclassement *m*
r переклассификация *f*

3047 reclassified image
d reklassifiziertes Bild *n*
f image *f* reclassifiée
r переклассифицированное изображение *n*

3048 recognition
d Erkennung *f*
f reconnaissance *f*
r распознавание *n*; различение *n*

3049 recoloring
d wiederholte Färbung *f*
f récoloriage *m*
r повторное раскрашивание *n*; повторная раскраска *f*

* **record** → 3184

3050 record ID
d Datensatz-Identifikator *m*;
Satz-Identifikator *m*
f identificateur *f* d'enregistrement
r идентификатор *m* записи

3051 rectangular facets
d rechteckige Facetten *fpl*
f facettes *fpl* rectangulaires
r прямоугольные фацеты *mpl*

3052 rectangular grid
d rechteckiges Gitter *n*
f grille *f* rectangulaire
r прямоугольная сетка *f*

3053 rectangular regular grid
d rechteckiges regelmäßiges Gitter *n*

f grille *f* rectangulaire régulière
r равносторонняя прямоугольная сетка *f*

3054 rectification
 d Rektifikation *f*; Rektifizierung *f*; Entzerrung *f*;
   Gleichrichtung *f*
 f rectification *f*; redressement *m*; redressage *m*
 r спрямление *n*; выпрямление *n*;
   исправление *n*; ректификация *f*

3055 rectifier
 d Gleichrichter *m*
 f redresseur *m*
 r выпрямитель *m*

3056 recurrence interval
 d Rekurrenzintervall *n*
 f intervalle *m* de récurrence
 r интервал *m* возврата

3057 recursion
 d Rekursion *f*
 f récursion *f*
 r рекурсия *f*

3058 recursive descendant method
 d rekursive Absteigemethode *f*
 f méthode *f* récurrente décroissante
 r метод *m* рекурсивного спуска

3059 redistricting
 d Redistricting *n*; Neueinteilung *f* der
   Wahlkreise; Neubestimmung *f* der Grenzen
   der Landkreise
 f redistribution *f*; remaniement *m* électoral
 r повторное деление *n* на округа

3060 redraw *v*
 d zeichnen im Weiterschlag; ziehen im
   Nachzug; aktualisieren der Bildschirmanzeige
 f redessiner; retracer
 r перечерчивать; перерисовывать

 * **reducing** → 3061

3061 reduction; reducing; foreshortening
 d Reduktion *f*; Reduzierung *f*; Verkleinerung *f*;
   Verringerung *f*
 f réduction *f*
 r редукция *f*; уменьшение *n*; понижение *n*;
   приведение *n*; сокращение *n*

3062 redundancy
 d Redundanz *f*
 f redondance *f*
 r избыточность *f*; резерв *m*

3063 reference
 d Bezug *m*
 f référence *f*

r эталон *m*; опора *f*

3064 reference
 d Referenz *f*; Verweis *m*
 f référence *f*
 r ссылка *f*; обращение *n*

3065 reference area
 d Bezugsgebiet *n*
 f domaine *m* de référence
 r эталонная область *f*

3066 reference edge
 d Bezugskante *f*; Bezugsrand *m*
 f marge *f* de référence
 r опорный край *m*

3067 reference ellipsoid
 d Referenz-Ellipsoid *n*
 f ellipsoïde *m* référentiel
 r эталонный эллипсоид *m*; ссылочный
   эллипсоид; референц-эллипсоид *m*

 * **reference frame** → 1475

 * **reference framework** → 1475

3068 reference grid
 d Referenzgitter *n*
 f quadrillage *m* de référence
 r модульная сетка *f*; масштабная сетка

3069 reference map
 d Referenzplan *m*
 f carte *f* référentielle
 r эталонная карта *f*

3070 reference meridian
 d Bezugsmeridian *m*
 f méridien *m* de référence
 r осевой меридиан *m*

3071 reference network
 d Referenznetz *n*
 f réseau *m* référentiel
 r эталонная сеть *f*

 * **reference point** → 2811

3072 reference receiver
 d Referenzempfänger *m*
 f récepteur *m* référentiel
 r эталонный приёмник *m*

 * **reference station** → 318

3073 reference system; referencing system
 d Bezugssystem *n*; Referenzsystem *n*

f système *m* de référence
r эталонная система *f*; система отсчёта;
референцная система

3074 referencing
d Bezugnahme *f*; Phasenregelung *f*
f référencement *m*
r эталонирование *n*; привязка *f*

* **referencing system → 3073**

3075 referential integrity
d referentielle Integrität *f*; referentielle
Konsistenz *f*
f intégrité *f* référentielle
r целостность *f* на уровне ссылок; ссылочная
целостность

* **refinement → 3076**

3076 refining; refinement
d Verfeinigung *f*; Verfeinerung *f*
f raffinement *m*; raffinage *m*
r измельчение *n*; усовершенствование *n*;
детализация *f*

* **reformat → 3077**

3077 reformat[ting]
d Umformatieren *n*; Formatänderung *f*
f reformatage *m*; modification *f* du format
r переформатирование *n*; изменение *n*
формата

3078 refresh *v*; freshen *v* up
d auffrischen; erneuern; Bild aktualisieren
f rafraîchir
r обновлять; освежать

3079 region
(contiguous areas with common or
complementary characteristics or linked by
intensive interaction or flows)
d Region *f*
f région *f*
r регион *m*; район *m*

3080 regional atlas
d regionaler Atlas *m*
f atlas *m* régional
r региональный атлас *m*

3081 regional database
d regionale Datenbasis *f*
f base *f* de données régionale
r региональная база *f* данных

3082 regional GIS
d regionales GIS *n*
f SIG *m* régional
r региональная ГИС *f*

3083 regionalism
d Regionalismus *m*
f régionalisme *m*
r регионализм *m*; местничество *n*

3084 regionalization
d Bereichsunterteilung *f*
f régionalisation *f*
r разбиение *n* на регионы; разбиение на
области

3085 region[al] map
d regionale Karte *f*; Gebietskarte *f*;
Regionkarte *f*
f carte *f* régionale; carte de région; carte
chorographique
r региональная карта *f*

3086 region fill
d Regionfüllung *f*
f remplissage *m* de régions
r закрашивание *n* регионов

* **region map → 3085**

3087 region of a map; map region
d Kartenbereich *m*
f région *f* de carte
r область *f* карты

3088 region specification
d Regionsspezifikation *f*
f spécification *f* de région
r спецификация *f* региона

3089 region style
d Regionstil *m*
f style *m* de régions
r стиль *m* регионов

* **register → 3091**

3090 registered table
d registrierte Tabelle *f*
f table *f* enregistrée
r регистрированная таблица *f*

3091 register[ing]; registration
d Registrierung *f*
f enregistrement *m*
r регистрирование *n*

* **register of real estate → 422**

* **registration → 3091**

* **registry map → 2652**

3092 regression
d Regression *f*

f régression *f*
r регрессия *f*

3093 regression curve; regression line
  d Regressionskurve *f*
  f courbe *f* de régression
  r кривая *f* регрессии

  * **regression line** → 3093

  * **regular error** → 3567

3094 regular grid
  d regelmäßiges Gitter *n*
  f grille *f* régulière
  r регулярная сетка *f*

3095 regularly spaced data
  d regelmäßiggezeigte Daten *pl*
  f données *fpl* régulièrement ramifiées
  r правильно расположенные данные *pl*;
    расположенные с равными интервалами
    данные; эквидистантные данные

3096 regularly spaced profiles
  d regelmäßiggezeigte Profile *npl*
  f profils *mpl* régulièrement ramifiés
  r эквидистантные профили *mpl*

3097 relate *v*
  d verbinden; synchronisieren
  f coupler; étalonner
  r устанавливать [co]отношение;
    устанавливать связь

3098 relational cartographic database
  d relationale kartografische Datenbasis *f*
  f base *f* de données cartographiques
    relationnelle
  r реляционная картографическая база *f*
    данных

3099 relational database
  d relationale Datenbasis *f*
  f base *f* de données relationnelle
  r реляционная база *f* данных

3100 relational database management system;
    RDBMS
  d relationales Datenbasis-Managementsystem *n*
  f système *m* de gestion de base de données
    relationnelle; SGDBR
  r система *f* управления реляционных баз
    данных

3101 relational indexing
  d Verknüpfungsindexierung *f*;
    Beziehungsindexierung *f*
  f indexation *f* relationnelle
  r реляционное индексирование *n*

3102 relational join
  d relationale Verbindung *f*
  f jonction *f* relationnelle
  r реляционная связь *f*

3103 relation key; index key
  (the common set of columns used to relate two
  attribute tables)
  d Indexschlüssel *m*
  f clé *f* d'index
  r ключ *m* отношения; ключ индекса

3104 relative accuracy
  d relative Genauigkeit *f*
  f précision *f* relative
  r относительная точность *f*

3105 relative height
  d relative Höhe *f*
  f hauteur *f* relative
  r относительная высота *f*

3106 relative location; relative position
  d relative Lokalisierung *f*
  f localisation *f* relative
  r относительное расположение *n*

3107 relative navigation
  d relative Navigation *f*
  f navigation *f* relative
  r относительная навигация *f*

3108 relative pointing device
  d relativer Positionsanzeiger *m*; relative
    Steuervorrichtung *f*; relatives Zeigegerät *n*
  f dispositif *m* de pointage relatif; dispositif de
    désignation relatif
  r относительное указательное устройство *n*

  * **relative position** → 3106

3109 relative positioning
  d relative Positionierung *f*
  f positionnement *m* relatif
  r относительное позиционирование *n*

3110 relaxation
  d Relaxation *f*
  f relaxation *f*
  r релаксация *f*

3111 relevance
  d Bedeutung *f*
  f pertinence *f*
  r релевантность *f*

3112 reliability of cartographic method of
    research
  d Zuverlässigkeit *f* der kartografischen
    Forschungsmethode

f fiabilité *f* de la méthode cartographique de recherche
r надёжность *f* картографического метода исследования

3113 reliability of map investigations
d Zuverlässigkeit *f* der Investitionen in der Kartografie
f fiabilité *f* d'investigations cartographiques
r надёжность *f* исследований по картам

* **relief** → 3605

3114 relief carte; relief map; feature map
d Reliefkarte *f*; Höhenschichtenkarte *f*
f carte *f* en relief
r рельефная карта *f*

* **relief effect** → 1260

* **relief lining** → 3137

* **relief map** → 3114

* **relief model** → 3614

3115 relief shading
d Reliefschattierung *f*
f ombrage *m* de relief
r оттенение *n* рельефа

3116 remote data collection
d Ferndatenerfassung *f*
f collection *f* de données éloignée
r дистанционный сбор *m* данных

3117 remote image processing
d Fernbildverarbeitung *f*
f traitement *m* d'images éloigné
r дистанционная обработка *f* данных

* **remotely sensed data** → 3121

3118 remote object evaluation
d Fernobjektsbewertung *f*
f évaluation *f* d'objets distante
r дистанционная оценка *f* объектов

3119 remote positioning unit
d Ferneinstellgerät *n*
f unité *f* de positionnement distante
r дистанционное позиционирующее устройство *n*

3120 remote sensing; remote surveying; RS
d Fernerkundung *f*; Fernablesung *f*; Fernabtastung *f*
f télédétection *f*; détection *f* à distance
r дистанционное зондирование *n*; ДЗ

3121 remote sensing data; remotely sensed data; remote surveying data; aerospace data
d Fernerkundungsdaten *pl*
f données *fpl* de télédétection
r данные *pl* дистанционного зондирования; данные аэрокосмического зондирования

3122 remote sensing generalization; optical generalization
d Fernerkundung-Generalisierung *f*
f généralisation *f* par télédétection
r дистанционная генерализация *f*

3123 remote sensing methods; distant methods
d Fernerkundungsmethoden *fpl*; Fernerkundungsverfahren *npl*
f méthodes *fpl* de télédétection
r методы *mpl* дистанционного зондирования

3124 remote sensor
d Fernsensor *m*
f télédétecteur *m*; télécapteur *m*
r дистанционный зонд *m*

* **remote surveying** → 3120

* **remote surveying data** → 3121

3125 remote table
d Ferntabelle *f*
f table *f* à distance
r дистанционная таблица *f*

3126 renaming
d Umbenennung *f*
f renomination *f*
r переименование *n*

3127 rendering
d Rendering *n*; Rendern *n*
f rendu *m*
r рендеринг *m*; тонирование *n*

3128 renumber *v*
d umnumerieren; neu numerieren
f rénuméroter
r перенумеровывать

3129 reordering; rearrangement
d Umordnung *f*; Neuordnung *f*
f réarrangement *m*
r переупорядочение *n*; перестройка *f*

3130 reordering of layers
d Schicht-Umordnung *f*
f réarrangement *m* de couches
r переупорядочение *n* слоев

3131 repeatability
d Wiederholbarkeit *f*

f répétitivité *f*
r повторяемость *f*

* **reporter** → 3132

* **report generation** → 3133

3132 **report generator; report writer; reporter; list generator**
   d Reportgenerator *m*; Listengenerator *m*
   f générateur *m* de rapports; générateur de listes
   r генератор *m* отчётов

3133 **reporting; report generation**
   d Reportgenerierung *f*
   f génération *f* de rapports
   r документирование *n*; генерация *f* отчётов

* **report writer** → 3132

3134 **repositioning**
   d Wiederpositionierung *f*; Verstellung *f*
   f repositionnement *m*
   r повторное позиционирование *n*

3135 **representational feature units**
   d repräsentative Merkmalseinheiten *fpl*
   f unités *fpl* caractéristiques descriptives
   r репрезентативные единицы *fpl* рельефа

3136 **representation of relief by contours**
   d Reliefdarstellung *f* durch Höhenlinien
   f représentation *f* de relief en courbes de niveau
   r изображение *n* рельефа горизонталями

3137 **representation of relief by hachures; relief lining**
   d Reliefdarstellung *f* durch Schraffen
   f représentation *f* de relief en hachures
   r изображение *n* рельефа штрихами

3138 **representation of terrain; representation of topographical surfaces**
   d Geländedarstellung *f*
   f représentation *f* de terrain; représentation de surfaces topographiques
   r представление *n* местности; представление топографических поверхностей

* **representation of topographical surfaces** → 3138

* **representative fraction** → 3139

3139 **representative fraction [scale]; RF scale; natural scale**
   d numerischer Maßstab *m*
   f échelle *f* numérique
   r численный масштаб *m*

* **reproduction scale** → 3218

3140 **request; interrogation; demand; inquiry; enquiry; query**
   d Abfrage *f*; Anfrage *f*; Auftrag *m*
   f requête *f* [logique]; interrogation *f*; question *f*; demande *f*
   r запрос *m*; опрос *m*; заказ *m*; справка *f*

3141 **request for information; RFI**
   d Abfrage *f* für Information
   f demande *f* d'information
   r запрос *m* на информацию

3142 **request for proposal; RFP**
   d Abfrage *f* für Vorschlag
   f demande *f* de proposition
   r запрос *m* на проектирование; запрос на утверждение; техническое задание *n*

3143 **resampling**
   d wiederholte Stichprobenauswahl *f*; Umrechnung *f*
   f rééchantillonnage *m*
   r повторная выборка *f*

3144 **rescale** *v*
   d Skale ändern
   f changer d'échelle
   r перемасштабировать

3145 **resection; trilinear surveying**
   (the locating of a single point by measuring horizontal angles from it to three known points)
   d Rückwärtseinschneiden *n*; Rückwärtseinschnitt *m*
   f relèvement *m*
   r [геодезическая] засечка *f*

3146 **reselection**
   d wiederholte Selektion *f*
   f resélection *f*
   r повторный отбор *m*; повторный выбор *m*

3147 **reservation**
   d Reservation *f*; Reservierung *f*
   f réservation *f*
   r резервирование *n*

3148 **reshape** *v*
   d neuprofilieren
   f reprofiler
   r придавать новую форму; приобретать новую форму

3149 **residential area; residential neighborhood; urban residential district**
   d Wohngebiet *n*; Wohnbezirk *m*

f zone *f* résidentielle; zone d'habitat
r жилой район *m*; жилая зона *f*

* **residential neighborhood** → 3149

3150 **residential quarter; quarter; square; block**
d Wohnviertel *n*; Wohnquartier *n*
f quartier *m* [d'habitation]; quartier résidentiel
r [жилой] квартал *m*; квартал города;
жилищный массив *m*

3151 **resizing**
d Größenänderung *f*; Vergrößern/Verkleinern *n*
f changement *m* de taille
r изменение *n* размера

3152 **resolution**
d Auflösung *f*
f résolution *f*
r разрешающая способность *f*; разрешение *n*

* **resolution cell** → 1764

3153 **resolving of ambiguity**
d Ambiguitätsauflösung *f*
f résolution *f* d'ambiguïté
r разрешение *n* неоднозначности

3154 **resource information system**
d Informationssystem *n* über die Ressourcen
f système *m* d'information sur les ressources
r справочная система *f* по информационным
ресурсам

* **responsivity** → 3263

3155 **restriction; constraint**
d Restriktion *f*; Einschränkung *f*;
Beschränkung *f*; Nebenbedingung *f*;
Grenzbedingung *f*
f restriction *f*; contrainte *f*
r ограничение *n*; рестрикция *f*

* **retardation** → 952

3156 **retrace line**
d Zeilenrücklauf *m*
f retour *m* ligne
r повторно трассируемая линия *f*

* **retrieval** → 3241

* **reverse azimuth** → 299

3157 **reverse polarity**
d inverse Polarität *f*
f polarité *f* reverse
r обратная полярность *f*

3158 **reverse surface**

d inverse Fläche *f*
f surface *f* reverse
r обратная поверхность *f*

3159 **revert** *v*
d zurückgeben; zurückkehren
f retourner; revenir
r возвращаться

* **revolution ellipsoid** → 1256

* **RFI** → 3141

* **RFP** → 3142

* **RF scale** → 3139

* **rhumb** → 2810

3160 **rhumb line; spherical helix; loxodrome;**
loxodromic spiral; loxodromic line;
loxodromic curve
(on a sphere or on the earth's surface)
d Rhumblinie *f*; Loxodrome *f* auf der Erdkugel;
Schieflaufende *f*
f loxodromie *f*; ligne *f* loxodromique; hélice *f*
sphérique; loxodrome *m*
r локсодромия *f*; локсодрома *f*; локсодромная
спираль *f*

3161 **ridge**
(line along which a surface diverges in two
different directions)
d Rücken *m*
f dorsale *f*; croupe *f*
r хребет *m*; гребень *m*

* **ridge-line** → 3975

* **rigid body** → 3339

* **RINEX format** → 3044

3162 **ring**
d Ring *m*
f anneau *m*; bague *f*
r кольцо *n*; замкнутая последовательность *f*
непересекающихся цепей или дуг

3163 **river channels; river network; stream**
network; river system
d Gewässernetz *n*; Gewässersystem *n*;
Flussnetz *n*
f réseau *m* hydrographique; hydrosystème *m*
r речная сеть *f*

* **river network** → 3163

* **river system** → 3163

*   RLE → 3189

*   RMSE → 3172

3164 **RMSE of unit weight; standard error of unit weight**
    d Gewichtseinheitsfehler *m*
    f erreur *f* moyenne quadratique de l'unité de poids
    r среднеквадратическая погрешность *f* единицы веса

*   road → 3177

3165 **road atlas**
    d Wegatlas *m*
    f atlas *m* de routes
    r дорожный атлас *m*

3166 **road classification**
    d Wegklassifizierung *f*
    f classification *f* de routes
    r классификация *f* дорог

3167 **road junction with overpass; road overpass; elevated road; elevated guide way; overhanging road**
    d aufgeständerte Straße *f*; Hochstraße *f*; hochgeführte Straße
    f voie *f* surélevée; route *f* surélevée; route en encorbellement; encorbellement *m*
    r транспортная развязка *f*

3168 **road map**
    d Autokarte *f*; Wegekarte *f*; Verkehrskarte *f*; Straßenkarte *f*
    f carte *f* de routes; carte routière
    r [авто]дорожная карта *f*; карта дорог; схема *f* дорог

*   road overpass → 3167

3169 **road symbology**
    d Wegsymbole *npl*; Wegsymbolsätze *mpl*
    f symboles *mpl* de routes
    r дорожная символика *f*; дорожные символические обозначения *npl*

3170 **rollback**
    d Rückkehr *f*
    f recul *m*; roulement *m* en arrière
    r откат *m*; отмена *f*

3171 **roll-feed plotter**
    d Walzenvorschubplotter *m*
    f traceur *m* à rouleaux
    r рулонный плоттер *m*; роликовый графопостроитель *m*; рулонный графопостроитель

*   rolling ball → 3777

3172 **root-mean-square error; RMSE**
    d quadratischer Mittelwert *m*
    f erreur *f* moyenne quadratique
    r среднеквадратическая погрешность *f*

3173 **root node; top node**
    d Wurzelknoten *m*; Wurzel *f*
    f nœud *m* racine; source *f* nœud
    r корневой узел *m*

3174 **root of a tree**
    d Wurzel *f* eines Baums
    f racine *f* d'un arbre
    r корень *m* дерева

3175 **rotated image**
    d gedrehtes Bild *n*; rotiertes Bild
    f image *f* en rotation
    r перевёрнутое изображение *n*

3176 **rotation**
    d Rotation *f*
    f rotation *f*
    r вращение *n*

*   rough error → 367

*   roughness → 3851

3177 **route; path; way; road; track**
    d Leitweg *m*; Route *f*; Weg *m*; Bahn *f*; Pfad *m*
    f route *f*; chemin *m*; voie *f*; cours *m*
    r маршрут *m*; путь *m*; дорога *f*; курс *m*

3178 **route attribute table**
    d Weg-Attributtabelle *f*
    f table *f* d'attributs de route
    r таблица *f* атрибутов маршрута

3179 **route measure**
    d Wegmessung *f*
    f mesure *f* de route
    r мера *f* маршрута

3180 **route network; route system**
    d Wegnetz *n*; Wegsystem *n*
    f réseau *m* de routes; système *m* de chemins; système d'itinéraire
    r система *f* маршрута

3181 **route number**
    d Wegnummer *f*
    f numéro *m* de route
    r номер *m* маршрута

*   route system → 3180

3182 **routing**
    d Leitweglenkung *f*; Leitwegsuchen *n*; Wegewahl *f*; Lenkung *f*; Routing *n*

f acheminement *m*; choix *m* d'itinéraire;
routage *m*

r маршрутизация *f*; выбор *m* пути;
трассировка *f*

3183 routing grid; placement grid
d Trassenführungsgitter *n*
f grille *f* d'acheminement
r координатная сетка *f* для трассировки;
сетка для трассировки

3184 row; data record; record; tuple
(of a table)
d Zeile *f*; Datensatz *m*; Satz *m*
f ligne *f*; enregistrement *m* [de données]
r строка *f*; запись *f*; кортеж *m*

* **row count** → **2580**

* **row number** → **2183, 2580**

* **RS** → **3120**

3185 R-tree
d R-Baum *m*
f R-arbre *m*
r R-дерево *n*

* **rubberband** → **3186**

3186 rubberband[ing]; rubbersheeting
(a process of stretching or shrinking line
lengths to fit between established reference
points)
d Einpassen *n* mit Gummibandfunktion;
Gummiband *n*
f étirement *m* par fil élastique
r эластичное соединение *n*; соединение
резиновой нитью

* **rubbersheeting** → **3186**

3187 ruler
d Zeilenlineal *n*; Lineal *n*
f rouleur *m*; règle *f*
r [измерительная] линейка *f*

3188 run length
d Lauflänge *f*
f longueur *f* de parcours
r длина *f* отрезка

* **run-length coding** → **3189**

**3189 run-length encoding; RLE; run-length
coding**
d Lauflängencodierung *f*
f codage *m* de répétitions; codage par plages
r групповое кодирование *n*; кодирование
группами отрезков

3190 running water information system
d Fliessgewässer-Informationssystem *n*;
Informationssystem *n* für die Fliessendwasser
f système *m* automatique d'information sur des
eaux courantes
r информационная система *f* для проточных
вод

* **rural area** → **784**

* **rural district** → **784**

S

* SA → 3260

3191 sample
 d Muster *n*; Stichprobe *f*; Probe *f*; Abtastwert *m*
 f échantillon *m* [d'essai]; modèle *m* d'essai;
    étalon *m*
 r образец *m*; выборка *f*; проба *f*; отсчёт *m*

3192 sample mean
 d Stichprobenmittel *n*; Stichprobenmittelwert *m*
 f moyenne *f* d'échantillon
 r выборочное среднее *n*

3193 samples analysis
 d Probenanalyse *f*
 f analyse *f* d'échantillons
 r анализ *m* выборок

3194 sampling
 d Stichprobenauswahl *f*; Stichprobenerhebung *f*
 f échantillonnage *m*; prélèvement *m*
    d'échantillons
 r процесс *m* выборки; опробирование *n*;
    отбор *m* выборок

3195 sampling error
 d Abtastfehler *m*
 f erreur *f* d'échantillonnage
 r погрешность *f* сканирования

* sampling time → 3227

3196 sampling transformation
 d Stichprobentransformation *f*
 f transformation *f* d'échantillonnage
 r преобразование *n* выборок

* SAR → 3562

* satellite-based position fixing → 3204

3197 satellite data
 d Satelliten[bild]daten *pl*
 f données *fpl* satellitaires
 r спутниковые данные *pl*

3198 satellite geodesy; celestial geodesy; space geodesy
 d Satellitengeodäsie *f*
 f géodésie *f* sur satellites; géodésie satellitaire
 r космическая геодезия *f*; спутниковая
    геодезия

3199 satellite geometry
 d Satellitengeometrie *f*
 f géométrie *f* satellitaire
 r спутниковая геометрия *f*

3200 satellite image
 d Satellitenbild *n*
 f image *f* satellitaire
 r спутниковое изображение *n*

3201 satellite image classification
 d Satellitenbildklassifikation *f*
 f classification *f* d'images satellitaires
 r классификация *f* спутниковых
    изображений

* satellite image maps → 3353

3202 satellite imagery
 d Satellitenaufnahmen *fpl*
 f imagerie *f* satellitaire; visionnique *m*
    satellitaire
 r спутниковые изображения *npl*

3203 satellite navigation system
 d Satellitennavigationssystem *n*
 f système *m* de navigation satellitaire
 r спутниковая навигационная система *f*

3204 satellite position fixing; satellite-based position fixing
 d satellitengestützte Ortung *f*;
    Standortbestimmung *f* per Satellit
 f détermination *f* de position par satellite;
    relèvement *m* de positions par satellite
 r определение *n* позиции спутником;
    фиксирование *n* позиции спутником

3205 satellite status display
    (a GPS screen which provides information
    about satellites)
 d Satellitenstatusanzeiger *m*
 f afficheur *m* d'état de satellite
 r экран *m* состояния спутников; индикатор *m*
    состояния спутников

3206 satellite triangulation
 d Satellitentriangulation *f*
 f triangulation *f* satellitaire
 r спутниковая триангуляция *f*

3207 saturation
 d Sättigung *f*
 f saturation *f*
 r насыщение *n*

3208 save *v*
 d aufbewahren; sicherstellen
 f conserver; sauvegarder
 r сохранять

3209 scalar field
(a surface whose value can be represented by
a single number)
d skalares Feld *n*
f champ *m* scalaire
r скалярное поле *n*

3210 scale
d Maßstab *m*
f échelle *f*
r масштаб *m*

3211 scale; graduation
d Skala *f*; Skale *f*; Tonleiter *f*
f échelle *f*; graduation *f*
r шкала *f* (*графическое изображение
последовательности изменения цвета,
насыщенности, количественных
характеристик условных знаков*)

3212 scale accuracy
d Maßstab-Genauigkeit *f*
f exactitude *f* d'échelle
r точность *f* масштаба [карты]

3213 scale accuracy limit
d Maßstab-Genauigkeitsgrenze *f*
f limite *f* d'exactitude d'échelle
r предельная точность *f* масштаба

3214 scale bar
d Maßstabslinie *f*; Maßstabsbalken *m*
f barre *f* d'échelle
r масштабная линия *f*

3215 scale error
d Maßstabsfehler *m*; Skalenfehler *m*
f erreur *f* d'échelle
r погрешность *f*, вносимая шкалой

**3216 scale factor; scaling factor; scaling
multiplier**
d Skalenfaktor *m*; Maßstabsfaktor *m*
f facteur *m* d'échelle
r масштабный коэффициент *m*; масштабный
множитель *m*

3217 scale of projection; projection scale
d Projektionsmaßstab *m*
f échelle *f* de projection
r масштаб *m* проекции

3218 scale of reproduction; reproduction scale
d Endmaßstab *m*; Abbildungsmaßstab *m*
f échelle *f* d'édition; échelle de reproduction
r масштаб *m* издания

3219 scale of survey
d Aufnahmemaßstab *m*
f échelle *f* du levé

r съёмочный масштаб *m*

3220 scale reduction factor
d Reduzierung-Maßstabsfaktor *m*
f coefficient *m* de réduction d'échelle
r коэффициент *m* уменьшения масштаба

3221 scaling
d Skalierung *f*; Maßstabsänderung *f*
f mise *f* à l'échelle; choix *m* d'échelle
r масштабирование *n*; выбор *m* масштаба;
пересчёт *m*

* **scaling factor** → 3216

* **scaling multiplier** → 3216

* **scan** → 3225

* **scan converter** → 3020

* **scan digitizing** → 3225

* **scan line** → 3226

3222 scanned data
d gescannte Daten *pl*
f données *fpl* scannées
r сканированные данные *pl*

3223 scanned map
d gescannte Karte *f*
f carte *f* scannée
r сканированная карта *f*

3224 scanner interface processor
d Scanner-Anpassungsprozessor *m*
f processeur *m* d'interface de scanner
r процессор *m* интерфейса сканера

3225 scan[ning]; scan digitizing
d Abtastung *f*; Abtasten *n*; Scanning *n*;
Scannen *n*; Scan *n*
f balayage *m*; scrutation *f*; exploration *f*;
scannérisation *f*; scannage *m*
r сканирование *n*

3226 scan[ning] line
d Abtastzeile *f*; Abtastlinie *f*
f ligne *f* de balayage; ligne d'exploration; ligne
de lecture
r строка *f* сканирования; строка развёртки;
полоса *f* сканирования

3227 scanning time; sampling time
d Abtastzeit *f*
f temps *m* d'échantillonnage; temps de
scannérisation
r время *n* выборки

* scan pattern → 3005

* scatter chart → 3228

* scatter diagram → 3228

3228 scatter graph; scatter chart; scatter plot;
scatter diagram; dispersion diagram; dot
chart
d Streu[ungs]diagramm *n*; Punktdiagramm *n*;
Haufendiagramm *n*
f graphique *m* de dispersion; diagramme *m* de
dispersion; nuage *m* de points; diagramme à
points; graphique par points
r диаграмма *f* разброса; диаграмма
рассеивания; точечная диаграмма

* scatter plot → 3228

3229 schematic map; sketch map
d schematische Karte *f*
f carte *f* schématique
r картосхема *f*; карта-схема *f*

3230 scientific-reference atlas
d wissenschaftlicher Atlas *m*
f atlas *m* scientifique
r научно-справочный атлас *m*

* scissoring → 571

* scratch file → 4003

3231 screen; display
d Bildschirm *m*
f écran *m*
r экран *m*; дисплей *m*

3232 screen capture; screen dump; screenshot
d Bildschirm-Sammeln *n*
f capture *f* d'écran; copie *f* d'écran
r экранный дамп *m*; скриншот *m*

* screen dump → 3232

* screenshot → 3232

3233 script[ing] language
d Skriptsprache *f*
f langage *m* de scripts
r язык *m* [описания] сценариев; язык
подготовки сценариев; сценарный язык

* script language → 3233

* scroll → 3235

* scrollbar → 3234

3234 scrollbar [slider]; scroll box; elevator;
slider box; thumb [mark]; slider [indicator]
d Rollbalken *m*; Bildlaufleiste *f*;
Schieberegler *m*
f barre *f* de défilement; case *f* de défilement;
curseur *m* de défilement; ascenseur *m*
r линейка *f* прокрутки; лента *f*
прокручивания; линейка просмотра;
лифт *m*

* scroll box → 3234

3235 scroll[ing]
d Rollen *n*; Blättern *n*;
Bild[schirmzeilen]verschiebung *f*; Bildlauf *m*;
Bildrollen *n*
f défilement *m*; décalage *m* vertical
r прокрутка *f*; прокручивание *n*; скроллинг *m*

3236 scrubbing
(the process of preparing data for input to a
GIS, intended to eliminate errors)
d Auswaschen *n*
f épuration *f*
r чистка *f*; очистка *f*

* SD → 3306

* SDTS → 3379

3237 sea level; zero level; ordnance datum
d Meeresspiegel *m*; Normalnull *f*;
Normal-Niveau *n*; NN
f niveau *m* de la mer; niveau zéro; zéro *m*
r уровень *m* моря; нуль *m* государственных
нивелировок

3238 seam
(the junction in the area of overlap between
raster objects combined by tiling or
mosaicing)
d Fuge *f*; Nacht *f*; Falte *f*; Narbe *f*; Flöz *n*;
Lager *n*
f filon *m*
r место *n* соединения; линия *f* сращения

* seamless data → 717

* seamless database → 3239

3239 seamless [geographical] database; logically
continuous database
d blattschnittfreie [geografische] Datenbasis *f*
f base *f* de données [géographiques] sans
soudure
r бесшовная база *f* данных

* seamless layer → 3240

* seamless map → 720

3240 **seamless [map] layer**
  d spaltfreie Schicht *f*
  f couche *f* sans soudure
  r бесшовный слой *m*; цельнокроенный слой

  * **search** → 3241

3241 **search[ing]; seek[ing]; retrieval**
  d Suche *f*; Suchen *n*; Untersuchung *f*
  f recherche *f*
  r поиск *m*; искание *n*

  * **search of optimum path** → 3259

3242 **secant**
  d Sekante *f*
  f sécante *f*
  r секанс *m*; секущая *f*

3243 **second**
  d Sekunde *f*
  f seconde *f*
  r секунда *f*

3244 **secondary data**
  d sekundäre Daten *pl*
  f données *fpl* secondaires
  r производные данные *pl*

3245 **second normal form**
  d zweite Normal[en]form *f*
  f deuxième forme *f* normale
  r вторая нормальная форма *f*

3246 **second order effect**
  d Effekt *m* zweiter Ordnung
  f effet *m* d'ordre secondaire
  r вторичный эффект *m*

  * **second projection** → 3940

3247 **section**
  (of a document)
  d Abschnitt *m*
  f section *f*
  r раздел *m*

3248 **section; link**
  d Sektion *f*; Abschnitt *m*; Schnitt *m*; Strecke *f*
  f section *f*; coupe *f*
  r секция *f*; отсек *m* [линии]; участок *m*;
  звено *n*

3249 **sectional aeronautical chart**
  d eingeteilte Luftfahrtkarte *f*
  f carte *f* aéronautique en coupe; carte
  aéronautique en plusieurs feuilles
  r аэронавигационная карта *f* в разрезе

  * **sector diagram** → 2753

  * **seek** → 3241

  * **seeking** → 3241

  * **seen area map** → 3954

3250 **segment**
  d Segment *n*
  f segment *m*
  r сегмент *m*; отрезок *m*

3251 **segmentation**
  d Segmentierung *f*
  f segmentation *f*
  r сегментирование *n*; сегментация *f*

3252 **segmented image classification**
  d Klassifizierung *f* von segmentiertem Bild
  f classification *f* d'images segmentées
  r классификация *f* сегментированных
  изображений

3253 **segment number**
  d Segmentnummer *f*
  f numéro *m* de segment
  r номер *m* сегмента

  * **seizing** → 2609

  * **seizure** → 2609

3254 **selectable layer**
  d wählbare Schicht *f*
  f couche *f* sélectable
  r выбираемый слой *m*

3255 **selected area**
  d ausgewählter Bereich *m*
  f zone *f* sélectionnée
  r выбранная область *f*

  * **selected entity set** → 1282

3256 **selected object**
  d ausgewähltes Objekt *n*
  f objet *m* sélectionné
  r выбранный объект *m*

3257 **selection**
  d Selektion *f*; Markierung *f*
  f sélection *f*
  r выбор *m*

  * **selection box** → 3258

3258 **selection marquee; marquee; selection box**
  d Zirkuszelt *n*; Partyzelt *n*

f rectangle *m* de sélection
r шатер *m*; отмеченная область *f*
(*охватывающая выделённые объекты или части изображения*)

3259 selection of optimum route; optimal path selection; search of optimum path
d Auswahl *f* der optimalen Route; Optimalwegauswahl *f*
f sélection *f* de route optimale; recherche *f* de route optimale
r выбор *m* оптимального маршрута; выбор оптимального пути; поиск *m* оптимального маршрута

3260 selective availability; SA
d selektive Verfügbarkeit *f*
f disponibilité *f* sélective
r селективная наличность *f*

* **SEM** → 3514

3261 semi-automated digitizing
d halbautomatische Digitalisierung *f*
f digitalisation *f* demi-auto; numérisation *f* semi-automatisée
r полуавтоматизированное оцифрование *n*; полуавтоматическое цифрование *n*

* **semiological factor** → 1749

3262 semi-variogram
d Halbvariogramm *n*
f semi-variogramme *m*
r полу-вариограмма *f*

3263 sensibility; sensitivity; responsivity
d Empfindlichkeit *f*
f sensitivité *f*
r чувствительность *f*

* **sensitivity** → 3263

3264 sensitivity analysis
d Empfindlichkeitsanalyse *f*
f analyse *f* de sensitivité
r анализ *m* чувствительности

3265 sensor
d Sensor *m*
f senseur *m*
r сенсор *m*; датчик *m*

* **SEP** → 3435

3266 sequential access; serial access; consecutive access
d sequentieller Zugriff *m*; serieller Zugriff; aufeinanderfolgender Zugriff
f accès *m* séquentiel; accès en série; accès

consécutif
r последовательный доступ *m*

3267 sequential file
d sequentielle Datei *f*
f fichier *m* séquentiel
r последовательный файл *m*

* **serial access** → 3266

3268 service area map
d Verkehrsgebietskarte *f*
f carte *f* de zone de service; carte de zone de couverture; carte de desserte
r карта *f* обслуживаемой области; карта зоны обслуживания

* **set** → 3269

3269 set[ting]; setup; establishment
d Einstellung *f*; Einstellen *n*
f établissement *m*; mise *f* au point
r установление *n*; установка *f*; настройка *f*

* **setup** → 3269

3270 sextant
d Sextant *m*
f sextant *m*
r шестая часть *f* окружности

* **shade** → 3276

3271 shade; hatch; hachure
d Schraffur *f*; Abdunkeln *n*
f hachure *f*
r штриховка *f*; штрихи *mpl* [для обозначения профилей местности]

3272 shaded relief image
d schattiertes Reliefbild *n*
f image *f* ombrée de relief
r затененное изображение *n* рельефа

3273 shaded symbol; shadow[ed] symbol
d schattiertes Symbol *n*
f symbole *m* ombré
r символ *m* с тенью; оттененный символ

* **shade of gray** → 1758

3274 shading
d Schattierung *f*; Schattieren *n*; Abschattung *f*; Beschattung *f*
f ombrage *m*; nuançage *m*
r оттенение *n*; затенение *n*; отмывка *f*; ретуширование *n*

3275 shading; hatching
d Schraffieren *n*; Schraffen *n*

f hachure *f*
r штрихование *n*

* **shading scale** → 1759

3276 shadow; shade
d Schatten *m*
f ombre *f*
r тень *f*; ретушь *f*; тон *m*

* **shadowed symbol** → 3273

3277 shadow map
d Shadow-Map *f*; Schattenkarte *f*
f schème *m* d'ombrage
r схема *f* затенения

3278 shadow matrix
d Schattenmatrix *f*
f matrice *f* fantôme
r матрица *f* тени

* **shadow symbol** → 3273

3279 shape; form; figure
d Form *f*; Figur *f*; Gestalt *f*
f forme *f*; figure *f*
r форма *f*; фигура *f*

3280 shape analysis
d Form[en]analyse *f*
f analyse *f* de formes
r анализ *m* форм

3281 shape distortion
d Formverzerrung *f*
f distorsion *f* de forme
r искажение *n* формы

3282 shapefile
d Formendatei *f*; Shape-Datei *f*
f fichier *m* de formes
r файл *m* графических форм

3283 shape recognition
d Formerkennung *f*
f reconnaissance *f* de formes
r распознавание *n* форм

* **sharpening filters** → 1731

* **sheet** → 2333

3284 sheet designation
d Kartenblattbezeichnung *f*
f désignation *f* de feuille
r обозначение *n* листа

* **sheet-fed scanner** → 1394

3285 sheet line
d Blattschnitt *m*; Kartenschnittlinie *f*;
Kartenschnitt *m*
f découpage *m* des feuilles; coupure *f*
r линия *f* резки на листы

* **sheet memory** → 2278

3286 sheet number
d Kartenblattnummer *f*; Blattnummer *f*
f numéro *m* de feuille
r номер *m* листа

3287 sheet numbering system; map numbering
d Blattzahlensystem *n*
f système *m* de numération des feuilles
r номенклатура *f* карт

* **shift** → 1126

* **shifting** → 1126

* **shoreline** → 594

3288 shortcut
d Shortcut *n*; schneller Zugang *m*; Kürzel *n*
f raccourci *m*; accès *m* rapide
r сокращённое наименование *n*;
укорачивание *n*; краткая форма *f*; быстрый
вызов *m*

3289 shortcut menu; context[-sensitive] menu
d Kontextmenü *n*; kontextbezogenes Menü *n*;
kontextsensitives Menü
f menu *m* contextuel
r контекстно-зависимое меню *n*; контекстное
меню; сокращённое меню

3290 shortest path
d kürzeste Strecke *f*
f chemin *m* le plus court
r наикратчайший пут *m*

3291 short-range navigation
d Kurzstreckennavigation *f*
f navigation *f* à courte distance
r ближняя навигация *f*

3292 short reference map
d kurzer Referenzplan *m*
f renvoi *m* d'appel abrégé; renvoi de courte
référence
r карта *f* близкого обращения

3293 side elevation
d Seitenaufriss *m*; Längsseite *f*
f élévation *f* latérale; vue *f* de côté
r вид *m* сбоку; боковая проекция *f*

* **side lap** → 2111

3294 side-looking airborne radar; SLAR
  d Flugzeug-Seitensichtradar *m*;
      Flugzeug-Schrägsichtradar *m*;
      Flugzeug-Seitwärtsradar *m*
  f radar *m* aéroporté à antenne latérale; RAAL;
      radar aéroporté à exploration latérale; radar à
      visée latérale
  r бортовая радиолокационная станция *f*
      бокового обзора

  * **SIF** → 3463

  * **sight-bar** → 93

  * **sight-rule** → 93

3295 sign; indicator
  d Vorzeichen *n*; Zeichen *n*
  f signe *m*
  r [дешифровочный] признак *m*

3296 signal-to-noise ratio; SNR
  d Signal-Rausch-Verhältnis *n*;
      Signal-Stör-Verhältnis *n*; Rauschabstand *m*
  f rapport *m* signal à bruit; rapport signal-bruit
  r отношение *n* сигнал-помеха; отношение
      сигнал-шум; коэффициент *m* помех

3297 signature
  d Signatur *f*; Unterschrift *f*; Untertitel *m*;
      Bildtext *m*
  f signature *f*
  r подпись *f*; надпись *f*

  * **significance features** → 3992

3298 sign of intersection; cap
  d Zeichen *n* der Durchschnittsbildung
  f signe *m* d'intersection
  r знак *m* пересечения

  * **sill** → 3671

3299 simple kriging
  d einfaches Kriging *n*
  f krigeage *m* simple
  r простой кригинг *m*

3300 simple object
  d einfaches Objekt *n*
  f objet *m* simple
  r простой объект *m*

3301 simple polygon
  d einfaches Polygon *n*
  f polygone *m* simple
  r простой полигон *m*

3302 simple rectification
  d einfache Rektifikation *f*; einfache

Gleichrichtung *f*
  f redressement *m* simple; rectification *f* simple
  r простое спрямление *n*

3303 simple regression; linear regression
  d einfache Regression *f*; lineare Regression
  f régression *f* simple
  r простая регрессия *f*; единичная регрессия;
      линейная регрессия

3304 simplification
  d Vereinfachung *f*
  f simplification *f*
  r упрощение *n*

3305 simulation
  d Simulation *f*; Simulierung *f*
  f simulation *f*; imitation *f*
  r симуляция *f*; [имитационное]
      моделирование *n*; имитация *f*

3306 single-difference; SD
  d Einzeldifferenz *f*
  f différence *f* unique
  r первая разность *f*; простая разность

3307 single line
  d Einzellinie *f*
  f ligne *f* unique
  r однопроводная линия *f*; отдельная линия

  * **single-look image** → 2620

3308 single-point intersection
  d Einzelpunkteinschaltung *f*
  f intersection *f* de point unique
  r одноточковое пересечение *n*

3309 single-valued image; single-valued mapping
  d eindeutige Abbildung *f*
  f projection *f* univoque
  r однозначное отображение *n*

  * **single-valued mapping** → 3309

3310 single-valued surface
      (single z (elevation) value for each coordinate
      pair)
  d einwertige Fläche *f*; eindeutige Fläche
  f surface *f* univoque
  r однозначная поверхность *f*

  * **site** → 1627

3311 site analysis; location[al] analysis
  d Standortanalyse *f*
  f analyse *f* de site; analyse de localisation
  r анализ *m* [место]положения; анализ
      расположения; позиционный анализ

3312 **site map**
   d Karte *f* der Lage; Standortkarte *f*; Ortsplan *m*
   f carte *f* de sites; plan *m* de sites
   r карта *f* узлов

   * **site planing** → 3885

3313 **site suitability analysis**
   d Standort-Tauglichkeitsanalyse *f*
   f analyse *f* d'aptitude; analyse de constructibilité
   r анализ *m* пригодности расположения

3314 **situs**
   d Situs *m*
   f situs *m*
   r ситус *m*

3315 **size of active area**
   d Größe *f* des aktiven Bereichs
   f taille *f* de zone active
   r размер *m* рабочего поля

   * **skeleton** → 1476

   * **sketch map** → 3229

   * **skewness** → 2606

   * **skew projection** → 2603

   * **slant** → 1921

   * **slant angle** → 126

3316 **slanting surface; sloping surface**
   d geneigte Fläche *f*
   f surface *f* oblique
   r наклонная поверхность *f*

   * **SLAR** → 3294

3317 **slice**
   d Scheibe *f*; Scheibenteil *m*
   f tranche *f*
   r слой *m*; вырезка *f*; разрез *m*

3318 **slice level**
   d Schnittpegel *m*; Scheibenpegel *m*
   f niveau *m* de tranche
   r уровень *m* вырезки; уровень слоя

   * **slicing** → 2121

   * **slider** → 3234

   * **slider box** → 3234

   * **slider indicator** → 3234

   * **sliver** → 3319

3319 **sliver [polygon]; splinter**
   (an artifact of polygon overlay usually created
   by overlay of two sources with different
   accuracy, different sources or different
   interpretations)
   d Splitterpolygon *n*; Splitter *m*
   f bûchette *f*; éclat *m*
   r [паразитный] иглообразный полигон *m*;
   сплинтер *m*

   * **slope** → 1921

3320 **slope calculation**
   d Neigungsberechnung *f*
   f calcul *m* d'inclinaison
   r вычисление *n* наклона

3321 **slope diagram**
   d Neigungsmaßstab *m*; Böschungsmaßstab *m*
   f échelle *f* de pente
   r шкала *f* заложений

   * **slope gradient** → 126

3322 **slope-line detection**
   d Böschungslinienerkennung *f*;
   Fallinienerkennung *f*
   f détection *f* de line de pente
   r поиск *m* линии склона

   * **slope map** → 569

3323 **slope stakes**
   (stakes placed to locate the top or bottom of a
   slope)
   d Böschungspickel *mpl*; Neigungspickel *mpl*
   f piquets *mpl* de pente; jalons *mpl* de pente
   r колышки *fpl* для разбивки откосов;
   откосные лекала *npl*

3324 **slope type**
   d Neigungstyp *m*
   f type *m* de pente
   r тип *m* склона

   * **sloping surface** → 3316

3325 **small area triangulation**
   d Kleinflächentriangulation *f*
   f triangulation *f* de petites surfaces
   r триангуляция *f* элементарной площадки

3326 **small circle**
   d Kleinkreis *m*
   f petit cercle *m*
   r мелкий круг *m*

3327 small-scale map
  d Karte *f* in verkleinertem Maßstab;
     kleinmaßstäbliche Karte; kleinmaßstäbige
     Karte
  f carte *f* à petite échelle
  r мелкомасштабная карта *f*

* **snap** → 3328

* **snap distance** → 3329

* **snap interval** → 3329

3328 snap[ping]
  d Fang *m*; Fangen *n*; Griff *m*
  f magnétisme *m*; accrochage *m*; croquage *m*
  r привязка *f*; привязывание *n*; защёлка *f*;
     прыжок *m*

**3329 snap[ping] distance; snapping tolerance;
       snap spacing; snap interval; edit-distance**
  d Fangabstand *m*; Fangradius *m*
  f intervalle *m* d'accrochage; espacement *m*
     d'accrochage
  r интервал *m* привязки

* **snapping tolerance** → 3329

3330 snapshot
  d Schnappschuss *m*
  f image *f* instantanée; photo[graphie] *f*
     instantanée
  r моментальный снимок *m*

* **snap spacing** → 3329

3331 snap *v* to grid
  d an dem Gitter fangen
  f accrocher à la grille
  r ухватываться за сетку

3332 snap *v* to node
  d an dem Knoten fangen
  f accrocher au nœud
  r привязывать к узлу; ухватываться за узел

* **SNR** → 3296

* **Sobel filters** → 1731

3333 socioeconomical data
  d sozioökonomische Daten *pl*
  f données *fpl* socio-économiques
  r социально-экономические данные *pl*

3334 socioeconomical map
  d sozioökonomische Karte *f*
  f carte *f* socio-économique
  r социально-экономическая карта *f*

3335 socioeconomical zone
  d sozioökonomische Zone *f*
  f zone *f* socio-économique
  r социально-экономическая зона *f*

* **SOE** → 3402

3336 software
  d Software *f*
  f logiciel *m*
  r программное обеспечение *n*

**3337 software for storing the information in the
       map**
  d Software *f* zur Speicherung der Informationen
     in der Karte
  f logiciel *m* de stockage de l'information
     cartographique
  r программное обеспечение *n* для
     сохранения информации в карте

* **soil** → 1784

* **soil lot** → 2089

3338 soil map; ground map
  d Bodenkarte *f*
  f carte *f* du sol; carte au sol; carte pédologique
  r почвенная карта *f*

* **solid** → 3339

3339 solid [body]; rigid body; body
  d Festkörper *m*; [fester] Körper *m*
  f corps *m* [solide]; solide *m*
  r [твёрдое] тело *n*; монолитное тело; жёсткое
     тело

3340 solid line
  d Vollinie *f*
  f ligne *f* solide
  r плотная линия *f*; монолитная линия

* **sort** → 3342

3341 sort *v*
  d sortieren
  f trier
  r сортировать

3342 sort[ing]
  d Sortierung *f*; Sortieren *n*
  f tri[age] *m*
  r сортирование *n*; сортировка *f*

3343 sounder
  d Klopfer *m*
  f sondeur *m*
  r зонд *m*

* soundex → 3344

3344 soundex [code]
(a phonetic spelling (up to six characters) of a
street name, used for address matching)
d Lautschrift *f*; Soundex-Code *m*
f code *m* Soundex; code phonétique
r фонетический код *m*

3345 source map
d Quellenkarte *f*
f carte *f* document de base
r исходная карта *f*

3346 source material
d Ausgangsmaterial *n*; Ursprungsmaterial *n*
f matière *m* brute
r исходный материал *m*

3347 source table; start table; original table
d Quelltabelle *f*
f tableau-source *m*; tableau *m* de départ; table *f*
d'origine
r исходная таблица *f*

3348 south
d Süden *m*
f sud *m*
r юг *m*

3349 southeast
d Südost *m*
f sud-est *m*
r юго-восток *m*

3350 South pole
d Südpol *m*
f pôle *m* Sud; pôle antarctique
r Южный полюс *m*

3351 southwest
d Südwest *m*
f sud-ouest *m*
r юго-запад *m*

3352 space
d Raum *m*
f espace *m*
r пространство *n*

* space coordinates → 3368

* spacecraft → 3356

* space geodesy → 3198

3353 space maps; satellite image maps
d Satelliten[bild]karten *fpl*
f spatiocartes *fpl*
r космические карты *fpl*; космокарты *fpl*

* space primitives → 3968

3354 space segment
(of GPS)
d Raumsegment *n*; Raumsektor *m*
f secteur *m* spatial
r космический сегмент *m*

* space unit → 3858

3355 space vector
d Raumvektor *m*
f vecteur *m* spatial
r пространственный вектор *m*

3356 space vehicle; spacecraft
d Weltraumfahrzeug *n*; Raumfahrzeug *n*;
Raumschiff *n*
f véhicule *m* spatial; engin *m* spatial; vaisseau *m*
spatial
r аппарат *m* для исследования космического
пространства; космический летательный
аппарат; КЛА

* spaghetti → 3358

3357 spaghetti code; spaghetti program
d Spaghetticode *m*
f code *m* spaghetti
r неструктурная программа *f*

3358 spaghetti [data]
(vector data composed of line segments which
are not topologically structured or organized
into objects)
d Spaghettidaten *pl*
f données *fpl* spaghetti
r данные *pl* [типа] "спагетти"

3359 spaghetti model
d Spaghetti-Modell *n*
f modèle *m* spaghetti
r модель *f* "спагетти"; векторное
нетопологическое представление *n*

* spaghetti program → 3357

3360 span; range
d Spannweite *f*; Variationsbreite *f*
f étendue *f*
r размах *m*; протяжение *n*

3361 spatial access
d raumbezogener Zugriff *m*; Raumzugriff *m*
f accès *m* spatial
r пространственный доступ *m*

3362 spatial access method
d räumliche Zugriffsmethode *f*

f méthode *f* d'accès spatial
r метод *m* пространственного доступа

3363 spatial address
  d räumliche Adresse *f*
  f adresse *f* spatiale
  r пространственный адрес *m*

3364 spatial analysis
  d räumliche Analyse *f*
  f analyse *f* spatiale
  r пространственный анализ *m*

3365 spatial attribute
  d räumliches Attribut *n*
  f attribut *m* spatial
  r пространственный атрибут *m*

3366 spatial autocorrelation
  d räumliche Autokorrelation *f*
  f autocorrélation *f* spatiale
  r пространственная автокорреляция *f*

  * **spatial control** → 3367

3367 spatial control [net]; three-dimensional net; 3D network
  d räumliches Netz *n*
  f contrôle *m* spatial
  r пространственная геодезическая сеть *f*

3368 spatial coordinates; space coordinates; 3D coordinates
  d Raumkoordinaten *fpl*
  f coordonnées *fpl* dans l'espace; coordonnées spatiaux
  r пространственные координаты *fpl*

3369 spatial correlation
  d räumliche Korrelation *f*
  f corrélation *f* spatiale
  r пространственная корреляция *f*

3370 spatial data
  d raumbezogene Daten *pl*; räumliche Daten; Raumbezugsdaten *pl*
  f données *fpl* spatiaux
  r пространственные данные *pl*

3371 spatial database
  d räumliche Datenbasis *f*
  f base *f* de données spatiale
  r пространственная база *f* данных; рассредоточенная база данных

  * **spatial data exchange standard** → 3379

3372 spatial data generalization
  d Raumbezugsdaten-Generalisierung *f*
  f généralisation *f* de données spatiaux

r генерализация *f* пространственных данных

3373 spatial data infrastructure
  d Raumbezugsdaten-Infrastruktur *f*
  f infrastructure *f* de données spatiaux
  r инфраструктура *f* пространственных данных

3374 spatial data manipulation language
  d räumliche Datenbearbeitungssprache *f*
  f langage *m* de traitement de données spatiaux
  r язык *m* обработки пространственных данных

3375 spatial data representation
  d Raumbezugsdatendarstellung *f*
  f représentation *f* de données spatiaux
  r представление *n* пространственных данных

3376 spatial data sources
  d Raumbezugsdatenquellen *fpl*
  f sources *fpl* de données spatiaux
  r источники *mpl* пространственных данных

3377 spatial data structure
  d Raumbezugsdatenstruktur *f*
  f structure *f* de données spatiaux
  r структура *f* пространственных данных

3378 spatial data transfer processor
  d Raumdatenübertragungsprozessor *m*
  f processeur *m* de transfert de données spatiaux
  r процессор *m* переноса пространственных координат

  * **spatial data transfer specification** → 3379

3379 spatial data transfer standard; spatial data transfer specification; SDTS; spatial data exchange standard
  d Raumbezugsdaten-Übertragungsstandard *m*
  f standard *m* de transfert de données spatiaux
  r стандарт *m* передачи пространственных данных; стандарт обмена цифровыми пространственными данными; стандарт на передачу географических данных

3380 spatial data transformation
  d Raumbezugsdatentransformation *f*
  f transformation *f* de données spatiaux
  r преобразование *n* пространственных данных

3381 spatial data warehouse; spatial DWH
  d Geodatenwarenhaus *n*; räumliches Datenbanksystem *n* zur Entscheidungsfindung
  f datawarehouse *m* spatial; informatique *f* décisionnelle spatiale
  r хранилище *n* пространственных данных

3382 spatial datum
  d räumliches Datum *n*
  f datum *m* spatial
  r пространственный репер *m*

3383 spatial digitizer; 3D digitizer
  d dreidimensionaler Digitalisierer *m*
  f numériseur *m* 3D
  r трёхмерный дигитайзер *m*

  * **spatial DWH** → 3381

3384 spatial equilibrium analysis
  d Analyse *f* des räumlichen Gleichgewichts
  f analyse *f* d'équilibre spatial
  r анализ *m* пространственного равновесия

3385 spatial filtering
  d spatiale Filterung *f*
  f filtrage *m* spatial
  r фильтрация *f* пространства

3386 spatial frequency
  d spatiale Frequenz *f*; Raumfrequenz *f*
  f fréquence *f* spatiale
  r пространственная частота *f*

3387 spatial index
  d räumlicher Index *m*; Raumindex *m*
  f indice *m* spatial
  r пространственный индекс *m*

3388 spatial indexing
  d Raumindizierung *f*
  f indexation *f* spatiale
  r пространственное индексирование *n*

3389 spatial information enquiry service
  d Auskunftsdienst *m* von räumlicher
    Information
  f service *m* de demande d'information spatiale
  r служба *f* исследования пространственной
    информации

3390 spatial information system
  d räumliches Informationssystem *n*;
    Rauminformationssystem *n*; RIS
  f système *m* d'information spatiale
  r пространственная информационная
    система *f*

3391 spatial interaction
  d räumliche Wechselwirkung *f*; räumliche
    Zusammenarbeit *f*
  f interaction *f* spatiale
  r пространственное взаимодействие *n*

3392 spatial interaction model
  d räumliches Wechselwirkungsmodell *n*
  f modèle *m* d'interaction spatiale

  r модель *f* пространственного
    взаимодействия

  * **spatial interpolation** → 170

3393 spatial intersection
  d Einzelbildorientierung *f*; räumlicher
    Rückwärtsschnitt *m*
  f intersection *f* spatiale
  r пространственное пересечение *n*

3394 spatial join
  d räumliche Verbindung *f*
  f jointure *f* spatiale
  r пространственное соединение *n*

3395 spatial location
  d räumliche Position *f*
  f location *f* spatiale
  r пространственное [рас]положение *n*

3396 spatial metadata
  d spatiale Metadaten *pl*
  f métadonnées *fpl* spatiaux
  r пространственные метаданные *pl*

3397 spatial modeling
  d räumliche Modellierung *f*
  f modelage *m* spatial
  r пространственное моделирование *n*

3398 spatial neighborhoods
  d räumliche Näherungen *fpl*
  f proximités *fpl* spatiaux
  r пространственные окрестности *fpl*

3399 spatial object
  d räumliches Objekt *n*; Raumobjekt *n*
  f objet *m* spatial
  r пространственный объект *m*

3400 spatial object controller
  d Raumobjektkontroller *m*
  f contrôleur *m* d'objets spatiaux
  r контроллер *m* пространственных объектов

3401 spatial object framework
  d Raumobjektstruktur *f*
  f structure *f* d'objets spatiaux
  r структура *f* пространственных объектов

3402 spatial occupancy enumeration; SOE
  d Aufzählung *f* von Raumbesetzung
  f énumération *f* d'occupance spatiale
  r перепись *f* пространственной занятости

3403 spatial order
  d räumliche Ordnung *f*
  f ordre *m* spatial
  r пространственный порядок *m*

3404 spatial organization
  d räumliche Organisierung *f*
  f organisation *f* spatiale
  r пространственная организация *f*

3405 spatial overlay
  d räumliches Overlay *n*
  f croisement *m* spatial
  r пространственное перекрытие *n*

3406 spatial pattern
  d räumliches Bild *n*
  f image *f* spatiale; profil *m* spatial
  r пространственная картина *f*;
        пространственное изображение *n*

  * **spatial polar coordinates** → **3437**

3407 spatial query
  d räumliche Abfrage *f*
  f interrogation *f* spatiale
  r пространственный запрос *m*

3408 spatial reasoning
  d räumliches Denken *n*; räumliches Schließen *n*
  f raisonnement *m* spatial
  r пространственное мышление *n*

3409 spatial reference
  d räumlicher Verweis *m*; räumliche Referenz *f*
  f référence *f* spatiale
  r пространственная ссылка *f*

3410 spatial reference information
  d räumliche Referenzdaten *pl*
  f information *f* de références spatiaux
  r информация *f* о пространственных ссылках

3411 spatial reference system
  d räumliches Bezugssystem *n*
  f système *m* de référence spatiale
  r эталонная пространственная система *f*

3412 spatial referencing
  d räumliche Referenzierung *f*
  f référencement *m* spatiale
  r пространственная привязка *f*

  * **spatial relation** → **3413**

3413 spatial relation[ship]
  d räumliche Beziehung *f*
  f relation *f* spatiale
  r пространственная взаимосвязь *f*

3414 spatial resolution
  d Raumauflösung *f*; räumliches
        Auflösungsvermögen *n*
  f résolution *f* spatiale
  r пространственное разрешение *n*

3415 spatial sampling
  d Flächenstichprobenverfahren *n*
  f échantillonnage *m* spatial
  r опробирование *n* пространства

3416 spatial smoothing
  d räumliche Glättung *f*
  f lissage *m* spatial
  r пространственное сглаживание *n*

3417 spatial statistics
  d räumliche Statistik *f*
  f statistique *f* spatiale
  r пространственная статистика *f*

3418 spatial structure
  d räumliche Struktur *f*
  f structure *f* spatiale
  r пространственная структура *f*

3419 spatial unit
  d räumliche Einheit *f*; Raumeinheit *f*
  f unité *f* spatiale
  r пространственная единица *f*

3420 spatiotemporal data
  d raum-zeitliche Daten *pl*
  f données *fpl* spatio-temporelles
  r пространственно-временные данные *pl*

3421 spatiotemporal database
  d raum-zeitliche Datenbasis *f*
  f base *f* de données spatio-temporelle
  r пространственно-временная база *f* данных

3422 spatiotemporal GIS
  d raum-zeitliches GIS *n*
  f SIG *m* spatio-temporel
  r пространственно-временная ГИС *f*

  * **SPCS** → **3470**

3423 special boundary
  d Sondergrenze *f*
  f limite *f* spéciale
  r специальная граница *f*

3424 special-purpose map
  d Sonderkarte *f*
  f carte *f* spéciale
  r карта *f* специального назначения;
        специальная карта

3425 speckle noise; modal noise
  d Modenrauschen *n*
  f bruit *m* de granulation; bruit modal
  r пятнистый шум *m*; гранулированный шум;
        модовый шум

3426 speckle noise filter
  d Modenrauschenfilter *m*

f filtre *m* de bruit de granulation
r фильтр *m* пятнистого шума

3427 spectral attribute
  d spektrales Attribut *n*
  f attribut *m* spectral
  r атрибут *m* спектра

3428 spectral band
  d spektrales Band *n*
  f bande *f* spectrale
  r спектральная полоса *f*; полоса спектра

3429 spectral brightness; spectral radiance
  d spektrale Helligkeit *f*
  f brillance *f* spectrale
  r спектральная яркость *f*

3430 spectral class
  d spektrale Klasse *f*
  f classe *f* spectrale
  r спектральный класс *m*

  * spectral correction → 2987

  * spectral radiance → 3429

3431 spectral resolution
  d Spektralauflösung *f*
  f résolution *f* spectrale
  r спектральная разрешающая способность *f*

3432 spectral signature
  d spektrale Unterschrift *f*
  f signature *f* spectrale
  r спектральная надпись *f*

3433 specular reflection; mirror reflection; mirroring
  d Spiegelreflexion *f*; Normalreflexion *f*; gerichtete Reflexion *f*; Spiegeln *n*; Spiegelung *f*
  f réflexion *f* spéculaire; réflexion dirigée
  r зеркальное отражение *n*; нормальное отражение

  * spherical coordinates → 3437

3434 spherical distortion
  d sphärische Verzerrung *f*
  f distorsion *f* sphérique
  r сферическое искажение *n*

3435 spherical error probability; SEP
  d Kugelfehlerwahrscheinlichkeit *f*
  f probabilité *f* d'erreur sphérique
  r вероятность *f* сферического отклонения

  * spherical helix → 3160

  * spherical map → 3436

3436 spherical map[ping]
  d sphärische Abbildung *f*
  f application *f* sphérique
  r сферическое отображение *n*

3437 spherical [polar] coordinates; spatial polar coordinates; polar coordinates in the space
  d sphärische Polarkoordinaten *fpl*; sphärische Koordinaten *fpl*; Kugelkoordinaten *fpl*; räumliche Polarkoordinaten
  f coordonnées *fpl* sphériques [polaires]
  r сферические координаты *fpl*; полярные координаты в пространстве

3438 spherical projections
  d sphärische Projektionen *fpl*
  f projections *fpl* sphériques
  r сферические проекции *fpl*

3439 spheroid
  d Drehellipsoid *n*
  f sphéroïde *m*
  r сфероид *m*

3440 spheroid[al] geodesy; geodesy on the ellipsoid
  d sphäroidische Geodäsie *f*
  f géodésie *f* sphéroïdale
  r сфероидическая геодезия *f*

  * spheroid geodesy → 3440

3441 spike
  (an error in which a line extends past a line it is supposed to join)
  d Spitze *f*
  f crête *f*
  r выброс *m*; остриё *n*

3442 spinner
  d Spinner *m*; Verteiler *m*
  f assiette *f*
  r спиннер *m*

3443 spline
  d Spline *m*
  f spline *m*
  r сплайн *m*

  * spline fitting → 3444

3444 spline smoothing; spline fitting
  d Splineglättung *f*; Spline-Anpassung *f*; Spline-Fitten *n*
  f lissage *m* de spline; déformation *f* sur un spline; ajustement *m* de points par fonction spline
  r сглаживание *n* сплайна; аппроксимация *f* сплайна

* **splinter** → 3319

* **split** → 3445

3445 split[ting]
   d Aufteilung *f*; Aufteilen *n*; Spaltung *f*;
       Splitting *n*; Splitten *n*
   f splittage *m*; scindement *m*; fractionnement *m*;
       scission *f*
   r расщепление *n*; дробление *n*; разбиение *n*

**3446 spoofing; deliberate data modification;
       deception of data**
   d Verfälschung *f* der Daten; wissentliche
       Täuschung *f* von Daten; absichtliches
       Unterlaufen *n* der Sicherheitseinrichtungen
   f falsification *f* des données; modification *f*
       apportée délibérément aux données
   r имитация *f* данных

3447 spotdepth
   d Tiefenpunkt *m*
   f point *m* de profondeur
   r отметка *f* глубин

* **spot elevation** → 1245

* **spotheight** → 1245

3448 spotheight symbols
   d Höhenpunktsymbole *npl*
   f symboles *mpl* de points cotés; symboles de
       cotes
   r символы *mpl* высотных отметок

3449 spot map
   d geografische Darstellung *f* aufgetretener
       Krankheitsfälle
   f carte *f* de répartition des sujets
   r спот-карта *f*

3450 spreadsheet
   d Rechenblatt *n*
   f feuille *f* de calcul électronique
   r электронная таблица *f*

3451 spread spectrum
   d Spread-Spektrum *n*
   f large spectre *m*; spectre étalé
   r широкополосный спектр *m*

* **SPS** → 3467

3452 spurious polygon
   d Parasitenpolygon *n*
   f polygone *m* parasite
   r паразитный полигон *m*; ложный полигон

* **SQL** → 3515

3453 SQL builder
   d SQL-Bilder *m*
   f bâtisseur *m* SQL
   r построитель *m* SQL запросов

3454 SQL query
   d SQL-Abfrage *f*
   f interrogation *f* SQL
   r SQL-запрос *m*

* **square** → 3150

3455 square; quadrat
   d Quadrat *n*
   f carré *m*
   r квадрат *m*

* **square grid** → 3456

3456 square [mesh] grid; standard grid
   d quadratisches Netz *n*; Quadratnetz *n*
   f réseau *m* quadratique; quadrillage *m*
   r километровая сетка *f*; квадратная сетка;
       квадратная сеть *f*

3457 squaring
   d Quadrat-Anlegung *f*
   f carroyage *m*
   r придание *n* квадратной формы;
       квадратура *f*

* **SRG** → 3464

3458 stack
   d Flugzeuge *npl* im Warteraum; Wartestapel *m*
   f étagement *m*; pile *f* d'attente
   r стек *m*; магазин *m*

3459 stadia
   d Messlatte *f*; Tachymeterlatte *f*
   f stadia *m*
   r дальномер *m* с окулярной сеткой;
       дальномерная рейка *f*; дальномерная
       съёмка *f*

* **stairstepping** → 92

* **staking** → 2122

3460 standard accuracy adjustment
   d Präzision *f* des Standards
   f ajustage *m* de précision d'un étalon
   r установка *f* точности стандарта

3461 standard deviation
   d Standardabweichung *f*
   f écart *m* type
   r стандартное отклонение *n*

* **standard error of unit weight** → 3164

3462 standard geodetic datum; geodetic datum; datum
 d geodätisches Datum *n*
 f datum *m* géodésique spatial; référentiel *m* géodésique
 r исходная геодезическая система *f*

 * **standard grid → 3456**

3463 standard interchange format; SIF
 d Standardaustauschformat *n*
 f format *m* d'échange standard
 r стандартный формат *m* обмена

3464 standardized raster graphics; SRG
 d Standardrastergrafik *f*
 f graphique *m* rastre standardisé
 r стандартизованная растровая графика *f*

3465 standard line
 d Standardlinie *f*
 f ligne *f* standard
 r стандартная линия *f*

3466 standard parallel
 d Standardbreitengrad *m*
 f parallèle *m* standard
 r стандартная параллель *f*

3467 standard positioning service; SPS
 d Standardortungsdienst *m*; GPS-Service *m* zur Standardpunktbestimmung für alle Nutzer
 f service *m* de positionnement étalon
 r служба *f* определения стандартного местоположения

 * **star map → 478**

3468 start node; beginning point; origin node
 d Anfangsknoten *m*
 f nœud *m* initial; nœud d'origine
 r начальный узел *m*; начальная точка *f*

3469 startnode ID
 d Anfangsknoten-Identifikator *m*
 f identificateur *m* de nœud initial
 r идентификатор *m* начального узла

 * **start table → 3347**

3470 state plane coordinate system; SPCS (US)
 d ebenes Koordinatensystem *n* von US
 f système *m* de coordonnées planes d'États-Unis
 r система *f* плоскостных координат США

3471 statics
 d Statik *f*
 f statique *f*
 r статика *f*

3472 stationarity
 d Stationarität *f*
 f stationnarité *f*
 r стационарность *f*; неизменность *f* во времени

3473 statistical area
 d statistischer Raum *m*; statistische Zone *f*
 f espace *m* statistique; zone *f* statistique
 r статистическая зона *f*

3474 statistical average
 d statistisches Mittel *n*
 f moyenne *f* statistique
 r статистическое среднее [значение] *n*

3475 statistical entity
 d statistische Entität *f*
 f entité *f* statistique
 r статистическая категория *f*

3476 statistical graphics
 d statistische Grafik *f*
 f graphique *m* statistique
 r статистическая графика *f*

3477 statistically equivalent entity
 d statistisch äquivalente Entität *f*
 f entité *f* équivalente statistique
 r статистически эквивалентная категория *f*

3478 statistical modeling
 d statistische Modellierung *f*
 f modelage *m* statistique
 r статистическое моделирование *n*

3479 statistical moment
 d statistisches Moment *n*
 f moment *m* statistique
 r статистический момент *m*

3480 statistical surface
 d statistische Fläche *f*
 f surface *f* statistique
 r статистическая поверхность *f*

3481 statistics window
 d Statistik-Fenster *n*
 f fenêtre *f* de statistique
 r окно *n* статистики

3482 steepness
 d Steilheit *f*
 f escarpement *m*; pente *f* raide
 r крутизна *f*; крутость *f*

 * **stellar chart → 478**

 * **stencil → 3595**

3483 stepwise linear classification
d schrittweise lineare Klassifikation f
f classification f linéaire pas à pas
r пошаговая линейная классификация f

3484 steradian
d Steradiant m
f stéradian m
r стерадиан m

3485 stereocompilation
d Stereokompilierung f
f stéréocompilation f
r стерео-оригинал m

3486 stereo digitizer
d Stereo-Digitalisierer m
f stéréodigitaliseur m
r стереодигитайзер m

3487 stereo elevation
(an elevation surface derived from stereo pairs
of remote sensing imagery)
d Stereoelevation f
f élévation f stéréo
r стереовысота f

* **stereograph → 2746**

**3488 stereographic [map] projection; azimuthal
orthomorphic projection; zenithal
orthomorphic projection**
d stereografische Projektion f; winkeltreue
Azimutalprojektion f; Kugelprojektion f;
stereografische Abbildung f
f projection f stéréographique; représentation f
stéréographique
r стереографическая проекция f

* **stereographic projection → 3488**

3489 stereomodel
d Stereomodell n
f stéréomodèle m
r стереомодель f

* **stereopair → 3491**

3490 stereophotogrammetry
d Stereomessbildverfahren n;
Stereofotogrammetrie f
f stéréophotogrammétrie f
r стереофотограмметрия f

* **stereoplotter → 3492**

* **stereoplotting → 3493**

3491 stereoscopic photograph; stereopair
d stereoskopisches Foto n; Stereofoto n;

Stereopaar n
f photo[graphie] f stéréoscopique
r стереоскопический аэрофотоснимок m;
стереопара f

**3492 stereoscopic plotter; stereoplotter; double
picture plotter**
d stereoskopischer Plotter m; Stereoplotter m;
Stereoauswertegerät n
f traceur m stéréoscopique; appareil m de
restitution stéréoscopique
r стереоплоттер m; стереоскопический
графопостроитель m;
стереообрабатывающий графопостроитель

3493 stereoscopic plotting; stereoplotting
d stereoskopisches Plotten n; Stereoplotten n
f traçage m stéréoscopique; stéréotracé m
r стереоскопическое вычерчивание n

3494 stereoscopic projection
d stereoskopische Projektion f
f projection f stéréoscopique
r стереоскопическая проекция f

3495 stock of maps; inventory of maps
d Bestandsverzeichnis n
f inventaire m de cartes
r картографический фонд m

* **stop → 3498**

3496 stop attributes
d Haltepunkt-Attribute npl
f attributs mpl de points d'arrêt
r атрибуты mpl перерывов; атрибуты
остановок

3497 stop impedance
d Haltepunkt-Impedanz f
f impédance f de point d'arrêt
r импеданс m перерывов

3498 stopping place; stop
(location visited in a path or tour)
d Haltepunkt m; Halteplatz m
f point m d'arrêt; lieu m d'arrêt
r промежуточная посадка f

3499 stops file
d Haltepunktsdatei f
f fichier m de points d'arrêt
r файл m остановок

3500 storage; store; memory
d Speicher m
f mémoire f
r запоминающее устройство n;
накопитель m; память f

* store → 3500

3501 **straight line**
   d Gerade f
   f ligne f droite; droite f
   r прямая [линия] f

3502 **straight line navigation**
   d geradlinige Navigation f
   f navigation f rectiligne
   r прямолинейная навигация f

3503 **strategic decision-making**
   d strategische Entscheidungsfindung f
   f prise f de décision stratégique
   r стратегическое принятие n решений

3504 **stratification**
   d Stratifikation f
   f stratification f
   r стратификация f; наслоение n

* **stratification plane** → 2777

3505 **stratified kriging**
   d stratifiziertes Kriging n
   f krigeage m stratifié
   r расслоённый кригинг m

3506 **stratigraphic unconformity; unconformity; disconformity**
   d [stratigrafische] Diskordanz f
   f discordance f [de stratification]
   r стратиграфическое несогласие n; непараллельное несогласие; угловое несогласие; несогласие

3507 **stratigraphy**
   d Stratigrafie f
   f stratigraphie f
   r стратиграфия f

* **stratum** → 732

* **stream line** → 1445

3508 **stream mode digitizing; dynamic mode digitizing**
   d dynamische Digitalisierung f
   f numérisation f [en mode] dynamique
   r цифрование n потоковым вводом

* **stream network** → 3163

3509 **street network**
   d Straßennetz n
   f réseau m filaire
   r уличная сеть f

3510 **street network database**

   d Straßennetz-Datenbasis f
   f base f de données de réseau filaire
   r база f данных уличной сети

* **string** → 527

* **strip** → 3511

3511 **strip[e]; band; bar**
   d Streifen m; Band n; Leiste f; Balken m
   f bande f; ruban m; feuillet m
   r лента f; полоса f

* **stroke** → 314

* **stroke marking** → 851

* **strong classification** → 1803

* **structural contour** → 732

* **structural contour map** → 3527

* **structural map** → 3527

3512 **structure**
   d Struktur f
   f structure f
   r структура f; строение n

3513 **structured data**
   d strukturierte Daten pl
   f données fpl structurées
   r структурированные данные pl

3514 **structured elevation model; SEM**
   d strukturiertes Höhenlinienmodell n
   f modèle m altimétrique structuré; modèle d'élévation structuré
   r структурированная модель f высотных точек

3515 **structured query language; SQL**
   d Sprache f der strukturierten Abfragen; Abfragesprache f
   f langage m de requête structurée
   r язык m структурированных запросов

* **structure map** → 3527

* **structure number** → 2242

3516 **subarea; subdomain; subregion**
   d Unterbereich m; Teilbereich m; Teilgebiet n
   f sous-domaine m; sous-région f
   r подобласть f

* **subassembly** → 585

3517 subblock; blockette
d Teilblock *m*
f sous-bloc *m*
r подблок *m*

3518 subclass
d Unterklasse *f*; Teilklasse *f*
f sous-classe *f*
r подкласс *m*

3519 subdivision
d Unterteilung *f*
f sous-division *f*; subdivision *f*
r подразделение *n*; подразбиение *n*

* subdomain → 3516

3520 sub-entity
(block attributes, segments of line etc.)
d Untereinheit *f*
f sous-entité *f*
r подобъект *m*

* submarine contour → 979

3521 subplane
(of a projective plane)
d Unterebene *f*
f sous-plan *m*
r подплоскость *f*

3522 subquadrant
d Unterquadrant *m*
f sous-quadrant *m*
r подквадрант *m*

* subregion → 3516

3523 subselects
d Unterselektionen *fpl*
f sous-sélections *fpl*
r подвыборки *fpl*

3524 subset
d Untermenge *f*; Teilmenge *f*
f sous-ensemble *m*; sous-multitude *f*
r подмножество *n*

3525 subsidence
d Grubensenkung *f*; Absinken *n*
f subsidence *f*
r ослабление *n*; оседание *n*

3526 substitution
d Substitution *f*; Ersetzung *f*; Ersetzen *n*
f substitution *f*
r подстановка *f*; субституция *f*

**3527 subsurface contour map; structural
[contour] map; structure map**
d [geologische] Strukturkarte *f*
f carte *f* structurale
r карта *f* подземного рельефа; структурная
карта

3528 subtotal; partial sum; partial total
d Zwischensumme *f*; Partialsumme *f*;
Teilsumme *f*
f sous-total *m*; somme *f* partielle
r промежуточная сумма *f*; частичная сумма

3529 subtractive primary colors
d subtraktive Primärfarben *fpl*
f couleurs *fpl* primaires soustractives
r вычитаемые основные цвета *mpl*

3530 subtree
d Subbaum *m*
f sous-arbre *m*
r поддерево *n*

3531 subwindow
d Unterfenster *n*
f sous-fenêtre *f*
r подокно *n*

3532 Suits-Wagner classification
d Suits-Wagner-Klassifikation *f*
f classification *f* de Suits-Wagner
r классификация *f* Сютса-Вагнера

3533 superimpose *v*
d übereinanderlegen
f surimposer
r накладывать; налагать

* superimposition → 3534

**3534 superposition; superimposition;
overlay[ing]**
d Überlagerung *f*; Superposition *f*
f surimposition *f*; superposition *f*; revêtement *m*
r суперпозиция *f*; операция *f* наложения;
наложение *n*

3535 supervised area
d Überwachungsbereich *m*
f zone *f* surveillée
r контролируемая область *f*

* supervised classification → 3536

3536 supervised [image] classification
d kontrollierte Klassifikation *f*; überwachte
Bildklassifizierung *f*
f classification *f* dirigée; classification contrôlée
r контролируемая классификация *f*

3537 supporting contour line
d Stützhöhenlinie *f*

f ligne *f* d'appui de contour
r опорная контурная линия *f*

3538 surface
d Fläche *f*; Oberfläche *f*
f surface *f*
r поверхность *f*

* **surface area** → 171

* **surface chart** → 160

* **surface contour map** → 3714

3539 surface derivative
d Flächenableitung *f*
f dérivative *f* de surface
r производная *f* поверхности

* **surface element** → 1242

3540 surface fitting
d Flächenanpassung *f*; Flächenausgleich *m*;
Flächenausgleichung *f*
f ajustage *m* de surface
r аппроксимация *f* поверхности;
сглаживание *n* поверхности

3541 surface geodesy
d Flächengeodäsie *f*
f géodésie *f* d'une surface
r геодезия *f* поверхности

* **surface graph** → 160

3542 surface in relief
d Relieffläche *f*
f surface *f* en relief
r рельефная поверхность *f*

3543 surface layer
d Oberflächenschicht *f*
f couche *f* superficielle
r поверхностный слой *m*; покрытие *n*

3544 surface of slope
d Böschungsfläche *f*
f surface *f* de pente
r поверхность *f* откоса

**3545 surface patch; area patch; patch; mesh
facet**
(piece of a surface bounded by a closed curve)
d Oberflächenstück *n*; Flächenstück *n*
f portion *f*; morceau *m* de surface limité par une
courbe fermée; fragment *m* surfacique
r фрагмент *m*; кусок *m* поверхности;
элемент *m* разрешения

* **surface source** → 165

3546 surface variable
(for example, elevation, mineral
concentration, human population density
etc.)
d Flächenvariable *f*
f variable *f* de surface
r переменная *f* поверхности

3547 surrogate data
d Ersatzdaten *pl*
f données *fpl* remplaçantes
r данные-заменители *pl*

* **surrounding points** → 2514

* **survey** → 1556, 3548

* **survey** *v* → 2249

3548 survey[ing]
d Überwachung *f*; Vermessung *f*
f arpentage *m*; surveillance *f*; mesurage *m*;
enquête *f*; opération *f* de triangulation et de
nivellement
r [текущий] контроль *m*; обзор *m*;
обследование *n*

* **surveying azimuth** → 1543

3549 surveying in; measuring in
d Punktaufnahme *f*; Punktvermessung *f*;
Einmessung *f*
f détermination *f* topographique de la position
d'un point fixe
r получение *n* точные размеры скважины

3550 surveying plane-table; plane-table [sheet]
d Erhebungsmesstisch *m*; Messtischblatt *n*
f planchette *f* [d'arpenteur]
r мензула *f*; [съёмочный] планшет *m*

3551 survey map; outline map; general map
d Übersichtskarte *f*
f plan *m* de surveillance; carte *f* muette;
croquis *m*
r обзорная карта *f*; немая карта

* **survey mark** → 333

* **survey of the relief** → 103

* **surveyor** → 2094

* **survey sheet** → 1405

3552 sustainable development
d nachhaltige Entwicklung *f*; dauerhafte
Entwicklung

f développement *m* soutenable; développement durable; écodéveloppement *m*
r устойчивое развитие *n*

* **swamp** → 368

3553 swath
d abgedecktes Gebiet *n*
f surface *f* balayée; zone *f* explorée
r полоса *f* захвата

* **symbol** → 525

3554 symbol environment
(defines the types of map symbols and their characteristics during a graphic display session)
d Symbolumgebung *f*
f environnement *m* de symboles
r [операционная] среда *f* символов

3555 symbol file
d Symboldatei *f*
f fichier *m* de symboles
r символьный файл *m*

* **symbol font** → 526

3556 symbolization
d Versinnbildlichung *f*
f symbolisation *f*
r символизация *f*; представление *n* в виде символов; совокупность *f* символов или знаков

3557 symbolized points
d symbolisierte Punkte *mpl*
f points *mpl* symbolisés
r символически отображённые точки *fpl*

3558 symbol overlay
d Symbolüberdeckung *f*
f couverture *f* de symboles
r символьное перекрытие *n*; наложение *n* символов

3559 symbol style
d Symbolstil *m*
f style *m* de symboles
r стиль *m* символов

3560 symbol table; character map
d Zeichentabelle *f*; Symboltabelle *f*
f table *f* de symboles; table de caractères
r таблица *f* символов

3561 synclinal axis; synclinal line; syncline; t[h]alweg
d Muldenlinie *f*; Talweg *m*
f ligne *f* de synclinal; ligne d'écoulement des

eaux d'une surface; t[h]alweg *m*
r синклинальная линия *f*; ось *f* долины; тальвег *m*

* **synclinal line** → 3561

* **syncline** → 3561

* **syntax analysis** → 2706

* **synthetical thematic map** → 3646

3562 synthetic aperture radar; SAR
d Radar *m* mit synthetischer Apertur; synthetischer Aperturradar *m*
f radar *m* à ouverture synthétique; radar à antenne synthétique; RAAS
r радиолокатор *m* с синтезированной апертурой; РСА; радиолокационная станция *f* с искусственным раскрывом антенны

3563 synthetic map
d synthetische Karte *f*
f carte *f* synthétique
r синтетическая карта *f*

3564 synthetic mapping
d synthetische Kartierung *f*
f cartographie *f* synthétique
r синтетическое картографирование *n*

3565 synthetic resolution
d synthetische Auflösung *f*
f résolution *f* synthétique
r синтетическая разрешающая способность *f*

3566 synthetic terrain data; artificial terrain data
d synthetische Daten *pl* über das Gelände
f données *fpl* de terrain artificielles; données synthétiques relatives au terrain
r искусственные данные *pl* о рельефе

3567 systematic error; constant error; fixed error; regular error
d systematischer Fehler *m*; konstanter Fehler; fester Fehler; regelmäßiger Fehler
f erreur *f* systématique; erreur constante; erreur fixe; erreur régulière
r систематическая погрешность *f*; систематическая ошибка *f*; постоянная ошибка; регулярная ошибка

3568 system mapping
d Systemkartierung *f*
f cartographie *f* système
r системное картографирование *n*

3569 system tables
  d Systemtabellen *fpl*
  f tables *fpl* de système
  r системные таблицы *fpl*

T

3570 table
d Tabelle f; Tafel f
f table f; tableau m
r таблица f

* **tableau** → 3571

* **table data** → 3575

3571 tableland; tableau
d Hochebene f; Hochterrasse f
f haut plateau m
r плоскогорье n; плато n; плоская
возвышенность f

3572 table plotter
d Tischplotter m
f traceur m tabulaire
r настольный плоттер m

3573 table structure
d Tabellenstruktur f
f structure f de table
r структура f таблицы

* **tablet** → 1071

3574 tablet menu area
d Gebiet n des Tablettmenüs
f zone f de menu de tablette
r область f меню планшета

3575 tabular data; table data
d Tabellendaten pl
f données fpl tabulaires
r табличные данные pl

* **tabular data representation** → 3576

**3576 tabular data view; tabular data
representation**
d Tabellendatendarstellung f
f représentation f de données tabulaires
r представление n табличных данных

3577 tabulating; tabulation
d Tabellierung f; Tabulation f; Darstellung f in
Tabellenform; tabellarische Darstellung
f tabulation f; mise f en table; représentation f
sous forme de tableau
r табулирование n; табуляция f;
составление n таблицы; расположение n
данных в виде таблицы; представление n
данных в виде таблиц

* **tabulation** → 3577

**3578 tacheometer; tachymeter; level theodolite;
transit theodolite**
d Tachymeter n; Höhennivellierinstrument n
f tachéomètre m
r тахеометр m

* **tachymeter** → 3578

* **tag** → 2061

* **tagged** → 2065

3579 tagged image file format; TIFF
d TIF-Format n
f format m TIF
r формат m TIF; формат файла с разметкой
для хранения изображений

* **tagging** → 2066

* **talus** → 1921

* **talweg** → 3561

* **tangent** → 3582

3580 tangent cone
(of a surface)
d Tangentenkegel m; Tangentialkegel m;
Berührungskegel m
f cône m tangentiel
r касательный конус m

3581 tangent conical projection
d Berührungskegelprojektion f
f projection f conique tangente
r тангенциальная коническая проекция f

* **tangential curvature** → 1546

3582 tangent [line]
d Tangente f; Berührende f
f tangente f; droite f tangente
r касательная f

* **tangent plane projection** → 3720

3583 target
d Ziel n
f cible f
r мишень f; цель f

* **target area** → 3584

3584 target district; target area
d Zielbezirk m; Zielgebiet n; Zielraum m

f district *m* de destination
r целевой район *m*

3585 targeting
d Ausrichtung *f* auf Zielgruppen; Zielung *f*;
Targeting *n*
f choix *m* des objectifs et des moyens de
traitement; ciblage *m*
r целеуказание *n*; таргетирование *n*

3586 target map
d Zielkarte *f*
f carte *f* d'objectifs
r целевая карта *f*

3587 target symbol
d Zielsymbol *n*
f symbole *m* de but
r символ *m* мишени; символ цели

* **TAT** → 3625

3588 taxonomic classification
d taxonomische Klassifizierung *f*
f classification *f* taxonomique
r таксономическая классификация *f*

3589 taxonomy
d Taxonomie *f*
f taxonomie *f*
r таксономия *f*

* **TAZ** → 3781

* **TCM** → 3606

* **TD** → 3827

* **tear-off menu** → 2953

3590 technical documentation
d technische Dokumentation *f*
f documentation *f* technique
r техническая документация *f*

3591 technical reliability
d technische Zuverlässigkeit *f*
f fiabilité *f* technique
r техническая надёжность *f*

3592 telecommunicational mapping
d Fernübertragungskartierung *f*
f cartographie *f* de télécommunication
r телекоммуникационное
картографирование *n*

3593 telemetry
d Telemetrie *f*; Fernmesstechnik *f*;
Fernmessung *f*; Fernwirktechnik *f*
f télémétrie *f*; télémesure *f*

r телеметрия *f*; телеизмерение *n*;
дистанционные измерения *npl*

3594 telescopic alidade
d Teleskopalhidade *f*
f alidade *f* holométrique; alidade *f* télescopique
r кипрегель *m*

3595 template; stencil; pattern
d Schablone *f*; Schrittschablone *f*; Vorlage *f*;
Formatbild *n*; Muster *n*
f maquette *f*; modèle *m*
r шаблон *m*; трафарет *m*; макет *m*; модель *f*

3596 temporal accuracy
d zeitliche Genauigkeit *f*
f exactitude *f* temporaire
r временная точность *f*

3597 temporal database
d zeitliche Datenbasis *f*
f base *f* de données temporaire
r временная база *f* данных

3598 temporal dimension
d zeitliche Dimension *f*
f dimension *f* temporaire
r временная размерность *f*

3599 temporal dimension of data
d zeitliche Datendimension *f*
f dimension *f* de données temporaire
r временная размерность *f* данных

3600 temporal extent
d zeitlicher Extent *m*
f étendue *f* temporaire
r временное расширение *n*

3601 temporal reference system
d zeitliches Referenzsystem *n*
f système *m* référentiel temporaire
r временная эталонная система *f*

* **temporal resolution** → 2149

3602 temporary column
d zeitliche Spalte *f*
f colonne *f* temporaire
r временная колонка *f*

3603 temporary topology
d zeitliche Topologie *f*
f topologie *f* temporaire
r временная топология *f*

3604 tensor
d Tensor *m*
f tenseur *m*
r тензор *m*

* **terminal point** → 1270

3605 terrain; landform; relief; ground features
  d Gelände *n*; Bodenform *f*; Erdform *f*; Terrain *n*;
     Relief *n*
  f terrain *m*; relief *m*
  r конфигурация *f* местности; формы *fpl*
     местности; топография *f* местности;
     рельеф *m*

* **terrain contour matching** → 3606

3606 terrain contour matching [system]; TCM
  d Geländelinienanpassungssystem *n*
  f système *m* de poursuite de terrain
  r радиолокационная система *f*
     обзорно-сравнительного метода наведения;
     система отслеживания рельефа местности

3607 terrain creation
  d Terrainerstellung *f*; Terrainschaffung *f*
  f création *f* de terrain
  r создание *n* конфигурации местности

3608 terrain data
  d Daten *pl* über das Gelände
  f information *f* de terrain; données *fpl* de terrain
  r данные *pl* о рельефе

* **terrain display** → 3618

* **terrain displaying** → 3618

3609 terrain display system
  d Terraindarstellungssystem *n*
  f système *m* d'affichage de terrain
  r система *f* визуализации конфигурации
     местности

3610 terrain edit system
  d Geländeeditiersystem *n*
  f système *m* d'édition de terrain
  r система *f* редактирования конфигурации
     местности

3611 terrain-following system
  d Geländefolgesystem *n*
  f système *m* suivi de terrain
  r система *f*, следящая рельефа местности

3612 terrain geometry
  d Terraingeometrie *f*
  f géométrie *f* de terrain
  r геометрия *f* форм местности

3613 terrain line interpolation; TLI
  d Geländelinieninterpolierung *f*
  f interpolation *f* de contour de terrain
  r интерполяция *f* линий поверхности земли;
     интерполяция рельефа местности

3614 terrain model; relief model
  d Geländemodell *n*; Terrainmodell *n*
  f modèle *m* de terrain
  r макет *m* [форм] местности

3615 terrain rendering
  d Terrain-Rendering *n*
  f rendu *m* de terrain
  r рендеринг *m* форм местности

**3616 terrain resource information management
     format; TRIM format** (Canada)
  d TRIM-Format *n*
  f format *m* TRIM
  r формат *m* TRIM (*обмена
     пространственными данными,
     используемый администрацией шт.
     Британская Колумбия*)

3617 terrain steepness
  d Geländesteilheit *f*
  f pente *f* de terrain
  r крутизна *f* местности

* **terrain-type map** → 2082

3618 terrain visualization; terrain display[ing]
  d Terrainvisualisierung *f*
  f visualisation *f* de terrain
  r визуализация *f* конфигурации местности

3619 terrestrial coordinates
  d terrestrische Koordinaten *fpl*
  f coordonnées *fpl* terrestres
  r наземные координаты *fpl*

* **terrestrial globe** → 1209

3620 terrestrial mapping
  d terrestrische Kartierung *f*
  f cartographie *f* terrestre
  r земное картографирование *n*

3621 territory
  d Geltungsbereich *m*
  f territoire *f*
  r территория *f*

* **territory search** → 178

* **tessellation** → 3684

* **test** → 532

3622 test area; test zone
  d Prüffläche *f*; Aufstellungsbereich *m*
  f zone *f* d'essais; polygone *m* d'essai; zone test
  r тестовой участок *m*

* **test zone** → 3622

* **tetragon** → 2959

* **texel** → 3634

3623 text
d Text *m*
f texte *m*
r текст *m*

3624 text attribute; character attribute
d Textattribut *n*
f attribut *m* textuel
r текстовый атрибут *m*

3625 text attribute table; TAT
d Textattributtabelle *f*
f table *f* d'attributs textuels
r таблица *f* текстовых атрибутов

3626 text box
d Textfeld *n*
f zone *f* de texte; boîte *f* de texte
r текстовая зона *f*; поле *n* текста; текстовая панель *f*

* **text data** → 3632

3627 text envelope
d Texthülle *f*
f enveloppe *f* de texte
r оболочка *f* текста

3628 text label
d Textetikett *n*
f étiquette *f* textuelle
r текстовая метка *f*

3629 text style
d Textstil *m*
f style *m* de texte
r текстовой стиль *m*

3630 text symbol
d Textsymbol *n*
f symbole *m* de texte
r текстовой символ *m*

3631 textual fidelity
d textuelle Genauigkeit *f*; Textrichtigkeit *f*
f fidélité *f* textuelle
r текстовая точность *f*; текстовая верность *f*

3632 textual information; text data
d textuelle Information *f*; Textdaten *pl*
f information *f* textuelle; données *fpl* textuelles
r текстовая информация *f*; текстовые данные *pl*

3633 texture
d Textur *f*

f texture *f*
r текстура *f*

3634 texture element; texel
d Textur-Element *n*; Texel *n*
f élément *m* de texture
r элемент *m* текстуры; тексел *m*

* **thalweg** → 3561

3635 thematic accuracy
d thematische Genauigkeit *f*
f précision *f* thématique
r тематическая точность *f*

3636 thematic atlas
d thematischer Atlas *m*
f atlas *m* thématique
r тематический атлас *m*

* **thematic cartography** → 3649

3637 thematic class
d thematische Klasse *f*
f classe *f* thématique
r тематический класс *m*

3638 thematic content
d thematischer Inhalt *m*
f contenu *m* thématique
r тематическое содержание *n*

3639 thematic data
d thematische Daten *pl*
f données *fpl* thématiques
r тематические данные *pl*

3640 thematic data structure
d Struktur *f* der thematischen Daten
f structure *f* de données thématiques
r структура *f* тематических данных

3641 thematic geo-field
d thematischer geografischer Bereich *m*
f champ *m* géographique thématique
r тематическая географическая область *f*

3642 thematic keyword
d thematisches Schlüsselwort *n*
f mot *m* clé thématique
r тематическое ключевое слово *n*

3643 thematic layer
d thematische Schicht *f*
f couche *f* thématique
r тематический слой *m*

3644 thematic map
d thematische Karte *f*; Themakarte *f*

f carte *f* thématique; mappe *f* thématique
r тематическая карта *f*; отраслевая карта

3645 thematic map of analytic representation; analytical thematic map
d analytische Themakarte *f*
f carte *f* thématique analytique
r аналитическая тематическая карта *f*

3646 thematic map of synthetic representation; synthetical thematic map
d synthetische Themakarte *f*
f carte *f* thématique synthétique
r синтетическая тематическая карта *f*

3647 thematic mapper
d thematischer Kartograph *m*
f dispositif *m* de cartographie thématique; cartographe *m* thématique; mappeur *m* thématique
r устройство *n* изготовления тематических карт; картопостроитель *m* с классификацией геологических районов

3648 Thematic Mapper; TM
(a satellite sensing system with resolution of 30 meters, 16 day repeat cycle, 185 km scene width and seven bands of spectral data)
d Thematic-Mapper *m*
f scanner *m* multibande pour thèmes multiples
r разметчик *m*

3649 thematic mapping; thematic cartography
d thematische Kartierung *f*; thematische Kartografie *f*
f cartographie *f* thématique
r тематическое картографирование *n*; тематическая картография *f*

3650 thematic modeling
d thematische Modellierung *f*
f modelage *m* thématique
r тематическое моделирование *n*

3651 thematic photomap
d thematische Fotokarte *f*
f photocarte *f* thématique
r тематическая фотокарта *f*

3652 thematic range
d thematisches Diapason *n*
f rangée *f* thématique
r тематический диапазон *m*

3653 thematic resolution
d thematische Auflösung *f*
f résolution *f* thématique
r тематическое разрешение *n*

3654 thematic shading

d thematische Schattierung *f*
f ombrage *m* thématique
r тематическое оттенение *n*

3655 thematic variable
d thematische Variable *f*
f variable *f* thématique
r тематическая переменная *f*

* **theme** → 1630

3656 theme-on-theme selection
d Thema-auf-Thema-Auswahl *f*; Thema-auf-Thema-Selektieren *n*
f sélection *f* thème sur thème
r выбор *m* объектов темы через объектов другой темы

3657 theme-specific symbology
d Thema-Symbolsätz *m*
f symbologie *f* thématique
r тематическая символика *f*

3658 theodolite
d Theodolit[e] *m*
f théodolite *m*
r угломер *m*; теодолит *m*

3659 theoretical circuit closure error
d theoretischer Schleifenschlussfehler *m*
f erreur *f* théorique de fermeture des polygones
r теоретическая невязка *f* контуров

3660 theory of cartographic communication
d Theorie *f* der kartografischen Kommunikation
f théorie *f* de communication cartographique
r теория *f* картографической коммуникации

3661 theory of cartography
d Theorie *f* der Kartografie
f théorie *f* de cartographie
r теория *f* картографии

3662 thermal infrared multispectral scanner
d thermischer Infrarot-Multispektralscanner *m*; thermischer Infrarot-Multispektral-Abtast-Radiometer *m*
f scanner *m* multibande infrarouge thermique; scanner multispectral infrarouge thermique
r тепловой многозональный сканер *m* ИК-диапазона

3663 thermal isoline map
d Thermoisoplethendiagramm *n*
f diagramme *m* thermo-isoplèthe
r термо-изолинейная карта *f*

* **thermal mapper** → 1947

3664 thermal plotter
d Thermalplotter *m*

f traceur *m* thermique
r термический графопостроитель *m*

* **theta angle** → 746

3665 **Thiessen polygons; Voronoi polygons;**
Voronoi diagrams; proximity polygons;
proximal polygons
d Thiessen-Polygone *npl*;
Voronoi-Diagramme *npl*;
Voronoi-Polygone *npl*
f polygones *mpl* de Thiessen; polygones de
Voronoi; diagrammes *mpl* de Voronoi;
polygones de proximité
r полигоны *mpl* Тиссена; полигоны
Вороного; диаграммы *fpl* Вороного;
ячейки *fpl* Вигнера-Зейтца;
многоугольники *mpl* близости

3666 **thinness ratio**
d Verhältnis *n*
f rapport *m* de ténuité
r коэффициент *m* неплотности

3667 **thinning**
(removing cells from wide line images in a
raster object or reducing the number of
coordinate pairs that describe a vector's line
and polygon elements)
d Verdünnung *f*; Ausdünnen *n*
f amenuisement *m*; amincissement *m*;
éclaircissage *m*
r выклинивание *n*; утонение *n*; утончение *n*

3668 **thin-plate splines; TPS**
d scharfkantige Spline *mpl*
f splines *mpl* en plaques minces; splines en
voile mince
r плоские сплайны *mpl*

3669 **third normal form**
d dritte Normal[en]form *f*
f troisième forme *f* normale
r третья нормальная форма *f*

3670 **three-dimensional; tri-dimensional; 3D**
d dreidimensional; 3D
f tridimensionnel; à trois dimensions; [en] 3D
r трёхмерный; 3D

* **three-dimensional net** → 3367

* **three-tuple** → 3828

3671 **throughput; sill**
d Schwelle *f*; Schwellenwert *m*
f seuil *m*
r порог *m*; пороговое значение *n*

* **thumb** → 3234

* **thumb mark** → 3234

* **tic** → 3676

3672 **tic coordinates**
d Gesichtszucken-Koordinaten *fpl*
f coordonnées *fpl* de tic
r координаты *fpl* отсчёта

3673 **tic file**
d Gesichtszucken-Datei *f*
f fichier *m* de tics
r файл *m* меток

* **tick** → 3674

3674 **tick [mark]**
(on the graph axis)
d Teilstrich *m*; Skalenstrich *m*
f coche *f*; amorce *f* [de carroyage]
r отметка *f* (*на оси графика*); птичка *f*;
галочка *f*; деление *n*

3675 **tic [match] tolerance**
(the maximum distance allowed between an
existing tic and a tic being digitized)
d Gesichtszucken-Toleranz *f*
f tolérance *f* de tics
r допуск *m* отсчётов

3676 **tic [point]**
(registration or geographic control points for a
coverage representing known locations on the
Earth's surface)
d Gesichtszucken *n*; Tic-Punkt *m*;
Registrierpunkt *m*
f tic *m*
r метка *f*; отсчёт *m*

* **tic tolerance** → 3675

3677 **tie-point; attachment point; junction point**
d Anschlusspunkt *m*; Verbindungspunkt *m*;
Verknüpfungspunkt *m*; Verbindungsstelle *f*;
Lötstelle *f*
f point *m* d'attachement; point de raccordement;
point de liaison
r связующая точка *f*; точка скрепления;
точка соединения или пересечения

* **TIFF** → 3579

3678 **TIGER-census tract street index;**
TIGER-CTSI
d Straßenindex *m* der TIGER-Erhebung
f index *m* TIGER-CTSI
r индекс *m* TIGER улиц переписных районов

* **TIGER-CTSI** → 3678

* **TIGER data format** → 3741

* **tight coupling** → 575

3679 **tile**
(the spatial unit by which geographic data is organized, subdivided, and stored in a map library)
d nebeneinander angeordnetes Muster *n*; mosaikartiges Muster; Musterkachel *f*; Kachel *f*; Fliese *f*
f pavé *m*
r фрагмент *m* [мозаичного изображения]; элемент *m* мозаичного изображения; плитка *f*; неперекрывающийся образец *m*; мозаичный шаблон *m*

3680 **tiled maps**
d nebeneinander angeordnete Karten *fpl*; mosaikartige Karten
f cartes *fpl* disposées en mosaïque
r неперекрывающиеся карты *fpl*

3681 **tiled pattern**
d Mosaik *n*
f mosaïque *f*
r мозаика *f*

3682 **tiled terrain model**
d nebeneinander angeordnetes Geländemodell *n*; mosaikartiges Geländemodell
f modèle *m* de terrain disposé en mosaïque
r модель *f* неперекрывающихся форм местности

3683 **tile indexing**
d Kachelindizierung *f*
f indexation *f* de mosaïque
r индексирование *n* элементов мозаичного изображения

3684 **tiling; tessellation; paving; covering of the plane with tiles**
d Nebeneinander-Anordnung *f*; Musterkachelung *f*; Parkettierung *f*
f pavage *m*; dallage *m*; tessellation *f*; disposition *f* en mosaïque
r фрагментирование *n*; разбиение *n* плоскости на многоугольники; покрытие *n* плоскости многоугольниками; мозаичное размещение *n*; неперекрывающееся размещение; тасселяция *f*

* **tilt** → 1921

* **tilt angle** → 126

* **tilt meter** → 1922

3685 **time accuracy**

d Präzision *f* der Zeit
f précision *f* de temps
r точность *f* [задания] времени

3686 **time-difference**
d Zeitunterschied *m*
f différence *f* de temps
r разновременность *f*; разность *f* времён

3687 **time dissemination**
d Zeitverbreitung *f*
f diffusion *f* de temps; diffusion de signaux horaires
r передача *f* размера единицы времени; передача сигналов точного времени

3688 **time/distance districting**
d Zeit/Entfernungsdistricting *n*; Districting *n* im Zeitabstand
f découpage *m* à la distance et à la durée; découpage hodochrone
r деление *n* на округа годографом

3689 **time error**
d Zeitfehler *m*
f erreur *f* instantanée de temps; erreur chronométrique
r сбой *m* счёта времени

* **time resolution** → 2149

3690 **time-section block-diagram**
d metachrones Blockdiagramm *n*
f diagramme-bloc *m* de coupe temps; diagramme-bloc métachronique
r метахронная блок-диаграмма *f*; блок-диаграмма временного разреза

3691 **time slice; time slot**
d Zeitscheibe *f*; Zeitschlitz *m*; Zeitlage *f*
f tranche *f* de temps; créneau *m* temporel
r квант *m* времени; отрезок *m* времени; временной канал *m*; интервал *m* временного канала

* **time slot** → 3691

3692 **time-space conceptualization**
d Raum-Zeit-Konzeptdarstellung *f*
f conceptualisation *f* spatio-temporelle
r пространственно-временная концептуализация *f*; пространственно-временное концептуальное представление *n*

3693 **time-space convergence**
d Zeit-Raum-Konvergenz *f*
f convergence *f* temps-espace
r пространственно-временная конвергенция *f*

3694 time-weighted average
  d zeitlich mittlerer Durchschnittswert *m*;
    zeitbezogener Durchschnittswert
  f moyenne *f* intégrée dans le temps; moyenne
    pondérée dans le temps
  r средневзвешенная во времени величина *f*

3695 time zone
  d Zeitzone *f*
  f zone *f* de temps
  r часовой пояс *m*

  * **TIN** → 3820

3696 TIN data structure
  d TIN-Datenstruktur *f*
  f structure *f* de données de réseau de triangles
    irréguliers
  r структура *f* данных сети неравносторонних
    треугольников

3697 TIN facets
  d TIN-Facetten *fpl*
  f facettes *fpl* de réseau de triangles irréguliers
  r фацеты *mpl* сети неравносторонних
    треугольников

  * **TIP** → 3713

  * **tip** → 3700

  * **Tissot's indicatrix** → 1253

  * **TLI** → 3613

  * **TLM** → 3714

  * **TM** → 3648

3698 tolerance
  d Toleranz *f*
  f tolérance *f*
  r допуск *m*; допустимый предел *m*;
    погрешность *f*

  * **tone value** → 617

3699 toolbar
  d Werkzeugstreifen *m*; Hilfsmittelsleiste *f*
  f barre *f* d'instruments; barre d'outils
  r инструментальная линейка *f*; планка *f*
    инструментов; инструментальная лента *f*;
    панель *f* инструментов

3700 tooltip; tip
  d Tip *m*
  f info-bulle *f*
  r сведение *n* для инструментальной кнопки

3701 top layer
  d obere Schicht *f*
  f couche *f* du haut
  r верхний слой *m*

  * **top node** → 3173

3702 topocentric angle
  d topozentrischer Winkel *m*
  f angle *m* topocentrique
  r топоцентрический угол *m*

3703 topocentric coordinates
  d topozentrische Koordinaten *fpl*
  f coordonnées *fpl* topocentriques
  r топоцентрические координаты *fpl*

3704 topocentric frame
  d topozentrischer Rahmen *m*
  f trame *f* topocentrique; repère *m* topocentrique
  r топоцентрический репер *m*;
    топоцентрическая основа *f*

  * **topographic** → 3705

3705 topographic[al]
  d topografisch
  f topographique
  r топографический

3706 topographic base; topographic basis; base map
  d Karten[unter]grund *m*; Basiskarte *f*;
    Arbeitskarte *f*; Grundkarte *f*; topografische
    Grundlage *f*
  f carte *f* de base; carte de travail; fond *m* de
    carte
  r топографическая основа *f* карты;
    географическая основа карты; топооснова *f*

3707 topographic base plate
  d topografische Basisplatte *f*
  f plaque *f* de base topographique
  r оригинал *m* географической основы

  * **topographic basis** → 3706

3708 topographic data
  d topografische Daten *pl*
  f données *fpl* topographiques
  r топографические данные *pl*

3709 topographic database
  d topografische Datenbasis *f*
  f base *f* de données topographique
  r топографическая база *f* данных

3710 topographic divide
  d oberirdische Scheide *f*
  f division *f* topographique
  r топографическая граница *f*;
    топографическое деление *n*

3711 **topographic drawing**
  d topografische Zeichnung *f*
  f dessin *m* topographique
  r топографический чертёж *m*;
     топографический рисунок *m*

3712 **topographic element; [topographic] feature**
  d topografisches Element *n*;
     Geländebedeckung *f*
  f élément *m* topographique
  r топографический элемент *m*

* **topographic feature** → **3712**

3713 **topographic information management; TIP**
  d topografisches Informationsmanagement *n*
  f gestion *f* d'information topographique
  r управление *n* топографической
     информации

3714 **topographic [line] map; TLM; topographic plan; plan; surface contour map**
  d topografische Karte *f*
  f carte *f* topographique
  r топографическая [контурная] карта *f*;
     топографический план *m*

* **topographic map** → **3714**

3715 **topographic map coverage**
  d Deckung *f* der topografischen Karte
  f couverture *f* de carte topographique
  r топографическая изученность *f* территории

* **topographic plan** → **3714**

3716 **topographic plane**
  d topografische Ebene *f*
  f plan *m* topographique
  r топографическая плоскость *f*

3717 **topographic planimetry**
  d topografische Planimetrie *f*; Grundrisstreue *f*
  f planimétrie *f* topographique
  r топографическая планиметрия *f*

* **topographic plotting** → **3723**

3718 **topographic point**
  d topografischer Punkt *m*
  f point *m* topographique
  r топографическая точка *f*

3719 **topographic profile**
  d topografisches Profil *n*
  f profil *m* topographique
  r топографический профиль *m*

3720 **topographic projection; tangent plane projection; projection with heights; coted projection**
  d topografische Projektion *f*; kongruente
     Projektion; kotierte Projektion
  f projection *f* topographique; projection sur un
     plan tangent; projection cotée
  r топографическая проекция *f*; проекция с
     числовыми отметками

* **topographic quadrangle** → **2318**

3721 **topographic relief model**
  d topografisches Reliefmodell *n*
  f modèle *m* de relief topographique
  r топографическая модель *f* рельефа

3722 **topographic surface**
  d topografische Fläche *f*; Geländefläche *f*
  f surface *f* topographique
  r топографическая поверхность *f*

* **topographic survey** → **3723**

3723 **topographic survey[ing]; land survey; cadastral survey; field mapping; topographic plotting**
  d topografische Geländeaufnahme *f*;
     topografische Aufnahme *f*; Fluraufnahme *f*;
     Katasteraufnahme *f*; Feldmessung *f*;
     Landesvermessung *f*; Katastervermessung *f*
  f levé *m* topographique; levé cadastrale;
     service *m* du cadastre; établissement *m* de
     plans cadastraux
  r топографическая съёмка *f*; землемерная
     съёмка; полевая съёмка

3724 **topographic view**
  d topografische Ansicht *f*
  f vue *f* topographique
  r топографический вид *m*

3725 **topography**
  d Topografie *f*
  f topographie *f*
  r топография *f*

* **topography texture map** → **3726**

3726 **topography texture map[ping]**
  d topografische Texturenabbildung *f*
  f application *f* de texture topographique
  r топографическое отображение *n* текстуры

3727 **topogrid interpolation**
  d Topogrid-Interpolation *f*
  f interpolation *f* topogrid
  r топогридная интерполяция *f*

3728 **topological analysis**
  d topologische Analyse *f*

f analyse *f* topologique
r топологический анализ *m*

3729 topological cell
d topologische Zelle *f*; krumme Zelle
f cellule *f* topologique
r топологическая клетка *f*; кривая клетка

* **topological circle** → 3730

3730 topological circumference; topological circle
d topologischer Kreis *m*
f circonférence *f* topologique
r топологическая окружность *f*

3731 topological cleaning
d topologische Reinigung *f*
f nettoyage *m* topologique
r топологическая очистка *f*

3732 topological closure
(as a operation)
d topologische Hüllenoperation *f*
f opération *f* de fermeture topologique
r топологическая операция *f* замыкания

3733 topological complexity
d topologische Komplexität *f*
f complexité *f* topologique
r топологическая сложность *f*

3734 topological consistency
d topologische Konsistenz *f*
f consistance *f* topologique
r топологическая согласованность *f*

* **topological display** → 3757

3735 topological error; topology error
d topologischer Fehler *m*
f erreur *f* topologique
r топологическая погрешность *f*

3736 topological function; topology function
d topologische Funktion *f*; Topologiefunktion *f*
f fonction *f* topologique
r топологическая функция *f*

3737 topological graph
d topologischer Graph *m*
f graphe *m* topologique
r топологический граф *m*

3738 topological group
d topologische Gruppe *f*
f groupe *m* topologique
r топологическая группа *f*

3739 topologically complete quadrangle

d topologisch vollständiges Quadrangel *n*
f quadrangle *m* topologiquement complet
r топологически полный четырёхугольник *m*

3740 topologically complex objects
d topologisch komplexe Objekte *npl*
f objets *mpl* topologiquement complexes
r топологически сложные объекты *mpl*

3741 topologically-integrated geographic encoding and referencing data format; TIGER data format
d TIGER-Format *n*
f format *m* TIGER
r формат *m* топологического интегрированного географического кодирования и стандартизации; формат TIGER (*используемый в Бюро переписей США с конца 80-х годов*)

3742 topologically-structured data
d topologisch strukturierte Daten *pl*
f données *fpl* topologiquement structurées
r топологически структурированные данные *pl*

* **topological map** → 3743

3743 topological map[ping]
d topologische Abbildung *f*
f mappage *m* topologique
r топологическое отображение *n*

3744 topological modeling
d topologische Modellierung *f*
f modelage *m* topologique
r топологическое моделирование *n*

3745 topological overlay
d topologisches Overlay *n*
f calque *m* topologique
r топологический оверлей *m*

3746 topological primitive
d topologisches Primitiv *n*
f primitive *f* topologique
r топологический примитив *m*

* **topological relation** → 3747

3747 topological relation[ship]
d topologische Beziehung *f*; topologische Verbindung *f*
f relation *f* topologique
r топологическая связь *f*

3748 topological sort
d topologisches Sortieren *n*
f tri[age] *m* topologique
r топологическая сортировка *f*

3749 topological space
d topologischer Raum *m*
f espace *m* topologique
r топологическое пространство *n*

3750 topological structure
d topologische Struktur *f*
f structure *f* topologique
r топологическая структура *f*

3751 topological transformation
d topologische Transformation *f*
f transformation *f* topologique
r топологическое преобразование *n*;
   топологическая трансформация *f*

3752 topological vector lattice
d topologisches Vektorgitter *n*
f lattis *m* vectoriel topologique
r топологическая векторная решётка *f*

3753 topological verification; topology auditing
d topologische Revision *f*
f inspection *f* de topologie
r топологическая ревизия *f*; контроль *m*
   топологии

3754 topologization
d Topologisation *f*
f topologisation *f*
r топологизация *f*

3755 topology
d Topologie *f*
f topologie *f*
r топология *f*

* **topology auditing** → 3753

3756 topology description
d topologische Beschreibung *f*
f description *f* de topologie
r топологическое описание *n*

3757 topology diagram; topological display
d Topologie-Diagramm *n*
f diagramme *m* de topologie
r диаграмма *f* топологии

3758 topology digitalization
d topologische Digitalisierung *f*
f numérisation *f* de topologie
r цифрование *n* топологии

3759 topology element
d topologisches Element *n*
f élément *m* de topologie
r элемент *m* топологии

* **topology error** → 3735

* **topology function** → 3736

3760 topology list
d Topologieliste *f*
f liste *f* topologique
r список *m* топологических схем

3761 topology name
d topologischer Name *m*
f nom *m* de topologie
r топологическое имя *n*

3762 topology object data
d topologische Objektdaten *pl*
f données *fpl* d'objets topologiques
r данные *pl* топологических объектов

3763 topology of convergence in measure
d Topologie *f* der Konvergenz nach Maß
f topologie *f* de convergence par mesure
r топология *f* сходимости по мере

3764 topology of network; network topology
d Netz[werk]topologie *f*
f topologie *f* d'un réseau
r топология *f* сети

3765 topology property
d topologische Eigenschaft *f*
f propriété *f* topologique
r свойство *n* топологии

3766 topology query
d topologische Abfrage *f*
f interrogation *f* topologique
r топологический запрос *m*

3767 topology statistics
d topologische Statistik *f*
f statistique *f* topologique
r топологическая статистика *f*

3768 topology type code
d topologischer Typcode *m*
f code *m* de type de topologie
r код *m* типа топологии

3769 topology variable
d topologische Variable *f*
f variable *f* topologique
r топологическая переменная *f*

3770 toponymy
d Toponymie *f*
f toponymie *f*
r топонимия *f*; топонимика *f*

3771 total station
   (an electronic theodolite combined into an
   EDM and an electronic data collector)
d Total-Station *f*

f station *f* totale
r электронный тахеометр *m*

3772 total variance
  d Gesamtvarianz *f*
  f variance *f* totale
  r полная вариантность *f*; общая вариантность

3773 tour
  d Tour *f*
  f tour *m*; voyage *m*
  r ход *m*; тур *m*; объезд *m*; путешествие *n*

3774 tourist's atlas
  d touristischer Atlas *m*; Atlas für Touristen
  f atlas *m* touristique
  r туристский атлас *m*

  * **town plan** → 548

  * **town planning** → 3885

  * **TPS** → 3668

3775 trace
  d Spur *f*; Ablauffolge *f*
  f trace *f*; poursuite *f* de déroulement; droite *f* de suite
  r след *m*; трасса *f*

  * **trace** *v* **a map** → 1180

  * **tracer** → 2174

3776 tracing
  d Verfolgung *f*; Trassieren *n*; Trassierung *f*; Tracing *n*; Aufzeichnung *f*; Durchzeichnung *f*
  f traçage *m*; tracement *m*
  r трассирование *n*; трассировка *f*

  * **track** → 3177

  * **trackball** → 3777

3777 trackball [mouse]; tracker ball; control ball; rolling ball
  d Trackball *m*; Rollkugel *f*
  f trackball *m*; souris *f* trackball; boule *f* roulante; boule de commande
  r [координатный] шар *m*; шаровой указатель *m*; трекбол *m*

  * **tracker ball** → 3777

3778 tracking station
  d Verfolgungsstation *f*
  f station *f* de poursuite
  r станция *f* слежения; пункт *m* слежения; измерительный пункт

3779 tract identification
  d Traktidentifikation *f*
  f identification *f* de tractus
  r идентификация *f* полосы [пространства]

3780 trade-area map; trading area map
  d Einzugsbereich-Karte *f*; Handelszone-Karte *f*; Gewerbeflächenkarte *f*
  f carte *f* de zone de chalandise
  r карта *f* торговой области

  * **trading area map** → 3780

3781 traffic analysis zone; TAZ
  d Verkehrs[fluss]analysebereich *m*
  f zone *f* d'analyse de trafic
  r зона *f* анализа трафика

  * **traffic attributes** → 3784

3782 traffic census
  d Verkehrszählung *f*
  f recensement *m* de trafic
  r учёт *m* трафика

3783 traffic network
  d Verkehrsnetz *n*
  f réseau *m* de trafic
  r сеть *f* трафика

3784 traffic[-related] attributes
  d Trafik-Attribute *npl*
  f attributs *mpl* de trafic
  r атрибуты *mpl* трафика

3785 traffic visualization
  d Trafik-Visualisierung *f*
  f visualisation *f* de trafic
  r отображение *n* трафика

3786 trajectory
  d Trajektorie *f*; Bahn *f*
  f trajectoire *f*
  r траектория

3787 trajectory space; orbit space
  d Raum *m* der Trajektorien; Bahnenraum *m*
  f espace *m* de trajectoires
  r пространство *n* траекторий

3788 transaction
  d Transaktion *f*
  f transaction *f*
  r транзакция *f*; информационный обмен *m*

3789 transect
  d Probestreifen *m*; Transekt *m*
  f transect *m*
  r трансект *m*

* **transfer format** → 1337

3790 transfer function; transmission function
d Übertragungsfunktion *f*;
übertragungstechnische Funktion *f*
f fonction *f* de transfert; fonction de
transmission
r передаточная функция *f*; функция передачи

3791 transfer medium
d Transferträger *m*
f milieu *m* de transfert
r средство *n* передачи

3792 transfer standard
d Übertragungsstandard *m*
f standard *m* de transfert
r стандарт *m* передачи

3793 transformation
d Umwandlung *f*; Umsetzung *f*;
Transformation *f*
f transformation *f*
r трансформация *f*; преобразование *n*

**3794 transformation operator; transformation
statement**
d Transformationsoperator *m*
f opérateur *m* de transformation
r оператор *m* преобразования

* **transformation statement** → 3794

3795 transit
d Transit *m*; Weiterleitung *f*
f transit *m*
r транзит *m*; ретрансляция *f*

3796 transition zone
d Steilabfallgebiet *n*; Sperrzone *f*;
Übergangszone *f*
f zone *f* de transition
r зона *f* перехода; переходная зона

* **transit theodolite** → 3578

3797 translation
d Translation *f*
f translation *f*
r перевод *m*; трансляция *f*

3798 translation curve
d Umsetzungskurve *f*
f courbe *f* de translation
r кривая *f* перемещения; кривая сдвига

* **transmission function** → 3790

3799 transmitter
d Sender *m*

f émetteur *m*
r передатчик *m*

3800 transparency
d Transparenz *f*
f transparence *f*
r прозрачность *f*

3801 transparent region
d transparente Zone *f*
f région *f* transparente
r прозрачный регион *m*

3802 transportation analysis
d Transportanalyse *f*
f analyse *f* de transport
r анализ *m* транспорта

3803 transportation forecasting
d Transport-Prognostizierung *f*
f prédiction *f* de transport
r прогнозирование *n* транспорта

3804 transportation planning
d Transport-Planung *f*
f planification *f* de transport
r составление *n* схемы уличного движения;
дорожно-транспортная планировка *f*

3805 transport geography
d Verkehrsgeografie *f*
f géographie *f* de transport
r география *f* транспорта

* **transversal** → 3807

**3806 transversal projection; transverse
projection; transverse aspect of a map
projection; meridional projection;
equatorial projection**
(a projection oriented at right angles to the
equator)
d querachsige Projektion *f*; transversale
Abbildung *f*; äquatorachsige Abbildung
f projection *f* transversale; projection
méridienne
r поперечная проекция *f*; меридиональная
проекция

3807 transverse; transversal; cross
d transversal; Quer-
f transversal; croisé
r поперечный; перекрёстный

* **transverse aspect of a map projection** →
3806

* **transverse projection** → 3806

* **transverse scanning** → 22

3808 travelling salesman problem
 d Problem *n* des Handlungsreisenden
 f problème *m* du représentant de commerce
 r задача *f* коммивояжера

 * **traverse → 2830**

3809 traverse *v*
 d durchlaufen
 f traverser; pointer en direction
 r пересекать; проходить

3810 traverse data
 d Polygonzugsdaten *pl*
 f données *fpl* de cheminement polygonal
 r данные *pl* о полигональных ходах

 * **traverse network → 2829**

3811 tree
 (a fully connected network without circuits)
 d Baum *m*
 f arbre *m*
 r дерево *n*

3812 tree[-line] topology
 d Baumstruktur *f*; Baum-Topologie *f*
 f topologie *f* arborescente
 r древовидная топология *f*; топология типа "дерево"

3813 tree network
 d Baumnetzwerk *n*
 f réseau *m* arborescente
 r древовидная сеть *f*

 * **tree topology → 3812**

 * **trend curve → 3814**

3814 trend line; trend curve
 d Trendlinie *f*
 f ligne *f* de tendance
 r линия *f* тренда; тренд-линия *f*; линия роста

3815 trend surface
 d Trendfläche *f*
 f surface *f* de tendance
 r поверхность *f* тренда; тренд-поверхность *f*

3816 trend surface analysis
 d Oberflächentrendanalyse *f*
 f analyse *f* de surface de tendance
 r анализ *m* тренд-поверхности

3817 triad configuration; triad geometry
 d Triade-Geometrie *f*
 f conformation *f* de triade
 r конфигурация *f* триады

 * **triad geometry → 3817**

3818 triangle
 d Dreieck *n*
 f triangle *m*
 r треугольник *m*

3819 triangular grid; triangular network
 d dreieckiges Netz *n*
 f réseau *m* triangulaire
 r треугольная сетка *f*

 * **triangular network → 3819**

3820 triangulated irregular network; TIN
 (a system of terrain representation that builds triangular facets to connect point heights)
 d trianguliertes nichtregelmäßiges Gitter *n*; trianguliertes irreguläres Netz[werk] *n*; TIN
 f réseau *m* de triangles irréguliers
 r сеть *f* неравносторонних треугольников

3821 triangulation
 d Triangulation *f*; Triangulierung *f*; Dreiecksaufnahme *f*
 f triangulation *f*
 r триангуляция *f*; триангулирование *n*; разбиение *n* на треугольники; треугольная декомпозиция *f*

3822 triangulation-based digitizer
 d Triangulation-basierter Digitalgeber *m*
 f numériseur *m* orienté à triangulation
 r дигитайзер *m*, базированный на триангуляции

 * **triangulation mark → 3823**

3823 triangulation mark[er]
 d Triangulationsmarker *m*
 f marqueur *m* de triangulation; marque *f* de triangulation
 r триангуляционный маркер *m*; триангуляционная марка *f*

 * **tri-dimensional → 3670**

3824 trigonometry
 d Trigonometrie *f*
 f trigonométrie *f*
 r тригонометрия *f*

3825 trilateration
 d Trilateration *f*
 f trilatération *f*
 r трилатерация *f*

3826 trilinear interpolator
 d Trilinear-Interpolator *m*

f interpolateur *m* trilinéaire
r трилинейный интерполятор *m*

* **trilinear surveying** → 3145

* **TRIM format** → 3616

3827 triple-difference; TD
  d Dreifachdifferenz *f*
  f différence *f* triple
  r третья разность *f*; строенная разность

3828 triplet; three-tuple
  d Trippel *n*; Drilling *m*
  f triplet *m*
  r тройка *f* [координат]; триплет *m*

3829 tripod data
  d Dreibein-Daten *pl*
  f données *fpl* trépiedes
  r треножные данные *pl*

3830 tri tree
  d Dreibaum *m*
  f tri-arbre *m*
  r трихотомическое дерево *n*

* **TRK** → 3039

3831 tropospheric errors
  d troposphärische Fehler *mpl*
  f erreurs *fpl* troposphériques
  r тропосферные задержки *fpl*

3832 true 3D view
  d wahre 3D-Darstellung *f*
  f vue *f* 3D vrai
  r истинное трёхмерное изображение *n*

3833 true latitude
  d wahre Breite *f*
  f latitude *f* vraie
  r истинная широта *f*

3834 true north
  d geografischer Norden *m*
  f nord *m* vrai; nord géographique
  r истинный север *m*; географический север

3835 true-to-scale representation
  d maßstabgetreue Darstellung *f*
  f représentation *f* en échelle vraie
  r точномасштабное представление *n*

* **tuple** → 3184

3836 turn
  (a transition from one network link to another
  at a network node)
  d Kurvenflug *m*; Windung *f*

f retournement *m*; renversement *m*; virage *m*
r поворот *m*; переход *m*

3837 turn impedance
  (the impedance or cost of making a turn at a
  network node)
  d Windung-Impedanz *f*
  f impédance *f* de retournement
  r импеданс *m* поворота; импеданс перехода

3838 turning point
  d Wendepunkt *m*
  f point *m* d'inflexion; point de virage; point de
  retournement
  r точка *f* поворота; поворотная точка

3839 two-point equidistant projection
  d äquidistante Zweipunkt-Projektion *f*;
  abstandstreue Zweipunkt-Projektion
  f projection *f* équidistante à deux points
  r двухточечная равнопромежуточная
  проекция *f*; двухточечная эквидистантная
  проекция

* **type font** → 526

U

* **UART** → 3859

* **UGA** → 3880

3840 unclipped
d nichtabgeschnitten
f non coupé
r необрезанный

3841 unclosed contour
d offene Höhenlinie f
f courbe f de niveau non fermée
r незамкнутая горизонталь f

3842 unclosed polygon
d offenes Polygon n
f polygone m non fermé
r незамкнутый полигон m

3843 unclosed [polygonal] traverse; open traverse
d offener Polygonzug m
f cheminement m polygonal non fermé
r висячий полигонный ход m

* **unclosed traverse** → 3843

* **uncompress** v → 943

* **unconformity** → 3506

3844 unconnected arc
d nichtverbundener Bogen m; unverbundener Bogen
f arc m non connecté
r несвязная дуга f

3845 unconnected node
d nichtverbundener Knoten m
f nœud m non connecté
r несвязная вершина f; несвязный узел m

* **uncouple** v → 3871

3846 underlay
d Unterlage f; Unterlegung f
f empiècement m; thibaude f; prise f de piquage
r подслаивание n

* **undershoot** → 2870

3847 undershoot inaccuracy
d Unterschwingen-Ungenauigkeit f
f inexactitude f due au sous-dépassement
r неточность f из-за недохода до заданной координаты

3848 undetected error
d Restfehler m; unentdeckter Fehler m
f erreur f non détectée
r необнаруженная ошибка f

3849 undocumented data
d undokumentierte Daten pl
f données fpl non documentées
r недокументированные данные pl

3850 unended line
d nichtbeendete Linie f
f ligne f sans fin
r незаконченная линия f

3851 unevenness; roughness
(of a surface)
d Ungleichmäßigkeit f; Unebenheit f; Gleisunebenheit f
f inégalité f; défaut m de planéité
r шероховатость f; неровность f

3852 unformatted data; nonformatted data
d nichtformatierte Daten pl; unformatierte Daten
f données fpl non formatées
r неформатированные данные pl

3853 ungeocoding
d Ungeocodierung f
f non-géocodage m
r дегеокодирование n; отмена f геокодирования

3854 uniformity; homogeneity
d Gleichmäßigkeit f; Homogenisierung f
f uniformité f; homogénéité f
r однородность f; равномерность f; однообразность f

3855 uniform regions; homogeneous regions
d homogene Regionen fpl
f régions fpl uniformes; régions homogènes
r однородные регионы mpl

* **union** → 2044

3856 unique property reference number
d eindeutige Bezugsnummer f einer Landparzelle; eindeutige Referenznummer f einer Landparzelle
f numéro m unique de référence d'une parcelle
r однозначный ссылочный номер m парцеллы

3857 United States Geological Survey; USGS
d geologischer Dienst m der Vereinigten Staaten

f Levé *m* géologique d'États-Unis
r Геологическая съёмка *f* США

3858 unit of space; space unit; area[l] unit
d Flächeneinheit *f*
f unité *f* de surface; unité de volume; surface *f* unitaire
r единица *f* площади

3859 universal asynchronous receiver/transmitter; UART
d universeller asynchroner Empfänger/Sender *m*
f récepteur/émetteur *m* asynchrone universel
r универсальный асинхронный приёмопередатчик *m*; УАПП

3860 universal decimal classification
d universale dezimale Klassifizierung *f*
f classification *f* décimale universelle
r универсальная десятичная классификация *f*

3861 universal geographic identity
d universale geografische Identität *f*
f identité *f* géographique universelle
r универсальная географическая идентичность *f*

3862 universal GIS
d universelles GIS *n*
f SIG *m* universel
r универсальная ГИС *f*

3863 universal polar stereographic projection; UPS
d universale polare stereografische Projektion *f*
f projection *f* stéréographique polaire universelle
r универсальная полярная стереографическая проекция *f*

3864 universal projection plotting system
d Plottensystem *n* in der universalen Projektion
f système *m* de tracement à projection universelle
r система *f* вычерчивания в универсальной проекции

3865 universal rectifier
d universeller Gleichrichter *m*
f redresseur *m* universel
r универсальный выпрямитель *m*

* **universal space rectangular → 3866**

3866 universal space rectangular [coordinate system]
d universelles rechteckiges Raumkoordinatensystem *n*
f système *m* de coordonnées rectangulaire spatial universel

r универсальная пространственная прямоугольная система *f* координат; прямоугольная координатная система универсального пространства

* **universal time → 758**

* **universal time coordinated → 758**

3867 universal transverse Mercator; UTM; Mercator['s] projection
(a spatial reference system using a set of transverse Mercator projections six degrees wide that cover the Earth)
d Mercator-Transversale *f*; Mercatorprojektion *f*; Mercator-Abbildung *f*
f Mercator *m* transverse universel; projection *f* Mercator [transverse universelle]
r [универсальная поперечно-цилиндрическая] проекция *f* Меркатора; проекция UTM

* **universe face → 3868**

3868 universe polygon; universe face
d Weltpolygon *n*
f polygone *m* d'univers
r универсальный полигон *m*

* **unlabeled → 3869**

3869 unlabel[l]ed; mislabel[l]ed
d nichtetikettiert
f sans étiquette; non étiqueté
r непомеченный

3870 unlayered sheet
d ungeschichtetes Blatt *n*
f feuille *f* sans courbes de niveau ni teintes hypsométriques
r бесслойный лист *m*

3871 unlink *v*; uncouple *v*
d trennen; spalten; entzweien
f délier
r разъединять; расщеплять; размыкать

3872 unlinked text
d nichtverbundener Text *m*
f texte *m* sans liens
r несвязанный текст *m*

3873 unorganized territory; UT
d nichterfasstes Gebiet *n*
f territoire *m* non organisé
r неорганизованная территория *f*

3874 unrectified image data
d nichtberichtigte Bilddaten *npl*

f données *fpl* [d']image non rectifiées
r невыверенные данные *pl* об изображении; неисправленные данные об изображении

* **unselecting** → 991

* **unsharp** → 1495

3875 unsmoothing; desmoothing
   d Desmoothing *n*
   f annulation *f* de lissage
   r отмена *f* сглаживания

3876 unsupervised classification
   d nichtkontrollierte Klassifikation *f*; unüberwachte Klassifizierung *f*
   f classification *f* non-dirigée; classification non-contrôlée
   r неконтролируемая классификация *f*; безусловная классификация

* **unzip** *v* → 943

3877 updating
   d Aktualisierung *f*
   f actualisation *f*
   r обновление *n*; актуализация *f*

3878 upland; highland
   d Erhebung *f*; Hochland *n*
   f terre *f* haute
   r возвышенность *f*; нагорье *n*; холмистая местность *f*

* **UPS** → 3863

3879 upstream
   d flussaufwärts; stromaufwärts; bergseits; bergwärts
   f [en] amont; à l'amont; vers l'amont
   r выше по течению; вверх по течению; против течения

* **urban cluster** → 3884

* **urban core** → 512

3880 urban development area; urban expansion area; urban growth area; UGA
   d städtebaulicher Entwicklungsbereich *m*
   f zone *f* d'extension urbaine
   r зона *f* градостроительства

* **urban expansion area** → 3880

3881 urban GIS
   d städtebauliches GIS *n*
   f SIG *m* urbain
   r городская ГИС *f*; муниципальная ГИС; МГИС

* **urban growth area** → 3880

3882 urban growth boundary
   d Entwicklungsbereichgrenze *f*
   f limite *f* de zone d'extension
   r граница *f* урбанизации

3883 urbanized area; urbanized region
   d verstädtertes Gebiet *n*
   f zone *f* urbanisée; région *f* urbanisée
   r городское население *n*

* **urbanized region** → 3883

3884 urban nebula; urban cluster
   d Städtegruppe *f*
   f nébuleuse *f* urbaine
   r группа *f* городов

* **urban nucleus** → 512

3885 urban planning; town planning; site planing
   d Stadtplanung *f*; städtebauliche Planung *f*; Städtebau *m*
   f planification *f* urbaine; urbanisme *m*
   r городская планировка *f*

* **urban residential district** → 3149

3886 Usenet map
   d Usenet-Karte *f*
   f carte *f* d'Usenet
   r карта *f* сети Usenet

3887 user-defined attributes
   d benutzerdefinierte Attribute *npl*; benutzerbestimmte Attribute
   f attributs *mpl* définis par utilisateur
   r атрибуты *mpl*, определяемые пользователем

3888 user-defined projection
   d benutzerdefinierte Projektion *f*
   f projection *f* définie par utilisateur
   r проекция *f*, определяемая пользователем

3889 user identification
   d Benutzerkennzeichnung *f*
   f identification *f* d'utilisateur
   r идентификация *f* пользователя

3890 user interface
   d Benutzerschnittstelle *f*; Anwenderschnittstelle *f*
   f interface *f* d'utilisateur
   r пользовательский интерфейс *m*

3891 user requirements
   d Benutzeforderungen *fpl*

f nécessités *fpl* d'utilisateur
r требования *npl* пользователя

3892 user segment
(of GPS)
d Nutzersegment *n*
f segment *m* d'utilisateurs
r сегмент *m* пользователей; сегмент потребителей; подсистема *f* аппаратуры пользователей

* **USGS → 3857**

* **UT → 3873**

* **UTC → 758**

3893 utility mapping
(a special class of GIS application for managing information about public utilities such as water piper, sewerage, telephone, electricity, and gas networks)
d Leitungskartierung *f*
f cartographie *f* d'utilités
r картографирование *n* утилит

3894 utility network
d Leitungsnetz *n*
f réseau *m* utilitaire
r сеть *f* утилит

* **UTM → 3867**

3895 UTM coordinates
d UTM-Koordinaten *fpl*
f coordonnées *fpl* UTM
r UTM координаты *fpl*

V

* **vagueness** → 108

3896 valency of a node; node valency; nodal degree
(the number of arcs at the node)
d Knotenvalenz *f*
f valence *f* de nœud
r валентность *f* узла

3897 valency table
d Valenztabelle *f*
f table *f* de valences
r таблица *f* валентностей

* **valid data** → 562

* **valuation** → 1321

3898 value attribute table; VAT; attribute-value table
d Attributwert-Tabelle *f*
f table *f* de valeurs d'attributs
r список *m* свойств; список атрибутов

3899 value set
d Wertesatz *m*
f ensemble *m* de valeurs
r набор *m* значений

* **vanishing point** → 2532

3900 variable
d Variable *f*; Veränderliche *f*
f variable *f*
r переменная *f*

3901 variable attribute
d veränderliches Attribut *n*
f attribut *m* variable
r переменный атрибут *m*

* **variable interval digitizing** → 2807

3902 variable length
d veränderliche Länge *f*
f longueur *f* variable
r переменная длина *f*

3903 variable length list
(a data structure that can store a flexible number of elements in an ordered sequence)
d Liste *f* mit variabler Länge
f liste *f* de longueur variable

r список *m* переменной длины

3904 variance
d Varianz *f*
f variance *f*
r [среднее] отклонение *n*; вариантность *f*

3905 variation of data
d Datenvariation *f*
f variation *f* de données
r вариация *f* данных

* **variations in color** → 609

3906 variogram
d Variogramm *n*
f variogramme *m*
r вариограмма *f*

* **VAT** → 3898

3907 vector
d Vektor *m*
f vecteur *m*
r вектор *m*

* **vector-based format** → 3910

3908 vector-based GIS
d vektorbasiertes GIS *n*
f SIG *m* vectoriel
r векторная ГИС *f*

3909 vector data
d Vektordaten *pl*
f données *fpl* vectorielles
r векторные данные *pl*

3910 vector [data] format; vector-based format
d Vektor[daten]format *n*
f format *m* [de données] vectoriel
r векторный формат *m* [пространственных] данных

3911 vector data model; vector data structure
d Vektordatenmodell *n*
f modèle *m* de données vectoriel
r векторная модель *f* данных

* **vector data structure** → 3911

3912 vector field
d Vektorfeld *n*
f champ *m* vectoriel; champ vecteur
r векторное поле *n*

* **vector format** → 3910

3913 vector function
d Vektorfunktion *f*

f fonction *f* vectorielle
r векторная функция *f*; вектор-функция *f*

3914 vector graphics
   d Vektorgrafik *f*
   f graphique *m* vectoriel
   r векторная графика *f*

3915 vector image format
   d Vektorbildformat *n*
   f format *m* image-vecteur
   r формат *m* векторных изображений

   * **vectorization → 3028**

3916 vectorized cell
   d vektorisierte Zelle *f*
   f cellule *f* vectorisée
   r векторизированная ячейка *f*

   * **vectorizer → 3029**

3917 vector layer
   d Vektorschicht *f*
   f couche *f* vectorielle
   r векторный слой *m*

3918 vector map
   d Vektorkarte *f*
   f carte *f* vectorielle
   r векторная карта *f*

3919 vector plotter
   d Vektorplotter *m*
   f traceur *m* vectoriel
   r векторный графопостроитель *m*

3920 vector product format; VPF
   d Format *n* des Vektorprodukts
   f format *m* de produit vectoriel
   r формат *m* векторного произведения

   * **vector-raster conversion → 3923**

3921 vector representation
   d Vektordarstellung *f*
   f représentation *f* vectorielle
   r векторное представление *n*

3922 vector structure
   d Vektorstruktur *f*
   f structure *f* vectorielle
   r векторная структура *f*

3923 vector-[to-]raster conversion; rasterization; gridding; grating
   d Vektor-Raster-Konvertierung *f*; Rasterung *f*
   f conversion *f* de vecteur en rastre
   r векторно-растровое преобразование *n*; растеризация *f*

3924 vector-to-raster processor; rasterizator
   d Vektor-Raster-Prozessor *m*
   f processeur *m* de vecteur en rastre
   r процессор *m* векторно-растрового преобразования; растеризатор *m*

3925 vegetation zone
   d Vegetationszone *f*
   f zone *f* de végétation
   r зона *f* растительности

   * **vehicle management system → 3926**

   * **vehicle tracking → 3926**

3926 vehicle tracking [system]; vehicle management system
   d Standortortungssystem *n* von Fahrzeugen
   f système *m* de localisation de véhicule [à distance]; système de repérage de voiture par satellite
   r [сателлитная] система *f* навигации транспортных средств; автомобильная навигационная система

3927 velocity
   d Geschwindigkeit *f*
   f vélocité *f*
   r скорость *f*

   * **verification → 532**

3928 vertex; apex
   (of a graph)
   d Spitze *f*
   f sommet *m*; apex *m*
   r вершина *f*

   * **vertex angles → 3930**

3929 vertical accuracy
   d vertikale Genauigkeit *f*
   f exactitude *f* verticale
   r вертикальная точность *f*

3930 vertical angles; vertex angles; apex angles; [vertically] opposite angles
   d Scheitelwinkel *mpl*; Spitzenwinkel *mpl*
   f angles *mpl* opposés par le sommet; angles au sommet
   r вертикальные углы *mpl*; углы при вершине

   * **vertical control → 2140**

3931 vertical datum
   d Vertikaldatum *n*
   f datum *m* vertical; repère *m* vertical
   r вертикальное начало *n* отсчёта

3932 vertical dilution of precision
   d vertikaler Genauigkeitsabfall *m*

f dilution *f* verticale de précision
r снижение *n* точности определения
положения в вертикальной плоскости

3933 vertical error
   d Vertikalfehler *m*
   f erreur *f* verticale
   r вертикальная погрешность *f*

3934 vertical error of closure; vertical misclosure; leveling error of closure
   d Vertikalabschlussfehler *m*; nivellierischer Schleifenschlussfehler *m*
   f erreur *f* verticale de fermeture; erreur de fermeture d'une maille en altimétrie
   r вертикальный зазор *m*; вертикальная невязка *f* [полигона]

3935 vertical exaggeration
   d vertikale Übertreibung *f*
   f exagération *f* verticale; surhaussement *m*
   r вертикальное утрирование *n* [размера или формы]

3936 vertical hill shading
   d Böschungsschummerung *f*, Böschungsschummer *m*
   f estompage *m* de pente
   r вертикальная отмывка *f* возвышений

3937 vertical interpolator
   d vertikaler Interpolator *m*
   f interpolateur *m* vertical
   r вертикальный интерполятор *m*

   * **vertical interval** → 730

3938 vertical line
   d vertikale Linie *f*
   f ligne *f* verticale
   r вертикальная линия *f*

   * **vertically opposite angles** → 3930

   * **vertical misclosure** → 3934

   * **vertical net** → 2140

3939 vertical position
   d senkrechte Schweissposition *f*
   f position *f* verticale
   r вертикальная позиция *f*

3940 vertical projection; second projection; elevation view
   d Vertikalprojektion *f*, Aufriss *m*; zweite Projektion *f*
   f projection *f* verticale; seconde projection; coupe *f* verticale
   r вертикальная проекция *f*; вторая проекция;

вертикальный разрез *m*

3941 vertical resolution
   d Vertikalauflösungsvermögen *n*
   f résolution *f* verticale
   r вертикальная разрешающая способность *f*

   * **videocasque** → 1810

3942 video digitizer
   d Video-Digitalisierer *m*; Video-Digitizer *m*
   f digitaliseur *m* vidéo
   r устройство *n* оцифровки видеоизображений

   * **video hard copy unit** → 2418

   * **view angle** → 127

3943 view direction
   d Ansichtrichtung *f*
   f direction *f* de vue
   r направление *n* взгляда

   * **viewer** → 3963

   * **viewing** → 3962

   * **viewing angle** → 127

   * **viewpoint** → 3957

3944 viewport; view window
   d Ansicht[s]fenster *n*; Darstellungsfeld *n*
   f clôture *f*, fenêtre *f* d'affichage; fenêtre de vue
   r область *f* просмотра; окно *n* просмотра; видовой экран *m*

3945 viewshed
   (the boundaries of sight from a single vantage point, assuming an unobstructed surface)
   d Sichtbereich *m*
   f aire *f* de visibilité
   r зона *f* видимости

3946 viewshed analysis; visibility/invisibility analysis
   d Sichtbereichanalyse *f*
   f analyse *f* de visibilité/non-visibilité
   r анализ *m* видимости/невидимости

   * **viewshed map** → 3954

   * **view window** → 3944

3947 village area; communal area
   d Dorfgebiet *n*
   f territoire *m* communal
   r сельская территория *f*

3948 **virtual active control point**
   d virtueller aktiver Kontrollpunkt *m*; virtuelle
      aktive Bedienungsstelle *f*; virtueller aktiver
      Orientierungspunkt *m*
   f point *m* directeur virtuel; point pilote virtuel
   r активная виртуальная опорная точка *f*;
      активная виртуальная реперная точка

3949 **virtual data**
   d Virtualdaten *pl*
   f données *fpl* virtuelles
   r виртуальные данные *pl*

3950 **virtual GIS**
   d virtuelles GIS *n*
   f SIG *m* virtuel
   r виртуальная ГИС *f*

3951 **virtual reality; VR**
   d virtuelle Realität *f*
   f réalité *f* virtuelle; RV
   r виртуальная реальность *f*

3952 **virtual road**
      (a street point in a ZIP centroid used for
      geocoding)
   d virtueller Weg *m*
   f route *f* virtuelle
   r виртуальный пут *m*

3953 **visibility**
   d Sichtbarkeit *f*
   f visibilité *f*
   r видимость *f*

3954 **visibility [area] map; visible area map; seen
      area map; viewshed map**
   d Sichtkarte *f*; Karte *f* der sichtbaren Flächen
   f carte *f* de visibilité[/non-visibilité]
   r карта *f* видимости[/невидимости]

   * **visibility/invisibility analysis** → 3946

   * **visibility map** → 3954

   * **visible area map** → 3954

3955 **visible feature**
   d sichtbares Feature *n*
   f détail *m* cartographique visible
   r видимый [топографический] элемент *m*

3956 **visible layer**
   d sichtbare Schicht *f*
   f couche *f* visible
   r видимый слой *m*

3957 **vista point; viewpoint; point of view**
   d Blickpunkt *m*; Durchblickpunkt *m*
   f point *m* de perspective

   r точка *f* обзора; точка перспективы

3958 **visual accumulation**
   d visuelle Akkumulation *f*
   f accumulation *f* visuelle
   r визуальная аккумуляция *f*

   * **visual angle** → 127

3959 **visual attribute**
   d visuelles Attribut *n*
   f attribut *m* visuel
   r визуальный атрибут *m*

   * **visual databank** → 1899

   * **visual field** → 1403

3960 **visual image interpretation**
   d visuelle Bilddeutung *f*
   f interprétation *f* visuelle d'images
   r визуальное дешифрирование *n*
      изображений

3961 **visual interface**
   d visuelle Schnittstelle *f*
   f interface *f* visuelle
   r визуальный интерфейс *m*

3962 **visualization; viewing; display[ing]**
   d Visualisierung *f*; Sichtbarmachung *f*
   f visualisation *f*; visionnement *m*
   r визуализация *f*; графическое
      воспроизведение *n*; отображение *n*

3963 **visualizer; viewer; browser; navigator**
   d Browser *m*; Suchprogramm *n*; Navigator *m*
   f logiciel *m* de visualisation; visionneur *m*;
      visionneuse *f*; module *m* de revue; butineur *m*;
      navigateur *m*
   r программа *f* просмотра; браузер *m*;
      просмотрщик *m*; визуализатор *m*;
      бродилка *f*; навигатор *m*; вью[в]ер *m*

3964 **visual line of position on land**
   d terrestrische Standlinie *f*
   f ligne *f* de position déterminée par observation
      visuelle
   r визуальная линия *f* наземной позиции

3965 **visual representation**
   d visuelle Darstellung *f*
   f représentation *f* visuelle
   r визуальное представление *n*

3966 **visual simulation of natural phenomena**
   d visuelle Simulation *f* der Naturerscheinungen
   f simulation *f* visuelle de phénomènes naturels
   r визуальная симуляция *f* естественных
      явлений

3967 volume calculation
- *d* Volumenberechnung *f*
- *f* calcul *m* de volume
- *r* исчисление *n* объёма

* **volume element** → 1243

* **volume pixel** → 1243

3968 volume primitives; space primitives
- *d* räumliche Primitivelemente *npl*;
  Raumprimitive *npl*; raumbezogene
  Grundelemente *npl*
- *f* primitives *fpl* volumétriques
- *r* объёмные примитивы *mpl*;
  пространственные примитивы

* **volumetric feature** → 1157

* **volumetric geoimage** → 1011

3969 volumetric image; 3D image
- *d* Volumenbild *n*; 3D-Bild *n*
- *f* image *f* volumétrique; image 3D
- *r* трёхмерное изображение *n*; объёмное
  изображение

* **Voronoi diagrams** → 3665

3970 Voronoi edge
- *d* Voronoi-Kante *f*
- *f* arête *f* de Voronoi
- *r* ребро *n* Вороного

3971 Voronoi network
- *d* Voronoi-Netz[werk] *n*
- *f* réseau *m* de Voronoi
- *r* сеть *f* Вороного

* **Voronoi polygons** → 3665

3972 voting district; electoral district
- *d* Wahlkreis *m*; Wahlbezirk *m*; Stimmbezirk *m*
- *f* circonscription *f* électorale
- *r* избирательный округ *m*

* **voxel** → 1243

* **VP** → 2532

* **VPF** → 3920

* **VR** → 3951

W

* WAC → 4006

3973 **wall map**
 d Wandkarte *f*
 f carte *f* murale
 r [на]стенная карта *f*

* WAN → 3996

* ward → 1148

* ward boundary → 1149

* warpage → 3974

3974 **warping; warpage**
 d Verbiegung *f*; Durchbiegung *f*;
    Verkrümmung *f*
 f gauchissement *m*; déformation *f* [non
    uniforme]
 r деформирование *n*; [неоднородное]
    искажение *n*

3975 **water divide; watershed; ridge-line**
    (area bounded by ridges that would converge
    (downhill) to a single exit point)
 d Wasserscheide *f*
 f ligne *f* de partage des eaux; crête *f* de partage
    des eaux
 r водораздел *m*

3976 **water line**
 d Wasserlinie *f*
 f ligne *f* d'eau; hauteur *f* du plan d'eau
 r горизонт *m* воды; уровень *m* воды

* watershed → 3975

3977 **waterway information system**
 d Wasserstraßen-Informationssystem *n*;
    Wasserweg-Informationssystem *n*
 f système *m* d'information des voies d'eau;
    système d'information des cours d'eau
 r информационная система *f* водотоков

3978 **wave-mode filter; mode filter**
 d Modenfilter *m*; Wellentypfilter *m*
 f filtre *m* de mode[s d'ondes]; filtre de modes
    d'oscillations
 r фильтр *m* мод; фильтр типов волн; фильтр
    типов колебаний; модовый фильтр

* way → 3177

3979 **waypoint**
 d Wegpunkt *m*; Wegmarke *f*
 f point *m* de route; point de cheminement
 r точка *f* маршрута

3980 **waypoint information**
 d Wegpunktdaten *pl*
 f information *f* de point de cheminement
 r информация *f* о точке маршрута

* WCS → 4008

* weather chart → 3982

3981 **weathering**
 d Eisverwitterung *f*; Bewitterung *f*; Wetterung *f*
 f érosion *f*; atmosphérisation *f*; vieillissement *m*
    climatique; altération *f* atmosphérique
 r выветривание *n*; разрушение *n* в
    атмосферных условиях; воздействие *n*
    атмосферных условий

3982 **weather map; weather chart;**
     meteorological map
 d Wetterkarte *f*
 f carte *f* météorologique
 r метеорологическая карта *f*

3983 **Web-based electronic atlas**
 d Web-basierter elektronischer Atlas *m*
 f atlas *m* électronique par le Web
 r атлас *m* в Web

3984 **Web-based GIS**
 d Web-[basiertes]GIS *n*; Internet-GIS *n*;
    Online-GIS *n*
 f SIG *m* [par le] Web
 r ГИС *f* в Web

3985 **Web[-based] map**
 d Web-basierte Karte *f*
 f carte *f* [par le] Web
 r карта *f* в Web

* Web map → 3985

3986 **Web mapping**
 d Webmapping *n*; Web-Kartografierung *f*
 f cartographie *f* Web
 r картографирование *n* в Web

3987 **Web-mapping system**
 d Web-Kartografierungssystem *n*
 f système *m* de cartographie [par le] Web
 r система *f* картографирования в Web

* weed → 3988

3988 **weed[ing]**
 d Jäten *n*; Weeding *n*

f émondage *m*; élagage *m* (*action de réduire un ensemble de données, de documents, de dossiers, etc. en éliminant les éléments les moins désirables*)
r разрядка *f*

3989 weed tolerance
(the minimum allowable distance between any two vertices along an arc)
d minimaler Abstand *m* zweier Vertices bei geraden Linien
f tolérance *f* d'émondage
r допуск *m* разрядки

* **weighted average** → 3991

3990 weighted buffering
d gewichtete Pufferung *f*
f tamponnage *m* pondéré
r буферизация *f* с взвешиванием

3991 weighted mean; weighted average
d gewogener Mittelwert *m*; gewogenes Mittel *n*
f moyenne *f* pondérée
r взвешенное среднее *n*

3992 weight features; significance features
(features of the importance of particular data elements)
d Gewicht-Charakteristiken *fpl*
f caractéristiques *fpl* de pondération
r признаки *mpl* весомости

3993 well-defined point
(a point-like (isolated) feature that can be distinguished on the source and on the ground to sufficient accuracy)
d klar abgegrenzter Punkt *m*; klar bestimmter Punkt; klarer Punkt
f point *m* bien défini
r вполне определённая точка *f*; чётко определённая точка

3994 western
d West *m*
f ouest *m*
r восток *m*

* **WGS** → 4010

3995 wide area differential GPS
d Weitverkehrs-Differenzial-GPS *n*
f GPS *m* différentiel global
r мировая дифференциальная ГПС *f*

3996 wide area network; WAN; long-haul network
d Weltdatennetz *n*; flächendeckendes Computernetz *n*; Weitverkehrsnetz *n*; weiträumiges Netzwerk *n*

f réseau *m* global; grand réseau; réseau étendu
r глобальная [вычислительная] сеть *f*; ГВС; мировая сеть

3997 wildlife map
d Wildfauna-Karte *f*; Wildtiere-Karte *f*
f carte *f* de faune sauvage; carte faunique
r карта *f* дикой природы; карта дикой флоры и фауны

3998 window configuration
d Fensterkonfiguration *f*
f configuration *f* de fenêtres
r конфигурация *f* окон

3999 windowing; fenestration
d Fensterung *f*; Fenstertechnik *f*; Ausschnittsdarstellung *f*
f fenêtrage *m*
r кадрирование *n* [изображения]; организация *f* окон

4000 wireframe image
d Drahtdarstellung *f*
f image *f* filaire; image fil de fer
r [проволочно-]каркасное изображение *n*

4001 wireframe model
d Drahtmodell *n*; Kantenmodell *n*
f modèle *m* fil de fer
r проволочно-каркасная модель *f*

4002 witness tree
(generally used in the US public land states, this refers to the trees close to a section corner)
d Zeugenbaum *m*
f arbre *m* témoin
r дерево-свидетель *n*

4003 work file; scratch file
(a temporary file holding intermediate data during an operation)
d Arbeitsfile *n*; Arbeitsdatei *f*; Notizdatei *f*
f fichier *m* de travail
r рабочий файл *m*; вспомогательный файл

4004 workspace
d Arbeitsbereich *m*
f espace *m* de travail; zone *f* de travail
r рабочая область *f*; рабочее пространство *n*

4005 workstation
d Arbeitsstation *f*
f station *f* de travail
r рабочая станция *f*; автоматизированное рабочее место *n*

4006 world aeronautical chart; WAC
d Weltluftfahrtkarte *f*

 f carte *f* aéronautique mondiale
 r мировая аэронавигационная карта *f*

4007 world coordinate
 d Weltkoordinate *f*
 f coordonnée *f* universelle
 r мировая координата *f*

4008 world coordinate system; WCS; global coordinate system
 d Weltkoordinatensystem *n*; globales Koordinatensystem *n*
 f système *m* de coordonnées mondial
 r мировая координатная система *f*; глобальная система координат

4009 world ellipsoid
 d Weltellipsoid *n*
 f ellipsoïde *m* mondial
 r общеземной эллипсоид *m*

4010 world geodetic system; WGS
 d geodätisches Weltsystem *n*
 f système *m* mondial géodésique
 r всемирная геодезическая система *f*

4011 world geographic reference system; GEOREF
 (used for aircraft navigation)
 d geografisches Weltreferenzsystem *n*
 f système *m* référentiel géographique mondial
 r всемирная система *f* географических координат

 * **world map → 2288**

X

4012 X axis
   d X-Achse *f*
   f axe *m* de X; axe des abscisses
   r ось *f* X

4013 X coordinate
   d X-Koordinate *f*
   f coordonnée *f* X
   r X-координата *f*

  * **x-shift** → 1367

  * **XTE** → 816

Y

4014 yaw *v*
  d gieren
  f faire dévier; faire une embardée
  r отклоняться от курса

4015 Y axis
  d Y-Achse *f*
  f axe *m* de Y; axe des ordonnées
  r ось *f* Y

  * **Y-azimuth** → **329**

4016 Y coordinate
  d Y-Koordinate *f*
  f coordonnée *f* Y
  r Y-координата *f*

  * **y-shift** → **1368**

Z

4017 Z axis
d Z-Achse *f*
f axe *m* de Z
r ось *f* Z

4018 Z coordinate
d Z-Koordinate *f*
f coordonnée *f* Z
r Z-координата *f*

4019 zenith
d Zenit *m*
f zénith *m*
r зенит *m*

* **zenithal orthomorphic projection** → 3488

* **zenithal projection** → 296

* **zenith angle** → 4020

4020 zenith distance; zenith angle
d Zenithdistanz *f*; Zenithwinkel *m*
f distance *f* zénithale; angle *m* zénithal
r зенитное расстояние *n*

* **zero** → 2889

* **zero level** → 3237

* **zero meridian** → 2889

4021 zero meridian plane
d Ebene *f* des Hauptmeridians
f plan *m* méridien origine
r плоскость *f* нулевого меридиана

* **ZIP code** → 4034

4022 ZIP code area
d Postleitzahl-Bereich *m*
f zone *f* de code postal
r область *f* почтового индекса; зона *f* почтового кода

4023 ZIP code boundary
d Postleitzahl-Grenze *f*
f limite *f* de code postal
r граница *f* почтового индекса; граница почтового кода

4024 zonal index
d zonaler Index *m*; Zonalindex *m*
f indice *m* zonal
r зональный индекс *m*

4025 zonal map
d zonale Karte *f*
f carte *f* zonale
r зональная карта *f*

4026 zonal operators
d zonale Operatoren *mpl*
f opérateurs *mpl* de zone
r зональные операторы *mpl*

4027 zonal polynomial
d zonales Polynom *n*
f polynôme *m* zonal
r зональный многочлен *m*

4028 zonal sampling
d zonale Probennahme *f*; zonale Abtastung *f*
f échantillonnage *m* par zones; échantillonnage géographique
r выборка *f* по зонам; выборка по слоям

4029 zonal structure
d Zonenstruktur *f*
f structure *f* zonée
r зонная структура *f*

* **zone** → 158

* **zone boundary** → 387

4030 zone character
d Zonenzeichen *n*
f caractère *m* de zone
r символ *m* зоны

* **zone code** → 161

4031 zone number
d Zonennummer *f*; Kennziffer *f* des Meridianstreifens
f numéro *m* de zone; repère *m* de fuseau
r номер *m* зоны

4032 zone of influence; influence zone; area of influence
d Einflussbereich *m*; Einflusszone *f*
f zone *f* d'influence
r зона *f* влияния

4033 zoning
d Unterteilung *f* in Zonen; Zoneneinteilung *f*; Zonenbildung *f*; Zonung *f*; Bodenordnung *f*
f zonification *f*; zonage *m*; division *f* en zones
r зонирование *n*; зональное распределение *n*; распределение по поясам

4034 zoning improvement plan code; ZIP code
(US)
d Zipcode *m*
f code *m* postal américain; code de zone
r почтовый код *m*; почтовый индекс *m*

* **zoom → 4036**

4035 zoom in
d Vergrößerung *f*
f zoom *m* [d'accompagnement] avant
r наезд *m*; приближение *n* [изображения];
увеличивание *n* [изображения]

4036 zoom[ing]
d Zoomen *n*; dynamisches Skalieren *n*
f variation *f* de focale; effet *m* de loupe;
zooming *m*; zoom *m*
r изменение *n* масштаба изображения;
масштабирование *n*; трансфокация *f*;
наводка *f* на резкость

4037 zoom layering
d Zoomen-Schichtenteilung *f*
f division *f* de zoom en couches
r расслоение *n* масштабирования

4038 zoom out
d Verkleinerung *f*
f zoom *m* [d'accompagnement] arrière
r отъезд *m* [от изображения]; уменьшение *n*
[изображения]

4039 Z-value
d Z-Wert *m*
f valeur *f* Z
r аппликата *f*

4040 Z-value data points
d Z-Wert-Datenpunkte *mpl*
f points *mpl* d'information de valeur Z
r знаки-символы *mpl* аппликаты

Deutsch

Abbild 1888
abbilden 2248
abbildete Variable 2296
Abbildung 2908
   durch Computer generierte ~ 668
Abbildungseinheit 2306
Abbildungsextent 2270
Abbildungsflächen 177
Abbildungsfunktion 2301
Abbildungsgleichung 2912
Abbildungsmaßstab 3218
Abbildungsniveau 2143
Abbildungsprogramm 2313
Abblendung 238
Abdeckband 2361
Abdeckklebeband 2361
Abdunkeln 3271
Abfall 1365
Abfertigung 531
Abfluss 1442
Abfrage 3140
Abfrage durch Beispiel 2973
Abfrage für Information 3141
Abfrage für Vorschlag 3142
Abfrage-Sprache 2974
Abfragesprache 3515
abgebildete Erscheinungen 2295
abgebildete Tabelle 2293
abgeblendete Selektion 1104
abgedecktes Gebiet 3553
abgeleitete Einheit 985
abgeleitete Karte 983
abgeleitetes Feld 984
abgeschlossene Feature 576
abgeschlossene Gerade 1348
abgeschlossener Objektbereich 395
abgeschnittene Region 570
abgetrennt 2615
Abgleich 2365
Abgleichstoleranz 2366
Abgrenzung 957
abhängige Datenstruktur 713
abhängige Variable 974
ablandig 2616
Ablaufdiagramm 1443
ablauffähiger Treiber 2205
Ablauffolge 3775
Ablauflinie 1445
Ablaufrichtung 1444
Ablaufschema 1443
Ablenkung 947, 1005
Ablenkungslinie 1006
Abmarkungsmethode 2407
Abnahmeprüfung 11
Abrisspunkt 333
Abschattung 3274
Abschätzung 1321
abschließen 574

Abschließung 582
Abschluss 582
Abschlussfehler 1319
Abschnitt 3247, 3248
Absenden 1125
Absendung 1125
absichtliches Unterlaufen der Sicherheitseinrichtungen 3446
Absinken 3525
absolute Exaktheit 2
absolute Genauigkeit 2
absolute Höhe 3
absolute Koordinaten 4
absolute Lokalisierung 5
absolute Messung 6
absoluter Positionsanzeiger 7
absolute Steuervorrichtung 7
absolutes Zeigegerät 7
Absorption 8
Abstammung 2151
Abstand 1131, 1510
abstandstreue Linien 1311
abstandstreue Projektion 1312
abstandstreue Zweipunkt-Projektion 3839
Abstandverzerrung 1133
Absteckung 2122
absteigender Index 986
absteigender Knoten 987
absteigende Sortierung 988
absteigendes Sortieren 988
Abstraktion 10
Abszisse 1
Abtasten 3225
Abtaster 1071
Abtastfehler 3195
Abtastlinie 3226
Abtastlupe 1084
Abtastung 3225
Abtastwert 3191
Abtastzeile 3226
Abtastzeit 3227
Abteilung 973
Abwählung 991
Abweichung 939, 1005
Abwicklung 1004
Abzählung 487
Abzählungsbezirk 499
adaptive dreiseitige Masche 27
addieren 28
Address-Matching 39
adiabatischer Prozess 44
adjazente Bögen 48
adjazente Fläche 50
adjazente Kante 49
adjazentes Kartenblatt 52
Adjazenz 45
Adjazenzanalyse 46
Adjunktion 53
administrative Division 57

Adressdatenverarbeitung 36
Adresse 29, 2242
Adressenbasislinie 32
Adressenbereich 42
Adressencodierung 33
Adressengitter 37
Adressengrundlinie 32
Adressenkorrektur 35
Adressenparität 40
Adressenraum 43
Adressenrichtigstellung 35
Adressenstandlinie 32
Adressenteil 34
Adressenverbesserung 35
Adressenverbuchung 41
Adressenversetzung 41
adressierbarer Punkt 30
Adressiersystem 38
Adressierung 31
Adresskonvertierung 39
Adressraum 43
Adresszuordnung 31
Aerofotografie 65
Aerofotografie-Quadratdatei 67
Aerotriangulation 71
   voll analytische ~ 1489
Aerotriangulierung 71
affine Abbildung 76
affine Transformation 77
AFIS 2489
Agglomeration 80
Aggregation 83
Akquisitionsintervall 21
aktiver Bereich 23
aktiver Sensor 25
aktive Schicht 24
aktives Ziel 822
aktualisieren der Bildschirmanzeige 3060
Aktualisierung 3877
Aktualität 26
algorithmische Generalisierung 91
Algorithmus 90
Alhidade 93
Alhidadenlibelle 1844
Aliasing 92
ALKIS 2488
allgemeine bathymetrische Meereskarte 1517
allgemeiner Atlas 1516
allgemeiner Kreislauf 1518
allgemeiner Satz 2952
allgemeines Zirkulationsmodell 1518
Allozierung 94
Almanach-Daten 96
Alphadaten 99
alphanumerisch 98
alphanumerische Daten 99

alphanumerisches Gitter 100
Altdaten 78
Alterspyramide 963
Altimeter 102
Amalgamation 106
Ambiguität 108
Ambiguitätsauflösung 3153
AM/FM-System 278
Amortisation 975
Amt 973
amtliche Karte 2614
amtliche Kartenbehörde 2651
amtliche Landvermessung 2651
amtliche Landvermessungskarte
    2652
amtliches
    Festpunktinformationssystem
    2489
amtliches Liegenschaftskataster-
    Informationssystem 2488
amtliche Stelle 1924
amtliches topografisch-
    kartografisches
    Informationssystem 423
Amtsbezirk 56
Anaglyphe 109
Anaglyphenbild 109
analog 112
analoge Karte 110
Analogplotter 111
Analyse 113
Analyse der Korrelation 779
Analyse des räumlichen
    Gleichgewichts 3384
Analyse-Operationen 114
Analyse von Nebenbedingungen
    706
analytische Karte 116
analytische Schummerung 115
analytisches Produkt 117
analytische Themakarte 3645
anamorphische Karte 118
Anamorphosis 119
Änderungsbild 523
Änderungserkennung 522
Anfangsbogen 331
Anfangsknoten 3468
Anfangsknoten-Identifikator 3469
Anfrage 3140
Angabe 922
angenäherte Interpolation 143
angepaßte Farbkarte 836
angepaßtes Etikett 838
angeschlossene Bögen 693
angeschlossene
    Erfassungsbereiche 694
angewandte Geodäsie 141
Angrenzende 45
angrenzende Facetten 51
Angrenzer 1487

Animation 131
animierte Karte 130
animiertes GIF-Bild 129
anisotrop 132
Anker 120
anklickbare Map 564
Anlage 194
Anlagendatei 195
anliegende Facetten 51
anliegender Winkel 47
Anlieger 1487
Anliegerung 236
Anmerkung 1454
Annäherungssuche 2931
Annotation 134
annotierendes Feature 133
annotierendes Merkmal 133
Anordnen 2645
Anordnung 2645
Anordnungswerte 2646
Anpassung 2365
Anschluss 2044, 2051
Anschlussgeräte 2727
Anschlusspunkt 3677
Anschriftsteil 34
Ansichtfenster 3944
Ansichtrichtung 3943
Ansichtsfenster 3944
ansteigend 188
antarktischer Polarkreis 135
anthropogene Karte 1857
anthropogenes Objekt 2243
Antialiasing 136
Antipoden 137
Anti-Täuschen 138
Anwenderschnittstelle 3890
Anwendung 139, 140
Anwendungssoftware 140
Anzeige 1934
Anziehungskraft 243
Apparatur 1314
Applikation 140
applikationsübergreifende
    Kommunikation 1981
Approach-Zone 142
Approximation 144
Äquator 1306
äquatorachsige Abbildung 3806
Äquatorebene 1308
äquatorialer Radius 1309
Äquatorialkoordinaten 1307
Äquideformate 1138
Äquidensiten 972
äquidistante Abbildung 1312
äquidistante Zweipunkt-
    Projektion 3839
äquidistante Zylinderprojektion
    1310
Äquidistanz 730
Äquipotentialfläche 2145

Arbeitsbereich 4004
Arbeitsdatei 4003
Arbeitsfile 4003
Arbeitskarte 3706
Arbeitsstation 4005
Arbeitsstättenzählung 493
Arc-Attributtabelle 147
Architektur 150
Archiv 151
Archivierung 152
Arc-Node-Modell 154
Arc-Node-Topologie 155
areolar 168
Array 183
Array-Prozessor 184
Artikel 2041
Artikelindizierung 2042
Assoziation 199
assoziierte Daten 197
assoziierte Zahl 198
astrogeodätisches Netz 201
Astronomie 212
astronomisch 203
astronomische Breite 205
astronomische Einheit 210
astronomische Kartierung 207
astronomische Länge 206
astronomischer Azimut 204
astronomischer Breitengrad 209
astronomischer Meridian 208
astronomischer Zenitwinkel 211
astronomische Zenitdistanz 211
Ästuar 1322
Asymmetrie 214
asymmetrische Verteilung von
    Information 213
asynchron 215
asynchrone Anforderung 216
asynchroner Transfermodus 218
asynchroner Übermittlungsmodus
    218
asynchrone Übertragung 217
ATKIS 423
Atlas 1575
Atlasentwurf 221
Atlas für Touristen 3774
Atmosphäre 224
atmosphärische Absorption 225
atmosphärische Effekte 228
atmosphärische Korrektur 227
atmosphärische
    Millimeterwellensonde 2422
atmosphärischer Schallweg 230
atmosphärisches Fenster 232
atmosphärisches Rauchenmodell
    229
atmosphärisches und
    ozeanografisches
    Bildverarbeitungssystem 226
atmosphärische Weglänge 231

atomarer Wert 234
Atomisierung 235
Atomwert 234
Atomzeit 233
Attitüde 239
Attraktivität 243
Attribut 244
Attributabfrage 265
Attributaktualisierung 263
Attributanzeige 257
Attributcode 248
Attributdaten 250
Attributdefinition 251
Attributdomäne 252
Attributetikett 260
Attributetikettierung 262
Attributinformation schleppen
   1356
Attributkennzeichen 248
Attributklasse 247
Attributobjekt 254
Attributoption 256
Attributsbefestigung 246
Attributsgenauigkeit 245
Attributsteuercode 249
Attributtabelle 259
Attributverknüpfung 253
Attributverweis 258
Attributwert 264
Attributwert-Tabelle 3898
Aufbereitung 1225
aufbewahren 3208
aufeinanderfolgender Zugriff
   3266
Aufenthalt 2125
auffrischen 3078
Auffüllen 1417
Aufgabe 140
aufgeständerte Straße 3167
Auflage 2680
Auflösung 3152
Auflösungsraumelement 1764
Aufmachung 2123
Aufnahme 1556, 2742
Aufnahmeblatt 1405
Aufnahmemaßstab 3219
Aufnahmeschwelle 12
aufnehmen 2249
Aufriss 1811, 3940
Aufseher 835
Aufspringfenster 2860
aufsteigend 188
aufsteigendes Sortieren 189
Aufstellung 2215
Aufstellungsbereich 3622
Aufstellungsplan 2123
Aufstellungsweise 2123
Auftauch-Menü 2859
Aufteilen 3445
Aufteilung 3445

Aufteilung eines
   Erhebungsbezirks 491
Auftrag 3140
Aufzählung 487
Aufzählung von Raumbesetzung
   3402
aufzeichnen 2248
Aufzeichnung 3776
Aufzug 1252
ausblenden 1822
Ausblenden 2360
Ausdruck 1347
Ausdünnen 3667
Ausfuhr 1345
Ausgabedaten 2673
Ausgang 1340
Ausgangsmaterial 3346
ausgewählter Bereich 3255
ausgewähltes Objekt 3256
Ausgleich durch die Methode der
   kleinsten Quadrate 2130
Ausgleichung 54
Ausgleichung des Polygonzugs
   55
Auskunftsdienst von räumlicher
   Information 3389
Auslassen 1185
Auslassung 1185
Auslaufbereich 932
Ausmaß 1101
Ausnutzung 2609
Ausrichtung auf Zielgruppen
   3585
Ausrüstung 1314
Ausscheidung 955
Ausschlag 947
Ausschlussflächen 1338
Ausschnitt 571
Ausschnittsdarstellung 3999
Außenbereich 2672
Außenpolygon 2675
äußere Datei 1353
äußere Datenbasisdatei 1352
äußere Dimension 2674
äußeres Attribut 1351
äußere Schicht 2670
äußeres Polygon 2675
Ausstattung 1314
Austesten 532
Austritt 1340
Auswahl der optimalen Route
   3259
Auswaschen 3236
Auswertung 1321
Auswertungskarte 1328
Auswirkung 1354
Auswirkung von Grenzen 381
Auszeichnen 2066
Autokarte 3168
Autokorrelation 266

automatisch 280
automatische Bildübertragung
   284
automatische Digitalisierung 274
automatische Etikettierung 270
automatische Fahrzeugortung 288
automatische Funkpeilung 275
automatische Generalisierung 277
automatische Geocodierung 283
automatische Kartografie 660
automatische
   Merkmalserkennung 276
automatische Namenplazierung
   279
automatische
   Netzwerkentdeckung 286
automatischer Abschluss 281
automatischer topologischer
   Fehler 287
automatisches Konturen 282
automatische
   Standortbestimmung von
   Fahrzeugen 288
automatische Teilnahme 269
automatische Umrisszeichnung
   282
automatische Verallgemeinerung
   277
automatische Verbindung 269
automatisierte Datenverarbeitung
   273
automatisierte
   Kartenproduktionssystem 272
automatisiertes
   Kartierungssystem 285
automatisiertes kartografisches
   System 285
automatisiertes System der
   Kartografie 285
automatisiertes System der
   Kartografie von
   Erhebungsdaten 271
autonom 2615
autonome Positionierung 289
Autorenkarte 2659
Autorenkartenmanuskript 639
Autorenmanuskript 639
Azimut 293
Azimutalauflösung 298
azimutale Korrektur 295
azimutales Eindrehen 295
Azimutalgrad 297
Azimutalprojektion 296
Azimutalwinkel 294
Azimutwinkel 294

Bahn 791, 3177, 3786
Bahnenraum 3787
bahntreue Abbildung 1551
Balken 3511

Balkendiagramm 315
Balkengrafik 315
ballistische Kamera 305
ballistische Meßkammer 305
Ballung 80
Ballungsgebiet 81
Ballungsraum 81
Band 3511
Bandauflösung 309
Bandbreite 311
Bandbreitenerweiterung 313
Bandbreitenkompression 312
Bandbreitenreduzierung 312
Bandbreitenverdichtung 312
Bandeinpassen 306
Band-Pixelüberlagerungsformat
    308
Band-Zeilenüberlagerungsformat
    307
Basishöhe-Verhältnis 316
Basiskarte 3706
Basislinie 317
Basisrahmen 1475
Basistabelle 319
Basisvermessung 321
Basiswetterkarte 1801
Bathymetrie 326
bathymetrische Karte 324
bathymetrische Vermessung 325
Baudrate 327
Baud-Zahl 327
Baufläche 414
Baugrenze 415
Baulinie 1487
Baum 3811
Baumnetzwerk 3813
Baumstruktur 3812
Baum-Topologie 3812
Bauplan 992
Bayes-Methoden 328
Bearbeitung 1225
Bearbeitungsprogramm 1226
Bebauungsplan 2101
Bedarfsschaltepunkt 401
Bedeutung 3111
Bedienungsgebiet 521
Bedienungspult 743
bedingte Linie 681
bedingter Operator 682
Befehlszeile 626
befestigen 1430
Befestigung 236
Befestigungsgrenze 237
begrenzte Koordinaten 394
begrenztes Kästchen 395
begrenztes Rechteck 395
Begrenzungslinie 385
Begriffsvermögen 653
Beikarte 1962
Belegung 2609

Belegungsdichte-Karte 2856
Beleuchtung 1886
Beleuchtungsgeometrie 1887
Belichtung 633
Beltramische Abbildung 1551
benachbarte Fläche 50
benachbarte Kante 49
Benachbartheit 712
Benchmark 333
Benchmarking 334
Benennung 2485
Benennung-Konvention 2486
Benutzeforderungen 3891
benutzerbestimmte Attribute 3887
benutzerdefinierte Attribute 3887
benutzerdefinierte Projektion
    3888
Benutzerkennzeichnung 3889
Benutzerschnittstelle 3890
Beobachtung 2607
Beobachtungsdichte 970
Beobachtungsfehler 2608
Berandung 728
Bereich 158
Bereich gemeinschaftlicher
    Nutzung 2049
Bereich mit Blocknummerierung
    366
Bereichsauswahl 3004
Bereichssuche 178
Bereichsunterteilung 3084
Bereich- und Perimeter-
    Berechnung 174
Bereich von deskriptive Daten
    883
bergige Region 2452
Bergrutsch 2091
Bergrutsch-Risiko-Modell 2092
Bergrutsch-Suszeptibilität 2093
bergseits 3879
bergwärts 3879
berichtigte Gesamtlänge 775
berichtigte Leitung 425
Berührende 3582
Berührungskegel 3580
Berührungskegelprojektion 3581
Beschaffungszeit 21
Beschattung 3274
Bescheinigung über den Titel 515
Beschlussfassung 934
Beschneidbereich 804
Beschneidenbereich 804
Beschränkung 3155
beschreibende Daten 990
Beschreibung 989
Beschriftung 453
Besetzung 2609
Besetzungsdichte-Karte 2856
Besitzstands- und
    Schätzungsnachweisdatei

1412
Beständigkeit 2004
Bestandsverzeichnis 3495
beste lineare nichtverzerrende
    Schätzfunktion 335
Bestimmung des Profils 2903
Beta-Index 336
Betrachtungswinkel 127
Betriebsforschung 2635
Betriebsgebiet 172
Betriebsgemeinschaft 2047
Betriebskonstellation 2634
Betriebssystem 2633
Bettgeometrie 524
Beugung 1939
Bevölkerung 2854
Bevölkerungsbaum 963
Bevölkerungscharakteristik 2855
Bevölkerungsdaten 960
Bevölkerungsdichte-Karte 2856
Bevölkerungsentwicklung 2858
Bevölkerungspyramide 963
Bevölkerungsumverteilung 2857
Bevölkerungsvorausberechnung
    962
Bevölkerungszählung 497
Bewertung 1321
Bewitterung 3981
Bezeichner 993, 1883
Beziehung
    1:1~ 2621
    m:1~ 2247
    m:n~ 2246
Beziehungsindexierung 3101
Bézier-Fläche 339
Bézier-Kurve 337
Bézier-Vieleck 338
bezifferte Karte 1080
Bezirk 1148
Bezirkfüllung 1418
Bezirkgrenze 1149
Bezug 3063
Bezugnahme 3074
Bezugsebene 925
Bezugsfläche 925
Bezugsgebiet 3065
Bezugshorizont 924
Bezugskante 3066
Bezugsmarke 333
Bezugsmeridian 3070
Bezugspunkt 2811
Bezugsrahmen 1475
Bezugsrahmenidentifikation 1474
Bezugsrand 3066
Bezugssystem 3073
Bhattacharya-Distanz 340
bibliografische Daten 341
bibliografische Indizierung 342
Bibliothek 2146
bikubische Fläche 344

bikubischer Interpolator 343
Bild 1888
   2D-~ 1098
   2,5D-~ 1099
   3D-~ 3969
   3D-~-Messung 1100
Bildaddition 1889
Bild aktualisieren 3078
Bildalgebra 1890
Bildanalyse 1891
Bildanreicherung 1903
Bildarithmetik 1892
Bildatlas 564
Bild-basiertes
   Informationssystem 1893
Bildbasis 2738
Bilddaten 1898
Bilddigitalisierung 1901
Bildelement 2758
Bilderdatenbank 1899
Bilderkennung 2717
Bildfeldwinkel 1403
Bildintegrator 1904
Bildkatalog 1894
Bildklassifizierung 1895
Bildlauf 3235
Bildlaufleiste 3234
Bildmessung 1905
Bildmustererkennung 2717
Bildneuzuordnung 1909
Bildpaar 1906
Bildpunkt 2758
Bildregistrierung 1910
Bildrollen 3235
Bildschirm 3231
Bildschirm-Digitalisierung 2627
Bildschirm-Sammeln 3232
Bildschirmzeilenverschiebung
   3235
Bildsegmentierung 1911
Bildsubtraktion 1912
Bildsummierung 1889
Bildsynthese 1897
Bildteilung 1902
Bildtext 3297
Bildungsatlas 222
Bildunterscheiden 1900
Bildverarbeitung 1907
Bildverarbeitungsystem 1908
Bildvergleich 1896
Bildverschiebung 3235
Bildwinkel 124
Bildzeichen 1714
bilineare Interpolation 345
bilinearer Interpolator 346
binäre lineare Generalisierung
   350
binäre Rasterdaten 353
binärer Code 347
binärer Raster 352

binäres Gitter 348
binäres Overlay 351
Binden 2199
Biom 355
Bioregion 356
Bisekante 535
Bit 357
Bitmap 3015
Bitmap-Bild 3015
Bitmap-Objekt 3019
Bit-Scheibe 358
Bit-Slice 358
Blasendiagramm 407
Blättern 3235
Blattnummer 3286
Blattschnitt 3285
blattschnittfreie Datenbasis 3239
blattschnittfreie geografische
   Datenbasis 3239
Blattzahlensystem 3287
Blickfeld 1403
Blickpunkt 3957
Blickwinkel 127
Blindendigitalisierung 361
blindes Digitalisieren 361
Blob 349
block-basierte Adresszuordnung
   362
Blockbild 1443
Blockdiagramm 1443
Blockgrenze 363
Blockierung 2222
Blockkennnummeringsgebiet 366
Blockname 364
Blocknummerierung 365
Blockschaltbild 1443
Blockschema 1443
Bodenaufnahme 1407
Bodenbearbeitung 2086
Bodenbedeckung 2077
Bodenbedeckungsdatenbasis
   2078
Bodenbedeckungsgenerator 796
Bodenbedeckungskarte 2079
Bodenbedeckungsstatistik 2080
Bodenbewirtschaftung 2086
Bodenform 3605
Bodenkarte 3338
Bodenkontrolle 1785
Bodenkontrollstation 1786
Bodenleitung 2084
Bodenleitungeinstellung 2085
Bodennutzung 2097
Bodennutzungsplan 2101
Bodenordnung 4033
Bodenstation 1786
Bodenwelle 1795
Bogen 146
Bogenattributtabelle 147
Bogendigitalisierungssystem 149

Bogen-Knoten-
   Kartendatenstruktur 153
Bogen-Knoten-Modell 154
Bogen-Knoten-Topologie 155
Bogenlänge 2136
Bogenversicherungsschutz 148
Boolesche Operationen 371
Boolescher Ausdruck 369
Boolescher Operator 372
Boolesches Bild 370
Boolesche Suche 373
Böschung 1921
Böschungsfläche 3544
Böschungslinienerkennung 3322
Böschungsmaßstab 3321
Böschungspickel 3323
Böschungsschummer 3936
Böschungsschummerung 3936
Böschungswinkel 126
Boundary-File 383
Bowditch-Regel 396
Boxcar-Klassifizierung 397
Boxplot 1443
Breite 205
Breite-Länge 2113
Breite-Linie 2184
Breitengrad 2699
Breitenkreis 2701
breitenorientierte Suche 399
Breitensuche 399
Browser 3963
Browser-Fenster 405
Browsing 406
Bruchkante 400
Bruchlinie 400
Brücke 402, 403
BSQ-Bildformat 310
buchstäbliche Konstante 2201
Bufferberechnung 408
Buffer-Distanz 409
Buffer-Zone 412
Bundesdatenverarbeitungs-
   standards 1393
Business-Mapping-System 416
Bussole 632
Byte 420

Check-in 531
Chi-Quadrat-Maßzahl 534
Chi-Quadrat-Verteilung 533
Chorologie 536
Choroplethe 538
Choroplethe-Linie 538
Choroplethenkarte 539
Choroplethe-Rahmenwerk 537
Choroplethe-Zonen 541
Chroma 542
Chrominanz 542
Clearinghaus 563
Clearing-House 563

Cluster 585
Cluster-Analyse 586
Clusterbildung 587
Cluster-Etikettieren 588
Cluster-Karte 589
Clusterung 587
Code 595
Code mit hoher Genauigkeit für
    autorisierte Nutzer 2879
Code von hydrologischer Einheit
    1871
Codieren 1264
Codierung 1264
Co-Kriging 600
Computer-Atlas 667
computergesteuerte
    Wiedergabetechnik 668
computergestützte Kartografie
    660
Computergrafik 669
Computergrafik-Schnittstelle 670
Computerkartografie 660
computerunterstützte
    Abschätzung 663
computerunterstützte Planung 664
computerunterstützte
    Reproduktion 666
computerunterstütztes Design 661
computerunterstütztes Entwerfen
    661
computerunterstütztes Entwerfen
    und Zeichnen 662
computerunterstütztes Zeichnen
    665
computerunterstützte Wiedergabe
    666
Constraint-Erfüllungsmethode
    707
Coregistrierung 772
Coverage-Geometrie 797
Cursor 823
Cursorpositioniertasten 824
Cursorsteuertasten 824
Cursortasten 824
Curvimeter 834
Cybergeografie 842
Cyberspace-Karte 843
Cyberspace-Kartierung 844

3D 3670
DAC 1073
Dachlatte 2112
Damm 2137
Dämpfung 238
Darstellung in Tabellenform 3577
Darstellungsattribut 1127
Darstellungsfeld 3944
Datamining 900
Data-Mining 900
Datei 1411

Dateiformat 1413
Dateiübertragung 1414
Daten 852
Datenalter 79
Datenalterung 856
Datenausgabe 903
Datenaustausch 888
Datenaustauschfehler 889
Datenauszug 890
Datenbank 858
Datenbanksicht 859
Datenbankverzeichnis 885
Datenbasis 860
Datenbasis des geografischen
    Registers 1604
Datenbasisindizierung 863
Datenbasisintegrator 864
Datenbasisintegrität 865
Datenbasiskatalog 862
Datenbasis-Managementsystem
    867
Datenbasis-Modell 868
Datenbasis-Portabilität 869
Datenbasisprojektierung 861
Datenbasisserver 870
Datenbasissperre 866
Datenbasis-Spezifikation 871
Datenbasistechnologie für GIS
    1692
Datenbasiswerkzeuge 872
Datenbehandlungssprache 898
Datenbeschreibung 881
Datenbeschreibungsdatei 884
Datenbeschreibungssprache 882
Datenbeschreibungszone 883
Datendefinition 881
Datendefinitionssprache 882
Dateneinführung 886
Dateneingabe 886
Dateneinstweiligkeit 917
Dateneinzelheit 922
Datenelement 922
Datenerfassung 875
Datenerfassungssystem
    im Flugzeug eingebautes ~ 85
Datenextraktion 890
Datenfehler 887
Datenfeld 891
Datenfluss 892
Datenformat 893
Datengeschichte 894
Datenglättung 915
Datengrenzen 873
Datengrundeinheit 922
Datenintegrator 895
Datenintegrität 896
Daten in unregelmäßigen
    Abständen 2017
Datenkalibrierung 874
Datenklassifizierung 876

Datenkonsistenz 878
Datenkonvertierung 879
Datenmaske 899
Datenmenge 911
Datenmengekarte 912
Datenmengename 913
Datenmodell 901
Datenmosaik 902
Datenpackbetrieb 904
Datenpunkt 905
Datenqualität 906
Datenqualitätsbericht 907
Datenreduktion 908
Datenreform 910
Datensammeln 875
Datensatz 3184
Datensatz-Identifikator 3050
Datenschicht 897
Datenstruktur 916
Datenteilung 914
Datentyp 919
Daten über das Gelände 3608
Datenüberdeckung 880
Datenübermittlung 918
Datenübertragung 918
Datenumorganisierung 910
Datenumstellung 910
Datenunterstützung 1944
Datenvariation 3905
Datenverbindung 877
Datenverfügbarkeit 857
Datenvermittlung 888
Datenverschmelzung 877
Datenverzeichnis 885
Datenzeitweiligkeit 917
Datenzugriff 853
Datenzugriffsicherheit 855
Datenzugriffsprache 854
Datenzusammenlegung 910
Datenzuverlässigkeit 909
Datum 920, 922, 2811
    3D~ 929
Datumkonversion 923
Datumsgrenze 921
Datumsverschiebung 926
dauerhafte Entwicklung 3552
D/A-Umwandler 1073
Deckerfolie 353
Deckung der topografischen
    Karte 3715
Decodieren 941
Decodierung 941
Defekt 944
Defektstruktur 945
degressive Abschreibung 975
Deich 1095
Deklination 939
Deklinationsdiagramm 940
Dekomposition 942
dekomprimieren 943

Delaunay-Triangulation 951
Delaunay-Zone 950
Demografie 965
demografische Analyse 959
demografische Daten 960
demografische Prognostizierung 961
demografischer Übergang 964
demografische Tendenz 2858
demografische Transition 964
Dendrogramm 966
Denormalisierung 967
Denormierung 967
Densifikation 968
Density-Slicing 972
Depression 976
Design 992
Deskription 989
deskriptive Daten 990
Desktop-GIS 995
Desktop-Kartografieren 996
Desmoothing 3875
Detektor 1000
Detektorabrustung 1001
Detektorungleichgewicht 1001
Deviation 1005
dezentrale kartografische Datenbasis 1139
Dezernat 973
dezimaler Grad 933
DGM-Format 1049
Diagonale 1017
diagonale Richtung 1016
Diagramm 1018
diagrammatische Ansicht 1019
diagrammatische Darstellung 1019
Diagramm-Fenster 1752
Diagrammkarte 539
Diagrammkarte ohne Klassenbreite 540
Dialogbetrieb 1979
Dialogverarbeitung 1980
Dialogverkehr 1979
Diameter 1020
Diapason 2999
Diapasonextraktion 3002
Diazokopie 1021
Diazokopierverfahren 1021
Dichte 969
Dichtheit 969
Dichtigkeit 969
Dienstkarte 2614
differentiale Entfernungsmessung 1026
differentiale Korrekturen 1022
differentiale Nivellierung 1023
Differentialentzerrung 1027
differentiale Positioniergenauigkeit 1024

differentiale Positionierung 1025
differentiale Positionsgenauigkeit 1024
diffuse Farbe 1028
diffuse Quelle 165
diffuse Reflektanz 1029
Diffusion-begrenzte Aggregation 1030
DIGEST-Standard 1047
digital 1031
Digital-Analog-Konverter 1073
Digitalbild 1051
Digitalbildanpassung 1052
Digitalbildverarbeitung 1053
digital darstellen 2969
digitale Bildzuordnung 1052
digitale Elevationsdaten 1041
digitale Höhenliniendaten 1041
digitale Höhenlinienmatrix 1042
digitale Höhenlinienmodell-Extraktion 958
digitale Kammer 1033
digitale Karte 1239
digitale Kartenbibliothek 1060
digitale Kartendaten 1059
digitale Katasterkarte 1032
digitale Katasterplankarte 1032
digitaler Erfassungsbereich 1039
digitaler Raum 1069
digitaler Streckengraph 1058
digitales Abbild 1051
digitales Bild 1051
digitales Foto 1064
digitales Geländemodell 1072
digitales geotopografisches Informationssystem 1050
digitales Höhenlinienmodell 1043
digitales kartografisches Modell 1034
digitales objektstrukturiertes Landschaftsmodell 1057
digitales Situationsmodell 1067
digitale Topografie 1075
digitale Weltkarte 1036
Digitalfläche 1070
Digitalgeber 1078
digitalisieren 1055
digitalisierendes Gerät 1078
Digitalisierer 1078
Digitalisierer-Menü 1085
Digitalisierer-Schablone 1086
Digitalisierlupe 1084
digitalisierte Daten 1077
digitalisierte Fotografie 1064
digitalisiertes Bild 1079
digitalisierte Wolkenaufnahme 1076
Digitalisierung 1054
3D-~ 1189
Digitalisierungsfehler 1087

Digitalisierungsfenster 1093
Digitalisierungshardware 1088
Digitalisierungsmodul 1089
Digitalisierungsoptionen 1090
Digitalisierungssession 1091
Digitalisierungssoftware 1092
Digitalisierung von Karten 2266
Digitalkamera 1033
Digitizer 1078
Digitizer-Konfiguration 1081
Digitizer-Treiber 1083
Dijkstra-Algorithmus 1094
Dilatationsfilter 1096
Dimension 1101
Dimensionalität 1102
dimensionlose Darstellung 1103
DIME-System 1193
Diopterlineal 93
direkte Befragung 1407
direkter DBMS-Fernzugang 2204
direktes 3D-Display 1114
direkte Zeichen 1113
direkte Zugriffsschicht 2202
Direktverbindung 1106
Dirichlet-Musterkachelung 1115
Disaggregation 1117
Diskontsatz 1119
Diskordanz 3506
diskrete Daten 1120
diskrete Erscheinungen 1122
diskrete Verteilung 1121
Diskretisierung 1123
Diskretisierungstechnik 3D-~ 930
Diskriminanzanalyse 1124
Dispachierungssystem für optimale Führung 2637
Distanz 1131
Distanzbereich 412
Distanzeinheiten 1135
Distanzvektor 1136
Distorsion 1137
Districting 1150
Districting im Zeitabstand 3688
Distrikt 1148
Diversity 1151
Division 1154
DLG-Format 1058
Dokument 1158
Domain 1159
Domain-Name 1160
Domain-Name-Kartierung 1161
Domain-Struktur 1162
Domäne 1159
Doppelbild 1172
Doppeldifferenz 1170
Doppeldigitalisierung 1171
Doppelfrequenz-Empfänger 1192
Doppelquadname 1173
Doppelsekante 535

Dopplereffekt 1165
DOQ-Standard 1062
Dorfgebiet 3947
Drahtdarstellung 4000
Drahtmodell 4001
Drapieren 1179
Drehellipsoid 3439
Drehungsellipsoid 1256
Dreibaum 3830
Dreibein-Daten 3829
dreidimensional 3670
dreidimensionaler Digitalisierer
    3383
Dreieck 3818
dreieckiges Netz 3819
Dreiecksaufnahme 3821
Dreieckskette 520
Dreifachdifferenz 3827
Drilling 3828
dritte Normalenform 3669
dritte Normalform 3669
Dropdown-Menü 2953
Druck 2886
Drucken 2895
Drucker 2894
Druckrahmen 2896
Duplikation 1196
duplizierte Daten 1194
duplizierte Linie 1195
Duplizierung 1196
Durchbiegung 3974
Durchblickpunkt 3957
Durchführbarkeitsanalyse 1373
Durchführbarkeitsstudie 1373
Durchführung 1915
durchgehende Datenstruktur 713
Durchlässigkeit 14
durchlaufen 3809
Durchmesser 1020
Durchprüfung 532
Durchsicht 406
Durchzeichnung 3776
DXF-Format 1182
dynamische Ansichten 1853
dynamische Digitalisierung 3508
dynamische Generalisierung
    1199
dynamischer Datenaustausch
    1198
dynamische Segmentierung 1201
dynamisches Geobild 1200
dynamisches Skalieren 4036
dynamisch gelinkte
    Datendarstellung 1197
dynamisch verknüpfte
    Datendarstellung 1197

Ebene 2774
Ebene des Hauptmeridians 4021
ebene Koordinaten 2775

Ebenenkoordinaten 2775
Ebenenkoordinatensystem 2776
ebenes Koordinatensystem von
    US 3470
ebene Überwachung 2778
Echo 1212
Echobild 1172
Echtzeit-Korrekturen 3037
Echtzeit-Positionierung 3040
Ecke 773
Ecken-Verbindung 774
Editieren 1225
Editierprogramm 1226
Editiertoleranzen 1227
Editor 1226
Effekt erster Ordnung 1429
effektiver Erdkrümmungsfaktor
    1228
effektiver momentaner
    Bildfeldwinkel 1229
effektives momentanes Bildfeld
    1229
Effekt zweiter Ordnung 3246
Eichung 426
Eigenkorrelation 266
Eigenschaft
    dem Geist zugeordnete ~
    2390
Eilinie 2676
Einbetten 1259
Einbettung 1259
Einbild 2620
Eindämmung 709
eindeutige Abbildung 3309
eindeutige Bezugsnummer einer
    Landparzelle 3856
eindeutige Fläche 3310
eindeutige Referenznummer einer
    Landparzelle 3856
eindimensional 2618
eindimensionales Datum 2619
Eindruck 1918
einfache Gleichrichtung 3302
einfache Regression 3303
einfache Rektifikation 3302
einfaches Kriging 3299
einfaches Objekt 3300
einfaches Polygon 3301
Einfallswinkel 126
Einfang 429
Einfassung 728
Einflussbereich 4032
Einflusszone 4032
einfügen 1961
Eingabe/Ausgabe-Geräte 1960
Eingabedaten 1958
Eingabegerät 1959
eingebettete Sprache der
    strukturierten Abfragen 1258
eingeordnete Karte 3000

eingeteilte Luftfahrtkarte 3249
Einheit 2041
Einheit-Auswahlsatz 1282
Einheitsbeziehungsdiagramm
    1279
Einheitsbeziehungsmodell 1280
Einheitsbeziehungsmodellierung
    1281
Einheitsdeformation 1276
Einheitsklasse 1275
Einheitsname 1278
Einheitstransformation 1284
Einheitstyp 1285
Einkapselung 1263
Einkreisung 1286
Einleiten 1952
Einmessung 3549
Einpassen mit
    Gummibandfunktion 3186
Einrahmung 1478
Einrichtung 1314
Einsatzdaten 1398
Einsatz der Ressourcen 95
Einschachtelung 1259
einschieben 1961
Einschluss 1923
Einschnitt- und Auftrag-Analyse
    841
Einschränkung 3155
Einsenkungskontur 977
Einsenkungsprofil 977
Einstellen 3269
Einstellung 3269
Einstellzeit 21
Einteilung
    in Bezirke ~ 1150
Einteilung der Wahlkreise 1150
einvariable Themakarte 2622
einwertige Fläche 3310
Einzelbild 1128
Einzelbildorientierung 3393
Einzeldifferenz 3306
Einzelheit 2041
Einzellinie 3307
Einzelpunkteinschaltung 3308
Einzugsbereich-Karte 3780
Einzugsscanner 1394
Eisverwitterung 3981
elastische Linie 1231
elastischer Kreis 1230
elastische Transformation 1232
elektromagnetische Forschung
    1235
elektromagnetisches Spektrum
    1234
elektromagnetische Strahlung
    1233
elektromechanischer Sensor 1236
elektronische
    Entfernungsmessung 1238

elektronische Karte 1239
elektronisches Planimeter 1240
elektronisch fühlbare Fläche 1237
elektrostatischer Plotter 1241
Element 2041
Elevation 1244
Elevationsbenchmark 1245
Elevationscomputer 1246
Elevationsdaten 1247
Elevationsdaten-Browser 1248
Elevationsdatenformat 1249
Elevationsmaske 1250
Elevationsruder 1252
Elevationswinkel 123
Elevator 1252
Ellipsoid 1254
elliptische Koordinaten 1255
elliptische Projektion 1257
elliptische Projektion von Donald
    1164
Emissionsquelle 165
Empfänger 3043
Empfänger-Rauschen 3045
empfängerunabhängiges
    Austauschformat 3044
Empfangstest 11
Empfindlichkeit 3263
Empfindlichkeitsanalyse 3264
Endknoten 1266
Endkoordinaten 1265
Endmaßstab 3218
Endpunkt 1270
Endpunkt des Bogens 1271
Endpunkt-Identifikator 1267
Endpunkt-Knoten 1266
Engpass 1913
Entaktivierung 931
Entfernung 1131, 2998
Entfernung der verdeckten Linien
    1821
Entfernung des Rauschens 2542
Entfernungsmesser 1134
Entfernungsmessgerät 1134
Entfernungsmessung 3004
Enthaltung 9
Enthüllungsdistanz 2431
Entität 1274
Entitätenintegrität 1277
Entitätsklasse 1275
entkomprimieren 943
Entscheidungsbaum 938
Entscheidungsfindung 934
Entscheidungsfindung-
    Informationssystem 935
Entscheidungshilfe-System 935
Entscheidungstheorie 937
Entscheidungstreffen 934
Entscheidungstreffen-Datenbasis
    936
Entscheidungsvorbereitung 934

Entschlüsseln 941
Entsendung 1125
entwickelbare Fläche 1002
Entwickler-Werkzeugsatz 1003
Entwicklung 1004
Entwicklungsbereichgrenze 3882
Entwurf 992
Entwurfsfehler 413
Entzerrung 3054
entzweien 3871
Enumeration 487
Enumerationsbezirk 499
Ephemeriden 2640
epipolar 1302
Erdauflösung 1790
Erdball 1209
Erdbeobachtung 1205
Erdbeobachtung-Satellit 1206
Erde 1784
Erdellipsoid 1203
erdfestes Achsensystem 1204
Erdfließen 2091
Erdform 3605
ER-Diagramm 1279
Erdimpedanz 1788
Erdkarte 2288
Erdkrümmung 825
Erdmagnetfeld 1648
Erdrutsch 2091
Erdsimulator 1208
Erdsteuerpunkt 1786
Erdsteuerung 1785
Erdteilkarte 714
Ereignis 1329
Ereigniskennzeichen 1331
Ereignislokalisierung 1330
Ereignisquelle 1332
Ereignistabelle 1333
Ereignisthema 1334
erfasster Bereich 1582
Erfassungsbereichsidentifikator
    798
Erfassungsbereichsname 799
Erfassungsextent 795
Erfassungsgenerator 796
Erfassungskarte 793
Erfassungskontur 794
Erfassungszeit 21
Ergibt-Anweisung 196
Erhebung 3878
Erhebungsmesstisch 3550
Erhebungswinkel 123
Erkennung 3048
erläuternder Maßstab 1343
Erledigung 1125
erneuern 3078
Erreichbarkeit 14
Erreichbarkeitsindex 15
Ersatzdaten 3547
Erscheinung 2737

Erscheinungen der realen Welt
    3041
Ersetzen 3526
Ersetzung 3526
erste Normalenform 1428
erste Normalform 1428
erste Projektion 1851
Erst-Schlüssel 2888
erweiterte Gerade 1348
erweitertes Objekt 1349
Erweiterung des DLG 1155
Erwerb von Grundstücken 20
Erzeugnis
    mit der Hand gefertigtes
    kartografisches ~ 1801
Etikett 2061
Etikettennummer 2067
Etikettieren 2066
etikettiertes Polygon 2062
Etikettierung 2066
Etikettierungsprozedur 2064
Euklidische Distanz 1323
Euklidische Ebene 1325
Euklidische Geometrie 1324
Euklidischer Raum 1326
Europäische Raumfahrtbehörde
    1327
Europäische
    Weltraumorganisation 1327
Exaggeration 1336
exakter Interpolator 1335
Exaktheit 17
Exemplar 769
Expertensystem 1341
explanatorische Beschriftungen
    1342
explanatorischer Maßstab 1343
explorative Datenanalyse 1344
Export 1345
Expression 1347
extensive Messung 1350
externer Raster 1355
Extrapolation 1357
Extrapolieren 1357
Extrapolierung 1357
Extrempunkte 1358
Exzentrizität 1211

Facette 1360
Facettennormale 1359
Facettenspitze 1361
Fach 473
Fachabteilung 973
Fachbezeichnung 2543
fachlicher Erhebungsbereich 494
Fachwerk 1476
Facility-Management 1362
Fadenkreuz 808
Fadenkreuz-Zeiger 1082
Fahrleitung 477

Faktoreneinsatz 95
Fall 1364
Fallinienerkennung 3322
falscher Hochwert 1368
falscher Rechtswert 1367
falscher Ursprung 1369
Falschfarbbild 2936
Falschfarbe 2934
Falschfarbenbild 2936
Falschfarbenkomposition 1366
Falte 3238
Faltung 753
Familienhaushalt 1371
Fang 3328
Fangabstand 3329
fangen
    an dem Gitter ~ 3331
    an dem Knoten ~ 3332
Fangen 429, 3328
Fangradius 3329
Farbaerofotografie 603
Farbauszug 616
Farbdiagramm 606
Farbe 619
Farbeinheit 542
farbempfindlich 2691
Farbeneinheit 542
Farbenkompression 608
Farbenschwankungen 609
Farbentrennung 616
Farbhintergrund 604
farbige Musterkachel 611
farbige Zone 1416
farbige Zonen 610
Farbinflexion 612
Farbinfrarotbild 613
Farbkarte 614
Farbkasten 605
Farbkomposit 607
Farbpalette 615
Farbqualität 542
Farbseparation 616
Farbskala 618
Farbtabelle 614
Farbtafel 606
Farbteilung 616
Färbungsproblem der Karten
    2258
Farbwert 617
Farbzuordnungstabelle 614
Feature 448
Feature-Attribut 1374
Feature-Attributtabelle 1375
Feature-Auswahl 1385
Feature-Code 1378
Feature-Identifikationsnummer
    1382
Feature-Klasse 1377
Feature-Raum 1388
Feature-Seriennummer 1387

Feature-Trennung 1386
Featuretyp 1390
fehlende Bildzeilen 2434
fehlender Knoten 2433
Fehler 1317
Fehleranalyse 1318
fehlerlose Daten 562
Fehler nach der Länge 3001
Fehler quer zur Bahn 816
Feingeneralisierung 1425
Feld 183
Feldaufnahme 1407
Felddaten 1398
Felddaten-Kollektion 1399
Felddaten-Sammlung 1399
Feld des Attributetiketts 261
Feldeinheit 1409
Felderdatenkollektor 1400
Feldkontrolle 1397
Feldlänge 1401
Feldmessung 3723
Feldname 1402
Feldprüf 1397
Feldstärke 1406
Feldstudie 1407
Feldterminator 1408
Feldtrennzeichen 1404
Feldvergleich 1397
Fensterkonfiguration 3998
Fenstertechnik 3999
Fensterung 3999
Fermeture 582
Fernablesung 3120
Fernabtastung 3120
Fernadresszuordnung 1132
Fernbildverarbeitung 3117
Ferndatenerfassung 3116
Ferneinstellgerät 3119
Fernerkundung 3120
Fernerkundung-Generalisierung
    3122
Fernerkundungsdaten 3121
Fernerkundungsmethoden 3123
Fernerkundungsverfahren 3123
Fernmesstechnik 3593
Fernmessung 3593
Fernobjektsbewertung 3118
Fernsensor 3124
Ferntabelle 3125
Fernübertragungskartierung
    3592
Fernwirktechnik 3593
fester Abstand 1431
fester Fehler 3567
fester Körper 3339
Festkörper 3339
Festland 2073
Festlegen eines
    Vermessungspunkts 2358
Festlegen von integrierten

Grenzen von
    Geländeeinheiten 1972
Festpunkt 740
Festpunktnetz 2530
Festwert 703
Figur 3279
Figur-Grund 1410
Filter 1422
Filterkriterien 1423
finden 1424
Finite-Element-Methode 1426
Finite-Element-Modell 1427
Fitten der Kurve 830
fixieren 1430
Flachbettplotter 1432
Flachbettscanner 1433
Fläche 157, 3538
    2D-~ 2773
flache Datei 1436
flache Karte 2769
flache Kartenprojektionen 2770
Flächen- 168
Flächenableitung 3539
Flächenanpassung 3540
Flächenausgleich 3540
Flächenausgleichung 3540
Flächenberechnung 659
Flächendeckel 169
flächendeckendes Computernetz
    3996
Flächendiagramm 160
Flächendigitalisierung
    3D-~ 1189
Flächeneinheit 3858
Flächenelement 1242
Flächenentzug 20
Flächengeodäsie 3541
Flächen-gewichtete
    Mittelauflösung 182
Flächeninhalt 157
Flächennutzung 2097
Flächennutzungserhebung 2103
Flächennutzungsplan 2101
Flächenobjekt 166
Flächenschwerpunkt 159
Flächenstichprobenverfahren
    3415
Flächenstück 3545
Flächenstückgeneralisierung 173
Flächensymbole 180
Flächensymbole-Methode 2408
flächentreu 1304
Flächentreue 2885
flächentreue Abbildung 1316
flächentreue azimutale Lambert-
    Projektion 2070
flächentreue Kartenprojektion
    1316
flächentreue konische Albers-
    Projektion 89

flächentreue Zylinderprojektion
    846
Flächentriangulation 181
Flächenvariable 3546
Flächenverschneidung 167
Flächenverzerrung 163
flacher Abfall 1434
flacher Durchschnitt 2768
flaches Erdmodell 1435
Flankendetektor 1220
Flankenerkennung 1219
Flau 1494
fliegender Teppich 1450
Fliegerkarte 73
Fliese 3679
Fliessgewässer-
    Informationssystem 3190
Flöz 3238
Fluchtpunkt 2532
Flugborddatenerfassungssystem
    85
Flugborddatenerfassung und
    -Registrierung 84
Fluglage 239
Fluglagemeßwertgeber 242
Fluglage- und
    Steuerkursreferenzsystem 240
Flugsimulator 88
Flugsimulator-Datenbasis 1437
Flugsimulatorsysteme 1438
Flugzeuge im Warteraum 3458
Flugzeugscanner 87
Flugzeug-Schrägsichtradar 3294
Flugzeug-Seitensichtradar 3294
Flugzeug-Seitwärtsradar 3294
Fluktuation 1449
Fluraufnahme 3723
Flurbuch 422
Flurkarte 2652
Flurnamen 1613
Flurstücksnummer 2652
flussabwärts 1174
Flussakkumulation 1442
flussaufwärts 3879
Flussdiagramm 1443
Flusslinie 1445
Flussnetz 3163
Flussrichtung 1444
Flussweg 1446
Flussweg-Raster 1447
Flutgrenze 1835
Fluttung 1439
Folge 516
Folgekarte 983
Folgerstufe 1452
Folgestufe 1452
Folie 353
Font 526
Form 3279
Formanalyse 3280

Formatänderung 3077
Format aufeinanderfolgenden
    Grauwerten pro Pixel 308
Formatbild 3595
Format des Vektorprodukts 3920
Formatfehler 1462
Format in aufeinanderfolgenden
    Zeilen 307
Formatkonvertierung 1461
Formblatt 1460
Formblatt-Interface 1464
Formblatt-Schnittstelle 1464
Formenanalyse 3280
Formendatei 3282
Formerkennung 3283
Formlinie 1463
Formverzerrung 3281
Forstrevier 1458
Forstwirtschaftskarte 1459
Fortführen einer Karte 2325
fortgeschrittene
    Kartografierungsausrüstung
    59
fortlaufende Datenstruktur 713
Fortpflanzung 2920
fortschrittliches System der
    Kartografie 60
Foto 2742
Fotografie 2742
fotografischer Film 2743
Fotogrammetrie 2741
fotogrammetrisch digitalisiert
    2740
Fotoindexschlüssel 2747
Fotokarte 2745
Fotokopierer 2739
fotomechanische Schummerung
    2744
Fotoplan 2748
Fotoplotter 2418
Fotosphäre 2749
Fototheodolitaufnahme 2750
Fourier-Polygone 1468
Fourier-Transformation 1469
Fractal 1470
fractale Fläche 1472
fractale Oberfläche 1472
fractales Bild 1471
Framing 1478
freier Knoten 1480
freie spatiale Bedarfskurve 1481
fremder Kennbegriff 1457
Fremdschlüssel 1457
Frequenzbereichsfilter 1485
Frequenzzerstreuung 1484
Frontseite 1488
Fuge 3238
Führungsanteil 2127
Füllen 1417
Füllfarbe 1415

Füllung 1417
Füllungsfarbe 1415
Füllungsmuster 1421
Füllungsname 1419
Füllungspalette 1420
Fundamentalebene der sphärische
    Koordinaten 1493
Fundort 1797
Funksignal 2993
funktionale Beziehung 1491
funktionale Region 1490
Funktionsfläche 1492
Funkzeichen 2993
Fusion 2393
Fußnote 1454
Fußpunkt 2482
Fuzzy- 1495
Fuzzy-Clusterung 1498
Fuzzylogik 1500
Fuzzylogik-Modell 1501
Fuzzy-Menge 1505
Fuzzy-Menge-Theorie 1506

Gadget 1508
Gall-Peters-Projektion 1509
Gang 782
Gauss-Krüger-Projektion 1514
Gausssche Approximation 1512
Gausssche Koordinaten 1513
Gaussscher Algorithmus 1511
Gausssche Verteilung 2556
Gaussverteilung 2556
Gebiet 158, 973
Gebietauswahl 179
Gebiet des Tablettmenüs 3574
Gebietsauswahl 179
Gebietseinteilung für eine
    Erhebung 491
Gebietsgrenze 387
Gebietskarte 3085
Gebietszerlegung 162
gebirgige Region 2452
Gebirgssattel 2708
gedrehtes Bild 3175
gefährliche Zone 1809
gefüllter Bereich 1416
Gegenazimut 299
gegenseitige Azimute 2481
gegenseitige Durchdringung 1997
Geisterbild 1172
gekennzeichnet 2065
gekoppelter Randsatz 790
gekreuztes Schraffen 809
Gelände 2072, 3605
Geländeanalyse 1407
Geländeausschnitt 1582
Geländebedeckung 3712
Geländedarstellung 3138
Geländedatenbasis 2081
Geländeeditiersystem 3610

Geländefläche 3722
Geländefolgesystem 3611
Geländeformenkarte 2082
Geländehöhe 1787
Geländelinie 2084
Geländelinienanpassungssystem
    3606
Geländelinieninterpolierung 3613
Geländemodell 3614
Geländesteilheit 3617
Geltungsbereich 3621
gemappte Erscheinungen 2295
gemeinsame Datenbenutzung 914
Gemeinsamgrenze 691
Gemeinsamobjekt 627
gemischtes Schattieren 625
Genauigkeit 17
Genauigkeit der Aufnahme 18
Genauigkeit der
    Kartenmessungen 2284
Genauigkeit der Messung 2382
Genauigkeitsabfall 1097
geneigte Fläche 3316
Generalisation 1520
generalisiert 1525
generalisierte Koordinaten 1526
Generalisierung 1520
Generalisierungsgrad 1522
Generalisierungsmaßstab 1524
Generalisierungsoperatoren 1523
Generationsdatengruppe 1521
generische Kartenwerkzeuge
    1527
genetische Algorithmen 1528
Geobild 1639
    3D-~ 1011
Geocode 1601
geocodierter Zensus 1535
geocodierte Zählung 1535
Geocodierung 1592
Geocodierungsfehler 1536
Geocodierungsschema 1600
Geodäsie 1542
Geodäte 1541
Geodaten 1584
Geodatenbasis 1585
Geodaten-Datei 1586
Geodatenentwicklung 1605
Geodateninfrastruktur 1677
Geodaten-Management 1587
Geodatenmodell 1678
Geodatennetz 1537
Geodatenserver 1538
Geodatensystem mit
    verschiedenen Verwendungen
    2468
Geodaten-Technologie 1589
Geodatenwarenhaus 3381
geodätisch 1540
Geodätische 1541

geodätische Abbildung 1551
geodätische Aufnahme 1556
geodätische Breite 205
geodätische Daten 1547
geodätische Grundlage 2530
geodätische Höhe 1548
geodätische Koordinaten 1545
geodätische Länge 1550
geodätische Linie 1541
geodätische Messungen 1552
geodätischer Azimut 1543
geodätischer Breitengrad 1554
geodätischer Kreis 1544
geodätischer Meridian 1553
geodätisches Datum 3462
geodätisches Dreieck 1559
geodätisches Grundnetz 2530
geodätisches Instrument 1549
geodätisches Netz 2530
geodätisches Referenzsystem
    1555
geodätisches Weltsystem 4010
geodätische Torsion 1558
geodätische Überwachung 1557
geodätische Windung 1558
geodätische Zenitdistanz 1560
Geodemografie 1539
Geodynamik 1562
Geofachdaten 1584
Geografie des Internet-
    Adressraums 1636
Geografie-Markierungssprache
    1635
geografisch 1569
geografisch codiert 1572
geografisch codierte Imagemap
    1672
geografische Abfrage 1609
geografische Adresse 1568
geografische Ansicht 1634
geografische Anwendung 1574
geografische Beziehung 1621
geografische Breite 205
geografische Darstellung 1622
geografische Darstellung
    aufgetretener Krankheitsfälle
    3449
geografische Daten 1584
geografische Datenbasis 1585
geografische Dateneingabesystem
    1596
geografische Datenmenge 1588
geografische Dimensionen 1561
geografische Einheit 1633
geografische Engine 1594
geografische Entität 1595
geografische Hierarchie 1599
geografische Identität 1602
geografische Information 1584
geografische Kartografie 1577

geografische Kategorie 1579
geografische Koordinaten 1581
geografische Länge 206
geografische Metapher 1612
geografische Namen 1613
geografische Namenorthografie
    2665
geografische Namenschreibweise
    2665
geografische Operatoren 1616
geografischer Abschnitt 1591
geografischer Atlas 1575
geografischer Bereich 1564
geografische Referenz 1619
geografischer Erfassungsbereich
    1582
geografische Ressource 1623
geografischer Grunddatensatz
    1576
geografischer Identifikator 1601
geografischer Katalog der
    politischen und statistischen
    Bereiche 1578
geografischer Kreis 1583
geografischer Norden 3834
geografischer Pol 1618
geografischer Zugriff 1566
geografisches Äquivalent 1597
geografisches Display 1590
geografische Sicherheit 1626
geografisches Informationssystem
    1608
geografisches Koordinatensystem
    1598
geografisches Lexikon 1515
geografisches Marketing 1650
geografisches Register 1603
geografisches Thema 1630
geografisches Traceroute-
    Programm 1632
geografische Suche 1625
geografisches
    Weltreferenzsystem 4011
geografisches wissensbasiertes
    System 1610
geografisches Zentrum 1580
geografische Teilung 1591
geografische Traceroute 1631
geografische Vermarktung 1650
geografische Visualisierung in
    Realzeit 3038
geografische Wissenschaften
    1624
geografische Zugriffsrechte 1567
geografisch verteilt 1570
geografisch verteiltes System
    1571
Geograph 1565
geographisch 1569
Geoid 1637

Geoinformatik 1607
Geoinformatik-Bildung 1640
Geoinformatik-Kartierung 1642
Geoinformatik-Konzeption 1641
Geoinformationssystem 1608
geologische Bildung 1645
geologische Daten 1644
geologische Formation 1645
geologische Karte 1646
geologische Kartografie 1647
geologischer Dienst der
   Vereinigten Staaten 3857
geologische Strukturkarte 3527
geomagnetische Überwachung
   1649
Geomarketing 1650
Geometriedaten 1655
geometrische Daten 1655
geometrische Datenbasis 1656
geometrische
   Erfassungsbereichsgrenzen
   1654
geometrische Fläche 1664
geometrische Genauigkeit 1651
geometrische Grafik 1660
geometrische Komplexität 1652
geometrische Korrektur 1653
geometrische Operation 1662
geometrischer Genauigkeitsabfall
   1657
geometrisches Grundelement
   1663
geometrisches Modell 1661
geometrisches Primitiv 1663
geometrische Transformation
   1665
geometrische Verzerrung 1658
Geomöglichkeiten-Datenbasis
   1563
Geomorphometrie 1666
Geoobjekt 1615
geophysikalische Ereignisse 1668
geophysikalisches Datenzentrum
   1667
Geoportal 1669
geopotentieller Wert 1670
geordnete Liste 2644
Georeferenz 1619
georeferenziert 1572
georeferenzierte Daten 1573
georeferenzierte Statistik 1628
Georeferenzierung 1593
Georeferenzierungspunkt 2323
georelationale Konzeption 1673
georelationales Datenmodell 1674
Georessource 1623
Georessourcen-Gateway 1676
geostationär 1679
geostationärer Orbit 1681
geostationäres Navigationssystem

   1680
Geostatistik 1628
Geostatistikprogramm 1629
geosynchron 1679
geosynchroner Orbit 1682
Geowissenschaften 1675
geozentrische Breite 1531
geozentrische Koordinaten 1529
geozentrische Länge 1532
geozentrischer Breitengrad 1534
geozentrischer Meridian 1533
geozentrisches Datum 1530
geozentrisches Kartesisches
   Greenwich-
   Koordinatensystem 1202
Gerade 3501
geradlinige Navigation 3502
geradlinig verbundene Struktur
   2159
Gerätekoordinaten 1007
Gerätekoordinatenraum 1010
Gerätekoordinatensystem 1008
Geräteraum 1010
Gerätetechnik 1804
Gerätetreiber 1009
gerichtete Reflexion 3433
gerichteter Graph 1105
gerichtetes Netz 1107
gerichtetes Netzwerk 1107
Gerichtsbarkeitsgrenzen 2052
Gerinnegeometrie 524
Gerüst 1476
Gesamtaufnahme 644
Gesamtheit 83, 2854
Gesamtvarianz 3772
gescannte Daten 3222
gescannte Karte 3223
Geschäftsgrafik 417
Geschäftsgrafikdaten 418
geschätzter Wert 1320
geschlossene Form 577
geschlossene Höhenlinie 578
geschlossene Polylinie 580
geschlossener Polygonzug 579
geschorene Region 570
Geschwindigkeit 3927
Gesichtsfeld 1403
Gesichtswinkel 124
Gesichtszucken 3676
Gesichtszucken-Datei 3673
Gesichtszucken-Koordinaten
   3672
Gesichtszucken-Toleranz 3675
Gespräch-Karte 747
Gestalt 3279
Gestalterkennung 2717
Gestell 1476
gestreckte Linie 1348
getrennt 2615
Gewässernetz 3163

Gewässersystem 3163
Gewebe 1179
Gewerbeflächenkarte 3780
Gewicht-Charakteristiken 3992
gewichtete Pufferung 3990
Gewichtseinheitsfehler 3164
Gewinn 194
gewogener Mittelwert 3991
gewogenes Mittel 3991
gewöhnliche Abbildung 2603
gewöhnliches Kriging 2649
gieren 4014
GIF-Format 1683
GIS 1608
   auf Schichten beruhendes ~
   2117
   2D-~ 1012
   3D-~ 1013
   4D-~ 1467
GIS-Anwendung 1684
GIS-basierte Analyse 1688
GIS-Basisschicht 1689
GIS-Benchmark 1690
GIS-Benutzer 1702
GIS-Datenbasis 1691
GIS-Funktionalität 1693
GIS-Funktions-Server 1694
GIS-gestütztes
   Bürgerinformationssystem
   1686
GIS-gestütztes
   Managementsystem 1687
GIS im Internet 1695
GIS-Internetlösungen 1695
GIS-Markt 1696
GIS-Produkt 1697
GIS-Projekt 1698
GIS-Server 1694
GIS-Shareware 1699
GIS-Software 1700
GIS-Systemarchitektur 1685
GIS-Technologie 1701
Gitter 2115
Gitter-Ähnlichkeitsfaktor 1778
Gitter-Extent 1767
Gitterfläche 1781
Gittergenerierung 1769
Gittergrenzen 1763
Gitterintervall 1771
Gitterkoordinaten 433
Gitterkreuz 1770
Gitter-Maßstabsfaktor 1778
Gittermodus 1773
Gitternetz 2274
Gitternetzanlegung 2336
Gitternorden 1774
Gitternummern 1768
Gitterpunktdaten 1775
Gitterreferenzen 1776
Gitterreferenzsystem 1777

Gitterstraßenmuster 1780
Gittersüden 1779
Gitterverfolgung 1766
Gitterzelle 1764
Gitterzone 1783
Gleichaufteilung 1313
Gleichmäßigkeit 3854
Gleichrichter 3055
Gleichrichtung 3054
Gleichwahrscheinlichkeitskreis 544
Gleissicherungspunkt 2384
Gleisunebenheit 3851
globale Daten 82
global einheitlicher Identifikator 1709
globale Ressourcendatenbasis 1711
globaler Koordinatenraum 1703
globales geografisches 3D-Modell 1014
globales GIS 1706
globales Indizierungssystem 1707
globales Koordinatensystem 4008
globales Positioniersystem 1710
globales Raumdatensystem 1712
globales Umweltsystem 1705
globale Topologie 1713
globale Umweltmonitoringssystem 1704
Globalisierung 1708
Globus 1209
Globuskugel 1209
Glyphe 1714
GMT 758
gnomonische Kartenprojektion 1715
gnosiologische Konzeption 2440
Goal 997
GPS 1710
GPS-basiertes Navigationssystem 1718
GPS-Empfänger 1722
GPS-Konstellation 1719
GPS-Kontrollmonitor 1720
GPS-Navigation 1721
GPS-Service zur Positionsbestimmung für autorisierte Nutzer 2877
GPS-Service zur Standardpunktbestimmung für alle Nutzer 3467
GPS-Software 1723
Grabber-Schaltfläche 1724
Grad 949
Gradabteilungsblatt 2318
Gradabteilungskarte 2318
Gradation 1725
Gradient 1726

Gradientabschätzung 1729
Gradientdichte 1728
Gradientenabschätzung 1729
Gradientenanalyse 1727
Gradientendichte 1728
Gradientenwinkel 125
Gradientfilterung 1730
Gradientwinkel 125
Gradnetz 2274
graduierte Punktsymbole 1732
Grafik 1018
Grafik-Datenstruktur 1745
Grafik von Linien und Koordinaten 761
grafisch 1736
grafische Benutzerschnittstelle 1748
grafische Darstellung 1743
grafische Elemente 1738
grafischer Maßstab 1744
grafisches Ausgabegerät 1741
grafisches Bild 1739
grafisches Eingabegerät 1740
grafische Seite 1746
grafisches Overlay 1742
grafisches Primitiv 1274
grafische Superponierung 1747
grafische Überlagerung 1747
grafische und analytische Methoden 1737
grafische Variable 1749
Grafschaft 786
Grafschaftsgrenze 787
Grain-Toleranz 1733
Graph 1735
Graph-Darstellung 1751
Graukarte 1762
Grauskala 1759
Graustufe 1758
Graustufenkarte 1762
Graustufenskala 1759
Graustufung 1759
Grauwertbild 1760
Gravimetrie 1757
greenwiche Normalzeit 758
Grenzänderungen 378
Grenzbedingung 3155
Grenzbeeinflussung 381
Grenzbereich 395
Grenzdaten 2350
Grenzdatendarstellung 2349
Grenze 377
Grenzen des Kartenblatts 2334
Grenzen-Kartenzeichen 386
Grenzextraktion 382
Grenzfolgenalgorithmus 384
Grenzlinie 385
Grenzmarke 386
Grenzpunkt 390
Grenzstein 2087

Grenzvermessung 393
Grenzzeichen 386
Grid 3005
Grid-Datenmenge 1765
Griff 3328
Griffel-Palette 2723
Größe des aktiven Bereichs 3315
Größenänderung 3151
großes binäres Objekt 349
Großflächentriangulation 2106
Großformat-Atlas 2107
Großkreis 1761
großmaßstäbige Karte 2108
Großmaßstab-Karte 2108
Großstadtbereich 2416
Ground-Truth 1793
Ground-Truth-Punkt 1794
Grübchen 2757
Grubensenkung 3525
Grund 1784
Grundbesitz-Analyse 3034
Grundbuch 422
Grunddatenbasis 2081
Grundeigentum-Analyse 3034
Grundelement 2890
Grunderwerb 20
Grundfarbe 301
Grundfläche 414
Grundkarte 3706
Grundkarteneinheit 320
Grundkataster 422
Grundkreis 1493
Grundlandeinheit 322
Grundlinie 317
Grund-Patch-Area 1789
Grundraumeinheit 323
Grundriss 1851
Grundrissaufnahme 2789
Grundrissdarstellung 2790
Grundrisskoordinaten 1845
Grundrisstreue 3717
grundsätzliche Überdeckung 474
Grundsteuerregister 422
Grundstück 2089
Grundtest 1793
Grundvermessungsdienst 321
Grundzug 448
Grundzugscodierung 2271
Grundzugsdatei 449
Gruppe 584
Gruppierung 587
Gruppierungsfunktion 1796
gültige Daten 562
Gummiband 3186

Habitatmodell 1798
Hafenplan 1802
halbautomatische Digitalisierung 3261
Halbebenenunterteilung 354

Halbkugel 1815
Halbmesser 2994
Halbton 1799
Halbtonaufnahme 724
Halbton-Diagrammkarte 540
Halbvariogramm 3262
Halteplatz 3498
Haltepunkt 3498
Haltepunkt-Attribute 3496
Haltepunkt-Impedanz 3497
Haltepunktsdatei 3499
Hamilton'sche Kontur 1800
Handelszone-Karte 3780
Handheld-Computer 2689
Handmikrorechner 2689
Handsteuergeber 2050
Hang 1921
hängende Linie 849
hängender Bogen 848
hängender Knoten 850
hängender Knotenpunkt 850
Hardcopy-Karte 2697
Hardware 1804
Hardwareplattform 1805
harte Klassifizierung 1803
Haufendiagramm 3228
Haupthöhenlinie 1929
Hauptkomponenten-Analyse 2891
Hauptmaßstab 2892
Hauptmeridian 2889
Haushalt 1371
Haushaltserhebung 1854
Hebewerk 1252
heimischer geografischer Name 1163
Heiterkeit 404
heliografische Länge 1814
Helligkeit 404
herausgefallene Abtastzeilen 2434
Heterogenität 1816
Heuristik 1818
heuristisch 1817
Hexagonalbaum 1819
Hierarchie 1828
hierarchische Datenbasis 1825
hierarchische geografische Darstellung 1827
hierarchische kartografische Datenbasis 1823
hierarchische Klassifizierung 1824
hierarchisches Datenmodell 1826
Hilfshöhenlinie 290
Hilfsmittelleiste 3699
Himmelsäquator 479
Himmelserdball 480
Himmelsrichtung 2810
Hinausschießen 2683

Hintergrund 300
Hintergrundfarbe 301
Hinterland 1838
Hinweis 2801
Hinweiskarte 1933
hinzufügen 28
Histogramm 315
historische Ansicht 1839
hochauflösendes Meeresboden-Kartografierungssystem 1833
Hochebene 3571
hochgeführte Straße 3167
Hochland 3878
Hochpassfilter 1831
hochpräzise Referenznetzwerke 1829
hochpräzises geodätisches Netz 1832
Hochsee- 2616
Hochseevermessung 2612
hochsteigend 188
Hochstraße 3167
Hochterrasse 3571
Hochwasser 1834
Hochwasserlinie 1835
Hochwasserschutzkarte 1440
Hochwasserüberschwemmungs-gebiet 1441
Hochwert 2568
Höhe 1811
Höhenaufnahme 103
Höhenbestimmung 104
Höhenebene 2144
Höhenfixpunkte 2140
Höhenkote 734
Höhenkurve 732
Höhenlage eines Fixpunktes in Bezug auf eine Horizontale 1813
Höhenlinie 732
Höhenlinie indizieren 1928
Höhenlinienabstand 730
Höhenliniendaten 1247
Höhenlinienetikettierung 735
Höhenlinienkarte 727
Höhenlinienmodell 1251
Höhenlinienzahl 734
Höhenmatrix 105
Höhenmesser 102
Höhenmesskunde 104
Höhenmessung 104
Höhennetz 2140
Höhennivellierinstrument 3578
Höhenplan 727
Höhenpunkt 1245
Höhenpunkt in der Karte 1245
Höhenpunktsymbole 3448
Höhenschichtenfarbe 1878
Höhenschichtenkarte 3114
Höhenschichtlinie 732

Höhenschichtlinienkarte 727
Höhenschichtlinienzahl 734
Höhenstufenfarbe 1878
Höhenunterschied 1812
Höhenzahl 1245
höhere Geodäsie 1830
Höhe über Meer 1811
Höhe über NN 1811
Höhe über Normal-Null 1811
homogene Regionen 3855
Homogenisierung 3854
Homologiezentrum 510
Horizont 1841
Horizontabschlussfehler 1319
Horizontaldatum 1846
Horizontale 924
horizontale Auflösung 1852
horizontale Ebene 1849
horizontale Genauigkeit 1842
horizontale Koordinaten 1845
horizontale Position 1850
horizontaler Genauigkeitsabfall 1847
horizontaler Interpolator 1848
horizontaler Winkel 1843
Horizontalkurve 732
Horizontalprojektion 1851
Horizontalwinkel 1843
Horizontierlibelle 1844
Hub 1856
Hülle 582
Hüter 835
Hybrideditor 1858
hybrides GIS 1859
Hydrografie 1865
hydrografisch 1861
hydrografische Karte 1862
hydrografische Namen 1863
Hydroisoplethenkarte 1866
hydroklimatisches Datennetz 1860
hydrologischer Atlas 1867
hydrologisches Benchmark 1868
hydrologisches Benchmark-Netz 1869
hydrologisches Informationssystem 1870
hyperbolischer 3D-Graph 1015
hyperspektrale Bildanalyse 1872
hyperspektraler Bildsensor 1873
Hypertext-Karte 1874
hypsografische Karte 1875
Hypsometrie 1880
hypsometrische Farbe 1878
hypsometrische Farbenskala 1879
hypsometrische Karte 1876
hypsometrische Methode 1877

Identifikationsnummer 1882

Identifikationsnummer für eine
  geografische Referenz 1620
Identifikator 1883
Identifizierung 2752
Identität 1884
Identitätsabbildung 1885
IGES-Vorschrift 1951
Ikone 1881
Illuminierung 1886
Impedanz 1914
Implementierung 1915
Importieren 1916
Impressum 1918
Inaktivierung 931
Index 1927
Indexkarte 1933
Indexschlüssel 3103
Index-Überdeckung 1930
Index-Überlagerung 1930
Index-Überlagerungsmodell 1932
Index von angrenzenden Blättern
  1933
Indikation 1934
Indikator-Kriging 1935
Indikatorpunkt 1936
indirekte Messung 1937
indirekte Zeichen 1938
Indirektmessung 1937
Indizieren 1931
Indizierung 1931
Inflexion 1939
Informationskarte 1943
Informationskartierung 1940
Informationskartierung-
  Technologie 1941
Informationssystem 1945
Informationssystem der
  Aerofotografie 66
Informationssystem der
  geografischen Namen 1614
Informationssystem für die
  Fliessendwasser 3190
Informationssystem für
  statistische Daten eines
  Landes 2083
Informationssystem über die
  Ressourcen 3154
Informationswiederauffindung
  1942
Informationswiedergewinnung
  1942
Infrarotmapper 1947
Infrarotscanner 1946
Infrastruktur 1948
Infrastruktur-Planung 1949
Inhalt-Grundsätze 710
Initialisieren 1952
Initialisierung 1952
Inklinationswinkel 126
Inklinometer 1922

Inklusion 1923
Inkrement 1925
Innengebiet 1983
Innenpunkt 1990
innere Konversion 1988
innere Orientierung 1984
innerer Bereich 1983
innerer Knoten 1955
innerer Punkt 1990
inneres Polygon 1956
innere Textur 1957
innerste Höhenlinie 1954
innerste Isohypse 1954
Insel 2020
Inselfläche 2020
installieren 1963
Instanz 1964
Instanziierung 1966
Instrument 1967
Instrumentfehler 1968
Integration 1973
integrierte Aufnahme 1971
integrierte regionale Datenbasis
  1970
integriertes GIS 1969
Integrierung 1973
intelligentes GIS 1974
Intensitätslinie 2031
Intensitätswertlinie 2031
interaktive Digitalisierung 1975
interaktive Geocodierung 1976
interaktive grafische Gestaltung-
  Software 1977
interaktive Kartierung 2625
interaktive Linienverfolgung
  1978
interaktiver Betrieb 1979
interaktiver Linienverlauf 1978
interaktive Verarbeitung 1980
internationale Karte 1992
interne Nummer 1989
interner Punkt 1990
interner Zeiger 1991
Internet-GIS 3984
Internet-Kartenserver 1994
Internet-Kartierung 1993
Internet-Topologie 1995
Interoperabilität 1996
Interpolation 1998
Interpolation in regelmässigen
  Intervallen 1305
Interpolationsmethode 1999
Interpretation 2000
Intervall 2002
Intervallarithmetik 2003
Invariante 2004
Inventarisierung der natürlichen
  Ressourcen 2501
inverse Abbildung 2009
inverse Distanzgewichtung 2007

inverse Distanzgewichtung-
  Interpolation 2006
inverse Fläche 3158
inverse Fourier-Transformation
  2008
inverse Polarität 3157
inverser Fehler 2010
Inzidentkarte 1920
Inzidenz 1919
ionosphärische Fehler 2013
ionosphärische Verzögerung 2012
Isarhythmenkarte 2019
Isobar 2021
isobare Form 2022
Isobathe 979
Isochronenkarte 727
Isochronenplan 727
Isodeformate 1138
isodemografische Karte 2023
Isogon 2024
Isohypse 732
Isohypsenvorhersagekarte 2904
Isolation 2028
isolierte Lage 2026
isoliertes System 2027
isolierte Stelle 2026
Isolierung 2028
Isolinie 2031
Isolinienkarte 2034
Isolinien-Methode 2410
isometrische Ansicht 2033
isometrische Projektion 2032
isometrisches Netz 2030
Isoplethe 2031
Isoplethenblockdiagramm 2029
Isoplethenkarte 2034
Isotherme 2036
isotherme Schicht 2035
Isothermkarte 2037
Isotropie 2040
isotropische Kurve 2038
isotropische Schicht 2039

Jackknifing 2043
Jäten 3988
Joystick 2050
JPEG-Format 2048
JPG-Format 2048
Justierung 54

Kachel 3679
Kachelindizierung 3683
kalibrierte Imagemap 424
kalibrierte Leitung 425
Kalibrierung 426
Kalibrierungsanalyse 427
Kaltstart 601
Kanalgeometrie 524
Kante 374, 1215
Kantenabbildung 1222

kantenbasierte
  Dreieckunterteilung 1217
Kantenextraktion 1221
Kantenglättung 136
Kantengraph 2176
Kantenklebeband 2361
Kantenmodell 4001
Kantensichtbarkeit 1224
Kantenübereinstimmung 1223
Kappa-Koeffizient 599
Kappen 571
Kapselung 1263
Kardinalpunkt 431
Kardinalrichtung 430
Karte 1611
  aus den digital verschlüsselten
  Informationen konstruierte ~
  2312
Karte der Lage 3312
Karte der sichtbaren Flächen
  3954
Karte in verkleinertem Maßstab
  3327
Karte mit flächentreue Projektion
  1840
Karte mittlerer Größe 2388
Kartenabdeckung 2262
Kartenabfrage 2320
Kartenabstimmung 2250
Kartenalterung 2251
Kartenänderung 2253
Kartenarchivierung 2254
Kartenausgabe 2317
Kartenausschnitt 2328
Kartenautorrecht 771
Kartenbearbeitungsfunktion 2268
Kartenbereich 3087
Kartenberichtigung 2253
Kartenbeschriftung 453
Kartenbeschriftungssprache 2279
Kartenbewertung 2269
Kartenbibliothek 2264
Kartenbildsatz 2275
Kartenblatt 2333
Kartenblattbezeichnung 3284
Kartenblattnummer 3286
Kartendarstellung 2267
Kartendatei 2272
Kartendaten 2263
Kartendatenbasis 442
Karten der Natur- und
  Sozialwechselwirkung 2335
Kartendruck 2311
Kartenebene 2307
Karteneinstellung 2250
Kartenentwerfen 2297
Kartenentwerfen-Ökonomik 1214
Kartenentwicklung 2265
Kartenerstellung 2310
Kartenfarboriginal 2259

Kartenfarbvorlage 2259
Kartenfeld 2339
Kartenformatierung 2273
Kartengitter 2274
Kartengrenzen 2280
Kartengrund 3706
Karteninformativität 2276
Karteninhaltsfläche 2339
Kartenkomponenten 2260
Kartenkoordinatensystem 2261
Kartenkorrektur 528
Kartenkunde 466
Kartenmaßstab 2326
Kartenmontage 2285
Kartennetz 2274
Kartennull 529
Kartennutzung 2346
Kartenorientierung 2289
Kartenpaar 2294
Kartenplotten 2308
Kartenpräzision 2298
Kartenprojektierung 444
Kartenprojektion 2314
Kartenprojektion-Verzerrung
  2315
Kartenprojektor 2316
Kartenqualität 2319
Kartenrand 376
Kartenrandangaben 2350
Kartenreproduktion 2901
Kartensammlung 2257
Kartensatz 2332
Kartenschnitt 3285
Kartenschnittlinie 3285
Kartenschriften 453
Kartensegment 2328
Kartenserie 2331
Kartenserver 1994
Kartensprache 2277
Kartentechnik 460
Kartenthema 2343
Kartentitel 2344
Kartentransformation 2345
Kartenübersetzung 2321
Kartenübersetzungssystem 2322
Kartenuntergrund 3706
Kartenurheberrecht 771
Kartenverbesserung 2253
Kartenvergleich 2283
Kartenverknüpfung 2281
Kartenversöhnung 2250
Kartenvorbereitung 2310
Kartenwerk 1575
Kartenwerkzeugsatz 2305
Kartenzeichen 459
Kartenzeichnen 445
Kartenzeichner 435
Kartenzeiger 2282
Kartenzuverlässigkeit 2324
Kartesische Koordinaten 433

Kartesisches Koordinatensystem
  434
Kartiereinheit 2306
kartieren 2290
Kartiergerät 2302
Kartierinstrument 2302
Kartierung 2297
Kartierung am Tag 927
Kartierung der planimetrischen
  Basis 2785
Kartierungsgenauigkeit 2298
Kartierungsstelle 2299
Kartierungssystem 2304
Kartodiagramm 539
Kartografie 465
kartografieren 2249
Kartografierungsausrüstung 1315
Kartografierungssystem 2304
kartografisch 436
kartografische Abbildung 2314
kartografische Algebra 2252
kartografische Ansicht 2347
kartografische Arbeitsstation 464
kartografische Auflösung 2303
kartografische Bibliographie
  2255
kartografische Bildung 446
kartografische Clipart 438
kartografische Darstellung 457
kartografische Daten 2263
kartografische Datenbank 441
kartografische Druckerzeugnisse
  2311
kartografische Einheit 2306
kartografische
  Forschungsmethode 454
kartografische Generalisierung
  450
kartografische Grundeinheit 320
kartografische Information 451
kartografische Kommunikation
  439
kartografische Methoden 2342
kartografische Mosaiking 2286
kartografische Netzdatenbasis
  2519
kartografische Pragmatik 2309
kartografische Projektierung 2265
kartografische Projektion 2314
kartografischer Browser 2348
kartografischer
  Darstellungsmodus 2244
kartografischer Datenbrowser 443
kartografischer Katalog 437
kartografischer Maßstab 2326
kartografischer Scanner 458
kartografischer Skript 2327
kartografisches Arbeitsformat 463
kartografisches Bild 456
kartografische Schicht 2291

kartografisches
  Datenbediensystem 452
kartografisches Design 444
kartografische Semantik 2329
kartografische Semiotik 2330
kartografisches
  Informationsrückgewinnungs-
  system 452
kartografisches
  Koordinatensystem 2261
kartografisches Modellierung 455
kartografisches Netz 2274
kartografisches Resümee 2338
kartografisches Schnittobjekt
  2256
kartografische Steuerung 440
kartografische Stilistik 2337
kartografische Symbole 459
kartografische Syntax 2341
kartografische Teilung 2300
kartografische Toponymie 461
kartografische Umgebung 447
kartografische Umwelt 447
kartografische Verallgemeinerung
  450
kartografische Visualisierung 462
kartografische Zeichnung 445
Kartogramm 539
Kartograph 435
Kartographie 465
kartographisch 436
Kartometrie 469
kartometrische Operationen 468
kartometrischer Index 467
kaskadiert 471
kaskadierte Fenster 472
Kataster 422
Katasteraufnahme 3723
Katasterkarte 2652
Katasterkarten-
  Übertragungsformat 2653
Katasterklasse 421
Kataster mit verschiedenen
  Verwendungen 2467
Katasterplankarte 2652
Katastervermessung 3723
Katastrierung 2090
Katastrophe-Kartografie 1118
Kategorie 476
kategorische Bodenbedeckung
  474
kategorische Daten 475
Kegelprojektion 690
Kegelprojektion nach Lambert
  2071
Keil 1716
Kelsh-Plotter 2054
Kennsatz 2061
Kennsatznummer 2067
Kennsatzpunkt 2068

Kennnummer 1882
Kennzeichnung 2066
Kennziffer des Meridianstreifens
  4031
Kern 1856
Kernachse 1303
Kette 516, 527
Kettencode 518
Kettencodierung 519
Kettendurchhang 477
Kettenfahrleitung 477
Kettenkurve 477
Kettenlinie 477
Kettenübertrager 517
Kettenzug 516
Kinematik 2057
kinematisches GPS 2056
Kippregel 93
Kippungswinkel 126
klar abgegrenzter Punkt 3993
klar bestimmter Punkt 3993
klarer Punkt 3993
Klasse 549
Klassenbreite 559
Klassenidentifikator 551
Klassenintervall 559
Klassenliste 560
Klassenzentrum 550
Klassifikation 552
Klassifikationsfehler 2429
Klassifikationsgenauigkeit 553
Klassifikator 558
Klassifikator mit minimaler
  Distanz 2425
Klassifizierer 558
klassifiziertes Bild 556
klassifiziertes
  Bildbearbeitungsprogramm
  557
Klassifizierung 552
Klassifizierung der größten
  Wahrscheinlichkeit 2374
Klassifizierungsmatrix 554
Klassifizierungsmethode 555
Klassifizierung von
  segmentiertem Bild 3252
Kleinflächentriangulation 3325
Kleinkreis 3326
kleinmaßstäbige Karte 3327
kleinmaßstäbliche Karte 3327
Kleinstcomputer 2689
kleinstes achsenparalleles
  umschließendes Rechteck
  2423
klickbare Imagemap 564
klickbare Karte 564
Klimadaten 107
Klimakurve 566
Klimanomogramm 566
klimatische Region 568

Klimaveränderung 567
Klimazone 568
Klimogramm 566
Klinometer 1922
Klippen 571
Klopfer 3343
Klumpen 584
Knoten 2533
  vom ~ 1486
Knoten-Adjazenz 2534
Knotenanpassung 2538
Knotenanpassungstoleranz 2539
Knotenattributtabelle 2535
Knotenidentifikator 2537
Knoten-Kanten-Inzidenz 2536
Knotenlinie 2186
Knoten-Nachbarschaft 2540
Knoten-Näherung 2540
Knotenvalenz 3896
koextensiv 597
kognitive Karte 598
kohärente Einheit 985
Kollaps 602
kollineare Abbildung 2918
Kombination 623
kombinieren 624
kombinierte Bildabschattung 625
kombinierte Gliederung 805
kommunikative Zuverlässigkeit
  628
Kompaktheit 629
Kompaktheitsverhältnis 630
Kompaktion 656
Kompass 632
Kompassmissweisung 635
Kompassrichtung 634
Kompassrose 636
Kompatibilität 637
Kompilieren 638
Kompilierung 638
Kompilierungsmaßstab 640
komplexe Karte 647
komplexe Kartierung 648
Komplexfläche 651
Komplexität 646
Komplexität der Kurve 827
Komplexpolygon 650
Komponentenmodell 652
kompressieren 654
kompressierte digitale
  Höhenliniendaten 655
Kompression 656
Kompressionsmethode 657
komprimieren 654
Komprimierung 656
Konfidenzintervall 684
Konfiguration 685
konforme Projektion 686
Konformität 687
Konglomerat 689

kongruente Projektion 3720
konische Abbildung 690
Konkatenation 672
Konkavität 673
Konnektivität 698
Konnektor 699
Konsistenz 700
Konsolidation 702
konsolidierte Stadt 701
Konstante 703
konstante geometrische
   Genauigkeit 704
konstanter Fehler 3567
Konstellation 705
konstruktive Körpergeometrie
   708
konstruktive Vollkörpergeometrie
   708
kontextbezogenes Menü 3289
Kontextmenü 3289
kontextsensitives Menü 3289
Kontiguität 712
Kontinentalschelf 715
Kontinentalsockel 715
kontinuierliche Daten 717
kontinuierliche Erscheinung 721
kontinuierliche Fläche 722
kontinuierliche Karte 720
kontinuierlicher Ton 723
kontinuierliches Feature 719
kontinuierliche Variable 725
kontinuierliche Verteilung 718
Kontinuität 716
Kontrastkoeffizient 596
Kontrastkorrektur 736
Kontraststrecken 737
Kontraststreckung 737
Kontrastverhältnis 596
kontrollierte Klassifikation 3536
Kontrollkästchen 530
Kontrollkasten 530
Kontrollsegment 742
Kontur 726
Konturattribute 2671
Konturen 728
Konturerkennung 382
Konturlinie 731
konventionelle Projektion 744
Konvergenz 745
Konverter 750
Konvertierung 748
konvexe Hülle 751
Konvexität 752
Konvolution 753
konzeptionelle Genauigkeit 675
konzeptionelle Karte 678
konzeptionelles Modell 679
konzeptionelles Rahmenwerk 677
Konzeptualbeschreibung 676
Konzeptualmodell 679

Koordinate 754
Koordinatenanfangspunkt 2660
koordinatenbezogene
   Dimensionierung 756
koordinatenbezogene
   Positionierung 756
Koordinatendarstellung 757
Koordinatendatei 759
Koordinatengefrierung 1482
Koordinatengeometrie 760
Koordinatengitter 762
Koordinaten in der Ebene 2775
Koordinatenkonvertierung 755
Koordinatenleser 764
Koordinatennetz 762
Koordinatenpaar 763
Koordinatenschreiber 768
Koordinatensystem 765
Koordinatentabelle 766
Koordinaten-Transformation 767
Koordinatenursprung 2660
koordinierte Weltzeit 758
Kopfbildschirm 1810
Kopie 769
Kopieren 770
Kopierrahmen 2896
Körnigkeit 1734
Körper 3339
Korrektheit 777
Korrektur 776
Korrektur der geometrischen
   Verzerrungen 1659
Korrelation 778
Korrelation-Matrix 780
Korrelationsanalyse 779
Korrelationspunkte 781
Korridor 782
korrigierte Zeilenlänge 775
kostengünstiger Leitweg 2128
kostenoptimierter Leitweg 2128
Kote 734
kotierte Projektion 3720
Kovarianz 792
kräftige Klassifikation 1803
Kränungsmesser 1922
Kreis 543, 1148
Kreisdiagramm 2753
Kreisfernaufnehmer 2641
Kreissehne 535
Kreuzkorrelation 806
Kreuzschattierung 814
Kreuzschraffen 809
Kreuzschraffur 809
Kriging 2059
Kriging-Varianz 2060
kritische Analyse 801
kritischer Bereich 803
kritische Stellen 802
krumme Zelle 3729
krummlinige Koordinaten 833

Kryptografie 817
kubische Abbildung der
   Umgebung 819
kubische Konvolution 818
kubischer Spline 820
kubischer Trend 821
Kugelfehlerwahrscheinlichkeit
   3435
Kugelkoordinaten 3437
Kugelprojektion 3488
kundenspezifische
   Produktgestaltung 837
kundenspezifisches Symbol 839
künstliche Grenze 186
künstliche Intelligenz 187
künstlich erzeugtes Objekt 2243
Kurs 791
Kursor 823
Kursrose 636
Kurs über Grund 1792
Kurve 826
Kurvenanpassung 830
Kurvenausgleich 830
Kurvenbild 1018
Kurvenblatt 1018
Kurvenfläche 828
Kurvenflug 3836
Kurvengenerierung 831
Kurvenleser 2174
Kurvenmesser 834
Kurvenmuster 832
Kurvenschreiber 2795
Kürzel 3288
kurzer Referenzplan 3292
kürzeste Strecke 3290
Kurzkupplung 575
Kurzstreckennavigation 3291
Küstengebiet 592
Küstenkartierung 590
Küstenlinie 594
Küstenstreifen 594
Küstenzonen-Farbscanner 593

ladbarer Treiber 2205
Ladefläche 330
Lage 1627, 2862
lagebezogener Verweis 2217
Lagefehler 2864
Lagegenauigkeit 2863
Lager 3238
Lagerbestandskarte 2005
Lagereferenz 2217
Lagereferenzsystem 2218
Lageregelungssystem 241
Lagerkarte 2005
Lagerung 330
Lagesymbol 2219
Land 2072
Landbedingungskarte 2076
Landeinheit 2096

Landesatlas 2487
Landeshorizont 2490
Landesinformationssystem 2083
landeskundlicher Atlas 2487
Landesplanung 785
Landesvermessung 3723
Landkarte 1611
Landkartenformat 223
Landkartensymbole 459
Landkartensymbolsätze 2340
Landkartentechnik 465
Landklassifizierung 2074
Landklassifizierungskarte 2075
Landkreis 784
Landkreisgrenze 787
Landkreismerkmale 788
Landkreisunterteilung 789
ländlicher Bezirk 784
Landmarke 2087
Landmarke-
   Identifikationsnummer 2088
Landmesser 2094
Landnutzung 2097
Landnutzung-
   Klassifizierungssystem 2099
Landnutzung-Managementsystem
   2100
Landnutzungsaussicht 2104
Landnutzungsgrenze 2098
Landnutzungskarte 2101
Landnutzungsplanung 2102
Landnutzungszone 2105
Landstrich 1148
Landsystem 2095
Landtauglichkeitskarte 2076
Landüberdeckung 2077
Länge 206
Länge-Linie 2185
Längenbtastung 97
Längenkreis 2395
längentreue Projektion 1312
Längen- und Winkel- 2152
Längen- und Winkelnetz 2153
Langläufer 2232
langsame Transaktion 2232
längslaufende Überprüfung 2231
Längsseite 3293
Längsüberdeckung 1465
langweiliger Abschluss 2232
Lärmkontur 2541
Laserplotter 2109
Latte 2112
Laufendhalten einer Karte 2325
Lauflänge 3188
Lauflängencodierung 3189
Lautschrift 3344
Layout 2123
Layoutfenster 2124
Lebensraum 1797
LED-Ploter 2147

leere Zelle 1262
Leerzeile 360
Legende 2278
Legendfenster 2135
Legendrahmen 2133
Legendtyp 2134
Leiste 3511
Leistungsvergleich 334
leitender Filter 1110
Leitfähigkeitskarte 683
Leitpflock 956
Leitpfosten 956
Leitung 2150
Leitungskartierung 3893
Leitungsnetz 3894
Leitweg 3177
Leitweglenkung 3182
Leitwegsuchen 3182
Lenkung 3182
Lichtmarke 823
Lichtquelle 2148
Liegenschaftskarte 2652
Liegenschaftskataster 422
Lineal 3187
lineare Anzeige 2177
lineare Datei 1436
lineare Interpolation 2157
lineare Optimierung 2163
lineare projektive Abbildung
   2918
lineare Referenzmethode 2164
lineare Regression 3303
linearer Fehler 2154
linearer Filter 2156
linearer Indikator 2177
linearer Interpolator 2158
linearer Maßstabsfaktor 2166
linearer Trend 2169
lineares Objekt 2162
lineares Referenzsystem 2165
lineare Streckung 2167
lineare Struktur 2168
lineare Triangulation 2170
Linie 2150
Linie gleicher magnetischer
   Mißwertung 2024
Linie-im-Polygon-Operation
   2178
Linienabschnitt 2189
Linienabstand 2191
Linienausdünnung 2194
Liniendurchlaß 2188
Linienereignis 2155
Linienerkennung 2172
Linienextraktion 2173
Linienglättung 2190
Liniengrafik 761
Linienjäten 2197
Linienkarte 2160
Linienkartierung 2181

Liniennetz 2161
Linienstil 2192
Liniensymbole 2193
Liniensymbol-Methode 2411
Linientyp 2196
Linienverdünnung 2194
Linienverfolgung 2195
Linienverfolgungsalgorithmus
   2175
Linienverlauf-Algorithmus 2175
Link 2199
Liste mit variabler Länge 3903
Listenfeld 2200
Listengenerator 3132
Literal 2201
Location-Allocation-Prozedur
   2216
logarithmische Skala 2224
logische Abfrage 2228
logische Auswahl 2230
logische Beziehung 2229
logische Genauigkeit 2226
logischer Anschluss 2227
logischer Verbinder 2227
lokale Datenbank 2210
lokale Entwässerungsrichtung
   2211
lokaler Gummiband 2207
lokales Einpassen mit
   Gummibandfunktion 2207
lokales GIS 2212
lokales Koordinatensystem 2209
lokales Netz 2206
lokales Netzwerk 2206
lokales Netzwerk von
   geologischer Aufnahme 1643
lokale Stetigkeit 2208
Lokalisierer 2221
Lokalisierung 2215
Lokalisierungsgitter 2214
Lokationsquotient 2220
Lookup-Tabelle 2233
löschen 953
Löschen 954
Löschung 954
lose Kopplung 2235
Lotabweichung 948
Lötauge 2073
Lötstelle 3677
Loxodrome auf der Erdkugel
   3160
Lücke 1510
Luftbild 72
Luftbildaufnahme 70
Luftbildkarte 2745
Luftbildtriangulation 71
Luftbildvermessung 64
Luftfahrtdaten 74
Luftfahrtkarte 73
Luftfotoplan 68

Luftperspektive 63
Luftprofilregistriergerät 86
Luftreihenaufnahme 62
Luftspektrofotometrie 69
Luftvermessung 70

Mäanderlinie 2376
Mäandervermessung 2377
Magnetfeld der Erde 1648
magnetische Abweichung 635
magnetische Deklination 635
magnetische Nordrichtung 2240
Mahalanobis-Distanz 2241
Mail-Adresse 2242
Makrobefehl 2239
Makroinstruktion 2239
Makrokommando 2239
Makros 2239
manuelle Digitalisierung 2245
Map-Algebra 2252
Map-Overlay 2291
mappen 2248
Mapping 2908
marginale Daten 2350
Marke 2061
Marker 2352
Markerform 2354
Markergröße 2355
Marker-Palette 2353
Markerzeichen 2356
Markierer 2352
markiert 2065
markierte Region 2063
markiertes Polygon 2062
Markierung 3257
Markierungsfeature 2357
Markscheidekunde 2427
Masche 2396
Masche-Knoten-Inzidenz 2398
Maschen-Adjazenz 2397
Maschinenintelligenz 187
Maske 2359
Maskieren 2360
Maskierung 2360
Massenberechnung 2363
Massenbilanz 2362
Massenmittelpunkt 509
Massenpunkt 2364
Maßstab 3210
Maßstab-Genauigkeit 3212
Maßstab-Genauigkeitsgrenze
    3213
maßstabgetreue Darstellung 3835
Maßstabsänderung 3221
Maßstabsbalken 3214
Maßstabsfaktor 3216
Maßstabsfehler 3215
Maßstabslinie 3214
Materialbilanz 2362
mathematische Funktion 2369

mathematische Kartenbasis 2367
mathematische Kartografie 2368
mathematische Modellierung
    2370
Matrix 2371
maximale Mutmaßlichkeit 2373
maximale Plausibilität 2373
maximiertes Fenster 2372
Maximum-Likelihood-
    Klassifizierung 2374
Mean-Filter 291
Mediane 2386
Median-Filter 2387
Meeresatlas 2610
Meereshöhe 1811
Meereskartografie 2351
Meeresspiegel 3237
Mehrbild 1172
Mehrdeutigkeit 108
mehrdimensionaler Raum 2460
mehrdimensionale Skalierung
    2459
Mehrfacheinordnung 805
mehrfache Regression 2465
Mehrfachregression 2465
Mehrkanal- 2455
Mehrkanalempfänger 2456
mehrkanalig 2455
Mehrkriterienanalyse 2457
Mehrkriterienentscheidungs-
    findung 2458
Mehrleitung 2464
Mehrmusterkachel 2478
Mehrschichtdarstellung 2119
Mehrskalendarstellung 2470
Mehrskalen-GIS 2469
Mehrstrecke 2462
Mehrstreckenfehler 2463
Mehrweg 2462
mehrwertige Fläche 2479
Melden 1934
Mengenanalyse von Gelände
    2968
Menü 2391
Menüelement 2392
Menüpunkt 2392
Mercator-Abbildung 3867
Mercatorprojektion 3867
Mercator-Transversale 3867
Meridian 2395
Meridian von Greenwich 2889
Merkmalgewinnung 1379
Merkmalsabbildung 1384
Merkmalsauswahl 1385
Merkmalseinheit 1391
Merkmalsextraktion 1379
Merkmalsextraktionsprozessor
    1380
Merkmalsextraktionssegment
    1381

Merkmalsidentifikator 1382
Merkmalskarte 1383
Merkmalsklasse 1377
Merkmalsraum 1388
Merkmalstrennung 1386
Merkmaltyp 1390
Messbildverfahren 2741
Messkunde 2389
Messlatte 3459
Messpunkt 2384
Messstelle 2384
Messtischaufnahme 2780
Messtischblatt 3550
Messunggitter 2383
Messung-Rahmenwerk 2380
Messwerterfassung 875
metachrones Blockdiagramm
    3690
Metadatei 2403
Metadaten 2401
Metadatenbasis 2402
Metainformation 2401
Metainformationssystem 2404
Metakartografie 2400
Metamorphismus 2405
meteorologisches Modell 2406
Methode der Bewegungssymbole
    2412
Methode der kartografischen
    Symbole 2409
Methode der kleinsten Quadrate
    2131
Methode des qualitativen
    Hintergrunds 2413
Methode des quantitativen
    Hintergrunds 2414
Methode von Erfüllen von
    Nebenbedingungen 707
metrisches System 2415
Mikrowellenträgersignal 2419
militärer Atlas 2420
militärer geografischer Atlas
    2420
militärgeografischer Dienst 946
Militärgitterreferenzsystem 2421
Millimeterpapier 1750
Minderung des Wertes 975
Minimal-Distanz-Abbildung 2426
minimale Distanz 2424
minimaler Abstand zweier
    Vertices bei geraden Linien
    3989
minimales begrenztes Rechteck
    2423
Minute 2428
Mischen 2399
Missklassifikation 2429
Missklassifikationsmatrix 2430
Missweisung 746
Mittel 2379

Mittelachse 507
Mittelachsedaten 508
Mittelfläche 2375
mittelgroße Karte 2388
Mittellinie 507
Mittelliniedaten 508
Mittelpunkt 511
Mittelpunkt der Kollineation 510
Mittelwert 2379
mittlere Greenwichzeit 758
mittlere Meereshöhe 2378
mittlere Seehöhe 2378
mobiles GIS 2437
Modell der Rasterdaten 3009
Modellierung der
    Naturerscheinungen 2500
Modellraum 2439
Modellversuch 2754
Modenfilter 3978
Modenrauschen 3425
Modenrauschenfilter 3426
Modentrennung 2438
modifizierte Projektion 2443
Mollweide-Projektion 2444
Monitorstation 2445
Montage 2453
Moor 368
Morast 368
Morgenempfang 2137
morphografische Karte 2082
Morphologie 2447
morphologischer Filter 2446
Morphometrie 2449
morphometrischer Parameter
    2448
Morton-Matrix 2450
Morton-Ordnung 2451
Mosaik 3681
mosaikartige Karten 3680
mosaikartiges Geländemodell
    3682
mosaikartiges Muster 3679
Mosaiken 2286
Mosaiking 2286
MOSS-Format 2292
Motiv 2716
Muldenlinie 3561
Multifunktionskataster 2467
Multi-Look-Radarbild 2978
Multimedia-GIS 2461
Multiplexempfänger 2466
Multisegment-Weg 2471
Multisensorbild 2472
Multiskalen-GIS 2469
multispektrale Daten 2473
multispektrale Fotografie 2475
multispektrale Pixel-zu-Pixel-
    Klassifizierung 2476
multispektraler Abtaster 2477
multispektraler elektrooptischer

Bildsensor 2474
Multispektralscanner 2477
multivariate Themakarte 2480
Mündungsgebiet 1322
Muster 3191, 3595
Mustererkennung 2717
Musterkachel 3679
Musterkachelung 3684

Nabe 1856
Nachbarpunkte 2514
Nachbarschaft 2929
Nachbarschaftsanalyse 2512
Nachbearbeitung 2874
Nachdigitalisierungsprobleme
    2873
Nacheilung 952
nachfolgende Facetten 51
nachhaltige Entwicklung 3552
Nachklassifizierung 2872
Nachschlagtabelle 2233
nächste Nachbarschaftsanalyse
    2508
nächster Knoten 2510
nächster Nachbar 2506
Nächster-Nachbar-Abtastung
    2509
Nächster-Nachbar-Algorithmus
    2507
Nacht 3238
Nachziehen 1178
NA-Code 2577
Nadir 2482
Nahaufnahme 581
Nähe 2929
Näherung 2929
Näherungsanalyse 2512
Näherungskarte 2930
Näherungsoperator 2513
Näherungszone 2932
Name 2483
Name der Karte 2287
Namengebung 2485
Namenoriginal 2484
Namensregister 2543
Narbe 3238
Nationalatlas 2487
nationale Geodatenbasis 2491
nationale
    Kartenpräzisionstandards
    2493
nationales Übertragungsformat
    2495
nationales Verbundnetz 2492
nativer Modus 2496
Naturbruchalgorithmus 2497
Naturlandschaft 2498
natürliche Kartenprojektion 2070
natürliche Netzwerke 2499
natürliche Projektion 1316

natürliches
    Ressourcenmanagement 2502
natürliche Umwelt 2498
Navigation 2503
Navigationssystem 2504
Navigator 2505, 3963
Nebel 1451
Nebenbedingung 3155
nebeneinander angeordnete
    Karten 3680
nebeneinander angeordnetes
    Geländemodell 3682
nebeneinander angeordnetes
    Muster 3679
Nebeneinander-Anordnung 3684
Nebenkarte 1962
Nebenwinkel 47
Neigung 1921
Neigungsberechnung 3320
Neigungskarte 569
Neigungsmaßstab 3321
Neigungsmesser 1922
Neigungspickel 3323
Neigungstyp 3324
Neigungswinkel 126
Nennwert 2544
Netzadresse 2515
Netzadresse-Translation 2516
Netzberechnung 2518
Netzerfassung 2520
Netzkarte 2528
Netzknoten 2529
Netzstruktur 2525
Netztopologie 3764
Netztrassieren 2531
Netzverbindung 2527
Netzverdichtung 2522
Netzverfolgung 2531
Netzwerkadresse 2515
Netzwerkanalyse 2517
Netzwerkdateisystem 2524
Netzwerkdatenmodell 2521
Netzwerkelement 2523
Netzwerkinformationssystem
    2526
Netzwerkkarte 2528
Netzwerkknoten 2529
Netzwerktopologie 3764
Neubestimmung der Grenzen der
    Landkreise 3059
Neueinteilung 3046
Neueinteilung der Wahlkreise
    3059
Neuordnung 3129
neuprofilieren 3148
Neuzuordnung 3046
Neuzuweisung 3042
nichtabgeschnitten 3840
nichtbeendete Linie 3850
nichtberichtigte Bilddaten 3874

nichterfasstes Gebiet 3873
nichtetikettiert 3869
nichtetikettiertes Polygon 2432
nichtexakter Interpolator 2549
nichtformatierte Daten 3852
nichtgroßstädtisches Gebiet 2552
nichthierarchische Datei 1436
nichtisometrische Linie 2550
nichtisotrop 132
nichtkontrollierte Klassifikation
     3876
nichtlineares Kriging 2551
nichtplanare Kreuzung 2553
nichtplanarer Knoten 2553
nichtplanarer Schnitt 2553
nichtplanares Netzwerk 2554
nichträumlich 190
nichträumliche Daten 192
nichträumliches Attribut 191
nichträumliches Objekt 193
Nichtstationarität 2555
nichtverbundener Bogen 3844
nichtverbundener Knoten 3845
nichtverbundener Text 3872
nichtvisuelle Digitalisierung 361
nichtzeitliche Daten 219
nichtzeitliche Datenbasis 220
Niederdrückung 976
niedriger Wasserspiegel 2237
Niedrigwasserlinie 2238
Niveau 2138
Niveaufläche 2145
Niveauliniendarstellung 2142
Nivellement 2141
Nivellierinstrument 2139
nivellierischer
     Schleifenschlussfehler 3934
Nivellierung 2141
Nivellierungsplan 727
NN 3237
NOAA 2494
Nomenklatur 2543
nominelle Orbitalebene 2545
Nomogramm 2546
Non-Earth-Koordinatensystem
     2548
Norden 2564
nördlich 2567
nördlicher Polarkreis 156
Nordosten 2566
Nordpfeil 2565
Nordpol 2569
Nordwest 2570
Normale 2560
normale Höhe 2558
Normalenform 2557
normale Polarität 2561
normales Schneiden 2563
Normalform 2557
Normalisierung 2559

Normal-Niveau 3237
Normalnull 3237
Normalprojektion 2562
Normalreflexion 3433
Normalverteilung 2556
Normierung 2559
Normung 2559
Notfallplanung 1261
Notizdatei 4003
NP-vollständiges Problem 2547
NTF-Format 2495
Nullhypothese 2573
Nullpunkt 2657
Nullpunkt des
     Koordinatensystems 2660
Nullwert 2574
Nullzelle 2572
numerieren
     neu ~ 3128
Numerierung 2575
Numerierungsbereich 2576
numerisch 1031
numerische Achsen 2582
numerische Daten 1040
numerische Farbteilung 1038
numerische Fotogrammetrie 1063
numerische Geochemie 1046
numerische Kartografie 1035
numerische Klassifizierung 1037
numerische Orthofotografie 1061
numerische Quelle 1068
numerischer Kurvenschreiber
     1065
numerischer Maßstab 3139
numerischer Plotter 1065
numerisches Austauschformat
     1044
numerisches Format 1045
numerisches Foto 1064
numerisches geografisches
     vektorbasiertes Format 1048
numerisches Landmaßsystem
     1056
numerisches Messbildverfahren
     1063
numerisches Tablett 1071
numerische Taxonomie 2583
numerische topografische Daten
     1074
numerische Unterschrift 1066
numerische Weltkarte 1036
Nutzersegment 3892
Nutzung 2609

obere Schicht 3701
Oberfläche 3538
Oberflächenbereich 171
Oberflächenelement 1242
Oberflächenschicht 3543
Oberflächenstück 3545

Oberflächentrendanalyse 3816
oberirdische Scheide 3710
Objekt 2584
     2D-~ 1156
     3D-~ 1157
     vom Menschen verursachtes ~
     2243
Objekt-Code 2586
Objektdarstellung 2599
Objektgeometrie 2589
Objektgestell 2588
Objektkatalogisierungs-
     methodologie 1376
Objektklasse 2585
Objektknoten 2600
Objektkombination 2587
Objektmanagement-Gruppe 2593
Objektmodell 2594
objektorientierte Datenbasis 2596
objektorientierte
     Datenmodellierung 2597
objektorientierte Verarbeitung
     von geografischen Daten 2598
Objektorientierung 2595
Objektschlüssel 2590
Objektschlüsselkatalog 2592
Observation 2607
Observationsdichte 970
Offenbarungsdistanz 2431
offene Datenbankschnittstelle
     2629
offene Fläche 2628
offene Höhenlinie 3841
offener Polygonzug 3843
offenes Polygon 3842
offenes System 2632
Offenlegungsdistanz 2431
öffentliche Information 2949
öffentliches
     Landvermessungssystem 2950
öffentliches terrestrisches System
     2951
Offline- 2615
Oktagonbaum 2613
Oktanenbaum 2613
Oktbaum 2613
OMG-Standard 2617
Online-Abfrage 2626
Online-GIS 3984
Online-Kartografie 2625
Online-Katalog 2624
Online-Zugriff 2623
Open-GIS-Konsortium 2631
Operationsforschung 2635
Operationssystem 2633
optimale Route 2636
optimaler Weg 2636
Optimalwegauswahl 3259
Optionenknopf 2638
Orbit 2639

Orbitalebene 2642
Ordinale 2648
Ordinalklasse 2647
Ordinalzahl 2648
Ordinate 2650
Ordnen 2645
Ordnung 2643, 2645
Ordnung der Einheitssortierung 1283
organisatorische Zuverlässigkeit 2654
orientierter Graph 1105
Orientierung 2655
Orientierung des Objekts 2595
Orientierungspunkt 2656
Originaldaten 3030
Originalkarte 2658
orografische Karte 2751
Ort 2762
orthodromische Linie 2661
Orthofoto 2666
Orthofotografie 2666
Orthofotokarte 2667
Orthofotoplan 2667
Orthofotoskop 2668
orthogonale Projektion 2663
orthogonaler Meridian 2662
Orthogonalprojektion 2663
Orthogonalraum 2664
orthometrische Höhe 1638
orthomorphe Projektion 686
Orthoprojektor 2668
Orthorektifikation 2669
Ortidentifizierung 2765
Ortleitung 2766
örtliche Konsistenz 2208
örtliche Lage 2213
Örtlichkeit 2213
Ortsbestimmung 2868
Ortscode 2764
Ortsgrenze 2763
Ortskennzahl 161
Ortslage 1627
Ortsnetzkennzahl 161
Ortsplan 3312
Ortsteil 1148
Ortung 2215
Oval 2676
Overlay 2680
ozeanografische Erhebung 2612
ozeanografisches Modell 2611

Paarigkeitsvergleich 2365
Packung 2685
Palette 2687
Palettedatei 2688
Palmtop 2689
Pampa 2690
panchromatischer Bereich 2692
Panning 2694

Pantograph 2695
Papierkarte 2697
papierlos 2696
Parallaxe 2698
Parallelflächner-Klassifizierung 2700
Parallelkreis 2701
Parallelprojektion 2702
parametrische Fläche 2704
parametrische Teste 2705
parametrisiertes Bild 2703
Parasitenpolygon 3452
Parkettierung 3684
Parsing 2706
Partialsumme 3528
partikulärer Maßstab 2707
Partyzelt 3258
Parzelle 2089
Passpunkt 1786
PC 2730
P-Code 2879
Peano-Hilton-Ordnung 2719
Peano-Kurve 2721
Peano-Nummern 2720
Peanosche Kurve 2721
Peanosche Nummern 2720
Pegel 2138
Perimeter 2726
periphere Einheiten 2727
Peripherie 2672
Peripheriegeräte 2727
persistente Sperre 2729
Personalcomputer 2730
Personalisierung 837
persönliche Datentransposition 2897
persönliche Kartografierung 2731
persönlicher Rechner 2730
Perspektive 690
perspektivische 3D-Ansicht 1175
perspektivische Aerofotografie 2601
perspektivische Projektion 690
perspektivisches Plotten 2733
perspektivisches Rendering 2734
perspektivisches Zeichengerät 2732
perspektivisches Zeicheninstrument 2732
Perspektivität 690
Pfad 3177
Pfadanalyse 2709
Pfadbestimmung 2710
Pfadbeurteilung 2710
Pfadname 2712
Pfadselektion 2715
Pfadsuchwahl 2710
Pfeil 185
Pflanzenweltkarte 2793
Phasenmehrdeutigkeit 108

Phasenmessungsmethode 2736
Phasenrastung 2735
Phasenregelung 3074
Piktogramm 1881
Pilotprojekt 2754
Pilotvorhaben 2754
Pin-Karte 2755
Pipeline 2756
Piquerloch 2757
Pixel 2758
Pixel-Interpolator 2760
Pixel-Klassifizierung 2759
Pixelraster 3015
Pixel-zu-Pixel-Klassifizierer 2761
Plan 992
planare Fläche 2773
Planargraph 2767
Planarnetz 2771
Planarnetzwerk 2771
Planarpolygon 2772
planen 2249
Planet 2779
planetarische Kartierung 2782
planetarischer Erdball 2781
planetarische Weltkugel 2781
Planimeter 2783
Planimetrie 2791
planimetrische Basis 2784
planimetrische Daten 2786
planimetrische Karte 2788
planimetrisches Bild 2787
Planquadrat 2318
Planung der Bodennutzung 2102
Planung-Informationssystem 2792
Planzeichen 459
Plate-Carree-Projektion 1310
platter Abfall 1434
Platz 2762
Platzcode 2764
Plazierung 2215
Plottausgabe 2794
Plottbereich 2796
Plottengeschwindigkeit 2797
Plottensystem in der universalen Projektion 3864
Plotter 2795
PLZ 2871
Pointer 2801
Polardistanz 2818
Polarität 2819
Polarkoordinaten 2816
Polarkoordinatensystem 2817
Polarprojektion 2820
Polhöhe 205
politische Grenzen 2821
Polyederabbildung 2851
polyedrische Abbildung 2851
Polygon 2823
    3D-~ 1176

Polygonabdeckung 2837
Polygon-Adjazenz 2824
polygonaler Fehler 2828
Polygonapproximation 2825
Polygonattributtabelle 2832
Polygonauflösung 2838
Polygonauflösung mit
   Vereinigung 2838
Polygon-basierte Modellierung
   2833
Polygonbildung 2844
Polygonbogen 2826
Polygon-Bogen-Topologie 2831
Polygon-Converter 2836
Polygonfüllblock 2840
Polygonfülleinheit 2840
Polygonfüllung 2839
Polygonfüllungsblock 2840
Polygonfüllungseinheit 2840
Polygongenerierung 2841
Polygongrenze 2834
Polygon-Identifikator 2842
Polygonierung 2844
Polygon-im-Polygon-Operation
   2843
polygonizieren 2845
Polygonkante 164
Polygonkappen 2827
Polygonmasche 2846
   3D-~ 1177
Polygonnetz 2829
Polygonschaltung 2835
Polygonschnitt 2827
Polygontopologie 2849
Polygon-Triangulation 2850
Polygonüberlagerung 2847
Polygonverarbeitung 2848
Polygonzerlegung 2844
Polygonzug 2830
Polygonzugsdaten 3810
polykonische Projektion 2822
Polylinie 2852
Polylinie mit Form der Kurve
   829
Polymorphismus 2853
POLYVRT 2836
Population 2854
Populationsdichte-Karte 2856
Popup-Menü 2859
Portabilität 2861
Position 2862
Positionieren 2869
Positioniergenauigkeit 2863
Positioniertoleranz 2865
Positionierung 2869
Positionsanzeiger 2804
Positionsbestimmung 2868
Positionsdarstellung 2867
Positionsfehler 2864
Positionsgenauigkeit 2863

Positionsgenauigkeitsabfall 2866
Positionstoleranz 2865
Postleitzahl 2871
Postleitzahl-Bereich 4022
Postleitzahl-Grenze 4023
Potentialfläche 2145
Potentialflächeberechnung 2875
Potentialwert 1670
Prädiktion 1455
Präzedenz 2876
Präzision 2878
Präzision der Zeit 3685
Präzision des Standards 3460
Präzisionortungsdienst 2877
Präzisionsdigitalisierung 2880
Präzisionsnormen 19
Präzisionstablett 2881
Präzisionstandards 19
Presseindruck 1918
Pressung 2886
primäres digitales
   Oberflächenmodell 2887
Primärschlüssel 2888
Primitiv 1274, 2890
Prinzip des geringsten Aufwands
   2893
Privathaushalt 1371
Probe 3191
Probenanalyse 3193
Probenzeichnung 2919
Probestreifen 3789
Problem der Gebietseinheit 2442
Problem der vermischten Pixel
   2436
Problem des Handlungsreisenden
   3808
Problem des kostengünstigen
   Leitweges 2129
Profil 812
Profilaufzeichnung 2903
Profillinie 732
Profilliniendarstellung 733
Profilllinieninterpolations-
   programm 729
Profilregistriergerät 2902
Prognostizierung 1455
Projekt 992
projektierte Toleranz 2906
Projektierung 994
Projektion 2907
Projektionsachse 292
Projektionsdatei 2913
Projektionsfläche 2915
Projektionskonvertierung 2909
Projektionskoordinaten 2910
Projektionslinie 2187
Projektionsmaßstab 3217
Projektionsparameter 2914
Projektionszentrum 510
projektive Algebra 2916

projektive Geometrie 2917
projektive Transformation 2918
Projektor 2911
Projizieren 2908
Propagierung 2920
proportionale Objekt-
   Größenänderung 2922
Proportionalität 2921
Proportionalsymbole 2923
Proportionieren 2924
Protokoll 2925
Prototyp 2926
Provinz 786
proximale Toleranz 2928
Proxydaten 2933
Prüffläche 3622
Prüfmarke 333
Pseudofarbbild 2936
Pseudofarbänderung 2937
Pseudofarbenänderung 2937
Pseudofarbanreicherung 2935
Pseudofarbenanreicherung 2935
Pseudofarbenerhöhung 2935
Pseudofarberhöhung 2935
Pseudokegelprojektion 2938
Pseudoknoten 2940
pseudokonische Abbildung 2938
Pseudoperspektive 1370
pseudoskopische Betrachtung
   2945
Pseudostatik 2946
Pseudostrecke 2942
Pseudostreckenmessung 2943
Pseudostreckennavigation 2944
Pseudo-Wahlkreis 2947
pseudozufälliger Rauschencode
   2941
Pseudozufallsstörungscode 2941
Pseudozylinderprojektion 2939
psychomotorischer Fehler 2948
Pufferberechnung 408
Pufferdistanz 409
Puffergröße 411
Puffer mit konzentrische Kreise
   674
Puffern 410
Pufferung 410
Pufferzone 412
Pulldown-Menü 2953
Punkt 1166, 2798
Punktabbildung 2806
Punktadresse 2799
Punkt-Aggregation 2800
Punktaufnahme 3549
Punktbeschriftung 2802
Punktdatei 2803
Punktdiagramm 3228
Punktdichte 971
Punktdichte-Karte 1167
Punkte pro Zoll 1169

Punkt-im-Polygon-Operation 2805
Punktmarke 1166
Punktmethode 1168
Punkt-Modus des Digitalisierens 2807
Punkt-Nachbarschaft 2808
Punkt-Näherung 2808
Punktobjekt 2809
Punktobjekt-Etikettenstellung 2802
Punktsymbole 2812
Punktübertragung 2815
Punktvermessung 3549
Punkt-zu-Punkt-Sichtbarkeit 2814
Punkt-zu-Zone-Konvertierung 2813
pyramidale Schichten 2955
pyramidales Diagramm 2954

Quadrangel 2959
Quadrant 2960
Quadrant-Bildzeile 2970
Quadrantfang 2961
Quadrat 3455
Quadrat-Anlegung 3457
Quadratdatei 2957
quadratischer Mittelwert 3172
quadratischer Spline 2963
quadratischer Trend 2964
quadratisches Netz 3456
quadratisches Polynom 2962
Quadratname 2958
Quadratnetz 3456
Quadtree 2965
Quadtree-Gebiete 2966
Quadtree-Struktur 2967
Quantelung 1123
Quantisierung 1123
quantitative Geländeanalyse 2968
Quasigeoid 2972
quasi-geozentrische Koordinaten 2971
Quellenkarte 3345
Quellentabelle 2975
Quelltabelle 3347
Quer- 3807
Querabtastung 22
querachsige Projektion 3806
Querkorrelation 806
Querkorrelation-Matrix 807
Querschnitt 812
Querschnitt-Flussdiagramm 813
Quertabulierung 815
Querüberdeckung 2111
Query-by-Example 2973

Radarfernabtastung 2979
Radarfernerkundung 2979
Radar mit synthetischer Apertur 3562
Radialfehler 2981
Radialleitung 2982
Radialverzerrung 2980
Radiometer 2985
Radiometer mit sehr hoher Auflösung 61
Radiometrie 2991
radiometrische Auflösung 2989
radiometrische Empfindlichkeit 2990
radiometrische Kalibrierung 2986
radiometrische Korrektur 2987
radiometrische Verzerrung 2988
Radiopositionierung 2992
Radioschaltfläche 2984
Radius 2994
Rahmen 374
Rahmenmarken 1396
Rahmenwerk 1476
Rahmung 1478
Rahmungssystem 1479
Rand 377, 1216
Randanpassung 1223
Randbedingung 1218
Randbögene 375
Randdarstellung 391
Rand einer Fläche 388
Rand eines Bereichs 387
Rand eines Gebiets 387
Rand eines partikulären Bereichs 389
Randerkennung 1219
Randerkennungsfilter 1220
Ränderung 728
Ränderzuordnung 379
Randkante 380
Randpunkt 390
Randsatz 392
Rangfolge 176
Ranggröße 2646
Rangordnung 176
Raster 3005
Rasterabtastung 3024
Raster-Algebra 3006
Rasterauffrischung 3022
Rasterbearbeitung 3011
Rasterbild 3015
Rasterbildprozessor 3017
Rasterbildverarbeitung 3016
Rasterdarstellung 3023
Rasterdatei 3012
Rasterdaten 3008
Rasterdichte 3010
Rasterelement 1764
Rasterfang 3025
Rasterformat 3013
Rasterhintergrund 3007
Rasterkarte 1772
Rasterlinie 3018

Rasterobjekt 3019
Rasterplotter 3020
Raster-Prozessor 3017
Rasterpunktabfühlung 3024
Rasterpunktdaten 1775
Rasterraum 3026
Rasterscannen 3024
Raster-Steuerungsprogramm 3014
Rasterstruktur 3027
Raster-Themakarte 1782
Rasterung 3923
Raster-Vektor-Editor 1858
Raster-Vektor-Konvertierung 3028
Raster-Vektor-Konvertor 3029
Rastervorverarbeitung 3021
Rasterzelle 1764
Raum 3352
Raumauflösung 3414
raumbezogene Daten 3370
raumbezogene Grundelemente 3968
raumbezogener Zugriff 3361
Raumbezugsdaten 3370
Raumbezugsdatendarstellung 3375
Raumbezugsdaten-Generalisierung 3372
Raumbezugsdaten-Infrastruktur 3373
Raumbezugsdatenquellen 3376
Raumbezugsdatenstruktur 3377
Raumbezugsdatentransformation 3380
Raumbezugsdaten-Übertragungsstandard 3379
Raumdatenübertragungsprozessor 3378
Raum der Trajektorien 3787
Raumeinheit 3419
Raumelement 1243
Raumfahrzeug 3356
Raumfrequenz 3386
Raumindex 3387
Raumindizierung 3388
Rauminformationssystem 3390
Raumkoordinaten 3368
räumliche Abfrage 3407
räumliche Adresse 3363
räumliche Analyse 3364
räumliche Autokorrelation 3366
räumliche Beziehung 3413
räumliche Daten 3370
räumliche Datenbasis 3371
räumliche Datenbearbeitungssprache 3374
räumliche Einheit 3419
räumliche Glättung 3416

räumliche Interpolation 170
räumliche Korrelation 3369
räumliche Modellierung 3397
räumliche Näherungen 3398
räumliche Ordnung 3403
räumliche Organisierung 3404
räumliche Polarkoordinaten 3437
räumliche Position 3395
räumliche Primitivelemente 3968
räumliche Referenz 3409
räumliche Referenzdaten 3410
räumliche Referenzierung 3412
räumlicher Index 3387
räumlicher Rückwärtsschnitt 3393
räumlicher Verweis 3409
räumliches Attribut 3365
räumliches Auflösungsvermögen 3414
räumliches Bezugssystem 3411
räumliches Bild 3406
räumliches Datenbanksystem zur Entscheidungsfindung 3381
räumliches Datum 3382
räumliches Denken 3408
räumliches Informationssystem 3390
räumliches Netz 3367
räumliches Objekt 3399
räumliches Overlay 3405
räumliches Schließen 3408
räumliche Statistik 3417
räumliche Struktur 3418
räumliches Wechselwirkungsmodell 3392
räumliche Verbindung 3394
räumliche Wechselwirkung 3391
räumliche Zugriffsmethode 3362
räumliche Zusammenarbeit 3391
Raum-Metadaten-Standards 711
Raummuster 1617
Raumobjekt 3399
Raumobjektkontroller 3400
Raumobjektstruktur 3401
Raumordnung 2086
Raumplanung 175
Raumprimitive 3968
Raumschiff 3356
Raumsegment 3354
Raumsektor 3354
Raumvektor 3355
Raum-Zeit-Konzeptdarstellung 3692
raum-zeitliche Daten 3420
raum-zeitliche Datenbasis 3421
raum-zeitliches GIS 3422
Raumzugriff 3361
Rauschabstand 3296
Ray-Casting 3031
Ray-Tracing 3033

R-Baum 3185
Realismus 3035
Realitätsvisualisierung 3036
Realzeit-Kinematik 3039
Realzeit-Ortsbestimmung 3040
Realzeit-Positionieren 3040
Rechenblatt 3450
Rechenfehler 658
Rechnungsfehler 658
rechteckige Facetten 3051
rechteckiges Gitter 3052
rechteckiges regelmäßiges Gitter 3053
Rechtswert 1210
Rechtwinkelkoordinaten 433
Redistricting 3059
Reduktion 3061
Reduktion der Daten 908
Redundanz 3062
Reduzierung 3061
Reduzierung-Maßstabsfaktor 3220
referentielle Integrität 3075
referentielle Konsistenz 3075
Referenz 3064
Referenzebene 925
Referenz-Ellipsoid 3067
Referenzempfänger 3072
Referenzgitter 3068
Referenznetz 3071
Referenzplan 3069
Referenzpunkt 2811
Referenzstation 318
Referenzsystem 3073
Referenzzahl 428
regelmäßiger Fehler 3567
regelmäßiges Gitter 3094
regelmäßiggezeigte Daten 3095
regelmäßiggezeigte Profile 3096
Regelwidrigkeit 2014
Regenkarte 2996
Regenmesser 2995
Regierungsbezirk 56
Region 3079
regionale Abdeckung 169
regionale Datenbasis 3081
regionale Karte 3085
regionaler Atlas 3080
regionales GIS 3082
regionales Koordinatensystem 2209
Regionalismus 3083
Regionenzahl 2579
Regionfüllung 3086
Regionkarte 3085
Regionsspezifikation 3088
Regionstil 3089
Register 473
Registrierpunkt 3676
registrierte Tabelle 3090

Registrierung 3091
Regression 3092
Regressionskurve 3093
Reichweite 2999
Reichweiten-Karte 3000
Reinzeichnung 1363
Reklassifikation 3046
reklassifiziertes Bild 3047
Rektifikation 3054
Rektifizierung 3054
rekurrenter Code 518
Rekurrenzintervall 3056
Rekursion 3057
rekursive Absteigemethode 3058
relationale Datenbasis 3099
relationale kartografische Datenbasis 3098
relationales Datenbasis-Managementsystem 3100
relationale Verbindung 3102
Relationsattribut 255
relative Genauigkeit 3104
relative Höhe 3105
relative Lokalisierung 3106
relative Navigation 3107
relative Positionierung 3109
relativer Positionsanzeiger 3108
relative Steuervorrichtung 3108
relatives Zeigegerät 3108
Relaxation 3110
Relief 3605
Reliefdarstellung durch Höhenlinien 3136
Reliefdarstellung durch Schraffen 3137
Reliefeffekt 1260
Relieffläche 3542
Relieffotografie 2744
Reliefkarte 3114
Reliefschattierung 3115
Relieftyp 1389
Relief-Vektor 1392
Rendering 3127
Rendern 3127
Reportgenerator 3132
Reportgenerierung 3133
repräsentative Merkmalseinheiten 3135
Reservation 3147
Reservefläche 2672
Reservierung 3147
Ressort 973
Restfehler 3848
Restriktion 3155
Rhumblinie 3160
Richtigkeit 777
Richtpunkt 1112, 2352
Richtung 1108
Richtungsableitung 1109
Richtungspunkt 1112

Richtungsverzerrung 1111
Richtungswinkel 329
RINEX-Format 3044
Ring 3162
RIS 3390
Rohdaten 3030
Rohfehler 367
Rollbalken 3234
Rollen 3235
Rollkugel 3777
Rollmenü 2953
Rotation 3176
Rotationsellipsoid 1256
rotiertes Bild 3175
Route 3177
Routenoptimierung 2713
Routing 3182
Rückblick 304
Rücken 3161
Rückenansicht 304
Rückkehr 3170
Rückland 1838
Rückstreuung 303
Rückwärtseinschneiden 3145
Rückwärtseinschnitt 3145
rundes Optionsfeld 2984

Sachdaten 250
Sammeln 429
Sammlung 429
Satellitenaufnahmen 3202
Satellitenbild 3200
Satellitenbilddaten 3197
Satellitenbildkarten 3353
Satellitenbildklassifikation 3201
Satellitendaten 3197
Satellitengeodäsie 3198
Satellitengeometrie 3199
satellitengestützte Ortung 3204
Satellitenkarten 3353
Satellitennavigationssystem 3203
Satellitenstatusanzeiger 3205
Satellitentriangulation 3206
Sättigung 3207
Satz 3184
Satz-Identifikator 3050
Scan 3225
Scannen 3225
Scanner-Anpassungsprozessor 3224
Scanning 3225
Schablone 3595
Schachtelung 1259
Schärfefilter 1731
scharfkantige Spline 3668
Schatten 3276
Schattenkarte 3277
Schattenmatrix 3278
Schattieren 3274
schattiertes Reliefbild 3272

schattiertes Symbol 3273
Schattierung 3274
Schätzwert 1320
Schätzung 1321
Schätzungswert 1320
Schaubild 1018
Scheibe 3317
Scheibenpegel 3318
Scheibenteil 3317
Scheide 1152
Scheitel 2718
Scheitelwinkel 3930
schematische Ansicht 1019
schematische Karte 3229
Schicht 2116
Schicht-basierendes GIS 2117
Schichtenlinienkarte 727
Schichtenteilung 2121
Schichtgefrierung 1483
Schichtindex 2120
Schichtlinie 732
Schichtlinienkarte 727
Schichtsteuerung 2118
Schicht-Umordnung 3130
Schichtung 2121
Schichtungsebene 2777
Schieben 1126
Schieberegler 3234
schiefe Ansicht 2605
schiefe Projektion 2603
Schiefheit 214
Schieflaufende 3160
schiefwinklige Projektion 2603
Schiftung 1126
Schlagen eines Kreises 546
Schleier 1451
Schleife 2234
schließen 574
Schlüssel 2055
Schnappschuss 3330
Schneiden 2121
Schnelldigitalisierung 2976
schneller Zugang 3288
Schnellplotten 2977
Schnellstatistik 1372
Schnitt 840, 3248
Schnittfenster 572
Schnittmenge 2001
Schnittpegel 3318
Schnittstelle 1982
Schnürbodenverfahren 2223
Schottenmuster 810
Schraffen 3275
Schraffennegativ 2182
Schraffieren 3275
schraffierter Hintergrund 1806
Schraffur 3271
Schraffurfrequenz 1807
Schraffurlinien 1808
Schräge 1921

Schrägheit 2606
Schrägprojektion 2603
Schrägschattierung 2604
Schrägschummerung 2602
Schranke 1216
Schreibmarke 823
Schrift 526
Schriftart 526
Schriftpalette 1453
Schrittschablone 3595
schrittweise Dimension 1473
schrittweise lineare Klassifikation 3483
Schummerung 1837
Schummerungsbild 1836
Schwankung 1449
Schwarzweiß-Aerofotografie 359
Schwelle 3671
Schwellenwert 3671
Schwenken 2694
Schwere-Abdeckung 1754
Schwerenetz 1755
Schwereoberfläche 1756
Schwerpunkt 159
Seekarte 1862
See- und Sumpfkarte 2069
Seevermessung 1864
Segment 3250
Segmentierung 3251
Segmentnummer 3253
Sehfeld 1403
Sehne 535
Sehweite 3003
Sehwinkel 127
Seilzug 2199
seismische 3D-Aufnahme 1188
Seitenaufriss 3293
Seitenextent 2686
Seitenhalbierende 2386
Seitenüberlappung 2111
Seitenverbindung 2110
seitenverkehrte Topologie 2132
Sekante 3242
Sektion 3248
Sektor- 168
Sektordiagramm 2753
sekundäre Daten 3244
Sekundärschlüssel 101
Sekunde 3243
Selbstindizierung 268
Selbstinkrement 267
Selbstkorrelation 266
Selektion 3257
selektive Verfügbarkeit 3260
Sender 3799
senkrechte Parallelprojektion 2663
senkrechte Schweissposition 3939
Sensor 3265
sequentielle Datei 3267

sequentieller Zugriff 3266
serieller Zugriff 3266
Sextant 3270
Shadow-Map 3277
Shape-Datei 3282
Shortcut 3288
Sicherheitsbehälter 709
Sicherheitsmaßnahme 1261
sicherstellen 3208
Sicherstellungsdatei 2225
sichtbare Schicht 3956
sichtbares Feature 3955
Sichtbarkeit 3953
Sichtbarmachung 3962
Sichtbereich 3945
Sichtbereichanalyse 3946
Sichtkarte 3954
Sieb 1422
Siebkettung 306
Siedlungskern 2571
Signal-Rausch-Verhältnis 3296
Signal-Stör-Verhältnis 3296
Signatur 3297
Signaturen 459
Simulation 3305
Simulation von Strömungen 1448
Simulierung 3305
Situationsdarstellung 2790
Situs 3314
Skala 3211
skalares Feld 3209
Skale 3211
Skale ändern 3144
Skalenfaktor 3216
Skalenfehler 3215
Skalenstrich 3674
Skalierung 3221
Skelettfläche 2375
Skriptsprache 3233
Software 3336
Software für geografische
    Koordinatentransformation
    1519
Software zur Speicherung der
    Informationen in der Karte
    3337
Sondergrenze 3423
Sonderkarte 3424
Sonnennähe 2725
Sortieren 3342
sortieren 3341
Sortieren in aufsteigender
    Reihenfolge 189
Sortierung 3342
Soundex-Code 3344
sozioökonomische Daten 3333
sozioökonomische Karte 3334
sozioökonomische Zone 3335
Spaghetticode 3357
Spaghettidaten 3358

Spaghetti-Modell 3359
Spalte 1510
Spalte 620
spalten 3871
Spaltenbilder 621
Spaltenindex 622
spaltfreie Schicht 3240
Spaltung 3445
Spannweite 3360
spatiale Filterung 3385
spatiale Frequenz 3386
spatiale Interpolation 170
spatiale Metadaten 3396
Speicher 3500
Speicheruferlinie 594
Spektralauflösung 3431
spektrale Helligkeit 3429
spektrale Klasse 3430
spektrales Attribut 3427
spektrales Band 3428
spektrale Unterschrift 3432
Sperrung 2222
Sperrzone 3796
Spezifikation für offene
    Geodaten-Interoperabilität
    2630
sphärische Abbildung 3436
sphärische Koordinaten 3437
sphärische Polarkoordinaten 3437
sphärische Projektionen 3438
sphärische Verzerrung 3434
sphäroidische Geodäsie 3440
Spiegeln 3433
Spiegelreflexion 3433
Spiegelung 3433
Spinner 3442
Spitze 2718, 3441, 3928
Spitzenwinkel 3930
Spitzzirkel 1153
Spline 3443
Spline-Anpassung 3444
Spline-Fitten 3444
Splineglättung 3444
Splitten 3445
Splitter 3319
Splitterpolygon 3319
Splitting 3445
Sprache der strukturierten
    Abfragen 3515
Spread-Spektrum 3451
Spur 3775
SQL-Abfrage 3454
SQL-Bilder 3453
staatliche Einheit 1717
Stab 314
Städtebau 3885
städtebauliche Planung 3885
städtebaulicher
    Entwicklungsbereich 3880
städtebauliches GIS 3881

Städtegruppe 3884
Stadtgebiet 2416
Stadtgrenze 2417
Stadtkern 512
Stadtmitte 512
Stadtparzelle 547
Stadtplan 548
Stadtplanung 3885
Stadtrandgebiet 2672
Stadtteil 1148
Stadtteilgrenze 1149
Stadtviertel 1148
Standardabweichung 3461
Standardaustauschformat 3463
Standardbreitengrad 3466
Standardlinie 3465
Standardortungsdienst 3467
Standardrastergrafik 3464
Standlinie 317
Standort 1627
Standortanalyse 3311
Standortbestimmung per Satellit
    3204
Standortfeststellung 2868
Standortkarte 3312
Standortortungssystem von
    Fahrzeugen 3926
Standort-Tauglichkeitsanalyse
    3313
Standregelung 2140
Statik 3471
Stationarität 3472
Statistik-Fenster 3481
statistisch äquivalente Entität
    3477
statistische Entität 3475
statistische Fläche 3480
statistische Grafik 3476
statistische Modellierung 3478
statistischer Klassparameter 561
statistischer Raum 3473
statistisches Mittel 3474
statistisches Moment 3479
statistische Zone 3473
Stechzirkel 1153
Stehachsenlibelle 1844
stehendes Haus 1855
Steigung 1921
Steigungsmesser 1922
Steilabfallgebiet 3796
Steilhangssymbole 565
Steilheit 3482
Stelle 2762
Stellung 2215
Steradiant 3484
Stereoauswertegerät 3492
Stereo-Digitalisierer 3486
Stereoelevation 3487
Stereofoto 3491
Stereofotogrammetrie 3490

stereografische Abbildung 3488
stereografische Projektion 3488
stereografischer Plotter 2746
Stereokompilierung 3485
Stereomessbildverfahren 3490
Stereomodell 3489
Stereopaar 3491
Stereoplotten 3493
Stereoplotten-Festpunktnetz 1477
Stereoplotter 3492
stereoskopische Projektion 3494
stereoskopischer Plotter 3492
stereoskopisches Foto 3491
stereoskopisches Plotten 3493
Stereotriangulation 71
Sternbild 705
Sternhöhenmesser 202
Sternkarte 478
stetige Variable 725
Stetigkeit 716
Steuerhebel 2050
Steuerknüppel 2050
Steuerpunkt 739
Steuerpunkt-Liste 741
Steuerpunktzahl 2578
Steuersegment 742
Steuerstand 743
Steuervorrichtung 2804
Steuerzeichen 738
Stichbreite 2678
Stichprobe 3191
Stichprobenauswahl 3194
Stichprobenerhebung 3194
Stichprobenmittel 3192
Stichprobenmittelwert 3192
Stichprobentransformation 3196
Stiftplotter 2724
stillschweigender
    Erbschaftsverzicht 9
Stimmbezirk 3972
Stimmenthaltung 9
Stoffbilanz 2362
Strahlenbahn 3032
Strahlengang 3032
Strahlenverlauf 3032
Strahlung 2983
Strahlverfolgung 3033
Strahlweg 3032
Straßenindex der TIGER-
    Erhebung 3678
Straßenkarte 3168
Straßennetz 3509
Straßennetz-Datenbasis 3510
Straßenüberführungs- und
    Unterführungsinformation
    2677
Straßenverkehrsgesetz 2053
strategische
    Entscheidungsfindung 3503
Stratifikation 3504

Stratifikationsfläche 2777
stratifiziertes Kriging 3505
Stratigrafie 3507
stratigrafische Diskordanz 3506
Strecke 3248
Streckenlänge 2711
Streckenmessung 3004
Streifen 3511
Streifendiagramm 315
3D-~ 928
Streudiagramm 3228
Streukreis-Ausrichtungsfehler
    545
Streukreis-Lagefehler 545
Streukreisradius 544
Streuungsdiagramm 3228
Strich 314
Strichlinie 2722
Strichmarkierung 851
Strichoriginal 999
Strichvorlage 999
String 527
Strom
    mit dem ~ 1174
stromabwärts 1174
stromabwärts gelegen 1174
stromabwärts gerichtet 1174
stromaufwärts 3879
Stromlinie 1445
Strömungsrichtung 1444
Struktur 3512
Struktur der thematischen Daten
    3640
strukturelle Höhenlinie 732
strukturierte Daten 3513
strukturiertes Höhenlinienmodell
    3514
Strukturkarte 3527
Strukturwerte 960
Studie über die Durchführbarkeit
    1373
Stützhöhenlinie 3537
Stützpunkt 2811
Subbaum 3530
Subpixel-Klassifizierer 2761
Substitution 3526
subtraktive Primärfarben 3529
Suche 3241
suchen 1424
Suchen 3241
Suchprogramm 3963
Süden 3348
südlicher Polarkreis 135
Südost 3349
Südpol 3350
Südwest 3351
Suits-Wagner-Klassifikation 3532
Sumpf 368
Sumpfgebiet an der Küste 591
Sumpfgebiet in Küstennähe 591

Superposition 3534
Symbol 525
Symboldatei 3555
symbolisierte Punkte 3557
Symbolstil 3559
Symboltabelle 3560
Symbolüberdeckung 3558
Symbolumgebung 3554
Symbolverschiebung 2454
synchronisieren 3097
syntaktische Analyse 2706
Syntaxanalyse 2706
synthetische Auflösung 3565
synthetische Daten über das
    Gelände 3566
synthetische Karte 3563
synthetische Kartierung 3564
synthetischer Aperturradar 3562
synthetische Themakarte 3646
systematischer Fehler 3567
System der
    Geodatenwiedergewinnung-
    und Analyse 1606
Systemkartierung 3568
Systemtabellen 3569
System zur automatischen
    Kartierung 285

tabellarische Darstellung 3577
Tabelle 3570
Tabelle mit direktem Zugriff
    2203
Tabellendaten 3575
Tabellendatendarstellung 3576
Tabellenstruktur 3573
Tabellierung 3577
Tablett 1071
Tabulation 3577
Tachymeter 3578
Tachymeterlatte 3459
Tafel 3570
Tafelchen 1071
Tagesdatum 920
Talweg 3561
Tangente 3582
Tangentenkegel 3580
Tangentialkegel 3580
Tangentialkrümmung 1546
Targeting 3585
Taste mit doppelter Funktion 101
Taxonomie 3589
taxonomische Klassifizierung
    3588
technische Dokumentation 3590
technische Karte 1272
technische Zuverlässigkeit 3591
Teilbereich 3516
Teilblock 3517
Teilgebiet 3516
Teilkarte 1962

Teilklasse 3518
Teilmenge 3524
Teilstrich 3674
Teilstriche 1753
Teilstück 2089
Teilsumme 3528
Teilung 1154
teilweise Dimension 1473
Telemetrie 3593
Teleskopalhidade 3594
Tensor 3604
Terrain 3605
Terraindarstellungssystem 3609
Terrainerstellung 3607
Terraingeometrie 3612
Terrainmodell 3614
Terrain-Rendering 3615
Terrainschaffung 3607
Terrainvisualisierung 3618
terrestrische Kartierung 3620
terrestrische Koordinaten 3619
terrestrischer Erdball 1209
terrestrische Standlinie 3964
Texel 3634
Text 3623
Textattribut 3624
Textattributtabelle 3625
Textdaten 3632
Textetikett 3628
Textfeld 3626
Texthülle 3627
Textrichtigkeit 3631
Textstil 3629
Textsymbol 3630
textuelle Genauigkeit 3631
textuelle Information 3632
Textur 3633
Textur-Element 3634
Thema 1630
Thema-auf-Thema-Auswahl 3656
Thema-auf-Thema-Selektieren 3656
Themakarte 3644
Thema-Symbolsätz 3657
Thematic-Mapper 3648
thematische Auflösung 3653
thematische Daten 3639
thematische Fotokarte 3651
thematische Genauigkeit 3635
thematische Karte 3644
thematische Kartierung 3649
thematische Kartografie 3649
thematische Klasse 3637
thematische Modellierung 3650
thematischer Atlas 3636
thematischer geografischer Bereich 3641
thematischer Inhalt 3638
thematischer Kartograph 3647

thematische Schattierung 3654
thematische Schicht 3643
thematisches Diapason 3652
thematisches Schlüsselwort 3642
thematische Variable 3655
Theodolit 3658
Theodolite 3658
theoretischer Schleifenschlussfehler 3659
Theorie der Kartografie 3661
Theorie der kartografischen Kommunikation 3660
Thermalplotter 3664
thermischer Infrarotmapper 1947
thermischer Infrarot-Multispektral-Abtast-Radiometer 3662
thermischer Infrarot-Multispektralscanner 3662
Thermoisoplethendiagramm 3663
Theta-Winkel 746
Thiessen-Polygone 3665
Tic-Punkt 3676
Tiefe 978
Tiefenkarte 982
Tiefenlinie 979
tiefenorientierte Suche 981
Tiefenpunkt 3447
Tiefenwirkung 980
Tiefpassfilter 2236
TIF-Format 3579
TIGER-Format 3741
TIN 3820
TIN-Datenstruktur 3696
TIN-Facetten 3697
Tintenstrahlplotter 1953
Tip 3700
Tischkartierung 996
Tischplotter 3572
Tissot-Indikatrix 1253
Toleranz 3698
Tonleiter 3211
Topografie 3725
topografisch 3705
topografische Anordnung 2123
topografische Ansicht 3724
topografische Aufnahme 3723
topografische Basisplatte 3707
topografische Daten 3708
topografische Datenbasis 3709
topografische Ebene 3716
topografische Fläche 3722
topografische Geländeaufnahme 3723
topografische Grundlage 3706
topografische Karte 3714
topografische Luftfahrtkarte 75
topografische Messtischaufnahme 2780
topografische Planimetrie 3717

topografische Projektion 3720
topografischer Punkt 3718
topografisches Element 3712
topografisches Informationsmanagement 3713
topografisches Profil 3719
topografisches Quadrangel 2318
topografisches Reliefmodell 3721
topografische Texturenabbildung 3726
topografische Zeichnung 3711
Topogrid-Interpolation 3727
Topologie 3755
2D-~ 1190
Topologie der Konvergenz nach Maß 3763
Topologie-Diagramm 3757
Topologiefunktion 3736
Topologieliste 3760
Topologisation 3754
topologische Abbildung 3743
topologische Abfrage 3766
topologische Analyse 3728
topologische Beschreibung 3756
topologische Beziehung 3747
topologische Digitalisierung 3758
topologische Eigenschaft 3765
topologische Funktion 3736
topologische Gruppe 3738
topologische Hüllenoperation 3732
topologische Komplexität 3733
topologische Konsistenz 3734
topologische Modellierung 3744
topologische Objektdaten 3762
topologische Reinigung 3731
topologische Revision 3753
topologischer Fehler 3735
topologischer Graph 3737
topologischer Kreis 3730
topologischer Name 3761
topologischer Raum 3749
topologischer Typcode 3768
topologisches Element 3759
topologisches Overlay 3745
topologisches Primitiv 3746
topologisches Sortieren 3748
topologische Statistik 3767
topologische Struktur 3750
topologisches Vektorgitter 3752
topologische Transformation 3751
topologische Variable 3769
topologische Verbindung 3747
topologische Zelle 3729
topologisch komplexe Objekte 3740
topologisch strukturierte Daten 3742

topologisch vollständiges
    Quadrangel 3739
Toponymie 3770
topozentrische Koordinaten 3703
topozentrischer Rahmen 3704
topozentrischer Winkel 3702
Tortengrafik 2753
Total-Station 3771
Tour 3773
touristischer Atlas 3774
Tracing 3776
Trackball 3777
Trafik-Attribute 3784
Trafik-Visualisierung 3785
Tragbarkeit 2861
Trägerphasen-Verfolgung 432
Trajektorie 3786
Traktidentifikation 3779
Transaktion 3788
Transekt 3789
Transferträger 3791
Transformation 3793
Transformation der Mittellinie
    2385
Transformationsoperator 3794
Transit 3795
Translation 3797
transparente Zone 3801
Transparenz 3800
Transportanalyse 3802
Transport-Planung 3804
Transport-Prognostizierung 3803
transversal 3807
transversale Abbildung 3806
Trassenführungsgitter 3183
Trassieren 3776
Trassierung 3776
Treiber 1009
Trendfläche 3815
Trendlinie 3814
trennen 3871
Trennwand 1153
Treppeneffekt 92
Triade-Geometrie 3817
Triangulation 3821
Triangulation-basierter
    Digitalgeber 3822
Triangulationsmarker 3823
trianguliertes irreguläres Netz
    3820
trianguliertes irreguläres
    Netzwerk 3820
trianguliertes nichtregelmäßiges
    Gitter 3820
Triangulierung 3821
Trigonometrie 3824
trigonometrisches Festpunktfeld
    2530
Trilateration 3825
Trilinear-Interpolator 3826

TRIM-Format 3616
Trippel 3828
Trommelplotter 1186
Trommelscanner 1187
troposphärische Fehler 3831

Überblendung 1130
Überblick 2684
Überdeckungseinheiten 800
Überdruckung 2681
übereinanderlegen 3533
Überflutungsgebiet 1441
Überführungs- und
    Unterführungsinformation
    2677
Übergabeprüfung 11
Übergangspunkt 811
Übergangszone 3796
übergreifende Funktionsfähigkeit
    1996
Überlagerung 3534
überlappend 471
überlappendes Menü 470
Überlappung 2679
Überoszillation 2683
Überregelung 2683
Überschreitung 2683
Überschwemmungsfläche 1441
Überschwemmungsgebiet 1441
Überschwingen 2683
Überschwingen-Ungenauigkeit
    2682
Überschwingweite 2683
übersetzte Kartenmenge 641
Übersetzung 2000
Übersicht 2684
Übersichtsbild 1933
Übersichtskarte 3551
Übersteuerung 2683
Übertragungsformat 1337
Übertragungsfunktion 3790
Übertragungsstandard 3792
übertragungstechnische Funktion
    3790
Übertreibung 1336
überwachte Bildklassifizierung
    3536
Überwachung 3548
Überwachung der Erde-
    Ressourcen 1207
Überwachungsbereich 3535
Überzug 2680
UIS 1293
Umbenennung 3126
Umbuchung 3046
Umfang 2726, 2999
Umformatieren 3077
Umformung 748
Umgebung 1286
Umgebungsabbildung 1300

Umgehung 419
Umgehungsstraße 419
Umgliederung 3046
Umgrenzungsmarker 386
Umland 1838
Umlaufbahn 2639
umnumerieren 3128
Umordnung 3129
Umrahmung 1478
Umrechnung 3143
Umriss 726
Umrissattribute 2671
Umrisszeichnung 728
Umschreibung 546
Umschrift 546
umsetzen auf 2249
Umsetzung 3793
Umsetzung in kartografische
    Darstellungen 749
Umsetzungskurve 3798
Umwandler 750
Umwandlung 748, 3793
Umwelt 1286
Umweltabschätzung 1287
Umweltabteilung 1299
Umweltdaten 107
Umweltdatenbasis 1288
Umweltdatenkatalog 1289
Umwelteinflussanweisung 1291
Umweltempfindlichkeitsindex
    1297
Umwelt-GIS 1290
Umweltinformationssystem 1293
Umweltinformationsverwaltung
    1292
Umweltmodellierung 1294
Umweltplanung 1295
Umweltressourcen-
    Informationssystem 1296
Umweltsatellit 1298
Umweltvariable 1301
Umweltverträglichkeitsanweisung
    1291
unabhängige Variable 1926
unabhängige Veränderliche 1926
unbearbeitete Daten 3030
unbefahrbare Straße 1913
unbestimmt 1495
unbestimmter Graph 1499
unbestimmter Knoten 1503
unbestimmtes Objekt 1504
unbestimmte Toleranz 1507
Unbestimmtheit 1494
undokumentierte Daten 3849
Unebenheit 3851
unentdeckter Fehler 3848
unformatierte Daten 3852
ungekennzeichnetes Polygon
    2432
Ungenauigkeit 1917

Ungeocodierung 3853
ungeschichtetes Blatt 3870
ungleiche Linie 2015
Ungleichmäßigkeit 3851
ungleichmäßig verteilte Punkte
2016
ungleich verteilte Punkte 2016
Unglückkartieren 1118
universale dezimale
Klassifizierung 3860
universale geografische Identität
3861
universale polare stereografische
Projektion 3863
universeller asynchroner
Empfänger/Sender 3859
universeller Gleichrichter 3865
universelles GIS 3862
universelles rechteckiges
Raumkoordinatensystem 3866
unmittelbares Sehfeld 1965
Unordnungsmatrix 688
unregelmäßiges Polygon 2018
Unregelmäßigkeit 2014
Unrichtigkeit 1917
unscharf 1495
unscharfe Grenze 1496
unscharfe Klassifizierung 1497
unscharfe Logik 1500
unscharfe Menge 1505
unscharfe Mitgliedschaft 1502
unsichtbare Schicht 2011
Unterbereich 3516
Unterbrechungsstelle 401
Unterdrückung 976
Unterebene 3521
Untereinheit 3520
Unterfenster 3531
Untergrund 300
Untergrundkarte 302
Unterklasse 3518
Unterlage 3846
Unterlassung 9
Unterlegung 3846
Untermenge 3524
Unternehmensintegration 106
Unternehmenskonzentration 106
Unternehmenszusammenschluss
2393
Unterquadrant 3522
Unterschied 1151
Unterschrift 3297
Unterschwingen 2870
Unterschwingen-Ungenauigkeit
3847
Unterselektionen 3523
Untersuchung 3241
Unterteilung 3519
Unterteilung in Zonen 4033
Untertitel 3297

unüberwachte Klassifizierung
3876
unverbundener Bogen 3844
unvermischte Kante 2511
Urbild 2926
Urdaten 3030
Ursprung 2657
Ursprungsbreite 2114
Ursprungsdaten 3030
Ursprungsmaterial 3346
Usenet-Karte 3886
UTM-Koordinaten 3895

Vagheit 108
Valenztabelle 3897
Variable 3900
Varianz 3904
Variationsbreite 3360
Variogramm 3906
Vegetationszone 3925
Vektor 3907
vektorbasiertes GIS 3908
Vektorbildformat 3915
Vektordarstellung 3921
Vektordaten 3909
Vektordatenformat 3910
Vektordatenmodell 3911
Vektorfeld 3912
Vektorformat 3910
Vektorfunktion 3913
Vektorgrafik 3914
vektorisierte Zelle 3916
Vektorkarte 3918
Vektorobjekt-Export 1346
Vektorplotter 3919
Vektorprozessor 184
Vektor-Raster-Konvertierung
3923
Vektor-Raster-Prozessor 3924
Vektorschicht 3917
Vektorstruktur 3922
verallgemeinerte Koordinaten
1526
Verallgemeinerung 1520
veraltete Daten 78
Veränderliche 3900
veränderliche Länge 3902
veränderliches Attribut 3901
Verarbeitungstoleranzen 2900
Verarbeitung von geografischen
Daten 1671
verbesserte Linienlänge 775
Verbiegung 3974
verbinden 3097
Verbinden 2199
Verbindung 2044
Verbindungspunkt 3677
Verbindungsstelle 3677
Verbreitung 2920
Verbund 2044

verbundene Bodenstrukturen
1791
verbundene Bögen 693
verbundene Geradensegmente
696
verbundener Graph 695
verbundener Schlüssel 671
Verbundenheit 698
verdecken 1822
verdeckte Bildkante 1820
verdeckte Linie 1820
Verdichten 968
verdichten 654
Verdichtung 968
Verdoppelung 1196
Verdopplungsfenster 573
Verdrängung 1126
verdunkelte Auswahl 1104
Verdünnung 3667
Verdünnung der Genauigkeit
1097
Vereinfachung 3304
vereinigte Knoten 2046
vereinigte Linien 2045
Vereinigung 236, 2044
Vereinigung für geografische
Informationen 200
Vererbung 1950
Verfälschung der Daten 3446
Verfeinerung 3076
Verfeinigung 3076
Verfestigung 702
Verflechtung 2199
Verfolgung 3776
Verfolgungsstation 3778
Vergleichbarkeit 631
Vergleichsebene 925
Vergleichspunkt 333
vergröbertes Bild 556
Vergrößern/Verkleinern 3151
Vergrößerung 4035
Verhaltensmatrix 332
Verhältnis 3666
Verirrung 2606
Verkapselung 1263
Verkehrsanalysebereich 3781
Verkehrsflussanalysebereich
3781
Verkehrsgebiet 521
Verkehrsgebietskarte 3268
Verkehrsgeografie 3805
Verkehrskarte 3168
Verkehrsmittelwahl 2438
Verkehrsnetz 3783
Verkehrssimulation 1448
Verkehrsteilung 2438
Verkehrszählung 3782
verketteter Schlüssel 671
Verkleidung 2693
Verkleinerung 3061, 4038

verknüpfte Tabellen 2198
Verknüpfung 2051, 2199
Verknüpfungsindexierung 3101
Verknüpfungspunkt 3677
verkoppelter Randsatz 790
Verkrümmung 3974
Verlagerung 1126
verlängerter Maßstab 1273
Vermarkung 2122
vermaschtes Netz 1598
Vermaß 1101
Vermessung 2389, 3548
vermischtes Pixel 2435
vermittelnde Abbildung 145
Verpackung 2685
Verriegelung 2222
Verringerung 3061
Versand 1345
Versatz 955
Verschachtelung 1259
Verschieben 2694
Verschiebung 1126
Verschiedenartigkeit 1151
Verschlüsseln 1264
Verschmelzen 2399
Verschmelzungsobjekt 2394
Verschwindungspunkt 2532
Versinnbildlichung 3556
Versorgungsbereich 521
verstädtertes Gebiet 3883
Verständnis 653
verstecken 1822
Verstellung 3134
Verstoß 2014
Verteidigungskartierungsstelle
    946
Verteiler 3442
verteilte Daten 1140
verteilte Datenbank 1141
verteilte Datenbasis 1142
verteilte geografische Information
    1144
verteilte kartografische
    Datenbasis 1139
verteiltes Datenbasis-
    Managementsystem 1143
verteiltes Informationssystem
    1145
verteilte Verarbeitung 1146
Verteilungsdichte 1147
Vertiefung 976
Vertikalabschlussfehler 3934
Vertikalauflösungsvermögen
    3941
Vertikaldatum 3931
vertikale Genauigkeit 3929
vertikale Linie 3938
vertikaler Genauigkeitsabfall
    3932
vertikaler Interpolator 3937

vertikale Übertreibung 3935
Vertikalfehler 3933
Vertikalprojektion 3940
Verträglichkeit 637
Vertrauensintervall 684
Vervielfältigen 770
verwaltende Abteilung 57
Verwaltung der natürlichen
    Ressourcen 2502
Verwaltung paralleler Prozesse
    680
Verwaltungsbezirk 56
Verwaltungsgliederung 57
Verweis 3064
Verwendung 139
Verwendungszweck 997
Verzerrung 1137
Verzerrungsellipse 1253
Verzögerung 952
Verzug 952
Video-Digitalisierer 3942
Video-Digitizer 3942
Videohelm 1810
Vieleck 2823
vielseitiger Atlas 645
vielseitiges Objekt 649
Viel-zu-Eins-Beziehung 2247
Viel-zu-Viel-Beziehung 2246
Vielzweckkataster 2467
vierdimensionales GIS 1467
Viereck 2959
Vier-Ecken-Gebiet 2956
Viererbaum 2965
Vierfarbenproblem 1466
Virtualdaten 3949
virtuelle aktive Bedienungsstelle
    3948
virtueller aktiver Kontrollpunkt
    3948
virtueller aktiver
    Orientierungspunkt 3948
virtuelle Realität 3951
virtueller Weg 3952
virtuelles GIS 3950
Visierlineal 93
Visualisierung 3962
visuelle Akkumulation 3958
visuelle Bilddeutung 3960
visuelle Darstellung 3965
visuelles Attribut 3959
visuelle Schnittstelle 3961
visuelle Simulation der
    Naturerscheinungen 3966
Volkszählung 497
Volkszählungsbereich 499
Volkszählungsblock 488
Volkszählungsblockgruppe 489
Volkszählungscode 490
Volkszählungsdaten 492
Volkszählungsgeografie 495

Volkszählungskarte 496
Volkszählungsregion 499
Volkszählungsunterbereich 498
Volldatum 642
Vollinie 3340
vollräumliche Wahllosigkeit 643
vollräumliche Zufälligkeit 643
vollständige Menge 1339
Volumenberechnung 3967
Volumenbild 3969
Volumenelement 1243
Volumenmodellierung 2441
Vorbild 2926
Vorderfläche 1488
Vordergrund 1456
Vorderkante 2126
Vorderseite 1488
Vordigitalisierungsprobleme
    2882
Vorgang 1329
Vorhergehen 2876
Vorhersage 1455
Vorhersagekarte 2905
Vorlage 3595
vorläufige Karte 2883
vorläufige Koordinaten 2927
vorläufige Topologie 2884
Voronoi-Diagramme 3665
Voronoi-Kante 3970
Voronoi-Netz 3971
Voronoi-Netzwerk 3971
Voronoi-Polygone 3665
Vorrang 2876
Vorrecht 2876
vorübergehender Datenzustand
    917
Vorwahl 161
Vorwahl-Code 2577
Vorwahlnummer 2577
Vorzeichen 3295
Voxel 1243

waagerechte Ebene 1849
wachsend 188
Wächter 835
wählbare Schicht 3254
Wahlbezirk 3972
Wahl des Pfades 2715
Wahlkreis 3972
wahre 3D-Darstellung 3832
wahre Breite 3833
wahrscheinlicher kreisförmiger
    Fehlerbereich 544
Wahrscheinlichkeit 2898
Wahrscheinlichkeitsverteilungs-
    funktion 2899
Wahrzeichen 2087
Walzenplotter 1186
Walzenschreiber 1186
Walzenvorschubplotter 3171

Wandbekleidung 2693
Wandkarte 3973
Wandler 750
Wartestapel 3458
Wasserlinie 3976
Wasserscheide 3975
Wasserstraßen-
    Informationssystem 3977
Wasserweg-Informationssystem
    3977
Web-basierte Karte 3985
Web-basierter elektronischer
    Atlas 3983
Web-basiertes GIS 3984
Web-GIS 3984
Web-Kartografierung 3986
Web-Kartografierungssystem
    3987
Webmapping 3986
Wechselformat 1337
Weeding 3988
Weg 3177
Wegatlas 3165
Weg-Attributtabelle 3178
Wegekarte 3168
Wegewahl 3182
Wegklassifizierung 3166
Weglänge 2711
Wegmarke 3979
Wegmessung 3179
Wegnetz 3180
Wegnummer 3181
Wegpunkt 3979
Wegpunktdaten 3980
Wegstrecke 2714
Wegsymbole 3169
Wegsymbolsätze 3169
Wegsystem 3180
Weiche 309
Weiterleitung 3795
weiträumiges Netzwerk 3996
weitspannende Punkte 692
Weitverkehrs-Differenzial-GPS
    3995
Weitverkehrsnetz 3996
Wellenauslaufbereich 932
Wellentypfilter 3978
Weltdatennetz 3996
Weltellipsoid 4009
Weltkarte 2288
Weltkoordinate 4007
Weltkoordinatensystem 4008
Weltkugel 1209
Weltluftfahrtkarte 4006
Weltpolygon 3868
Weltraumfahrzeug 3356
Wende 2125
Wendepunkt 3838
Werkzeugstreifen 3699
Wertebereich 2999

Wertebereichsextraktion 3002
Wertefeldlinie 2031
Wertesatz 3899
Wertgleiche 2031
Wertlinie 2031
Wertminderung 975
West 3994
Wetterkarte 3982
Wetterung 3981
Widerspruchsfreiheit 700
Wiederholbarkeit 3131
wiederholte Färbung 3049
wiederholte Selektion 3146
wiederholte Stichprobenauswahl
    3143
Wiederpositionierung 3134
Wildfauna-Karte 3997
Wildtiere-Karte 3997
willkürliche Projektion 145
Windrose 636
Windung 3836
Windung-Impedanz 3837
Winkel 122, 773
Winkelabstand 128
Winkeldistanz 128
Winkelentfernung 128
Winkelmaß 2381
winkeltreue Abbildung 686
winkeltreue Azimutalprojektion
    3488
winkeltreue Karte 2025
Wirtschaftsgeografie 1213
Wissensbasis 2058
wissenschaftlicher Atlas 3230
wissentliche Täuschung von
    Daten 3446
Wohnbezirk 3149
Wohneinheit 1855
Wohngebiet 3149
Wohnquartier 3150
Wohnviertel 3150
Wolkenkarte 583
Wurzel 3173
Wurzel eines Baums 3174
Wurzelknoten 3173

X-Achse 4012
X-Koordinate 4013

Y-Achse 4015
Y-Koordinate 4016

Z-Achse 4017
Zählbezirk 499
Zahl der Regionen 2579
Zahl der Steuerpunkte 2578
Zahl der Zeilen 2580
Zahlenattribut 2581
Zähllinie 1929
Zählpfeil 185

Zählung 783
Zaunlinie 1395
Zeichen 525, 3295
Zeichen der Durchschnittsbildung
    3298
Zeichenerklärung 2278
Zeichengerät 1181
Zeicheninstrument 1183
Zeichenkette 527
Zeichenordnung 1184
Zeichenschriftart 526
Zeichentabelle 3560
Zeichen von Linien 2193
zeichnen
    eine Karte ~ 1180
zeichnen im Weiterschlag 3060
Zeichnungsaustausch-Format
    1182
Zeiger 2801
Zeigerarm 93
Zeile 3184
Zeilenanzahl 2580
Zeilenende 1268
Zeilenende-Code 1269
Zeilenlänge 2179
Zeilenlänge-Code 2180
Zeilenlineal 3187
Zeilennummer 2183
Zeilenrücklauf 3156
Zeilenzahl 2580
zeiliger Schablonentisch 2171
Zeitauflösung 2149
zeitbezogener Durchschnittswert
    3694
Zeit/Entfernungsdistricting 3688
Zeitfehler 3689
Zeitlage 3691
zeitliche Auflösung 2149
zeitliche Datenbasis 3597
zeitliche Datendimension 3599
zeitliche Dimension 3598
zeitliche Genauigkeit 3596
zeitlicher Extent 3600
zeitliche Spalte 3602
zeitliches Referenzsystem 3601
zeitliche Topologie 3603
zeitlich mittlerer
    Durchschnittswert 3694
Zeit-Raum-Konvergenz 3693
Zeitscheibe 3691
Zeitschlitz 3691
Zeitunterschied 3686
Zeitverbreitung 3687
Zeitzone 3695
Zelle 481
Zellenbreite 486
Zellengröße 483
Zellengrößekalibrierung 484
Zellenhöhe 482
Zellwert 485

Zenit 4019
Zenithdistanz 4020
Zenithwinkel 4020
Zensus 487
Zensusgebiet 499
Zensusgebietsbericht 503
Zensusgebietsbeziehung 501
Zensusgebietsbeziehung-Datei
    502
Zensusgebietsgrenze 500
Zensusgebiet-Straßenindex 504
Zensuskarte 496
zentrale Leitung 507
zentralisierte kartografische
    Datenbank 513
Zentralprojektion 690
Zentralstadt 512
Zentraltendenz 514
Zentrierung 506
Zentroid 159
Zentrum 505
Zentrum der Perspektive 510
Zerfall 1117
zerfallen 1116
zerlegen 1116
Zerlegung 942
Zerstäubung 235
zerstreuen 1116
Zerstreuung 1129
zerteilen 1116
Zeugenbaum 4002
Zeugnis 515
Ziehen 1178
ziehen im Nachzug 3060
Ziel 997, 3583
Zielbezirk 3584
Zielfunktion 2591
Zielgebiet 3584
Zielkarte 3586
Zielknoten 998
Zielraum 3584
Zielsymbol 3587
Zielung 3585
Zipcode 4034
Zirkuszelt 3258
Z-Koordinate 4018
zonale Abtastung 4028
zonale Karte 4025
zonale Operatoren 4026
zonale Probennahme 4028
zonaler Index 4024
zonales Polynom 4027
Zonalindex 4024
Zonenbildung 4033
Zoneneinteilung 4033
Zonennummer 4031
Zonenstruktur 4029
Zonenzeichen 4030
Zonung 4033
Zoomen 4036

Zoomen-Schichtenteilung 4037
zufällige Daten 2997
zufälliger Fehler 16
Zug 448
Zugang 13
Zugänglichkeit 14
Zugänglichkeitsindex 15
Zugriff 13
zulässiger Adressenbereich 2728
zulässiger Fehler 58
zunehmend 188
Zuordnung 94
zurückgeben 3159
zurückkehren 3159
Zusammenarbeit 2047
Zusammenbruch 602
Zusammenführung 877
zusammengelegte Innenstadt 701
zusammenhängende Fläche 697
zusammenhängender Graph 695
Zusammenmischen 2399
Zusatzdaten 121
Zuverlässigkeit der Investitionen
    in der Kartografie 3113
Zuverlässigkeit der
    kartografischen
    Forschungsmethode 3112
Zuverlässigkeit einer Karte 2324
Zuwachs 1925
Zuweisung 94
Zuweisungsanweisung 196
Zwei-Achsen-Fourier-
    Formalyse 1191
Zweige der thematischen
    Kartografierung 398
zweite Normalenform 3245
zweite Normalform 3245
zweite Projektion 3940
Z-Wert 4039
Z-Wert-Datenpunkte 4040
Zwickel 1716
Zwischenknoten 1986
Zwischenkonturlinie 1985
Zwischenmaßstab 1987
Zwischenraum 2002
Zwischensumme 3528
zyklische Laufenthalten 845
Zyklus 2234
Zylinderabbildung 847
Zylinderprojektion 847

Français

abornement 957
abscisse 1, 1210
abscisse d'un quadrillage 1210
abscisse fausse 1367
absorption 8
absorption atmosphérique 225
abstention 9
abstraction 10
accès 13
accès à distance au SGBD 2204
accès consécutif 3266
accès de données 853
accès en ligne 2623
accès en série 3266
accès géographique 1566
accès rapide 3288
accès séquentiel 3266
accessibilité 14
accès spatial 3361
accident géographique 1615
accrochage 3328
accrochage au quadrant 2961
accrochage au rastre 3025
accrocher à la grille 3331
accrocher au nœud 3332
accroissement 1925
accumulation en coulée 1442
accumulation visuelle 3958
acheminement 3182
acquisition de données 875
acquisition de terre 20
actif 194
actualisation 3877
actualité 26
adaptation d'images numériques
    1052
addition d'images 1889
additionner 28
adjacence 45
adjacence de mailles 2397
adjacence de nœuds 2534
adjacence de polgones 2824
adjonction 53
administrateur séquestre 835
Administration nationale des
    océans et de l'atmosphère
    2494
adressage 31
adressage éloignée 1132
adressage par blocs 362
adresse 29
adresse d'envoi 2242
adresse de point 2799
adresse de réseau 2515
adresse géographique 1568
adresse spatiale 3363
aérophoto 65
aérophoto couleur 603
aérophoto noire et blanche 359
aérophoto perspective 2601

aérophotoplan 68
aérotriangulation 71
aérotriangulation entièrement
    analytique 1489
affaiblissement 238
affaissement 602
affaissement brusque 602
affichage cartographique 2267
affichage de coordonnées 757
affichage de position 2867
affichage géographique 1590
afficheur d'état de satellite 3205
afficheur volumétrique à
    visualisation directe 1114
âge des données 79
agence de cartographie 2299
Agence de cartographie de
    défense 946
Agence spatiale européenne 1327
agglomération 80
agrégation 83
agrégation avec fusion de
    polygones 2838
agrégation de points 2800
agrégation de polygones 2838
agrégation limitée par diffusion
    1030
aire 157
    à ~ équivalente 1304
    d'~ 168
aire de service 521
aire de surface 171
aire de traçage 2796
aire-de-vent 2810
aire de visibilité 3945
aire métropolitaine 2416
aire non métropolitaine 2552
ajouter 28
ajustage 54
ajustage de cartes 2250
ajustage de précision d'un étalon
    3460
ajustage de surface 3540
ajustage par méthode des
    moindres carrés 2130
ajustement 54
ajustement de ligne terrestre 2085
ajustement de points par fonction
    spline 3444
ajustement d'une courbe 830
algèbre cartographique 2252
algèbre d'images 1890
algèbre projective 2916
algèbre rastre 3006
algorithme 90
algorithme de Dijkstra 1094
algorithme de filage de courbe
    2175
algorithme de filage de limite 384
algorithme de Gauss 1511

algorithme de pause naturelle
    2497
algorithme de poursuite de ligne
    2175
algorithme du plus proche voisin
    2507
algorithmes génétiques 1528
alidade 93
alidade holométrique 3594
alidade télescopique 3594
alignement 1487
allocation 94
allocation de ressources 95
alphanumérique 98
altération atmosphérique 3981
altimètre 102
altimétrie 104
altitude 1811
altitude absolue 3
amalgamation 106
ambiance 1286, 2929
ambiguïté 108
amélioration de pseudocouleur
    2935
amélioration d'image 1903
aménagement du territoire 2086
amendement dans une carte 2253
amenuisement 3667
amer terrestre 2087
amincissement 3667
amincissement de ligne 2194
amont 3879
    à l'~ 3879
    en ~ 3879
    vers l'~ 3879
amorçage à froid 601
amorce 3674
amorce de carroyage 3674
amorces de canevas 1753
amorces de réseau géographique
    1753
amortissement 238, 975
anaglyphe 109
analogique 112
analogue 112
analyse 113
analyse à base de SIG 1688
analyse adjacence 46
analyse cluster 586
analyse comparative 334
analyse critique 801
analyse d'aptitude 3313
analyse de calibration 427
analyse d'échantillons 3193
analyse de chemins 2709
analyse de constructibilité 3313
analyse de contraintes 706
analyse de corrélation 779
analyse de déblai-remblai 841
analyse de formes 3280

analyse de formes de Fourier à
    deux axes 1191
analyse de gradient 1727
analyse de localisation 3311
analyse démographique 959
analyse de propriété foncière
    3034
analyse d'équilibre spatial 3384
analyse de réseaux 2517
analyse d'erreurs 1318
analyse de sensitivité 3264
analyse de site 3311
analyse de surface de tendance
    3816
analyse de terrain 1407
analyse de transport 3802
analyse de visibilité/non-visibilité
    3946
analyse de voisinage 2512
analyse de voisinage le plus
    proche 2508
analyse d'images 1891
analyse d'images hyperspectraux
    1872
analyse discriminante 1124
analyse en composantes
    principales 2891
analyse exploratoire de données
    1344
analyse lexicale 2706
analyse multicritère 2457
analyse quantitative de terrain
    2968
analyse spatiale 3364
analyse syntaxique 2706
analyse topologique 3728
anamorphose 119
ancre 120
angle 122
angle adjacent 47
angle azimutal 294
angle de champ 124
angle de champ visuel 124
angle de déclivité 126
angle de déviation de la verticale
    948
angle de gradient 125
angle de hauteur 123
angle d'élévation 123
angle de site positif 123
angle de talus 126
angle de vision 127
angle d'incidence 126
angle d'inclinaison 126
angle d'inclinaison de talus 126
angle directeur 329
angle horizontal 1843
angles au sommet 3930
angles opposés par le sommet
    3930

angle topocentrique 3702
angle visuel 127
angle zénithal 4020
animation 131
anisotrope 132
anneau 3162
annotation 134
annulation de lissage 3875
annulation de sélection 991
anti-arnaque 138
anti-brouillage 138
anticrénelage 136
anti-déception 138
anti-leurrage 138
antipodes 137
anti-tromperie 138
aperçu 2684
apex 3928
appareil de mesure de distance
    1134
appareil de restitution
    stéréoscopique 3492
appareillage 1314
appariement 2365
appariement de nœuds 2538
appartenance floue 1502
application 139, 2908
application affine 76
application de Beltrami 1551
application d'environnement 1300
application de routage
    géographique 1632
application de texture
    topographique 3726
application géodésique 1551
application géographique 1574
application identique 1885
application inverse 2009
application ponctuelle 2806
application projective linéaire
    2918
application SIG 1684
application sphérique 3436
appréciation 1321
appréciation de masse par
    ordinateur 663
appréciation d'environnement
    1287
approximation 144
approximation de Gauss 1512
approximation polygonale 2825
arbre 3811
arbre à six branches 1819
arbre de décisions 938
arbre d'octants 2613
arbre quaternaire 2965
arbre témoin 4002
arc 146
architecture 150
architecture de SIG 1685

archivage 152
archivage en cartographie 2254
archives 151
arc initial 331
arc non connecté 3844
arc non saturé 848
arc pendant 848
arc polygonal 2826
arcs adjacents 48
arcs connectées 693
arcs de frontière 375
aréolaire 168
arête 1215
arête adjacente 49
arête de polygone 164
arête de Voronoi 3970
arête frontière 380
arithmétique d'images 1892
arithmétique d'intervalle 2003
armature 1476
arpentage 3548
arpenteur 2094
arrière-pays 1838
arrière-plan 300
arrière-plan couleur 604
arrière-plan hachuré 1806
arrivée 531
arrondissement 786
ascendant 188
ascenseur 3234
aspect aléatoire spatial complet
    643
assemblage d'aérophotographies
    62
assemblage de carte 2285
assiette 3442
association 199
association de type 1:1 2621
association de type m:1 2247
association de type n:m 2246
association d'information
    géographique 200
astrolabe 202
astronomie 212
astronomique 203
asymétrie 214
asynchrone 215
atlas 1575
atlas complexe 645
atlas d'éducation 222
atlas de gros format 2107
atlas de routes 3165
atlas électronique 667
atlas électronique par le Web
    3983
atlas général 1516
atlas géographique 1575
atlas hydrologique 1867
atlas militaire 2420
atlas national 2487

atlas océanographique 2610
atlas régional 3080
atlas scientifique 3230
atlas thématique 3636
atlas touristique 3774
atmosphère 224
atmosphérisation 3981
atomisation 235
attachement 236
attachement d'attribut 246
attelage serré 575
atténuation 238
atténuation de ligne 2194
attitude 239
attractivité 243
attribut 244
attribut affichable 1127
attribut de caractéristique 1374
attribut de relation 255
attribut d'objet 1374
attribut externe 1351
attribut mental 2390
attribut non spatial 191
attribut numérique 2581
attributs de contour 2671
attributs définis par utilisateur
    3887
attributs de points d'arrêt 3496
attributs de trafic 3784
attribut spatial 3365
attribut spectral 3427
attribut textuel 3624
attribut variable 3901
attribut visuel 3959
autocorrélation 266
autocorrélation spatiale 3366
auto-étiquetage 270
auto-incrément 267
automatique 280
autonome 2615
aval
    d'~ 1174
avantage du lieu 2766
avantage indirect 1354
avant plan 1456
axe 507
axe de projection 292
axe des abscisses 4012
axe des ordonnées 4015
axe de X 4012
axe de Y 4015
axe de Z 4017
axe nucléique 1303
axes numériques 2582
azimut 293
azimut astronomique 204
azimut géodésique 1543
azimut inverse 299
azimuts mutuels 2481

background 300
bague 3162
bâillement 1510
balayage 3024, 3225
balayage de trame 3024
balayage ligne par ligne 3024
balayage récurrent 3024
balayage trame 3024
balise de délimitation 386
balise du terrain 386
banc 330
banc d'essai 333
bande 3511
bande adhésive 2361
bande côtière 594
bande passante 311
bande spectrale 3428
banque de données 858
banque de données
    cartographiques 441
banque de données
    cartographiques centralisée
    513
banque de données distribuée
    1141
banque de données locale 2210
banque d'images 1899
barre 314
barre d'échelle 3214
barre de défilement 3234
barre d'instruments 3699
barre d'outils 3699
bas
    de ~ en haut 188
basculement 2125
base à l'échelle des photographies
    2738
base de connaissance 2058
base de données 860
base de données cartographiques
    442
base de données cartographiques
    distribuée 1139
base de données cartographiques
    hiérarchique 1823
base de données cartographiques
    relationnelle 3098
base de données décisionnelles
    936
base de données de couverture du
    sol 2078
base de données de la terre 2081
base de données de moyens
    géographiques 1563
base de données de réseau
    cartographique 2519
base de données de réseau filaire
    3510
base de données de ressources
    globale 1711

base de données de simulateur de
    vol 1437
base de données d'indices
    géographiques 1604
base de données distribuée 1142
base de données
    environnementaux 1288
base de données géographiques
    1585
base de données géographiques
    sans soudure 3239
base de données géométriques
    1656
base de données géospatiaux
    nationale 2491
base de données hiérarchique
    1825
base de données localisées 2083
base de données non temporelles
    220
base de données orientée objet
    2596
base de données régionale 3081
base de données régionale
    intégrée 1970
base de données relationnelle
    3099
base de données sans soudure
    3239
base de données SIG 1691
base de données spatiale 3371
base de données spatio-
    temporelle 3421
base de données temporaire 3597
base de données topographique
    3709
base de métadonnées 2402
base mathématique de cartes 2367
base photographique 2738
base planimétrique 2784
basses eaux 2237
bathymétrie 326
bâtisseur de colonne 621
bâtisseur SQL 3453
batterie 585
batterie de dérouleurs 585
berceau 330
bibliographie cartographique
    2255
bibliothèque 2146
bibliothèque de cartes 2264
bibliothèque de cartes numériques
    1060
bien 194
bien immeuble 2073
bilan massique 2362
bilan matière 2362
biome 355
biorégion 356
biotope 1797

bi-passe 419
bit 357
bitmap 3015
blocage 2222
blocage de coordonnées 1482
bloc de recensement 488
bloc de remplissage de polygones
    2840
bogue 413
boîte de légende 2133
boîte délimitée 395
boîte délimitée minimale 2423
boîte de liste 2200
boîte d'essai 530
boîte de texte 3626
boîte en couleurs 605
bord 374, 1216
bord avant 2126
bord de côte 594
bord de dessin 2511
bord d'une surface 388
bordurage 728
bordure 374
bordure de finition 2511
bordure pure 2511
boréal 2567
bornage 957
borne frontière 386
borne-repère 333
bornes de données 873
boucle 2234
boule de commande 3777
boule roulante 3777
bouquet 584, 585
boussole 632
bouton accrocheur 1724
bouton d'option 2984
bouton d'options 2638
bouton radio 2984
branches de cartographie
    thématique 398
brillance spectrale 3429
brouillard 1451
bruit de granulation 3425
bruit de récepteur 3045
bruit modal 3425
bûchette 3319
bureau local de recensement 491
but 997
butineur 3963
butineur cartographique 2348

cacher 1822
cadastrage 2090
cadastration 2090
cadastre 422
cadastre à objectifs multiples
    2467
cadastre d'immeubles 422
cadastre multifonction 2467

cadrage 1478
cadre 1128
cadre conceptuel 677
cadre d'objet 395
cadre-porteur 1476
cadre théorique 677
calage 2141
calage d'une carte 1593
calcul d'aire/périmètre 174
calcul d'aires 659
calcul de masse 2363
calcul de réseau 2518
calcul de surface de niveau 2875
calcul de surface équipotentielle
    2875
calcul de tampon 408
calcul de volume 3967
calcul d'inclinaison 3320
calibrage de données 874
calibration 426
calibration de taille de cellule 484
calibration radiométrique 2986
calque 2680
calque binaire 351
calque graphique 1742
calque topologique 3745
camembert 2753
caméra balistique 305
caméra numérique 1033
caméra numérisante 1033
canal de passage 1446
canevas 2530
canevas altimétrique 2140
canevas de traçage stéréoscopique
    1477
canevas géodésique 2530
CAO 661
cappa de Cohen 599
capsulage 1263
captage 429
capteur actif 25
capteur de luminescence 25
capteur digital 1078
capteur d'imagerie hyperspectral
    1873
capture 429
capture d'écran 3232
caractère 525
caractère de contrôle 738
caractère de zone 4030
caractère qualitatif 244
caractères directs 1113
caractères indirects 1938
caractéristique annotée 133
caractéristiques de département
    788
caractéristiques de la population
    2855
caractéristiques de pondération
    3992

carré 3455
carrefour 2051
carroyage 3457
carroyage cartographique 2336
carroyage de méridiens et
    parallèles 2274
carroyage de référence 1777
carroyage de référence militaire
    2421
carte 1611
carte à courbes de niveau 2142
carte à échelle sous-synoptique
    2388
carte aéronautique 73
carte aéronautique en coupe 3249
carte aéronautique en plusieurs
    feuilles 3249
carte aéronautique mondiale 4006
carte à grande échelle 2108
carte à moyenne échelle 2388
carte analogique 110
carte analytique 116
carte anamorphique 118
carte animée 130
carte à petite échelle 3327
carte à points 1167
carte à projection équivalente
    1840
carte au sol 3338
carte au trait 727
carte bathymétrique 324
carte-cadre 678
carte céleste 478
carte chorographique 3085
carte clinométrique 569
carte cognitive 598
carte complexe 647
carte conceptuelle 678
carte continentale 714
carte continue 720
carte couleur personnalisée 836
carte d'aménagement 2101
carte de base 3706
carte de bruit 2541
carte de caractéristiques 1383
carte de classification de biens
    immeubles 2075
carte de conductivité 683
carte de contrôle de l'étendue
    3000
carte de conversations 747
carte de couleurs 606
carte de couverture 793
carte de couverture du sol 2079
carte de demi-teintes gris 1762
carte de densité de la population
    2856
carte de densité démographique
    2856
carte de desserte 3268

carte de disposition 2123
carte de faune sauvage 3997
carte de fond 302
carte de foresterie 1459
carte de géomorphologie 2082
carte d'ensembles de données 912
carte d'entraînement à sec 2076
carte de nuages 583
carte de papier 2697
carte de plan cadastral 2652
carte de profondeurs 982
carte de proximité 2930
carte de recensement 496
carte de région 3085
carte de répartition des sujets 3449
carte de réseau 2528
carte dérivative 983
carte dérivée 983
carte de routes 3168
carte des isoplèthes de la nappe phréatique 1866
carte de sites 3312
carte des lacs et marais 2069
carte d'espace informationnel 1943
carte des pluies 2996
carte des possibilités d'exploitation des terres 2076
carte de travail 3706
carte d'évaluation 1328
carte de végétaux 2793
carte de visibilité 3954
carte de visibilité/non-visibilité 3954
carte de zone de chalandise 3780
carte de zone de couverture 3268
carte de zone de service 3268
carte de zones d'inondation 1440
carte diagramme 539
carte diagramme en demi-teintes 540
carte diagramme en tonalité continue 540
carte d'incidents 1920
carte d'inventaire 2005
carte d'isarhythmes 2019
carte d'isohypses prévue 2904
carte d'isoplèthes 2034
carte d'isothermes 2037
carte d'objectifs 3586
carte document de base 3345
carte d'Usenet 3886
carte du sol 3338
carte d'utilisation du sol 2101
carte électronique 1239
carte en courbes de niveau 727
carte en rangées 3000
carte en relief 3114
carte faunique 3997

carte générale bathymétrique des océans 1517
carte géographique 1611
carte géologique 1646
carte humaine 1857
carte hydrographique 1862
carte hypertexte 1874
carte hypsographique 1875
carte hypsométrique 1876
carte-image 564
carte imagée 564
carte imprimée 2697
carte-index de photographies 2747
carte infographique 1239
carte informatique 1239
carte informatisée 1239
carte internationale 1992
carte isochronique 727
carte isodémographique 2023
carte isogone 2025
carte limitée par des méridiens et des parallèles 2318
carte linéaire 2160
carte maillée 1167
carte météorologique 3982
carte mondiale 2288
carte mondiale numérique 1036
carte muette 3551
carte murale 3973
carte numérique 1239
carte numérisée 1080
carte officielle 2614
carte originale 2658
carte orographique 2751
carte orthophotographique 2667
carte par le Web 3985
carte par maille 1167
carte pédologique 3338
carte pente 569
carte planaire 2769
carte planimétrique 2788
carte préliminaire 2883
carte prévue 2905
carte produite à partir d'information digitalisée 2312
carte rastre 1772
carte réalisée à la main 1801
carte référentielle 3069
carte régionale 3085
carte répertoire 1933
carte routière 3168
carte scannée 3223
carte schématique 3229
cartes d'interaction naturelle et sociale 2335
cartes disposées en mosaïque 3680
carte sensible 564
carte socio-économique 3334

carte spéciale 3424
carte structurale 3527
carte sur papier 2697
carte synthétique 3563
carte thématique 3644
carte thématique analytique 3645
carte thématique à variables multiples 2480
carte thématique à variable unique 2622
carte thématique rastre 1782
carte thématique synthétique 3646
carte topographique 3714
carte topographique aéronautique 75
carte topologique 2123
carte vectorielle 3918
carte Web 3985
carte zonale 4025
cartographe 435
cartographe thématique 3647
cartographie 465
cartographie astronomique 207
cartographie automatisée 660
cartographie complexe 648
cartographie côtière 590
cartographie d'avaries 1118
cartographie de noms de domaines 1161
cartographie de télécommunication 3592
cartographie d'information 1940
cartographie d'Internet 1993
cartographie diurne 927
cartographie d'utilités 3893
cartographie géographique 1577
cartographie géo-informatique 1642
cartographie géologique 1647
cartographie intégrée d'unités terrestres 1972
cartographie interactive 2625
cartographie maritime 2351
cartographie mathématique 2368
cartographie numérique 1035
cartographie par ordinateur 660
cartographie personnelle 2731
cartographie planétaire 2782
cartographie planimétrique de base 2785
cartographier 2249
cartographie sur ordinateur 996
cartographie synthétique 3564
cartographie système 3568
cartographie tabulaire 996
cartographie terrestre 3620
cartographie thématique 3649
cartographie Web 3986
cartographique 436

cartologie 466
cartomatique 660
cartométrie 469
carton 539
carton annexe d'une carte 1962
cartothèque 1060
cartothèque cartographique 2257
cartouche 1962
cascade
    en ~ 471
case 473
case à cocher 530
case de défilement 3234
case délimitée 395
case d'option 530
casque de vision 3D 1810
casque de visualisation 1810
caste de cyberespace 843
catalogue cartographique 437
catalogue de clés d'objets 2592
catalogue de données
    environnementaux 1289
catalogue des noms de lieux 1515
catalogue d'images 1894
catalogue en ligne 2624
catalogue géographique de
    régions politiques et
    statistiques 1578
catégorie 476
catégorie géographique 1579
caténaire 477
cellule 481
cellule de grille 1764
cellule nulle 2572
cellule topologique 3729
cellule vectorisée 3916
cellule vide 1262
centrage 506
centre 505
centre d'échange 563
centre d'échange et de
    compensation 563
centre de classe 550
centre de collinéation 510
centre de données géophysiques
    1667
centre de masse 509
centre de perspective 510
centre de projection 510
centre d'homologie 510
centre géographique 1580
centre perspectif 510
centroïde 159
cercle 543
cercle de latitude 2701
cercle d'erreur probable 544
cercle d'option 2984
cercle élastique 1230
cercle géodésique 1544
cercle polaire antarctique 135

cercle polaire arctique 156
certificat de titre 515
certificat de titre de propriété 515
chaîne 516, 527
chaîne de caractères 527
chaîne de pilotage 241
chaîne de triangles 520
chaînette 477
chambre de compensation 563
champ 183, 1582
champ angulaire visuel 1403
champ de données 891
champ dérivé 984
champ d'étiquette d'attribut 261
champ de visée 1403
champ de vision 1403
champ de vision instantané 1965
champ de vision instantané
    effectif 1229
champ d'inondation 1441
champ géographique 1564
champ géographique thématique
    3641
champ géomagnétique 1648
champ scalaire 3209
champ vecteur 3912
champ vectoriel 3912
champ visuel 1403
changement de taille 3151
changer d'échelle 3144
châssis 1476
châssis de copie 2896
châssis-presse 2896
chemin 3177
chemin au plus bas coût 2128
chemin du moindre coût 2128
cheminement fermé 579
cheminement polygonal 2830
cheminement polygonal non
    fermé 3843
chemin le plus court 3290
chemin multisegment 2471
chemin optimal 2636
chercher 1424
chevauchement 2679
chiffre-référence 428
choix d'échelle 3221
choix des objectifs et des moyens
    de traitement 3585
choix d'itinéraire 3182
chorologie 536
chroma 542
chrominance 542
ciblage 3585
cible 3583
cible active 822
cinématique 2057
cinématique en temps réel 3039
circonférence topologique 3730
circonscription 546, 1148

circonscription électorale 3972
circonscription pseudo-électorale
    2947
circuit de Hamilton 1800
cité compactée 701
classe 549
classe cadastrale 421
classe d'attribut 247
classe de caractéristiques 1377
classe de formes 1377
classe d'entité 1275
classe d'objets 2585
classement 552
classement de zones 176
classe ordinale 2647
classe spectrale 3430
classe thématique 3637
classeur 558
classeur de thèmes composant
    seulement des fractions de
    pixels 2761
classificateur 558
classificateur à distance minimale
    2425
classificateur spectral avancé
    permettant de détecter de
    thèmes composant seulement
    des fractions de pixels 2761
classification 552
classification contrôlée 3536
classification croisée 805
classification de biens immeubles
    2074
classification d'échantillonneur
    monocanal 397
classification décimale
    universelle 3860
classification de données 876
classification de pixels 2759
classification de routes 3166
classification de Suits-Wagner
    3532
classification de terres 2074
classification d'images 1895
classification d'images
    satellitaires 3201
classification d'images
    segmentées 3252
classification dirigée 3536
classification du maximum de
    vraisemblance 2374
classification dure 1803
classification erronée 2429
classification floue 1497
classification hiérarchique 1824
classification linéaire pas à pas
    3483
classification multispectrale pixel
    à pixel 2476
classification non-contrôlée 3876

classification non-dirigée 3876
classification numérique 1037
classification parallélépipède
    2700
classification rigide 1803
classification selon le maximum
    de vraisemblance 2374
classification taxonomique 3588
clé 2055
clé d'index 3103
clé en chaîne 671
clé étrangère 1457
clé primaire 2888
climagramme 566
climatogramme 566
climogramme 566
clinomètre 1922
clipart cartographique 438
clôture 582, 3944
cluster 585
coalescence 587
coche 3674
cocher 530
co-crigeage 600
codage 1264
codage d'adresse 33
codage de chaîne 519
codage d'éléments
    cartographiques 2271
codage de répétitions 3189
codage par plages 3189
code 595
code binaire 347
code d'attribut 248
code de bruit pseudo-aléatoire
    2941
code de caractéristique 1378
code de commande d'attribut 249
code de fin de ligne 1269
code de forme 1378
code de juridiction 2053
code de longueur de ligne 2180
code d'enchaînement 518
code de place 2764
code de recensement 490
code de trait 1378
code de type de topologie 3768
code de zone 4034
code d'identification de nœud
    2537
code d'objet 2586
code d'unité hydrologique 1871
code en chaîne 518
code géographique 1601
code P 2879
code phonétique 3344
code postal 2871
code postal américain 4034
code Soundex 3344
code spaghetti 3357

code spécifique 551
code spécifique de classe de
    recensement 494
codification 1264
coefficient de réduction d'échelle
    3220
co-enregistrement 772
coexploitation 2047
coextensif 597
coin 773
coïncidence 2365
coïncidence d'arêtes 1223
coïncidence des cartes 2283
collapsus 602
collecteur de données
    d'exploitation 1400
collection de données 875
collection de données
    d'exploitation 1399
collection de données éloignée
    3116
colonne 620
colonne temporaire 3602
coloriage hypsométrique 1878
combinaison 623
combinaison d'objets 2587
combiner 624
communauté urbaine 1924
communication cartographique
    439
communication entre applications
    1981
compacité 629
compaction 656
comparabilité 631
comparaison d'images 1896
compas à pointes sèches 1153
compatibilité de géo-images 637
compensation de cheminement
    polygonal 55
compilation 638
compilation cartographique 2297
compléteur 2094
complexité 646
complexité de la courbe 827
complexité géométrique 1652
complexité topologique 3733
composante de couleur 607
composantes d'une carte 2260
composition à couleurs fausses
    1366
composition d'écritures 453
composition d'image 1897
compréhension 653
compresser 654
compression 656
compression de couleurs 608
compression de largeur de bande
    312
compression des signaux 312

comprimer 654
comptage 783
compte 783
comté 786
concaténation 672
concavité 673
concentrateur 1856
conception 994
conception assistée par ordinateur
    661
conception et dessin assisté par
    ordinateur 662
conception géo-informatique
    1641
conception géorelationnelle 1673
conception gnoséologique 2440
conceptualisation spatio-
    temporelle 3692
concession 20
condition aux bornes 1218
condition limite 1218
conditionneur de données 904
conduite d'écoulement 2127
cône tangentiel 3580
configuration 685
configuration de digitaliseur 1081
configuration de fenêtres 3998
confinement 709
conformation de triade 3817
conformité 687
congélation de coordonnées 1482
congélation de plan 1483
conglomérat 689
connectabilité 698
connecteur 699
connecteur logique 2227
connectivité 698
connexion aux bases de données
    ouverte 2629
connexion en polygone 2835
connexion latérale 2110
connexité 698
conservateur 835
conservateur de titres 835
conservation de l'aire 2885
conserver 3208
consigne d'attribut 257
consistance 700
consistance de données 878
consistance locale 2208
consistance topologique 3734
consolidation 702
Consortium de SIG ouverts 2631
constante 703
constante littérale 2201
constellation 705
constellation GPS 1719
constellation opérationnelle 2634
contenant 709
contenu thématique 3638

contiguïté 712
continuité 716
contour 726
contour de côtes 594
contour de couverture 794
contour isopsophique 2541
contrainte 3155
contrôle 532
contrôle au sol 1785
contrôle cartographique 440
contrôle de couches 2118
contrôle d'exploitation 1397
contrôle spatial 3367
contrôleur d'objets spatiaux 3400
convention de nommage 2486
convergence 745
convergence de méridiens 746
convergence temps-espace 3693
conversion 748
conversion de coordonnées 755
conversion de datum 923
conversion de données 879
conversion de formats 1461
conversion de projections 2909
conversion de rastre en vecteur 3028
conversion de repère 923
conversion de vecteur en rastre 3923
conversion en forme cartographique 749
conversion interne 1988
conversion point à zone 2813
convertisseur 750
convertisseur de rastre en vecteur 3029
convertisseur numérique-analogique 1073
convertisseur polygone 2836
convexité 752
convolution 753
convolution cubique 818
coordinatographe 768
coordonnée 754
coordonnée rectangulaire dans la direction ouest-est 1210
coordonnée rectangulaire dans la direction S-N 2568
coordonnées absolues 4
coordonnées cartésiennes 433
coordonnées curvilignes 833
coordonnées dans le plan 2775
coordonnées dans l'espace 3368
coordonnées d'appareil 1007
coordonnées de Gauss 1513
coordonnées délimitées 394
coordonnées de projection 2910
coordonnées de tic 3672
coordonnées ellipsoïdaux 1255
coordonnées équatoriaux 1307

coordonnées finales 1265
coordonnées généralisées 1526
coordonnées géocentriques 1529
coordonnées géodésiques 1545
coordonnées géographiques 1581
coordonnées horizontaux 1845
coordonnées planaires 2775
coordonnées planes 2775
coordonnées planimétriques 1845
coordonnées polaires 2816
coordonnées provisoires 2927
coordonnées quasi-géocentriques 2971
coordonnées rectangulaires 433
coordonnées spatiaux 3368
coordonnées sphériques 3437
coordonnées sphériques polaires 3437
coordonnées terrestres 3619
coordonnées topocentriques 3703
coordonnées UTM 3895
coordonnée universelle 4007
coordonnée X 4013
coordonnée Y 4016
coordonnée Z 4018
copiage 770, 2903
copie 769
copie au trait 2171
copie d'écran 3232
copyright en cartographie 771
corde 535
corner 773
corps 3339
corps solide 3339
correction 776
correction atmosphérique 227
correction d'adresse 35
correction de contraste 736
correction de distorsions géométriques 1659
correction d'une carte 528
correction géométrique 1653
correction radiométrique 2987
corrections en temps réel 3037
corrections numériques 1022
corrélation 778
corrélation avec décalage 806
corrélation avec retards 806
corrélation croisée 806
corrélation spatiale 3369
correspondance des frontières 379
correspondre 2248
corridor 782
cote 1245
côte 1216
cote de courbe 734
cote de courbe de niveau 734
côte frontière 380
cote orthométrique 1638
couche 2116

couche à accès en direct 2202
couche active 24
couche de base SIG 1689
couche du haut 3701
couche externe 2670
couche isotherme 2035
couche isotrope 2039
couche non visible 2011
couche sans soudure 3240
couche sélectable 3254
couches pyramidaux 2955
couche superficielle 3543
couche thématique 3643
couche vectorielle 3917
couche visible 3956
couleur 619
couleur d'arrière-plan 301
couleur de fond 301
couleur de remplissage 1415
couleur diffuse 1028
couleurs primaires soustractives 3529
couloir 412, 782
coupage 571
coup-arrière 304
coupe 840, 3248
coupe horizontale 1851
coupe transversale 812
coupe verticale 3940
couplage lâche 2235
coupler 3097
coupure 2333, 3285
courbe 826
courbe à l'effet 1463
courbe de Bézier 337
courbe de cuvette 977
courbe de demande spatiale libre 1481
courbe de la chaînette 477
courbe de niveau 732
courbe de niveau bathymétrique 979
courbe de niveau de dépression 977
courbe de niveau fermée 578
courbe de niveau la plus interne 1954
courbe de niveau maîtresse 1929
courbe de niveau non fermée 3841
courbe de Peano 2721
courbe de régression 3093
courbe de translation 3798
courbe d'isovaleurs 2031
courbe figurative 1463
courbe isobathe 979
courbe isohypse 732
courbe isotrope 2038
courbe maîtresse 1929
courbe représentative 1018

courbure de la Terre 825
courbure géodésique 1546
courbure tangentielle 1546
couronne de gisement 1922
cours 791, 3177
couverture catégorielle 474
couverture d'arcs 148
couverture de carte topographique 3715
couverture de levé gravimétrique 1754
couverture de polygone 2837
couverture de réseau 2520
couverture de symboles 3558
couverture d'indice 1930
couverture du sol 2077
couverture numérique 1039
couverture régionale 169
couvertures connectées 694
covariance 792
création de terrain 3607
créneau temporel 3691
crénelage 92
crête 2718, 3441
crête de partage des eaux 3975
critères de filtrage 1423
croisé 3807
croisée de fils 808
croisement spatial 3405
croisillon 1770
croisillon de repérage 808
croissant 188
croquage 3328
croquis 3551
croupe 3161
cryptographie 817
curseur 823
curseur de défilement 3234
curseur réticulaire 1082
curseur réticule 1082
curvimètre 834
cybergéographie 842
cycle géographique 1583

3D 3670
    en ~ 3670
dallage 3684
datamining 900
datawarehouse spatial 3381
date 920
datum 922, 2811
datum à une dimension 2619
datum cartographique 529
datum complet 642
datum 3D 929
datum géocentrique 1530
datum géodésique spatial 3462
datum horizontal 1846
datum spatial 3382
datum vertical 3931

datum vertical géodésique national 2490
déactivation 931
débit en bauds 327
décalage 1126
décalage de datum 926
décalage vertical 3235
décalque 2680
déchiffrage 941
déclaration relative aux incidences sur l'environnement 1291
déclinaison 939
déclinaison magnétique 635
déclivité 1921
décodage 941
décompacter 943
décomposition 942
décomposition de domaine 162
décomposition de zone 162
décompresser 943
découpage 571
découpage à la distance et à la durée 3688
découpage cartographique 2328
découpage des circonscriptions 1150
découpage des feuilles 3285
découpage électoral 1150
découpage hodochrone 3688
découpage polygonal 2827
découverte de décision 934
découverte de réseau automatique 286
découvrir 1424
décryptage 941
défaut 944
défaut de conception 413
défaut de planéité 3851
défilement 3235
défilement de couleur 609
définition d'attribut 251
définition de données 881
déflexion 947
déflexion de la verticale 948
déformation 3974
déformation d'entité 1276
déformation non uniforme 3974
déformation sur un spline 3444
degré 949
degré azimutal 297
degré carré 2318
degré décimal 933
degré de généralisation 1522
délai 952
délai ionosphérique 2012
délier 3871
délimitation 955
délimitation des circonscriptions 1150

délinéateur 956
demande 3140
demande de proposition 3142
demande d'information 3141
démarcation 957
demi-teinte 1799
demi-ton 1799
démographie 965
dendrogramme 966
dénivelée 1812
dénivellation 1812
dénombrement 487
dénomination 2485
dénormalisation 967
densification 968
densification de réseau 2522
densité 969
densité de distribution 1147
densité de points 971
densité de rastre 3010
densité de trame 3010
densité d'observation 970
densité graduelle 1728
département 786, 973
dépassement 2683
dépassement de niveau de réglage 2683
dépassement de réglage 2683
déplacement 1126
déplacement de symbole 2454
dépositaire 835
dépositaire des informations 835
dépréciation 975
dépression 976
dérivation 419
dérivative de surface 3539
dérivative directionnelle 1109
désagrégation 1117
désagréger 1116
désalignement 955
descente 1365
descente plate 1434
description 989
description conceptuelle 676
description de données 881
description de topologie 3756
désélection 991
déséquilibre de détecteur 1001
désignation 2485, 2752
désignation de feuille 3284
dessin assisté par ordinateur 665
dessin au net 1363
dessin cartographique 445
dessin de base de données 861
dessin définitif 1363
dessin de la configuration des broches 2755
dessin d'épreuve 2919
dessin en grisé 1410
dessiner une carte 1180

dessin topographique 3711
destinataire 3043
destination 997
destination active 822
détail 2041
détail cartographique 448
détail cartographique visible 3955
détail culturel 2243
détail de données 922
détecteur 1000
détecteur d'orientation 242
détection à distance 3120
détection de changements 522
détection de contours 1219
détection de côtes 1219
détection de lignes 2172
détection de line de pente 3322
détermination de position 2868
détermination de position par
    satellite 3204
détermination topographique de
    la position d'un point fixe
    3549
détourage 571
deuxième forme normale 3245
devanture 1487
développement 1004
développement de carte 2265
développement d'information
    géographique 1605
développement durable 3552
développement soutenable 3552
déversement radar 2125
déviation 1005
déviation de la verticale 948
dézipper 943
diagonale 1017
diagramme 1018
diagramme à colonnes 315
diagramme à colonnes 3D 928
diagramme à points 3228
diagramme à secteurs 2753
diagramme à tuyaux d'orgue 315
diagramme-bloc de coupe temps
    3690
diagramme bloc d'isolignes 2029
diagramme-bloc métachronique
    3690
diagramme circulaire 2753
diagramme climatique 566
diagramme de blocs 1443
diagramme de chromaticité 606
diagramme de déclinaison 940
diagramme de dispersion 3228
diagramme de flux 1443
diagramme de relations entre
    entités 1279
diagramme de topologie 3757
diagramme en aires 160
diagramme pyramidal 2954

diagrammes de Voronoi 3665
diagramme surfacique 160
diagramme thermo-isoplèthe
    3663
diamètre 1020
diapason d'adresses permissif
    2728
diazocopie 1021
dictionnaire de données 885
différence de temps 3686
différence triple 3827
différence unique 3306
différenciation d'images 1900
diffusion de signaux horaires
    3687
diffusion de temps 3687
diffusion en arrière 303
digitalisation 1054
digitalisation automatique 274
digitalisation d'écran 2627
digitalisation demi-auto 3261
digitalisation d'image 1901
digitalisation rapide 2976
digitaliser 1055
digitaliseur 1078
digitaliseur vidéo 3942
digue 2137
dilution de précision 1097
dilution de précision de
    positionnement 2866
dilution géométrique de précision
    1657
dilution horizontale de précision
    1847
dilution verticale de précision
    3932
dimension 1101
    à trois ~s 3670
    à une ~ 2618
dimension de données temporaire
    3599
dimension fractionnaire 1473
dimension hors-tout 2674
dimensionnalité 1102
dimensionnement de coordonnées
    756
dimension temporaire 3598
direction 1108
direction cardinale 430
direction de compas 634
direction d'écoulement 1444
direction de vue 3943
direction diagonale 1016
direction locale cours d'eau 2211
direction magnétique 634
directoire de base de données 862
discordance 3506
discordance de stratification 3506
discrétisation 1123
disponibiiité de données 857

disponibilité sélective 3260
dispositif de cartographie
    thématique 3647
dispositif de cartographie
    thermique 1947
dispositif de désignation 2804
dispositif de désignation absolu 7
dispositif de désignation relatif
    3108
dispositif de dessin 1181
dispositif de localisation 2221
dispositif de mesure de distance
    1134
dispositif d'entrée 1959
dispositif d'entrée graphique 1740
dispositif de pointage 2804
dispositif de pointage absolu 7
dispositif de pointage relatif 3108
dispositif de sortie graphique
    1741
dispositif digitalisé 1078
dispositifs d'entrée/sortie 1960
dispositifs périphériques 2727
disposition 2123
disposition en mosaïque 3684
disposition topologique 2123
dissémination 1129
dissémination de fréquence 1484
dissolution avec fusion de
    polygones 2838
dissolution de polygones 2838
dissymétrie 214
distance 1131
distance angulaire 128
distance astronomique zénithale
    211
distance d'amortissement 932
distance de Bhattacharya 340
distance d'écart de fermeture 2431
distance de dégénérescence 932
distance de Mahalanobis 2241
distance d'Euclide 1323
distance euclidienne 1323
distance minimale 2424
distance polaire 2818
distance zénithale 4020
distance zénithale géodésique
    1560
distorsion 1137
distorsion de direction 1111
distorsion de distance 1133
distorsion de forme 3281
distorsion de projection
    cartographique 2315
distorsion de zone 163
distorsion géométrique 1658
distorsion radiale 2980
distorsion radiométrique 2988
distorsion sphérique 3434
distribution chi carré 533

distribution continue 718
distribution de Gauss 2556
distribution de Laplace-Gauss 2556
distribution de X^2 533
distribution discrète 1121
distribution gaussienne 2556
distribution normale 2556
district 1148
district administratif 56
district communal 784
district de destination 3584
district de recensement 499
district forestier 1458
diversité 1151
division 1152, 1154
division administrative 57
division cartographique 2300
division de l'environnement 1299
division de recensement 491
division de zoom en couches 4037
division d'images 1902
division en couches 2121
division en zones 4033
division géographique 1591
division topographique 3710
DLG amélioré 1155
document 1158
documentation technique 3590
domaine 1159
domaine d'attribut 252
domaine de référence 3065
domaine intérieur 1983
domaine rempli 1416
donnée 922
données 852
données à application 1398
données aéronautiques 74
données aléatoires 2997
données alphanumériques 99
données ancillaires 121
données à référence géospatiale 1573
données associées 197
données auxiliaires 121
données bibliographiques 341
données brutes 3030
données cartographiques 2263
données cartographiques numériques 1059
données cartographiques numérisées 1059
données catégoriques 475
données climatiques 107
données continues 717
données crues 3030
données d'almanach 96
données d'attribut 250
données de cheminement

polygonal 3810
données de couverture 880
données de graphique d'entreprise 418
données d'élévation 1247
données d'élévation numériques 1041
données de ligne de centre 508
données démographiques 960
données d'entrée 1958
données de points de grille 1775
données de recensement 492
données descriptives 990
données de télédétection 3121
données de terrain 3608
données de terrain artificielles 3566
données de terrain numériques compressées 655
données de trame 3008
données de trame binaires 353
données d'exploitation 1398
données d'image 1898
données d'image non rectifiées 3874
données discrètes 1120
données distribuées 1140
données d'objets topologiques 3762
données dupliquées 1194
données espacées irrégulièrement 2017
données géodésiques 1547
données géographiques 1584
données géologiques 1644
données géométriques 1655
données géoréférencées 1573
données image 1898
données image non rectifiées 3874
données localisées 1573
données marginaux 2350
données multispectraux 2473
données non documentées 3849
données non formatées 3852
données non spatiaux 192
données non temporelles 219
données numériques 1040
données numérisées 1077
données obtenues par agrégation 82
données orbitaires 2640
données planimétriques 2786
données proximaux 2933
données pures 562
données rastrées 3008
données régulièrement ramifiées 3095
données remplaçantes 3547
données satellitaires 3197

données scannées 3222
données secondaires 3244
données socio-économiques 3333
données sortantes 2673
données spaghetti 3358
données spatiaux 3370
données spatio-temporelles 3420
données structurées 3513
données synthétiques relatives au terrain 3566
données tabulaires 3575
données textuelles 3632
données thématiques 3639
données topographiques 3708
données topographiques numériques 1074
données topologiquement structurées 3742
données trépiedes 3829
données valables 562
données valides 562
données vectorielles 3909
données vieilles 78
données virtuelles 3949
dorsale 3161
dossier d'impact sur l'environnement 1291
double différence 1170
double digitalisation 1171
double nom de quad 1173
doublon de fenêtre 573
drapage 1179
driver 1009
driver de digitaliseur 1083
driver de dispositif 1009
droit d'auteur en cartographie 771
droit d'exploitation en cartographie 771
droite 3501
droite achevée 1348
droite de suite 3775
droite tangente 3582
droite tendue 1348
droits d'accès géographique 1567
duplication 1196
dyke 1095

éboulement de terrain 2091
écart 1510
écart circulaire probable 544
écart de commune 784
écart de fermeture 1319
écart de hauteur 1812
écart type 3461
échange climatique 567
échange de données 888
échange de données dynamique 1198
échanges de limite 378
échantillon 3191

échantillon d'essai 3191
échantillonnage 3194
échantillonnage de cheminement
      longitudinal 97
échantillonnage de routes
      longitudinal 97
échantillonnage de trajectoire 97
échantillonnage du plus proche
      voisin 2509
échantillonnage géographique
      4028
échantillonnage longitudinal 97
échantillonnage par zones 4028
échantillonnage spatial 3415
échantillonnage sur des routes
      convergentes 22
échelle 3210, 3211
échelle cartographique 2326
échelle de carte 2326
échelle de coloriage
      hypsométrique 1879
échelle de compilation 640
échelle de couleurs 618
échelle d'édition 3218
échelle de généralisation 1524
échelle de gris 1759
échelle de pente 3321
échelle de projection 3217
échelle de reproduction 3218
échelle du levé 3219
échelle élargie 1273
échelle explicative 1343
échelle graphique 1744
échelle intermédiaire 1987
échelle logarithmique 2224
échelle nominale 2892
échelle numérique 3139
échelle particulière 2707
échelle principale 2892
écho 1212
écho-image 1172
éclaircissage 3667
éclairement 404
éclat 3319
écodéveloppement 3552
économie de production
      cartographique 1214
écran 3231
écran monté sur la tête 1810
écrasement de boisage 2043
écrêtage 571
écriture cartographique 453
éditeur 1226
éditeur d'images classifiées 557
éditeur hybride 1858
édition 1225
édition de rastre 3011
effaçage 954
effacement 954
effacer 953

effet au joint 381
effet de bande 306
effet de loupe 4036
effet de premier ordre 1429
effet de profondeur 980
effet Doppler 1165
effet d'ordre secondaire 3246
effet externe 1354
effet relief 1260
effets atmosphériques 228
effondrement 602
élaboration 1004
élagage 3988
élagage de ligne 2197
élément 2041
élément d'aire 1242
élément de données 922
élément de menu 2392
élément de réseau 2523
élément de surface 1242
élément de texture 3634
élément de topologie 3759
élément de volume 1243
élément d'image 2758
élément mixte 2435
élément pictural 2758
élément sensible d'orientation 242
éléments graphiques 1738
élément topographique 3712
élément volumétrique 1243
élévateur 1252
élévation 1244, 1811
élévation du sol 1787
élévation latérale 3293
élévation stéréo 3487
élimination dâ bruit 2542
élimination de lignes cachées
      1821
ellipse de distorsions 1253
ellipsoïde 1254
ellipsoïde de la Terre 1203
ellipsoïde de révolution 1256
ellipsoïde mondial 4009
ellipsoïde référentiel 3067
emballeur de données 904
embase 2199
emboîtage 1259
emboîtement 1259
émetteur 3799
émission de données 903
émondage 3988
émondage de ligne 2197
empiècement 3846
empiètement 2679
empilement 2685
emplacement 2215
empreinte 1918
encapsulage 1263
encastrement 1259
enceinte 582

enclos 582
encombrement 2609
encorbellement 3167
endroit 2762
engagement 2609
engin spatial 3356
enquête 3548
enquête auprès des ménages 1854
enquête sur les ménages 1854
enquête sur le terrain 1407
enquête sur l'utilisation du sol
      2103
enregistrement 3091, 3184
enregistrement cartographique
      2275
enregistrement d'adresses 41
enregistrement de données 3184
enregistrement d'images 1910
enregistrement public 2952
enregistreur de profils 2902
enregistreur de profils aéroporté
      86
enrobage 1263
enseignement cartographique 446
enseignement géo-informatique
      1640
ensemble compilé de cartes 641
ensemble de données 911
ensemble de données carroyés
      1765
ensemble de données
      géographiques 1588
ensemble de frontières 392
ensemble de frontières liées 790
ensemble de sélection d'entités
      1282
ensemble de valeurs 3899
ensemble exhaustif 1339
ensemble flou 1505
entité 1274
entité cartographique 448
entité équivalente statistique 3477
entité géographique 1595
entité graphique 1274
entité linéaire 2162
entité ponctuelle 2809
entités de marquage 2357
entités fermées 576
entité statistique 3475
entraînement 1178
entrée de données 886
énumération 487
énumération d'occupance spatiale
      3402
enveloppe convexe 751
enveloppe de texte 3627
environnement 1286
environnement cartographique
      447
environnement de symboles 3554

environnement naturel 2498
éphémérides 2640
épipolaire 1302
épuration 3236
équateur 1306
équateur céleste 479
équation de projection 2912
équidensitométrie 972
équidistance 730
équidistance des courbes de
    niveau 730
équidistribution 1313
équipement 1314
équipement cartographique
    avancé 59
équipement de cartographie 1315
équipement de mesure 2380
équivalent géographique 1597
érosion 3981
erreur 1317
erreur admissible 58
erreur aléatoire 16
erreur chronométrique 3689
erreur circulaire de position 545
erreur constante 3567
erreur de calcul 658
erreur d'échange de données 889
erreur d'échantillonnage 3195
erreur d'échelle 3215
erreur de classification 2429
erreur de digitalisation 1087
erreur de distance 3001
erreur de données 887
erreur de fermeture 1319
erreur de fermeture d'une maille
    en altimétrie 3934
erreur de format 1462
erreur de géocodage 1536
erreur de position 2864
erreur de positionnement 2864
erreur de trajets multiples 2463
erreur d'instrument 1968
erreur d'inversion 2010
erreur d'observation 2608
erreur en travers de la trajectoire
    816
erreur fixe 3567
erreur grossière 367
erreur instantanée de temps 3689
erreur latérale 816
erreur linéaire 2154
erreur moyenne quadratique 3172
erreur moyenne quadratique de
    l'unité de poids 3164
erreur non détectée 3848
erreur polygonale 2828
erreur psychomotrice 2948
erreur radiale 2981
erreur régulière 3567
erreurs ionosphériques 2013

erreurs troposphériques 3831
erreur systématique 3567
erreur théorique de fermeture des
    polygones 3659
erreur topologique 3735
erreur topologique automatique
    287
erreur transversale 816
erreur verticale 3933
erreur verticale de fermeture 3934
escalier de demi-teintes gris 1759
escarpement 3482
espace 3352
espace à bâtir 414
espace clos 582
espace d'adresses 43
espace d'appareil 1010
espace de caractéristique 1388
espace de coordonnées global
    1703
espace découvert 2628
espace de trait 1388
espace de trajectoires 3787
espace de travail 4004
espace d'Euclide 1326
espace euclidien 1326
espacement d'accrochage 3329
espacement des lignes du
    quadrillage 1771
espacement entre lignes 2191
espacement fixe 1431
espace multidimensionnel 2460
espace numérique 1069
espace orthogonal 2664
espace périphérique 1010
espace rastre 3026
espace statistique 3473
espace topologique 3749
essai 532
essai de réception 11
essai de recette 11
essai longitudinal 2231
essais paramétriques 2705
estimation 1321
estompage 1837
estompage analytique 115
estompage de pente 3936
estompage d'ombre oblique 2602
estompage numérique 115
estompage photographique 2744
estuaire 1322
établissement 3269
établissement de cartes de
    cyberespace 844
établissement de plans cadastraux
    3723
établissement de profil 2903
établissement d'un projet 994
étagement 3458
étalement de la largeur de bande

313
étalon 3191
étalonnage concurrentiel 334
étalonnage de performances 334
étalonnage de SIG 1690
étalonner 3097
étendue 3360
étendue d'application 2270
étendue de couverture 795
étendue de grille 1767
étendue de page 2686
étendue d'une erreur 3001
étendue temporaire 3600
étiage 2237
étiquetage 2066
étiquetage d'attributs 262
étiquetage de grappes 588
étiquetage de lignes de niveau
    735
étiqueté 2065
étiquette 2061
    sans ~ 3869
étiquette d'attribut 260
étiquette d'utilisateur 838
étiquette textuelle 3628
étirage linéaire 2167
étirement 306
étirement de contraste 737
étirement par fil élastique 3186
étirement par fil élastique local
    2207
étude de faisabilité 1373
étude de viabilité 1373
étude sur le terrain 1407
euristique 1817, 1818
évaluation 1321
évaluation de carte 2269
évaluation de gradient 1729
évaluation de performance 334
évaluation d'objets distante 3118
événement 1329
événement linéaire 2155
événements géophysiques 1668
évolution démographique 2858
exactitude 17
exactitude absolue 2
exactitude cartographique 2298
exactitude conceptuelle 675
exactitude d'attribut 245
exactitude d'échelle 3212
exactitude de classification 553
exactitude de mesure 2382
exactitude de mesure
    cartographique 2284
exactitude de position 2863
exactitude de position
    différentielle 1024
exactitude géométrique 1651
exactitude géométrique constante
    704

exactitude horizontale 1842
exactitude logique 2226
exactitude temporaire 3596
exactitude verticale 3929
exagération 1336
exagération verticale 3935
excentricité 1211
exemplaire 769
exploitation conjointe 2047
exploitation du terre 2097
explorateur multispectral 2477
exploration 3225
exportation 1345
exportation d'objets vectoriels
    1346
exposition 633
expression 1347
expression booléenne 369
expression de Boole 369
extension du spectre 313
externalité 1354
extraction d'arêtes 1221
extraction d'attributs 1379
extraction de caractéristiques
    1379
extraction de données 890
extraction de formes 1379
extraction de frontière 382
extraction de modèle altimétrique
    digital 958
extraction de primitives de
    reconnaissance 1379
extraction de rangées 3002
extraction de traits géométriques
    2173
extraction de traits pertinents
    1379
extraire une information
    d'attributs 1356
extrapolation 1357
extrémité 1270

face 1488
face adjacente 50
face frontale 1488
facette 1360
facettes adjacentes 51
facettes de réseau de triangles
    irréguliers 3697
facettes rectangulaires 3051
facteur d'échelle 3216
facteur d'échelle de grille 1778
facteur d'échelle linéaire 2166
facteur de contraste 596
facteur de courbure équivalente
    de la Terre 1228
faille-pli 1812
faire dévier 4014
faire une embardée 4014
falsification des données 3446

fausse origine 1369
fenêtrage 3999
fenêtre à liste directe 2860
fenêtre atmosphérique 232
fenêtre d'affichage 3944
fenêtre de butineur 405
fenêtre de découpage 572
fenêtre de graphique 1752
fenêtre de légende 2135
fenêtre de numérisation 1093
fenêtre de statistique 3481
fenêtre de vue 3944
fenêtre du layout 2124
fenêtre en incrustation 2860
fenêtre flash 2860
fenêtre maximisée 2372
fenêtres en cascade 472
fente 1510
fermer 574
fermeture 582
fermeture automatique 281
feuille d'assemblage 1933
feuille de calcul électronique
    3450
feuille de compilation 2659
feuille de levé 1405
feuille d'une carte 2333
feuille d'une carte adjacente 52
feuille sans courbes de niveau ni
    teintes hypsométriques 3870
feuillet 3511
feuilletage 406
fiabilité de communication 628
fiabilité de données 909
fiabilité de la méthode
    cartographique de recherche
    3112
fiabilité d'investigations
    cartographiques 3113
fiabilité d'une carte 2324
fiabilité organisationnelle 2654
fiabilité technique 3591
fichier 1411
fichier carte 2272
fichier d'actifs 195
fichier d'aérophotos de quads 67
fichier de base de données
    externe 1352
fichier de coordonnées 759
fichier de détails cartographiques
    449
fichier de données géographiques
    1586
fichier de formes 3282
fichier de limites 383
fichier d'enregistrement 2225
fichier de palette 2688
fichier de points 2803
fichier de points d'arrêt 3499
fichier de projection 2913

fichier de quads 2957
fichier de relations entre secteurs
    de recensement 502
fichier des données descriptives
    884
fichier de surfaces et valeurs de
    parcelles 1412
fichier de symboles 3555
fichier de tics 3673
fichier de travail 4003
fichier externe 1353
fichier non hiérarchique 1436
fichier plat 1436
fichier rastre 3012
fichiers de base géographique
    1576
fichier séquentiel 3267
fichier trame 3012
fidélité textuelle 3631
figure 3279
figure fermée 577
figure fond 1410
filage interactif d'une courbe 1978
filage interactif d'une courbe de
    niveau 1978
filet 2396
filet à border 376
filet de cadre 376
filon 3238
filon rocheux 1095
fils croisés 808
filtrage de gradient 1730
filtrage spatial 3385
filtre 1422
filtre de bruit de granulation 3426
filtre de détection de contours
    1220
filtre de dilatation 1096
filtre de domaine fréquentiel
    1485
filtre de mode 3978
filtre de modes d'ondes 3978
filtre directionnel 1110
filtre linéaire 2156
filtre médian 2387
filtre morphologique 2446
filtre moyen 291
filtre passe-bas 2236
filtre passe-haut 1831
filtres distincts 1731
fin de ligne 1268
finesse de la chronologie 2149
fixer 1430
flèche 185
flèche d'orientation 2565
flèche du Nord 2565
flou 1494, 1495
fluctuation 1449
flux de données 892
fonction d'application 2301

fonction de distribution de probabilité 2899
fonction d'édition cartographique 2268
fonction de groupement 1796
fonction de l'ordre de l'observation 2646
fonction d'estimation linéaire sans biais 335
fonction de transfert 3790
fonction de transmission 3790
fonction d'objectif 2591
fonction mathématique 2369
fonctionnalité de SIG 1693
fonctionnement en réciprocité 1996
fonction objective 2591
fonction topologique 3736
fonction vectorielle 3913
fond 300
fond de carte 3706
fond rastre 3007
fondu 1130
fonte 526
fontis 2091
force de champ 1406
formatage de carte 2273
format d'atlas 223
format d'échange 1337
format d'échange numérique 1044
format d'échange standard 3463
format de données 893
format de données d'élévation 1249
format de données vectoriel 3910
format de fichier 1413
format de produit vectoriel 3920
format de transfert de plan cadastral 2653
format de transfert national 2495
format de travail cartographique 463
format d'images à lignes entrelacées 307
format d'images à pixels entrelacés 308
format d'images BSQ 310
format DLG 1058
format DXF 1182
format GIF 1683
format JPEG 2048
format JPG 2048
format image-vecteur 3915
formation géologique 1645
format MOSS 2292
format NTF 2495
format numérique 1045
format numérique géographique vectoriel 1048
format rastre 3013

format RINEX 3044
format TIF 3579
format TIGER 3741
format TRIM 3616
format vectoriel 3910
forme 2716, 3279
forme de courbe 832
forme de marqueur 2354
forme fermée 577
forme géométrique régulière 2716
forme isobare 2022
forme normale 2557
formulaire 1460
fractale 1470
fractionnement 3445
fractionnement modal 2438
fragment surfacique 3545
fréquence de hachure 1807
fréquence spatiale 3386
frontière 377
frontière d'un domaine particulier 389
frontière d'une zone particulière 389
fuseau 1716
fusion 2393
fusion d'information multisource 877
fusion multisource 877
fusionnement 2393

gadget 1508
gamme 2999
garnissage 1417
gauchissement 3974
généalogie 2151
généralisation 1520
généralisation algorithmique 91
généralisation automatique 277
généralisation cartographique 450
généralisation de données spatiaux 3372
généralisation de morceaux de surface 173
généralisation dynamique 1199
généralisation fine 1425
généralisation linéaire binaire 350
généralisation par télédétection 3122
généralisé 1525
générateur de listes 3132
générateur de rapports 3132
génération de courbes 831
génération de polygones 2841
génération de rapports 3133
génération de volume 2223
génération d'un quadrillage 1769
géobase 1585
géochimie numérique 1046
géocodage 1592

géocodage automatique 283
géocodage interactif 1976
géocodage par adresse postale 39
géoconception 1641
géodémographie 1539
géodésie 1542
géodésie appliquée 141
géodésie de mines 2427
géodésie d'une surface 3541
géodésien 2094
géodésie satellitaire 3198
géodésie sphéroïdale 3440
géodésie sur satellites 3198
géodésique 1540, 1541
géodimensions 1561
géodynamique 1562
géographe 1565
géographie de recensement 495
géographie d'espace des adresses Internet 1636
géographie de transport 3805
géographie économique 1213
géographique 1569
géographiquement dispersé 1570
géographiquement distribué 1570
géographiquement spécifique 1572
géoïde 1637
géo-image 1639
géo-image 3D 1011
géo-image dynamique 1200
géo-informatique 1607
geomarketing 1650
géomatique 1607
géomètre 2094
géométrie coordonnée 760
géométrie de lit 524
géométrie de recouvrement 797
géométrie de solides constructive 708
géométrie de terrain 3612
géométrie d'Euclide 1324
géométrie d'illumination 1887
géométrie d'objet 2589
géométrie euclidienne 1324
géométrie plane 2791
géométrie projective 2917
géométrie satellitaire 3199
géomorphométrie 1666
géonémie 536
géoportail 1669
géoprocessing 1671
géoréférence 1619
géoréférencé 1572
géoréférencement 1593
géo-ressource 1623
géostationnaire 1679
géosynchrone 1679
gestion de données géographiques 1587

gestion de ressources naturelles 2502
gestion de sites 1362
gestion des terres 2086
gestion d'information environnementale 1292
gestion d'information topographique 3713
gestion externe du système informatique 1362
gestion par impartition 1362
gisement 329
glissement 1178
glissement de terrain 2091
globalisation 1708
globe 1209
globe céleste 480
globe planétaire 2781
globe terrestre 1209
glyphe 1714
godet 1716
gouvernail d'élévation 1252
GPS 1710
GPS cinématique 2056
GPS différentiel global 3995
gradation 1725
grade 949
gradient 1726
graduation 3211
graduation de zones 176
grand cercle 1761
grandeur de cellule 483
grandeur dérivée 985
grand objet binaire 349
grand réseau 3996
granularité 1734
graphe 1735
graphe connexe 695
graphe dirigé 1105
graphe flou 1499
graphe hyperbolique 3D 1015
graphe orienté 1105
graphe plan 2767
graphe plat 2767
graphe topologique 3737
graphique 1018, 1736
graphique à bulles 407
graphique circulaire 2753
graphique de dispersion 3228
graphique de surface 160
graphique en segments 2176
graphique en tarte 2753
graphique géométrique 1660
graphique linéaire 2176
graphique par coordonnées et traits 761
graphique par ordinateur 669
graphique par points 3228
graphique prédessiné cartographique 438

graphique rastre standardisé 3464
graphique sectoriel 2753
graphique statistique 3476
graphique vectoriel 3914
grappe 585
graticule cartographique 2274
gravimétrie 1757
grille 2274
grille alphanumérique 100
grille binaire 348
grille cartographique 2274
grille d'acheminement 3183
grille d'adresses 37
grille de localisation 2214
grille de mesure 2383
grille rectangulaire 3052
grille rectangulaire régulière 3053
grille régulière 3094
groupage 587
groupage flou 1498
groupe 584
groupe de blocs de recensement 489
groupe de données de généralisation 1521
groupe de gestion des objets 2593
groupement 587
groupe topologique 3738

habitat 1797
hachure 3271, 3275
hachure croisée 809
hardware 1804
haute géodésie 1830
hautes eaux 1834
hauteur 1811
hauteur de cellule 482
hauteur d'un repère par rapport à un plan 1813
hauteur du plan d'eau 3976
hauteur géodésique 1548
hauteur normale 2558
hauteur orthométrique 1638
hauteur relative 3105
haut plateau 3571
hélice sphérique 3160
hémisphère 1815
hétérogénéité 1816
heure de Greenwich 758
heure du méridien de Greenwich 758
heuristique 1817, 1818
hiérarchie 1828
hiérarchie géographique 1599
histogramme 315
historique 151
historique de données 894
HO 1638
homogénéité 3854
horizon 1841

hydrographie 1865
hydrographique 1861
hydrosystème 3163
hypothèse nulle 2573
hypsométrie 1880

IA 187
icône 1881
identifiant numérique 1882
identificateur 1883
identificateur de caractéristique 1382
identificateur de classe 551
identificateur de couverture 798
identificateur de forme 1382
identificateur de nœud 2537
identificateur de nœud initial 3469
identificateur de nœud périphérique 1267
identificateur d'enregistrement 3050
identificateur de repère terrestre 2088
identificateur de trait 1382
identificateur d'objet 2590
identificateur unique globalement 1709
identification de place 2765
identification de repère 1474
identification de structure 1379
identification de tractus 3779
identification d'utilisateur 3889
identité 1884
identité géographique 1602
identité géographique universelle 3861
île 2020
illumination 1886
image 1888
image à points 3015
image booléenne 370
image cartographique 456
image changée 523
image classifiée 556
image cliquable 564
image 2D 1098
image 2,5D 1099
image 3D 3969
image d'estompage 1836
image double 1172
image en coordonnées 564
image en demi-teintes de gris 1760
image en rotation 3175
image fantôme 1172
image filaire 4000
image fil de fer 4000
image fractale 1471
image GIF animée 129

image graphique 1739
image infrarouge couleur 613
image instantanée 3330
image matricielle 3015
image multicapteur 2472
image numérique 1051
image numérisée 1079
image ombrée de relief 3272
image paramétrée 2703
image pixélisée 3015
image planimétrique 2787
image plurielle 2978
image plurielle par radar 2978
image pseudocouleur 2936
image rastre 3015
image réactive 564
image réactive calibrée 424
image réactive géoréférencée 1672
image reclassifiée 3047
imagerie satellitaire 3202
image satellitaire 3200
image simple 2620
image spatiale 3406
image volumétrique 3969
imagination générée par ordinateur 668
imbrication 1259
imbriquer 1961
imitation 3305
impasse 1913
impédance 1914
impédance de point d'arrêt 3497
impédance de retournement 3837
impédance de terre 1788
implémentation 1915
importation 1916
imprécision 1917
impression 1918, 2895
impression cartographique 2311
imprimante 2894
incidence 1919
incidence maille-arête 2398
incidence nœud-arête 2536
inclinaison 1921
inclinomètre 1922
inclusion 1923
incrément 1925
indexage 1931
indexage bibliographique 342
indexage d'article 2042
indexage de base de données 863
indexation 1931
indexation automatique 268
indexation de mosaïque 3683
indexation relationnelle 3101
indexation spatiale 3388
index CTSI 504
index de bibliographies

géographiques 1603
index descendant 986
indexer une ligne de niveau 1928
index géographique 1603
index ponctuel 1166
index TIGER-CTSI 3678
indicateur de pente 1922
indicateur linéaire 2177
indicatif 993
indicatif de zone 161
indicatif de zone de numérotage 2577
indication 1934
indice 1927
indice bêta 336
indice cartométrique 467
indice d'accessibilité 15
indice de colonne 622
indice de couche 2120
indice de position 2220
indice de sensibilité environnementale 1297
indice morphométrique 2448
indice spatial 3387
indice zonal 4024
inégalité 3851
inexactitude 1917
inexactitude de dépassement 2682
inexactitude due au sous-dépassement 3847
inflexion 1939
inflexion de couleur 612
influence de la limite 381
info-bulle 3700
infographie d'entreprise 417
infographie par coordonnées 761
information cartographique 451
information de passages supérieurs et inférieurs 2677
information de point de cheminement 3980
information de références spatiaux 3410
information de terrain 3608
information géographique 1584
information géographique distribuée 1144
information publique 2949
information textuelle 3632
informatique décisionnelle spatiale 3381
informativité de carte 2276
infrastructure 1948
infrastructure de carte diagramme 537
infrastructure de données géospatiaux 1677
infrastructure de données spatiaux 3373
infrastructure d'information

géographique 1677
inhéritance 1950
initialisation 1952
inondation 1439
inscription cartographique 453
inscriptions explicatives 1342
insérer 1961
insertion de données 886
inspection de topologie 3753
installer 1963
instance 1964
instanciation 1966
instruction d'affectation 196
instrument 1967
instrument de cartographie 2302
instrument de dessin 1183
instrument de dessin perspectif 2732
instrument de mesure de niveau 2139
instrument géodésique 1549
instruments de base de données 872
instruments de mappage génériques 1527
intégrateur de bases de données 864
intégrateur de données 895
intégrateur d'images 1904
intégration 1973
intégration de données multisource 877
intégrité de base de données 865
intégrité de données 896
intégrité d'entité 1277
intégrité référentielle 3075
intelligence artificielle 187
intensité de champ 1406
interaction spatiale 3391
interclassement 2399
interface 1982
interface de formes 1464
interface d'infographie 670
interface d'utilisateur 3890
interface graphique utilisateur 1748
interface visuelle 3961
Internet-SIG 1695
interopérabilité 1996
interpénétration 1997
interpolateur bicubique 343
interpolateur bilinéaire 346
interpolateur de pixels 2760
interpolateur exact 1335
interpolateur horizontal 1848
interpolateur linéaire 2158
interpolateur non exact 2549
interpolateur trilinéaire 3826
interpolateur vertical 3937
interpolation 1998

interpolation à intervalles égaux 1305
interpolation approchée 143
interpolation bilinéaire 345
interpolation d'aire 170
interpolation de contour de terrain 3613
interpolation linéaire 2157
interpolation pondérée inversement par les distances 2006
interpolation spatiale 170
interpolation topogrid 3727
interprétation 2000
interprétation de carte 2321
interprétation visuelle d'images 3960
interrogation 3140
interrogation cartographique 2320
interrogation géographique 1609
interrogation interactive 2626
interrogation logique 2228
interrogation spatiale 3407
interrogation SQL 3454
interrogation topologique 3766
intersection 2001
intersection de point unique 3308
intersection de zones 167
intersection du quadrillage 1770
intersection non planaire 2553
intersection plane 2768
intersection spatiale 3393
interstice 1510
intervalle 2002
intervalle d'accrochage 3329
intervalle de classe 559
intervalle de confiance 684
intervalle de récurrence 3056
introduction de données 886
invariance 2004
inventaire de cartes 3495
inventaire des ressources naturelles 2501
irrégularité 2014
isobare 2021
isobathe 979
isocartographie 733
isodensitométrie 972
isohypse 732
isolation 2028
isolignage 728
isolignage automatique 282
isoligne 2031
isoplèthe 2031
isotherme 2036
isotropie 2040

jalonnement différentiel 1026
jalons de pente 3323
joint 2044

jointure 2044
jointure spatiale 3394
jonction 2051
jonction relationnelle 3102
justesse 17, 777

kit d'instruments cartographiques 2305
kit d'instruments de réalisateur 1003
krigeage 2059
krigeage d'indicateur 1935
krigeage non linéaire 2551
krigeage ordinaire 2649
krigeage simple 3299
krigeage stratifié 3505

label 2061
lacet 2234
lâchage 1185
lacune 1510
laisse de basse mer 2238
laisse des hautes eaux 1835
lambrissage 2693
lambrissement 2693
lancé de rayons 3033
langage d'accès de données 854
langage de balisage de géographie 1635
langage de carte 2277
langage de définition de données 882
langage de description de données 882
langage de manipulation de données 898
langage de requête structurée 3515
langage de scripts 3233
langage de traitement de données spatiaux 3374
langage d'interrogation 2974
langage encastré de requêtes structurées 1258
large
    au ~ 2616
    au ~ des côtes 2616
large spectre 3451
largeur de bande 311
largeur de cellule 486
latitude 205
latitude astronomique 205
latitude d'origine 2114
latitude géocentrique 1531
latitude géodésique 205
latitude géographique 205
latitude-longitude 2113
latitude vraie 3833
latte 2112
lattis 2115

lattis vectoriel topologique 3752
LAV 288
layout 2123
lecteur de coordonnées 764
lecteur de courbes 2174
lecture de carte 2321
légende 2278
légende cartographique 2278
levant 188
levé 1556
levé aérien 70
levé altimétrique 103
levé bathymétrique 325
levé cadastrale 3723
levé d'ensemble 644
levé du fond 325
levée 2137
levé électromagnétique 1235
levé géodésique 1556
levé géodésique de base 321
Levé géologique d'États-Unis 3857
levé gravimétrique 1757
levé hydrologique 1864
levé intégré 1971
levé océanographique 2612
levé par cheminement 2377
levé photothéodolite 2750
levé planimétrique 2789
lever 2249
levé sismique en 3D 1188
levé terrestre 1407
levé topographique 3723
levé topographique à la planchette 2780
levier 2050
liaison 2199
liaison d'attributs 253
liaison de réseau 2527
liaison des cartes 2281
liaison orientée 1106
libellé 2065, 2201
lien 2199
lieu d'arrêt 3498
lieu découvert 2628
lieu géographique 1619
ligature 2199
ligne 2150, 3184
    hors ~ 2615
ligne à casser 400
ligne axiale 507
ligne blanche 360
ligne brisée 2722
ligne cachée 1820
ligne calibrée 425
ligne caténaire 477
ligne centrale 507
ligne conditionnelle 681
ligne corrigée 425
ligne d'appui de contour 3537

ligne d'eau 3976
ligne de balayage 3226
ligne de balayage de quadrant
    2970
ligne de base 317
ligne de base d'adresse 32
ligne de carte diagramme 538
ligne de cassure 400
ligne de centre 507
ligne de changement de date 921
ligne de cloison 1395
ligne de commande 626
ligne de comptage 1929
ligne de contact à suspension
    caténaire 477
ligne de contour 731
ligne de contour intermédiaire
    1985
ligne d'écoulement des eaux d'une
    surface 3561
ligne de déviation 1006
ligne de flux 1445
ligne de frontière 385
ligne de grille 3018
ligne de guidage 2127
ligne de latitude 2184
ligne de lecture 3226
ligne de longitude 2185
ligne de méandre 2376
ligne de niveau 732
ligne de niveau auxiliaire 290
ligne de nœuds 2186
ligne de partage 1152
ligne de partage des eaux 3975
ligne de position déterminée par
    observation visuelle 3964
ligne de projection 2187
ligne de rastre 3018
ligne de sonde 2127
ligne de synclinal 3561
ligne de tendance 3814
ligne de trame 3018
ligne d'exploration 3226
ligne droite 3501
ligne dupliquée 1195
ligne élastique 1231
ligne en traits 2722
ligne géodésique 1541
ligne isogone 2024
ligne loxodromique 3160
ligne non isométrique 2550
ligne non régulière 2015
ligne orthodromique 2661
ligne pendante 849
ligne radiale 2982
lignes 2193
ligne sans fin 3850
lignes de balayage manquantes
    2434
lignes d'égale altération 1138

lignes de hachure 1808
lignes équidistantes 1311
lignes jointes 2045
ligne solide 3340
ligne standard 3465
ligne terrestre 2084
ligne unique 3307
ligne verticale 3938
ligne vide 360
limite 377
limite à bâtir 415
limite artificielle 186
limite conjointe 691
limite d'arrondissement 363
limite d'attachement 237
limite de circonscription 1149
limite de code postal 4023
limite de comté 787
limite de département 787
limite de la superficie 387
limite de place 2763
limite de secteur 1149
limite de secteur de recensement
    500
limite d'exactitude d'échelle 3213
limite de zone d'extension 3882
limite d'un polygone 2834
limite d'utilisation du sol 2098
limite floue 1496
limite métropolitaine 2417
limites de carte 2280
limites de couverture
    géométriques 1654
limites de feuille de carte 2334
limites de grille 1763
limites de juridiction 2052
limite spéciale 3423
limites politiques 2821
limite unie 691
linéaire-angulaire 2152
lingot 584
lissage 136
lissage de données 915
lissage de ligne 2190
lissage de spline 3444
lissage d'une courbe 830
lissage spatial 3416
liste de classes 560
liste de longueur variable 3903
liste de points de contrôle 741
liste ordonnée 2644
liste topologique 3760
liteau 2112
littéral 2201
littoral 594
livre foncier 422
localisant 2068
localisateur 2221
localisateur de carte 2282
localisation 2215

localisation absolue 5
localisation automatique de
    véhicules 288
localisation d'événement 1330
localisation isolée 2026
localisation relative 3106
localisé 1572
localité 2213
localité centrale 512
location 2215
location spatiale 3395
logiciel 3336
logiciel appliqué 140
logiciel de dessin graphique
    interactif 1977
logiciel de génération de
    couverture 796
logiciel de GPS 1723
logiciel de numérisation 1092
logiciel de stockage de
    l'information cartographique
    3337
logiciel de transformation de
    coordonnées géographiques
    1519
logiciel de visualisation 3963
logiciel de visualisation
    cartographique 2348
logiciel SIG 1700
logique floue 1500
longitude 206
longitude astronomique 206
longitude géocentrique 1532
longitude géodésique 1550
longitude géographique 206
longitude héliographique 1814
longueur d'arc 2136
longueur de champ 1401
longueur de chemin 2711
longueur de ligne 2179
longueur de ligne corrigée 775
longueur de ligne révisée 775
longueur de parcours 3188
longueur de parcours
    atmosphérique 231
longueur de trajectoire 2711
longueur de trajectoire
    atmosphérique 231
longueur variable 3902
lot de données 911
lotissement 2089
loupe de digitaliseur 1084
loxodrome 3160
loxodromie 3160
luminosité 404

macro 2239
macrocommande 2239
macro-instruction 2239
magnétisme 3328

maillage 2396
maillage par polygones 2846
maillage par polygones 3D 1177
maille 2396
maille triangulaire adaptative 27
management de concurrence 680
manche à balai 2050
manette 2050
manuscrit de minute d'auteur 639
mappage 2908
mappage de caractéristiques 1384
mappage de côtes 1222
mappage de dépannages 1118
mappage de distance minimale
   2426
mappage de lignes 2181
mappage d'environnement
   cubique 819
mappage de traits 1384
mappage topologique 3743
mappemonde 2288
mapper 2290
mappe thématique 3644
mappeur thématique 3647
mapping 2908
maquette 3595
maquette d'atlas 221
marais 368
marais côtier 591
marche de rayons 3032
marché de SIG 1696
marée basse 2237
marge de référence 3066
marge de tampon 409
marquage 2066, 2122
marquage en trait 851
marquage par piquets 2122
marqué 2065
marque d'attribut 248
marque de canevas 740
marque de jalonnement 333
marque de triangulation 3823
marque d'événement 1331
marque ponctuelle 1166
marqueur 2352
marqueur de triangulation 3823
marqueur routier 956
masquage 2360
masque 2359
masque de données 899
masque d'élévation 1250
masquer 1822
massif 584
matérialisation d'un repère 2358
matériel 1804
matériel de digitalisation 1088
matière brute 3346
matrice 2371
matrice d'altitudes 105
matrice de classification 554

matrice de classifications
   erronées 2430
matrice de comportement 332
matrice de confusion 688
matrice de corrélations 780
matrice de corrélations croisées
   807
matrice d'élévation numérique
   1042
matrice de Morton 2450
matrice fantôme 3278
médiane 2386
meilleur estimateur linéaire sans
   biais 335
mémoire 3500
ménage 1371
mensuration 2389
mensuration pour le
   rétablissement des limites 393
menu 2391
menu à liste directe 2859
menu cascade 470
menu contextuel 3289
menu de digitaliseur 1085
menu déroulant 2953
menu en cascade 470
menu en incrustation 2859
menu relevant 2859
menu superposable 2859
mer
   en ~ 2616
Mercator transverse universel
   3867
méridien 2395
méridien astronomique 208
méridien de référence 3070
méridien d'origine 2889
méridien géocentrique 1533
méridien géodésique 1553
méridien orthogonal 2662
mesurage 3548
mesurage absolu 6
mesurage indirect 1937
mesurages géodésiques 1552
mesure angulaire 2381
mesure d'angle 2381
mesure de distance 3004
mesure de distance différentielle
   1026
mesure de pseudo-distance 2943
mesure de route 3179
mesure d'images 1905
mesure d'images 3D 1100
mesure électronique des distances
   1238
mesure étendue 1350
mesure extensive 1350
métacartographie 2400
métadonnées 2401
métadonnées spatiaux 3396

métafichier 2403
métamorphisme 2405
métaphore géographique 1612
méthode d'accès spatial 3362
méthode d'arrière-plan qualitatif
   2413
méthode d'arrière-plan quantitatif
   2414
méthode de classification 555
méthode de compression 657
méthode de description par
   tenants et aboutissants 2407
méthode de la projection
   horizontale 2407
méthode d'éléments finis 1426
méthode de lignes 2411
méthode de mesure de phase
   2736
méthode de présentation de lignes
   2411
méthode de présentation de
   symboles cartographiques
   2409
méthode de présentation de
   symboles de mouvement 2412
méthode de présentation de
   symboles de régions 2408
méthode de recherche
   cartographique 454
méthode de référencement
   linéaire 2164
méthode de régions 2408
méthode de satisfaction de
   contraintes 707
méthode des moindres carrés
   2131
méthode des points 1168
méthode de symboles
   cartographiques 2409
méthode de symboles de
   mouvement 2412
méthode d'interpolation 1999
méthode d'isolignes 2410
méthode hypsométrique 1877
méthode récurrente décroissante
   3058
méthodes de Bayesian 328
méthodes de télédétection 3123
méthodes graphiques et
   analytiques 1737
méthodologie de catalogage
   d'objets 1376
milieu de transfert 3791
minute 2428
minute d'auteur 2659
mire à dessin écossais 810
miroitement combiné 625
mise à échelle
   multidimensionnelle 2459
mise à l'échelle 3221

mise au net 1363
mise au point 3269
mise en ciseaux 2043
mise en contour 728
mise en forme 1225
mise en jour d'attributs 263
mise en page 2123
mise en table 3577
mode de dialogue 1979
mode de représentation
    cartographique 2244
mode de transfert asynchrone 218
mode-grille 1773
mode interactif 1979
modelage de données orienté
    objet 2597
modelage d'environnement 1294
modelage de phénomènes
    naturels 2500
modelage de relations entre
    entités 1281
modelage de volumes 2441
modelage mathématique 2370
modelage spatial 3397
modelage statistique 3478
modelage thématique 3650
modelage topologique 3744
modèle 3595
modèle altimétrique 1251
modèle altimétrique digital 1043
modèle altimétrique structuré
    3514
modèle arc-nœud 154
modèle cartographique numérique
    1034
modèle conceptuel 679
modèle de base de données 868
modèle de bruit atmosphérique
    229
modèle de circulation générale
    1518
modèle de components 652
modèle de courbe 832
modèle de couverture d'indice
    1932
modèle de données 901
modèle de données de trame 3009
modèle de données de type réseau
    2521
modèle de données géorelationnel
    1674
modèle de données géospatiaux
    1678
modèle de données hiérarchique
    1826
modèle de données vectoriel 3911
modèle d'élément fini 1427
modèle d'élévation 1251
modèle d'élévation numérique
    1043

modèle d'élévation structuré 3514
modèle de logique floue 1501
modèle de numériseur 1086
modèle de relations entre entités
    1280
modèle de relief topographique
    3721
modèle de risque d'éboulements
    2092
modèle d'essai 3191
modèle de terrain 3614
modèle de terrain disposé en
    mosaïque 3682
modèle d'habitat 1798
modèle d'interaction spatiale 3392
modèle d'objet 2594
modèle fil de fer 4001
modèle géographique global 3D
    1014
modèle géométrique 1661
modèle hiérarchique 1826
modèle météorologique 2406
modèle numérique de paysage
    structuré objet 1057
modèle numérique de situation
    1067
modèle numérique de terrain
    1072
modèle océanographique 2611
modèle plat de la Terre 1435
modèle primaire numérique de
    surface 2887
modèle quadrillé de rues 1780
modèle rastre 3009
modèle spaghetti 3359
modélisation cartographique 455
modélisation orientée à polygones
    2833
modélisation polygonale 2833
mode naturel 2496
modification apportée
    délibérément aux données
    3446
modification du format 3077
module de numérisation 1089
module de revue 3963
moins-value 975
moment statistique 3479
moniteur de contrôle GPS 1720
montage 2453
montant 188
monument 386
morceau de surface limité par une
    courbe fermée 3545
morcellement vertical 1117
morphologie 2447
morphométrie 2449
mosaïquage 2286
mosaïquage cartographique 2286
mosaïque 62, 3681

mosaïque colorée 611
mosaïque de données 902
mosaïque de prises de vue
    aériennes 62
mosaïque numérique de nuages
    1076
mot clé thématique 3642
moteur géographique 1594
motif 2716
motif de remplissage 1421
mouvement panoramique 2694
moyenne 2379
moyenne d'échantillon 3192
moyenne intégrée dans le temps
    3694
moyenne pondérée 3991
moyenne pondérée dans le temps
    3694
moyenne statistique 3474
multicanal 2455
multiligne 2464
multi-mosaïque 2478

nadir 2482
navigateur 2505, 3963
navigation 2503
navigation à courte distance 3291
navigation à pseudo-distance
    2944
navigation par GPS 1721
navigation rectiligne 3502
navigation relative 3107
nébuleuse urbaine 3884
nécessités d'utilisateur 3891
négatif de trait 2182
nettoyage topologique 3731
niveau 2138
niveau d'application 2143
niveau de gris 1758
niveau de l'alidade 1844
niveau de la mer 3237
niveau de référence 924
niveau de tranche 3318
niveau de verticalité 1844
niveau moyen de la mer 2378
niveau zéro 3237
nivelle de l'alidade 1844
nivellement 2141
nivellement différentiel 1023
nœud 2533
    de ~ 1486
nœud de destination 998
nœud de réseau 2529
nœud descendant 987
nœud d'extrémité 1266
nœud d'origine 3468
nœud flou 1503
nœud initial 3468
nœud intermédiaire 1986
nœud interne 1955

nœud le plus proche 2510
nœud libre 1480
nœud manquant 2433
nœud non connecté 3845
nœud pendant 850
nœud périphérique 1266
nœud racine 3173
nœuds de grille 1776
nœuds d'objet 2600
nœuds joints 2046
nom 2483
nombre associé 198
nombre de lignes 2580
nombre de points de contrôle 2578
nombre de rangées 2580
nombre de régions 2579
nombre de série de caractéristique 1387
nombre de série de trait 1387
nombre d'identification de polygone 2842
nombre d'identification de repère terrestre 2088
nombre géopotentiel 1670
nombre interne 1989
nombre ordinal 2648
nombres de Peano 2720
nom de bloc 364
nom de carte 2287
nom de champ 1402
nom de chemin 2712
nom de couverture 799
nom de domaine 1160
nom d'ensemble de données 913
nom d'entité 1278
nom de quad 2958
nom de remplissage 1419
nom de topologie 3761
nomenclature 2543
nomenclature toponymique 1515
nom géographique domestique 1163
nominal 2544
nommage 2485
nomogramme 2546
noms géographiques 1613
noms hydrographiques 1863
non coupé 3840
non distinct 1495
non étiqueté 3869
non-géocodage 3853
non isotrope 132
non-précision 1917
non spatial 190
non-stationnarité 2555
nord 2564
    du ~ 2567
nord de la grille 1774
nord d'un carroyage 1774

nord du quadrillage 1774
nord-est 2566
nord géographique 3834
nord magnétique 2240
nord-ouest 2570
nord vrai 3834
normale 2560
normale à une facette 1359
normalisation 2559
norme d'interopérabilité de systèmes d'information géographique 2630
note en bas de page 1454
noyau 1856
noyau de peuplement 2571
noyau urbain 512
nuage de points 3228
nuançage 3274
nuançage combiné 625
nucléique 1302
numérique 1031
numérisation 1054
numérisation cachée 361
numérisation cartographique 2266
numérisation 3D 1189
numérisation de carte 2266
numérisation de surfaces 3D 1189
numérisation de topologie 3758
numérisation dynamique 3508
numérisation en mode dynamique 3508
numérisation interactive 1975
numérisation manuelle 2245
numérisation par points 2807
numérisation précise 2880
numérisation semi-automatisée 3261
numérisé photogrammétriquement 2740
numériser 1055
numériseur 1078
numériseur 3D 3383
numériseur orienté à triangulation 3822
numéro d'acheminement postal 2871
numéro de feuille 3286
numéro de ligne 2183
numéro de route 3181
numéro de segment 3253
numéro d'étiquette 2067
numéro de zone 4031
numéro d'identification 1882
numéro d'identification du lieu géographique 1620
numéro postal 2871
numérotage 2575
numérotage de bloc 365
numérotage de quartier 365
numérotation du quadrillage 1768

numéro unique de référence d'une parcelle 3856

objectif 997
objet 2584
objet 2D 1156
objet 3D 1157
objet attribut 254
objet bitmap 3019
objet commun 627
objet complexe 649
objet de découpage cartographique 2256
objet d'origine artificielle 2243
objet d'origine humaine 2243
objet dû à l'homme 2243
objet étendu 1349
objet flat 1156
objet flou 1504
objet fusionné 2394
objet géographique 1615
objet linéaire 2162
objet non spatial 193
objet planimétrique 1156
objet point 2809
objet polygonal 166
objet ponctuel 2809
objet rastre 3019
objet sélectionné 3256
objet simple 3300
objet spatial 3399
objets topologiquement complexes 3740
objet surfacique 166
objet volumétrique 1157
objet zonal 166
obliquité 2606
observation 2607
observation de la Terre 1205
observation de ressources terrestres 1207
observation écographique 1207
occupation 2609
occupation du sol 2097
octet 420
offline 2615
off-shore 2616
ombrage 3274
ombrage croisé 814
ombrage de relief 3115
ombrage oblique 2604
ombrage thématique 3654
ombre 3276
omission de ligne 2188
onde de sol 1795
onde directe 1795
opérateur booléen 372
opérateur conditionnel 682
opérateur de transformation 3794
opérateur de voisinage 2513

opérateurs de généralisation 1523
opérateurs de zone 4026
opérateurs géographiques 1616
opération de définition de ligne à
    polygone 2178
opération de définition de point à
    polygone 2805
opération de définition d'un
    polygone à polygone 2843
opération de fermeture
    topologique 3732
opération de triangulation et de
    nivellement 3548
opération géométrique 1662
opérations booléennes 371
opérations cartométriques 468
opérations d'analyse 114
optimisation de route 2713
optimisation linéaire 2163
option d'attribut 256
options de numérisation 1090
orbite 2639
orbite géostationnaire 1681
orbite géosynchrone 1682
ordinal 2648
ordinateur d'élévations 1246
ordinateur de poche 2689
ordinateur personnel 2730
ordinateur qui tient dans la main
    2689
ordonnancement 2645
ordonnée 2568, 2650
ordonnée d'un quadrillage 2568
ordonnée d'un quadrillage fausse
    1368
ordonnée fausse 1368
ordre 2643
ordre de dessin 1184
ordre de Morton 2451
ordre de Peano-Hilton 2719
ordre de triage d'entités 1283
ordre spatial 3403
ordre symbolique 527
organigramme 1443
organisation spatiale 3404
orientation 633, 2655
orientation de carte 2289
orientation d'objet 2595
orientation interne 1984
original au trait 999
original couleur 2259
original couleur de carte 2259
origine 2657
origine de coordonnées 2660
origine de système de
    coordonnées 2660
orthodromie 2661
orthographie des noms
    géographiques 2665
orthophoto 2666

orthophotocarte 2667
orthophotographie 2666
orthophotographie numérique
    1061
orthophoto numérique 1061
orthophotoplan 2667
orthophotoscope 2668
orthoprojecteur 2668
orthoredressement 2669
ossature 1476
ouest 3994
ovale 2676
overlay 2680

page graphique 1746
paire de cartes 2294
paire de coordonnées 763
paire d'images 1906
palette 2687
palette de marqueur 2353
palette de plumes 2723
palette de polices 1453
palette de remplissage 1420
palette des couleurs 615
pampa 2690
pan 2694
panchromatique 2691
panoramique 2694
pantographe 2695
papier
    sans ~ 2696
papier millimétré 1750
papillon 1962
paquet contributif SIG 1699
parallaxe 2698
parallèle 2699
parallèle astronomique 209
parallèle de latitude 2701
parallèle géocentrique 1534
parallèle géodésique 1554
parallèle standard 3466
paramètres de projection 2914
paramètre statistique de classe
    561
parcelle 2089
parcelle de ville 547
parcours atmosphérique 230
parité d'adresse 40
partage 1154
partage de données 914
part d'un terrain 2089
partie adresse 34
partition binaire d'espace 354
passage 782, 1446
passerelle de ressources
    géographiques 1676
pavage 3684
pavé 3679
pellicule photographique 2743
pente 1364, 1921

pente de terrain 3617
pente raide 3482
périhélie 2725
périmètre 2726
périphériques 2727
permutation identique 1885
personnalisation 837
perspective 690
perspective aérienne 63
perte persistante 2729
pertinence 3111
petit cercle 3326
PF 2532
phénomène 2737
phénomène continu 721
phénomènes de monde réel 3041
phénomènes discrets 1122
phénomènes mappés 2295
photo 2742
photo aérienne 65
photocarte 2745
photocarte thématique 3651
photocomposition à couleurs
    fausses 1366
photocopieur 2739
photocopieuse 2739
photo en tonalité continue 724
photogrammétrie 2741
photogrammétrie aérienne 64
photogrammétrie numérique 1063
photographie 2742
photographie aérienne 65
photographie instantanée 3330
photographie multispectrale 2475
photographie numérique 1064
photographie numérisée 1064
photographie stéréoscopique
    3491
photo instantanée 3330
photo numérique 1064
photo numérisée 1064
photoplan 2748
photosphère 2749
photo stéréoscopique 3491
phototraceur 2418
pic 2718
pictogramme 1881
pile d'attente 3458
pilote 1009
pilote chargeable 2205
pilote de dispositif 1009
pipeline 2756
piquets de pente 3323
piqûre 2757
pixel 2758
pixel 3D 1243
pixel mixte 2435
place 2762
place centrale 512
place de recensement 493

place destinée de recensement 493
placement 2215
placement de noms automatisé 279
placement d'étiquettes d'objets ponctuels 2802
plage 2999
plaine alluviale 1441
plaine inondable 1441
plan 992, 2774
plan cadastral 2652
plan cadastral numérique 1032
planche d'écritures 2484
planchette 1071, 3550
planchette d'arpenteur 3550
plan d'arrière 300
plan d'assemblage 1933
plan de carte 2307
plan de configuration 589
plan de couches 2777
plan de données 897
plan de l'équateur 1308
plan de niveau 2144
plan de nivellement 727
plan de port 1802
plan de référence 925
plan de sites 3312
plan de stratification 2777
plan de surveillance 3551
plan de ville 548
plan d'occupation du sol 2101
plan d'orbite nominal 2545
plan équatorial 1308
planète 2779
plan euclidien 1325
plan fondamental de coordonnées sphériques 1493
plan horizontal 1849
planification des mesures d'urgence 1261
planification de transport 3804
planification d'exploitation du terre 2102
planification de zone 175
planification d'infrastructure 1949
planification environnementale 1295
planification par ordinateur 664
planification par pays 785
planification urbaine 3885
planimètre 2783
planimètre électronique 1240
planimétrie 2791
planimétrie topographique 3717
plan méridien origine 4021
plan orbitaire 2642
plan orbital 2642
plan topographique 3716
plan topométrique 1272

plan topométrique du génie civil 1272
plaque de base topographique 3707
plateau continental 715
plate-forme continentale 715
plate-forme en matériel 1805
plate-forme matériel 1805
plissement 1179
plissement moulant 1179
pluviomètre 2995
poignée 2050
point 1166, 2798
point adressable 30
point bien défini 3993
point cardinal 431
point central 511
point coté 1245
point d'appui 2811
point d'appui terrestre 1786
point d'arc final 1271
point d'arrêt 3498
point d'attachement 3677
point de canevas 740
point de cheminement 3979
point de contrôle 739
point de contrôle terrestre 1786
point de crossover 811
point de départ 2811
point de direction 1112
point de fuite 2532
point de géoréférencement 2323
point de liaison 3677
point de masse 2364
point de mesure 2384
point de perspective 3957
point de profondeur 3447
point de raccordement 3677
point de référence 2811
point de référence relatif 1369
point de repère 2811
point de réseau 2529
point de résolution 1764
point de retournement 3838
point de route 3979
point d'étiquette 2068
point de trame 1764
point de virage 3838
point d'indicateur 1936
point d'inflexion 3838
point d'information 905
point d'interruption 401
point directeur virtuel 3948
point d'orientation 2656
pointer en direction 3809
pointeur 2801
pointeur de lien 120
pointeur en croix 1082
pointeur en fils croisés 1082
pointeur interne 1991

point final 1270
point frontière 390
point géodésique 743
point intérieur 1990
point interne 1990
point limite 390
point neutre 2532
point permanent marqué 740
point pilote 740
point pilote virtuel 3948
points au pouce 1169
points conjugués 692
points critiques 802
points de corrélation 781
points d'information de valeur Z 4040
points distribués irrégulièrement 2016
points extrêmes 1358
points symbolisés 3557
points voisins 2514
point topographique 3718
point visé 1112
polarité 2819
polarité normale 2561
polarité reverse 3157
pôle antarctique 3350
pôle boréal 2569
pôle géographique 1618
pôle Nord 2569
pôle Sud 3350
police 526
police de caractères 526
polygone 2823
polygone 3D 1176
polygone complexe 650
polygone de Bézier 338
polygone d'essai 3622
polygone d'univers 3868
polygone en enveloppe convexe 751
polygone étiqueté 2062
polygone extérieur 2675
polygone externe 2675
polygone interne 1956
polygone irrégulier 2018
polygone isolé 2020
polygone non étiqueté 2432
polygone non fermé 3842
polygone parasite 3452
polygone plan 2772
polygones de Fourier 1468
polygones de proximité 3665
polygones de Thiessen 3665
polygones de Voronoi 3665
polygone simple 3301
polygonisation 2844
polygoniser 2845
polyligne 2852
polyligne fermée 580

polyligne lissée en courbe 829
polymorphisme 2853
polynôme quadratique 2962
polynôme zonal 4027
poncifs 180
pondération de distances inverses 2007
pont 402, 403
pontil à griffes 1508
population 2854
portabilité 2861
portabilité de base de données 869
portée 2998
portée de la vue 3003
portée de visibilité 3003
portée géographique 1582
porteuse du signal de micro-ondes 2419
portion 3545
portion de ligne 2189
position 2862
position de référence 1475
position horizontale 1850
positionnement 2869
positionnement autonome 289
positionnement différentiel 1025
positionnement en temps réel 3040
positionnement relatif 3109
position verticale 3939
post-classification 2872
poste de contrôle 2445
post-traitement 2874
poursuite de déroulement 3775
poursuite de phase de l'onde porteuse 432
poursuite de phase de porteuse 432
poursuite interactive de ligne 1978
pouvoir d'attraction 243
pragmatique cartographique 2309
précédence 2876
précision 2878
précision cartographique 2298
précision de levé 18
précision de mesure 2382
précision de mesure cartographique 2284
précision de position 2863
précision de positionnement 2863
précision de temps 3685
précision relative 3104
précision sémantique 245
précision thématique 3635
prédiction 1455
prédiction démographique 961
prédiction de transport 3803
prélèvement d'échantillons 3194

première forme normale 1428
premier plan 1456
préparation de carte 2310
présentation en image par ordinateur 668
présentation géographique 1622
présentation géographique hiérarchique 1827
pression 2886
prétraitement de rastre 3021
prévision 1455
primitive 1274, 2890
primitive géométrique 1663
primitives volumétriques 3968
primitive topologique 3746
principe du moindre effort 2893
prise 2609
prise de décision 934
prise de décision multicritère 2458
prise de décision stratégique 3503
prise de piquage 3846
probabilité 2898
probabilité d'erreur sphérique 3435
problème de chemin au plus bas coût 2129
problème de coloriage des cartes 2258
problème de l'unité territoriale modifiable 2442
problème des pixels mixtes 2436
problème des quatre couleurs 1466
problème du représentant de commerce 3808
problème non déterministe polynomial complet 2547
problème NP complet 2547
procédé de décision 934
procédé diazo 1021
procédé diazoïque 1021
procédure de localisation-allocation 2216
procédure d'étiquetage 2064
processeur de tableaux 184
processeur de transfert de données spatiaux 3378
processeur de vecteur en rastre 3924
processeur d'extraction de caractéristiques 1380
processeur d'images rastres 3017
processeur d'interface de scanner 3224
processeur matriciel 184
processeur vectoriel 184
processus adiabatique 44
production à la demande 837
production cartographique 2901

produit analytique 117
produit de SIG 1697
profil en travers 812
profil géographique 1617
profil spatial 3406
profils régulièrement ramifiés 3096
profil topographique 3719
profondeur 978
progiciel 140
programmation par pays 785
programme d'application 2313
programme d'édition 1226
programme de statistique géographique 1629
programme d'interpolation de lignes de niveau 729
programme traiteur de rastre 3014
projecteur 2911
projecteur cartographique 2316
projection 2907, 2908
projection à aire équivalente 1316
projection arbitraire 145
projection azimutale 296
projection azimutale équivalente de Lambert 2070
projection cartographique 2314
projection centrale 690
projection conforme 686
projection conique 690
projection conique conforme de Lambert 2071
projection conique équivalente d'Albers 89
projection conique tangente 3581
projection conventionnelle 744
projection cotée 3720
projection cylindrique 847
projection cylindrique équivalente 846
projection définie par utilisateur 3888
projection de Gall-Peters 1509
projection de Gauss-Krüger 1514
projection de la population 962
projection démographique 962
projection de Mollweide 2444
projection des cartes plate carrée 1310
projection d'ombre combiné 625
projection elliptique 1257
projection elliptique de Donald 1164
projection équiangle 686
projection équidistante 1312
projection équidistante à deux points 3839
projection équidistante cylindrique 1310
projection équivalente 1316

projection gnomonique 1715
projection horizontale 1851
projection isocylindrique 846
projection isométrique 2032
projection Mercator 3867
projection Mercator transverse
    universelle 3867
projection méridienne 3806
projection modifiée 2443
projection normale 2562
projection oblique 2603
projection orthogonale 2663
projection orthographique 2663
projection orthomorphique 686
projection parallèle 2702
projection perspective 690
projection plate carrée 1310
projection polaire 2820
projection polyconique 2822
projection polyédrique 2851
projection pseudo-conique 2938
projection pseudo-cylindrique
    2939
projections cartographiques
    planes 2770
projections sphériques 3438
projection stéréographique 3488
projection stéréographique polaire
    universelle 3863
projection stéréoscopique 3494
projection sur un plan tangent
    3720
projection topographique 3720
projection transversale 3806
projection univoque 3309
projection verticale 3940
projet 992
projet cartographique 444
projet pilote 2754
projet SIG 1698
propagateur de chaînes 517
propagation 2920
proportionnalité 2921
proportionnément 2924
propriétaire des informations 835
propriété topologique 3765
protocole 2925
prototype 2926
proximité 2929
proximités spatiaux 3398
pseudocouleur 2934
pseudo-distance 2942
pseudonœud 2940
pseudo-perspective 1370
pseudostatique 2946
publication de cartes 2317
puissance d'attraction 243
pyramide de population 963

quad 2959

quadrangle 2959
quadrangle topologiquement
    complet 3739
quadrant 2960
quadrilatère 2959
quadrillage 3456
quadrillage de référence 3068
quadrillage géographique 2274
quad-tree 2965
qualité de carte 2319
qualité des données 906
quantification 1123
quantifier 2969
quart 2810
quartier 3150
quartier d'habitation 3150
quartier résidentiel 3150
quasi-géoïde 2972
question 3140

RAAL 3294
RAAS 3562
raccordement 1223
raccourci 3288
racine d'un arbre 3174
radar à antenne synthétique 3562
radar aéroporté à antenne latérale
    3294
radar aéroporté à exploration
    latérale 3294
radar à ouverture synthétique
    3562
radar à visée latérale 3294
radiation 2983
radiation électromagnétique 1233
radiomètre 2985
radiomètre à résolution très
    grande 61
radiomètre à très grand pouvoir
    séparateur 61
radiométrie 2991
radiopositionnement 2992
radiosignal 2993
raffinage 3076
raffinement 3076
rafraîchir 3078
rafraîchissement de rastre 3022
raie 314
raisonnement spatial 3408
rangée 2999
rangée d'adresses 42
rangée thématique 3652
rapport base-éloignement 316
rapport base/hauteur 316
rapport de qualité des données
    907
rapport de secteurs de
    recensement 503
rapport de ténuité 3666
rapport prospectif d'ambiance

1291
rapport signal à bruit 3296
rapport signal-bruit 3296
R-arbre 3185
rastre 3005
rastre binaire 352
rastre de passage 1447
rastre externe 1355
rayon 2994
rayon équatorial 1309
rayonnement 2983
raytracing 3033
réalisation 1915
réalisme 3035
réalité 26
réalité du terrain 1793
réalité sur le terrain 1793
réalité virtuelle 3951
réarrangement 3129
réarrangement de couches 3130
réattribution 3042
recensement de population 497
recensement de trafic 3782
recensement géocodé 1535
récepteur 3043
récepteur de double fréquence
    1192
récepteur de multiplexage 2466
récepteur/émetteur asynchrone
    universel 3859
récepteur GPS 1722
récepteur multicanal 2456
récepteur référentiel 3072
recherche 3241
recherche booléenne 373
recherche de chemins 2710
recherche de proximité 2931
recherche de route optimale 3259
recherche d'information 1942
recherche d'opérations 2635
recherche en largeur 399
recherche en largeur d'abord 399
recherche en profondeur 981
recherche en profondeur d'abord
    981
recherche géographique 1625
recherche opérationnelle 2635
recherche par opérateurs de
    proximité 2931
rechercher 1424
recherche sélective 178
recherche sur zone 178
reclassement 3046
reclassement d'images 1909
récoloriage 3049
reconnaissance 3048
reconnaissance automatique de
    caractéristiques 276
reconnaissance de formes 3283
reconnaissance de modèles 2717

reconnaissance d'images 2717
recouvrement 2679
recouvrement de polygone 2847
recouvrement latéral 2111
recouvrement longitudinal 1465
rectangle de sélection 3258
rectification 3054
rectification simple 3302
recul 3170
récupération d'information 1942
récursion 3057
rédaction 1225
rédaction définitive 1363
redessiner 3060
redimensionnement
    proportionnellement d'objet
    2922
redistribution 3059
redistribution de la population
    2857
redondance 3062
redressage 3054
redressement 3054
redressement différentiel 1027
redressement simple 3302
redresseur 3055
redresseur universel 3865
réduction 3061
réduction de données 908
rééchantillonnage 3143
référence 3063, 3064
référence d'attribut 258
référence de position 2217
référencement 3074
référencement spatiale 3412
référence spatiale 3409
référenciation 334
référentiel géodésique 3462
réflectance diffuse 1029
réflexion dirigée 3433
réflexion spéculaire 3433
reflux 2237
reformatage 3077
région 3079
régionalisation 3084
régionalisme 3083
région climatique 568
région colorée 1416
région coupée 570
région critique 803
région de carte 3087
région de "quatre coins" 2956
région de recensement 499
région étiquetée 2063
région fonctionnelle 1490
région métropolitaine 2416
région montagneuse 2452
région non métropolitaine 2552
région périphérique 2672
régions colorées 610

régions d'exclusion 1338
régions homogènes 3855
régions uniformes 3855
région transparente 3801
région urbanisée 3883
registre 473
registre du cadastre 422
registre hypothécaire 422
registre terrien 422
règle 3187
règle de Bowditch 396
régression 3092
régression multiple 2465
régression simple 3303
regroupement 587
relation entre secteurs de
    recensement 501
relation fonctionnelle 1491
relation géographique 1621
relation logique 2229
relation plusieurs-à-un 2247
relation spatiale 3413
relation topologique 3747
relaxation 3110
relévateur 2221
relévateur de coordonnées 2221
relèvement 3145
relèvement de positions 2868
relèvement de positions par
    satellite 3204
relevé océanographique 2612
relief 3605
reliure 2199
remaniement électoral 3059
remplissage 1417
remplissage de district 1418
remplissage de polygones 2839
remplissage de régions 3086
rendu 3127
rendu de terrain 3615
rendu perspectif 2734
renomination 3126
rénuméroter 3128
renversement 3836
renvoi d'appel abrégé 3292
renvoi de courte référence 3292
réorganisation de données 910
répartition 1125
répartition d'information
    asymétrique 213
répartition du travail 1125
répartition entre les modes 2438
répartition entre les modes de
    transport 2438
repérage 3004
repérage des directions
    automatique 275

repère 333, 1475
repère de fuseau 4031
repère d'élévation 1245
repère de nivellement 333
repère hydrologique 1868
repères du cadre d'appui 1396
repères du fond de chambre 1396
repère terrestre 2087
repère topocentrique 3704
repère vertical 3931
répétitivité 3131
repliement 92
repositionnement 3134
représentation cartographique 457
représentation de cyberespace
    844
représentation de données
    marginaux 2349
représentation de données
    spatiaux 3375
représentation de données
    tabulaires 3576
représentation de limites 391
représentation de relief en
    courbes de niveau 3136
représentation de relief en
    hachures 3137
représentation de surfaces
    topographiques 3138
représentation de terrain 3138
représentation d'objet 2599
représentation en couches 2119
représentation en échelle vraie
    3835
représentation graphique 1743
représentation multi-échelle 2470
représentation planimétrique 2790
représentation sans dimensions
    1103
représentation schématique 1019
représentation sous forme de
    tableau 3577
représentation stéréographique
    3488
représentation stratifiée 2119
représentation trame 3023
représentation vectorielle 3921
représentation visuelle 3965
reproductibilité 631
reproduction d'un original au trait
    2171
reproduction par ordinateur 666
reprofiler 3148
requête 3140
requête alphanumérique 265
requête asynchrone 216
requête attributaire 265
requête définie par un exemple
    2973
requête logique 3140

réseau 2115, 2530
réseau arborescente 3813
réseau astrogéodesique 201
réseau cartographique 2274
réseau de données d'hydroclimate 1860
réseau de données géographiques 1537
réseau de données hydroclimatiques 1860
réseau de facettes 2396
réseau de points d'appui 2530
réseau de polygones 2829
réseau de quadrillage de coordonnées 762
réseau de repères hydrologiques 1869
réseau de routes 3180
réseau de trafic 3783
réseau de triangles irréguliers 3820
réseau de Voronoi 3971
réseau d'isolignes 2030
réseau étendu 3996
réseau filaire 3509
réseau géodésique 2530
réseau géodésique à haute précision 1832
réseau géographique 2274
réseau global 3996
réseau gravimétrique 1755
réseau hydrographique 3163
réseau linéaire 2161
réseau linéaire-angulaire 2153
réseau local 2206
réseau local géologique 1643
réseau maillé 1598
réseau maillé national 2492
réseau non planaire 2554
réseau orienté 1107
réseau plane 2771
réseau quadratique 3456
réseau référentiel 3071
réseau triangulaire 3819
réseau utilitaire 3894
réseaux naturels 2499
réseaux référentiels à haute exactitude 1829
resélection 3146
réservation 3147
résolution 3152
résolution azimutale 298
résolution cartographique 2303
résolution d'ambiguïté 3153
résolution de temps 2149
résolution d'une carte 2303
résolution horizontale 1852
résolution moyenne pondérée par aire 182
résolution radiométrique 2989

résolution spatiale 3414
résolution spectrale 3431
résolution synthétique 3565
résolution temporaire 2149
résolution terrestre 1790
résolution thématique 3653
résolution verticale 3941
restriction 3155
résumé cartographique 2338
retard 952
retardement 952
réticule en croix 808
retour ligne 3156
retournement 3836
retourner 3159
retracer 3060
rétrodiffusion 303
revenir 3159
revêtement 3534
révision cyclique 845
révision d'une carte 2325
revue 406
rhumb 2810
rose des vents 636
rose du compas 636
rotation 3176
rotation en azimut 295
rotondité de la Terre 825
roulement en arrière 3170
rouleur 3187
routage 3182
routage géographique 1631
route 3177
route en encorbellement 3167
route surélevée 3167
route virtuelle 3952
ruban 3511
ruban à masquer 2361
ruban-cache 2361
ruban de papier-cache 2361
rue impassible 1913
RV 3951

saisie 429
saisie de données 875
saisie et registration des données aériennes 84
satellite d'étude du milieu 1298
satellite d'observation de la Terre 1206
satellite pour l'étude du milieu 1298
saturation 3207
sauf-conduit 2708
sauvegarder 3208
scannage 3225
scanner à défilement 1394
scanner aérien 87
scanner à plat 1433
scanner à tambour 1187

scanner cartographique 458
scanner couleur de marais côtier 593
scanner en infrarouge 1946
scanner infrarouge 1946
scannérisation 3225
scanner multibande infrarouge thermique 3662
scanner multibande pour thèmes multiples 3648
scanner multispectral 2477
scanner multispectral infrarouge thermique 3662
schéma-bloc 1443
schéma-bloc de profil 813
schéma d'assemblage de cartes 1933
schème de codage géographique 1600
schème d'ombrage 3277
sciences de la Terre 1675
sciences géographiques 1624
scindement 3445
scission 3445
script cartographique 2327
scrutation 3225
sécante 3242
seconde 3243
seconde projection 3940
secteur 1148
secteur de contrôle 742
secteur de dépouillement 499
secteur de recensement 499
secteur spatial 3354
section 3247, 3248
sectionnement géographique 1591
section transversale 812
sécurité d'accès de données 855
sécurité géographique 1626
segment 3250
segmentation 3251
segmentation d'image 1911
segmentation dynamique 1201
segment de ligne 2189
segment de recouvrement 2680
segment de route 2714
segment d'extraction de caractéristiques 1381
segment d'utilisateurs 3892
segments linéaires connectés 696
sélection 3257
sélection de caractéristique 1385
sélection de chemin 2715
sélection de forme 1385
sélection de gamme 3004
sélection d'entités 1282
sélection de route optimale 3259
sélection des couleurs 616
sélection de trait 1385
sélection de voie 2715

sélection de zone 179
sélection logique 2230
sélection non accessible 1104
sélection thème sur thème 3656
sémantique cartographique 2329
sémiotique cartographique 2330
semis 2115
semi-variogramme 3262
sens
    dans le ~ descendant 1174
sens d'écoulement 1444
sens du courant 1444
senseur 3265
senseur actif 25
senseur électromécanique 1236
senseur électrooptique
    multispectral 2474
sensitivité 3263
sensitivité radiométrique 2990
séparateur de champs 1404
séparation de modes 2438
séparation de bandes 309
séparation de caractéristiques
    1386
séparation de formes 1386
séparation des couleurs 616
séparation des couleurs
    numérique 1038
séparation de traits 1386
septentrional 2567
série de cartes 2331
serveur de bases de données 870
serveur de cartes 1994
serveur de cartes d'Internet 1994
serveur de données
    géographiques 1538
serveur de SIG 1694
service de demande d'information
    spatiale 3389
service de géodésie 1556
Service de la cartographie
    d'artillerie 2651
service de positionnement étalon
    3467
service de positionnement précise
    2877
service du cadastre 3723
service géodésique de base 321
session de numérisation 1091
seuil 3671
seuil de réception 12
sextant 3270
SGBD 867
SGDBR 3100
SIG 1608
SIG 2D 1012
SIG 3D 1013
SIG 4D 1467
SIG à base de couches 2117
SIG bureautique 995

SIG commercial 416
SIG environnemental 1290
SIG global 1706
SIG hybrid 1859
SIG intégré 1969
SIG intelligent 1974
SIG local 2212
SIG mobil 2437
SIG multi-échelle 2469
SIG multimédia 2461
signal convoyeur de micro-ondes
    2419
signalé 2065
signature 3297
signature numérique 1066
signature spectrale 3432
signe 3295
signe de démarcation 386
signe d'intersection 3298
signe graphique 1714
signes symboliques 2812
signes symboliques dégradés
    1732
signet 120
SIG par le Web 3984
SIG régional 3082
SIG spatio-temporel 3422
SIG universel 3862
SIG urbain 3881
SIG vectoriel 3908
SIG virtuel 3950
SIG Web 3984
simili 1799
simplification 3304
simulateur de la Terre 1208
simulateur de vol aérien 88
simulation 3305
simulation de flux 1448
simulation visuelle de
    phénomènes naturels 3966
site géographique 1627
site témoin 1794
situs 3314
sol 1784
solide 3339
sommaire cartographique 2338
sommation d'images 1889
somme partielle 3528
sommer 28
sommet 3928
sommet de facette 1361
sommet pendant 850
sondeur 3343
sondeur atmosphérique d'onde
    millimétrique 2422
sortie 1340
sortie de données 903
sortie de traçage 2794
source d'éclairage 2148
source de lumière 2148

source de pollution étendue 165
source d'événements 1332
source diffuse 165
source dispersée 165
source étendue 165
source lumineuse 2148
source nœud 3173
source numérique 1068
sources de données spatiaux 3376
souris trackball 3777
sous-arbre 3530
sous-bloc 3517
sous-classe 3518
sous-correction 2870
sous-dépassement 2870
sous-division 3519
sous-division de département 789
sous-division de triangle à base
    d'arêtes 1217
sous-domaine 3516
sous-ensemble 3524
sous-entité 3520
sous-fenêtre 3531
sous-multitude 3524
sous-plan 3521
sous-quadrant 3522
sous-région 3516
sous-région de recensement 498
sous-sélections 3523
sous-total 3528
soustraction d'images 1912
soutien d'information 1944
spatiocartes 3353
spécification de base de données
    871
spécification de région 3088
spécification d'interopérabilité
    ouverte de géodonnées 2630
spectre électromagnétique 1234
spectre étalé 3451
spectre panchromatique 2692
spectrophotométrie aérienne 69
sphéricité de la Terre 825
sphéroïde 3439
spline 3443
spline cubique 820
spline quadratique 2963
splines en plaques minces 3668
splines en voile mince 3668
splittage 3445
stadia 3459
standard de communication
    graphique 1951
standard de transfert 3792
standard de transfert de données
    spatiaux 3379
standard DGM 1049
standard DIGEST 1047
standard DOQ 1062
standard IGES 1951

standard OMG 2617
standards de contenu 710
standards de contenu de
    métadonnées spatiaux 711
standards de précision 19
standards de précision
    cartographique 2493
standards d'exactitude 19
standards fédérales de traitement
    d'information 1393
station de base 318
station de contrôle 743
station de poursuite 3778
station de référence 318
station de travail 4005
station de travail cartographique
    464
station de triangulation 743
stationnarité 3472
station totale 3771
statique 3471
statistique de couverture du sol
    2080
statistique d'ordre 2646
statistique géographique 1628
statistique géoréférencée 1628
statistique rapide 1372
statistique spatiale 3417
statistique topologique 3767
stéradian 3484
stéréocompilation 3485
stéréodigitaliseur 3486
stéréographe 2746
stéréomodèle 3489
stéréophotogrammétrie 3490
stéréotracé 3493
strate 732
stratification 3504
stratigraphie 3507
structure 3512
structure arc-nœud de données
    cartographiques 153
structure d'arbre quaternaire 2967
structure de domaine 1162
structure de données 916
structure de données contiguë 713
structure de données de réseau de
    triangles irréguliers 3696
structure de données graphiques
    1745
structure de données spatiaux
    3377
structure de données thématiques
    3640
structure de réseau 2525
structure des défauts 945
structure de table 3573
structure d'objets 2588
structure d'objets spatiaux 3401
structure liée rectilignement 2159

structure linéaire 2168
structure rastre 3027
structure spatiale 3418
structures terrestres liées 1791
structure topologique 3750
structure vectorielle 3922
structure zonée 4029
style de ligne 2192
style de régions 3089
style de symboles 3559
style de texte 3629
style de types 526
stylistique cartographique 2337
subdivision 3519
subsidence 3525
substitution 3526
sud 3348
sud de la grille 1779
sud-est 3349
sud-ouest 3351
suite de cartes 2332
suiveur 1452
suiveur de courbes 2174
superposition 3534
surdépassement 2683
surface 3538
surface 2D 2773
surface à plusieurs valeurs 2479
surface balayée 3553
surface bicubique 344
surface cartographiée 2339
surface complexe 651
surface connexe 697
surface continue 722
surface courbée 828
surface couverte 2262
surface de Bézier 339
surface de gravité 1756
surface de grille 1781
surface de modèle 2439
surface de niveau 2145
surface de pente 3544
surface de projection 2915
surface de référence 925
surface de tendance 3815
surface développpable 1002
surface en relief 3542
surface équipotentielle 2145
surface fonctionnelle 1492
surface fractale 1472
surface géométrique 1664
surface médiane 2375
surface moyenne 2375
surface multi-valuée 2479
surface numérique 1070
surface oblique 3316
surface paramétrique 2704
surface plane 2773
surface reverse 3158
surfaces de projection 177

surface sensible électroniquement
    1237
surface statistique 3480
surface topographique 3722
surface unitaire 3858
surface univoque 3310
surhaussement 3935
surimposer 3533
surimposition 3534
surimposition graphique 1747
surimpression 2681
surjet 2678
suroscillation 2683
surposition graphique 1747
surréglage 2683
surrégulation 2683
surveillance 3548
surveillance géodésique 1557
surveillance géomagnétique 1649
surveillance plane 2778
survol 406
susceptibilité d'éboulements 2093
symbole 525
symbole de but 3587
symbole de marqueur 2356
symbole de position 2219
symbole de texte 3630
symbole graphique 1714
symbole ombré 3273
symbole personnalisé 839
symboles cartographiques 459
symboles de berge à forte
    déclivité 565
symboles de cotes 3448
symboles de falaise 565
symboles de points cotés 3448
symboles de routes 3169
symboles proportionnels 2923
symbolisation 3556
symbologie cartographique 2340
symbologie thématique 3657
syntactique cartographique 2341
syntaxe cartographique 2341
synthèse d'image 1897
système à haute résolution de
    cartographie des fonds marins
    1833
système automatique
    d'information sur des eaux
    courantes 3190
système automatisé de
    cartographie de recensement
    271
système automatisé de production
    cartographique 272
système cartographique 2304
système cartographique avancé
    60
système d'acquisition de données
    aériennes 85

système d'adressage 38
système d'affichage de terrain 3609
système d'arpentage public 2950
système de business-mapping 416
système de cartographie automatique 285
système de cartographie automatisé 285
système de cartographie par le Web 3987
système de cartographie Web 3987
système de chemins 3180
système de classification selon l'utilisation du sol 2099
système de commande d'orientation 241
système de connaissance géographique 1610
système de contrôle d'attitude 241
système de coordonnées 765
système de coordonnées cartésiennes 434
système de coordonnées d'appareil 1008
système de coordonnées de carte 2261
système de coordonnées géocentrique cartésienne de Greenwich 1202
système de coordonnées géographiques 1598
système de coordonnées local 2209
système de coordonnées mondial 4008
système de coordonnées non terrestres 2548
système de coordonnées plane 2776
système de coordonnées planes d'États-Unis 3470
système de coordonnées polaires 2817
système de coordonnées rectangulaire spatial universel 3866
système d'écriture cartographique 2279
système de digitalisation d'arcs 149
système d'édition de terrain 3610
système de données géographiques à objectifs multiples 2468
système de fichiers de réseau 2524
système de gestion assisté par

SIG 1687
système de gestion de base de données distribuée 1143
système de gestion de base de données relationnelle 3100
système de gestion de bases de données 867
système de gestion d'utilisation du sol 2100
système de lecture de cartes 2322
système de localisation de véhicule 3926
système de localisation de véhicule à distance 3926
système de masses continentales en code numérique 1056
système de méta-information 2404
système de navigation 2504
système de navigation à base de GPS 1718
système de navigation par recouvrement géostationnaire 1680
système de navigation satellitaire 3203
système d'entrée de données géographiques 1596
système de numération des feuilles 3287
système de numérisation d'arcs 149
système de phototraçage 2418
système de positionnement à capacité globale 1710
système de positionnement par satellites 1710
système de poursuite de terrain 3606
système de recherche d'information cartographique 452
Système de recherche et analyse d'information géographique 1606
système de référence 3073
système de référence de cap et d'attitude 240
système de référence de carroyage 1777
système de référence géodésique 1555
système de référencement linéaire 2165
système de référence spatiale 3411
système de repérage de voiture par satellite 3926
système de tracement à projection universelle 3864

système de traitement d'images 1908
système de traitement d'images atmosphériques et océanographiques 226
système de tramage 1479
système d'exploitation 2633
système DIME 1193
système d'indexage global 1707
système d'information 1945
système d'information à base d'images 1893
système d'information cartographique-topographique cadastral 423
système d'information civil assisté par SIG 1686
système d'information d'aérophotos 66
système d'information décisionnel 935
système d'information de noms géographiques 1614
système d'information de planification 2792
système d'information de ressources environnementaux 1296
système d'information des cours d'eau 3977
système d'information des voies d'eau 3977
système d'information distribué 1145
système d'information environnementale 1293
système d'information géographique 1608
système d'information géotopographique numérique 1050
système d'information hydrologique 1870
système d'information national cadastral 2488
système d'information national de points d'appui 2489
système d'information réseau 2526
système d'information spatiale 3390
système d'information sur les ressources 3154
système d'information sur le territoire 2083
système d'itinéraire 3180
système du type dispatching de routage optimal 2637
système expert 1341
système géodésique 2530

système géographiquement distribué 1571
système global de données spatiaux 1712
système global de l'environnement 1705
système global de monitorage de l'environnement 1704
système isolé 2027
système métrique 2415
système MF-MA 278
système mondial de radiorepérage 1710
système mondial de repérage 1710
système mondial géodésique 4010
système ouverte 2632
système public terrestre 2951
système référentiel de position 2218
système référentiel géographique mondial 4011
système référentiel temporaire 3601
systèmes de simulateurs de vol 1438
système suivi de terrain 3611
système terrestre 2095

table 3570
table à distance 3125
table attributaire 259
tableau 183, 3570
tableau d'assemblage 1933
tableau de départ 3347
tableau-source 3347
table d'accès vive 2203
table d'attributs 259
table d'attributs d'arcs 147
table d'attributs de nœuds 2535
table d'attributs de polygones 2832
table d'attributs de route 3178
table d'attributs d'objets 1375
table d'attributs textuels 3625
table de base 319
table de caractères 3560
table de coordonnées 766
table de couleurs 614
table de localisation 1333
table de numérisation 1071
table de numérisation de précision 2881
table de référence 2233
table de symboles 3560
table de valences 3897
table de valeurs d'attributs 3898
table d'interrogations 2975
table d'origine 3347

table enregistrée 3090
table mappé 2293
table numérisante 1071
tables de système 3569
tables liées 2198
table traçante à plat 1432
tablette 1071
tablette à numériser 1071
tabulation 3577
tabulation croisée 815
tabulation transversale 815
tache d'analyse 1789
tache de prise de vue 1789
tache élémentaire 1789
tachèle 1789
tachéomètre 3578
taches après numérisation 2873
taches avant numérisation 2882
taille de cellule 483
taille de marqueur 2355
taille de tampon 411
taille de zone active 3315
talus 1921
talweg 3561
tamponnage 410
tamponnage pondéré 3990
tampons en anneaux concentriques 674
tangente 3582
tapis volant 1450
taux d'actualisation 1119
taux de compacité 630
taux d'escompte 1119
taxonomie 3589
taxonomie numérique 2583
technique cartographique 460
technique de digitalisation 3D 930
techniques cartographiques 2342
technologie de base de données géospatiaux 1692
technologie de base de données SIG 1692
technologie de cartographie d'information 1941
technologie de données géographiques 1589
technologie de SIG 1701
teinte hypsométrique 1878
télécapteur 3124
télédétecteur 3124
télédétecteur orbitaire 2641
télédétecteur orbital 2641
télédétection 3120
télédétection radar 2979
télémesure 3593
télémétrie 3593
temporalité de données 917
temps atomique 233
temps d'acquisition 21

temps d'échantillonnage 3227
temps de scannérisation 3227
temps moyen de Greenwich 758
temps universel 758
temps universel coordonné 758
tendance centrale 514
tendance cubique 821
tendance démographique 2858
tendance linéaire 2169
tendance quadratique 2964
tenseur 3604
terminateur de champ 1408
terrain 3605
terrain enclavé 1716
terre 1784, 2072
terre haute 3878
territoire 3621
territoire communal 3947
territoire d'estuaire 1322
territoire non organisé 3873
tessellation 3684
tessellation de Dirichlet 1115
test 532
test de réception 11
tétrarbre 2965
texte 3623
texte sans liens 3872
texture 3633
texture interne 1957
thalweg 3561
thème 1630
thème actif 24
thème cartographique 2343
thème d'événement 1334
thème géographique 1630
théodolite 3658
théorie de cartographie 3661
théorie de communication cartographique 3660
théorie de la décision 937
théorie des ensembles flous 1506
thibaude 3846
tic 3676
tireté 2722
titre de carte 2344
tolérance 3698
tolérance d'appariement de nœuds 2539
tolérance de coïncidence 2366
tolérance de grain 1733
tolérance de localisation 2865
tolérance d'émondage 3989
tolérance de position 2865
tolérance de tics 3675
tolérance floue 1507
tolérance projective 2906
tolérance proximale 2928
tolérances d'édition 1227
tolérances de traitement 2900
tonalité continue 723

ton continu 723
top 2718
topogramme binaire 3015
topographe 2094
topographie 3725
topographie minière 2427
topographie numérique 1075
topographique 3705
topologie 3755
topologie 2D 1190
topologie arborescente 3812
topologie arc-nœud 155
topologie de convergence par
    mesure 3763
topologie de gauche à droite 2132
topologie de polygone 2849
topologie d'Internet 1995
topologie d'un réseau 3764
topologie globale 1713
topologie polygone-arc 2831
topologie préliminaire 2884
topologie temporaire 3603
topologisation 3754
toponymie 3770
toponymie cartographique 461
torsion géodésique 1558
touche à double fonction 101
touches de curseur 824
touches de gestion de curseur
    824
tour 3773
tourbière 368
traçage 3776
traçage de carte 2308
traçage de lignes 2195
traçage de réseau 2531
traçage de réseau de coordonnées
    1766
traçage immédiat 2977
traçage obtenu en première
    lecture 2977
traçage perspective 2733
traçage rapide 2977
traçage stéréoscopique 3493
trace 1792, 3775
tracé automatique des courbes de
    niveau 282
tracé de carte 2308
tracé des courbes de niveau 728
tracement 3776
tracement de réseau 2531
traceur 2795
traceur à bulle d'encre 1953
traceur à diodes
    électroluminescentes 2147
traceur à jet d'encre 1953
traceur analogique 111
traceur à plat 1432
traceur à rouleaux 3171
traceur à tambour 1186

traceur à trame 3020
traceur de courbes 2795
traceur de Kelsh 2054
traceur de plumes 2724
traceur électrostatique 1241
traceur graphique 2795
traceur laser 2109
traceur numérique 1065
traceur par ligne 3020
traceur stéréoscopique 3492
traceur tabulaire 3572
traceur thermique 3664
traceur vectoriel 3919
trackball 3777
traducteur 750
traînage 1178
traînement 1178
trait 314, 448
trait commenté 133
trait continu 719
trait de contour intermédiaire
    1985
traitement de données
    automatique 273
traitement de données d'adresse
    36
traitement de données
    géographiques 1671
traitement de données
    géographiques orienté objet
    2598
traitement de polygones 2848
traitement d'image numérique
    1053
traitement d'images 1907
traitement d'images éloigné 3117
traitement d'images rastres 3016
traitement distribué 1146
traitement interactif 1980
traits de département 788
trait tireté 2722
trajectoire 3786
trajectoire atmosphérique 230
trajectoire au sol 1792
trajectoire de rayons 3032
trajectoire de rayons optiques
    3032
trajet de lumière 3032
trajet de rayons 3032
trajet multiple 2462
tramage 1478
trame 3005
trame topocentrique 3704
tranchage 2121
tranchage normal 2563
tranche 3317
tranche de bits 358
tranche de temps 3691
transaction 3788
transaction longue 2232

transect 3789
transfert asynchrone 217
transfert de données 918
transfert de fichiers 1414
transfert de points 2815
transformation 3793
transformation affine 77
transformation colinéaire 2918
transformation d'axe médian 2385
transformation de cartes 2345
transformation d'échantillonnage
    3196
transformation de coordonnées
    767
transformation de données
    spatiaux 3380
transformation de Fourier 1469
transformation de Fourier inverse
    2008
transformation d'entité 1284
transformation élastique 1232
transformation géométrique 1665
transformation identique 1885
transformation projective 2918
transformation pseudocouleur
    2937
transformation topologique 3751
transit 3795
transition démographique 964
translation 3797
translation d'adresse de réseau
    2516
transmission de données 918
transmission d'images
    automatique 284
transparence 3800
transparent cartographique 2291
transposition d'information privée
    2897
transtypage de rayons 3031
transversal 3807
traverser 3809
treillis 2115
tri 3342
triage 3342
triage ascendant 189
triage croissant 189
triage de ligne 2197
triage descendant 988
triage topologique 3748
triangle 3818
triangle géodésique 1559
triangulation 3821
triangulation aérienne 71
triangulation d'aire 181
triangulation de Delaunay 951
triangulation de grandes surfaces
    2106
triangulation de petites surfaces
    3325

triangulation de polygones 2850
triangulation linéaire 2170
triangulation satellitaire 3206
tri-arbre 3830
tri décroissant 988
tridimensionnel 3670
trièdre terrestre 1204
trier 3341
trigonométrie 3824
trilatération 3825
triplet 3828
tri topologique 3748
troisième forme normale 3669
troncation 1185
tronçon 2189
trou 2757
trouver 1424
TU 758
type de caractéristique 1390
type de données 919
type de légende 2134
type de ligne 2196
type d'entité 1285
type de pente 3324
type de relief 1389

uniformité 3854
union 2044
union angulaire 774
union automatique 269
unité 2041
unité astronomique 210
unité caractéristique 1391
unité cartographique 2306
unité cartographique de base 320
unité de champ 1409
unité de logement 1855
unité de mesure cartographique
        2306
unité de positionnement distante
        3119
unité dérivée 985
unité de surface 3858
unité de terrain 2096
unité de terrain de base 322
unité de volume 3858
unité d'habitation 1855
unité géographique 1633
unité gouvernementale 1717
unités caractéristiques
        descriptives 3135
unités de couverture 800
unités de distance 1135
unités de mesure de contenu 710
unité spatiale 3419
unité spatiale de base 323
unités périphériques 2727
urbanisme 3885
utilisateur de SIG 1702
utilisation de carte 2346

utilisation des terres 2097
utilisation du sol 2097

vaisseau spatial 3356
valence de nœud 3896
valeur atomique 234
valeur attributaire 264
valeur d'attribut 264
valeur de chromaticité 617
valeur de couleur 617
valeur de l'ordonnée dans le
        système Gauss-Krüger 2568
valeur d'estimation 1320
valeur d'une cellule 485
valeur moyenne 2379
valeur Z 4039
valeur zéro 2574
variable 3900
variable chi carré 534
variable continue 725
variable d'environnement 1301
variable dépendante 974
variable de surface 3546
variable graphique 1749
variable indépendante 1926
variable mappée 2296
variable thématique 3655
variable topologique 3769
variable X^2 534
variance 3904
variance de crigeage 2060
variance totale 3772
variation de données 3905
variation de focale 4036
variogramme 3906
vecteur 3907
vecteur de distance 1136
vecteur de relief 1392
vecteur spatial 3355
vectorisation 3028
véhicule spatial 3356
vélocité 3927
vérification 532
vérité-terrain 1793
verrou de base de données 866
verrouillage 2222
verrouillage de phase 2735
vide 1510
vieillissement climatique 3981
vieillissement de carte 2251
vieillissement de données 856
ville centrale 512
ville-centre 512
virage 3836
visée arrière 304
visée inverse 304
visée rétrograde 304
viseur 1082
viseur tête-haute 1810
visibilité 3953

visibilité d'arête 1224
visibilité mutuelle de deux points
        2814
visière stéréoscopique 1810
visiocasque 1810
visionnement 3962
visionneur 3963
visionneur cartographique 2348
visionneur de données
        cartographiques 443
visionneur de données d'élévation
        1248
visionneuse 3963
visionnique satellitaire 3202
visualisation 3962
visualisation cartographique 462
visualisation de réalité 3036
visualisation de terrain 3618
visualisation de trafic 3785
visualisation géographique en
        temps réel 3038
visualisation par graphique 1751
vitesse de tracé 2797
vitesse de tracement 2797
vocabulaire de données 885
voie 3177
        en ~ descendante 1174
voie de passage 1446
voie surélevée 3167
voile 1451
voisinage 2929
voisinage de nœud 2540
voisinage de point 2808
voisin le plus proche 2506
voxel 1243
voyage 3773
vraisemblance maximale 2373
vue aérienne 72
vue cartographique 2347
vue de banque de données 859
vue de côté 3293
vue de données chaînées
        dynamiquement 1197
vue de plan rapprochée 581
vue d'utilisation du sol 2104
vue 3D vrai 3832
vue géographique 1634
vue historique 1839
vue isométrique 2033
vue matricielle 3023
vue oblique 2605
vue perspective 3D 1175
vue planimétrique 2790
vue pseudoscopique 2945
vue rapprochée 581
vue schématique 1019
vues dynamiques 1853
vue synoptique 2684
vue topographique 3724

zénith 4019
zéro 3237
zonage 4033
zone 158
zone à bâtir 414
zone active 23
zone adresse 34
zone approche 142
zone constructible 414
zone d'agglomération 81
zone d'analyse de trafic 3781
zone dangereuse 1809
zone de code postal 4022
zone de construction 414
zone de coupage 804
zone découverte 2628
zone de couverture 521
zone de Delaunay 950
zone de délimitation 395
zone de grille 1783
zone de liste déroulante 2200
zone de menu de tablette 3574
zone de numérotage 2576
zone de numérotage de blocs 366
zone de proximité 2932
zone des données descriptives
    883
zone de service 521
zone d'essais 3622
zone de temps 3695
zone de texte 3626
zone de traçage 2796
zone de transition 3796
zone de travail 4004
zone de végétation 3925
zone d'extension urbaine 3880
zone d'habitat 3149
zone d'influence 4032
zone d'inondation 1441
zone d'opérations 172
zone d'usage conjoint 2049
zone d'utilisation du sol 2105
zone d'utilisation en commun
    2049
zone estuarienne 1322
zone explorée 3553
zone géographique 1582
zone intérieure 1983
zone littorale 592
zone non affectée 2672
zone périphérique 2672
zone remplie 1416
zone réservée 2672
zone résidentielle 3149
zone rurale 784
zones d'arbres quaternaires 2966
zones de cartes diagrammes 541
zone sélectionnée 3255
zone socio-économique 3335
zone source 165

zone source de pollution 165
zone statistique 3473
zone surveillée 3535
zone tampon 412
zone test 3622
zone urbanisée 3883
zonification 4033
zoom 4036
zoom arrière 4038
zoom avant 4035
zoom d'accompagnement arrière
    4038
zoom d'accompagnement avant
    4035
zooming 4036

Русский

абсолютная высота 3
абсолютная высотная отметка 3
абсолютная отметка 3
абсолютная точность 2
абсолютное измерение 6
абсолютное размещение 5
абсолютное расположение 5
абсолютное указательное устройство 7
абсолютные координаты 4
абстракция 10
абсцисса 1
авиасимулятор 88
авиасимуляционные системы 1438
автодорожная карта 3168
автокорреляция 266
автоматизированная картографическая система 285
автоматизированная картография 660
автоматизированная массовая оценка 663
автоматизированная обработка данных 273
автоматизированная планировка 664
автоматизированная система картографирования переписи 271
автоматизированная система картографического производства 272
автоматизированное воспроизведение 666
автоматизированное изготовление чертежей 665
автоматизированное оцифрование 274
автоматизированное проектирование 661
автоматизированное проектирование и изготовление чертежей 662
автоматизированное рабочее место 4005
автоматизированное размещение надписей 279
автоматическая генерализация 277
автоматическая запись меток 270
автоматическая картографическая система 285
автоматическая отмывка возвышений 115
автоматическая передача изображений 284

автоматическая топологическая погрешность 287
автоматический 280
автоматический анализ конфигурации сети 286
автоматический координатограф 2795
автоматическое вычерчивание контуров 282
автоматическое геокодирование 283
автоматическое дигитализирование 274
автоматическое замыкание 281
автоматическое индексирование 268
автоматическое картографирование 660
автоматическое оконтуривание 282
автоматическое определение азимута 275
автоматическое определение местоположения подвижного объекта 288
автоматическое определение направления 275
автоматическое присваивание меток 270
автоматическое присваивание обозначений 270
автоматическое распознавание признаков 276
автоматическое соединение 269
автоматическое увеличивание 267
автоматическое увеличивание на единицу 267
автомобильная навигационная система 3926
автономное позиционирование 289
автономный 2615
авторский оригинал карты 639
авторское право в картографии 771
агломерация 80
агрегация 83
агрегация, ограниченная диффузией 1030
агрегирование 83
агрегированные данные 82
адаптивная треугольная сетка 27
адиабатический процесс 44
адиабатный процесс 44
административная единица 56
административная область 56
административное деление 57
административный округ 56

административный район 56
адрес 29, 2242
адресат 3043
адресная базисная линия 32
адресная привязка позиционно-неопределённых наборов данных 39
адресная система 38
адресная часть 34
адресное пространство 43
адресование 31
адрес точки 2799
адресуемая позиция 30
адресуемое отображение 564
азимут 293
азимутальная поправка 295
азимутальная проекция 296
азимутальная разрешающая способность 298
азимутальный градус 297
азимутальный угол 294
АКС 285
активная виртуальная опорная точка 3948
активная виртуальная реперная точка 3948
активная мишень 822
активная цель 822
активный сенсор 25
активный слой 24
активный центр реакции 517
актуализация 3877
актуализирвание атрибутов 263
алгебра изображений 1890
алгебра обработки растровых изображений 3006
алгебра растровых изображений 3006
алгоритм 90
алгоритм ближайшего соседа 2507
алгоритм Дийкстра 1094
алгоритм естественного перерыва 2497
алгоритмическая генерализация 91
алгоритм прослеживания горизонтали 2175
алгоритм слежения границы 384
алгоритм слежения контура 384
алидада 93
аллювиальная равнина 1441
альманах 96
альтернативный ключ 101
альтиметрия 104
амальгамирование 106
амортизационные отчисления 975
анаглиф 109

анаглифическая карта 109
анализ 113
анализ близости 2512
анализ видимости/невидимости 3946
анализ выборок 3193
анализ выёмки с закладкой 841
анализ гиперспектральных изображений 1872
анализ главных компонентов 2891
анализ градуировки 427
анализ данных об исследованиях 1344
анализ двуосных фигур Фурье 1191
анализ изображений 1891
анализ калибровки 427
анализ местоположения 3311
анализ недвижимости 3034
анализ ограничений 706
анализ окрестностей 2512
анализ основных компонентов 2891
анализ ошибок 1318
анализ положения 3311
анализ положительных и отрицательных объёмов 841
анализ по множеству критериев 2457
анализ пригодности расположения 3313
анализ пространственного равновесия 3384
анализ профиля 2903
анализ путей 2709
анализ расположения 3311
анализ сетей 2517
анализ смежности 46
анализ соседства 2512
анализ траекторией 2709
анализ транспорта 3802
анализ тренд-поверхности 3816
анализ форм 3280
анализ чувствительности 3264
аналитическая карта 116
аналитическая тематическая карта 3645
аналитический продукт 117
аналогический 112
аналоговая карта 110
аналоговые данные 717
аналоговый 112
аналоговый графопостроитель 111
анаморфированная карта 118
анаморфоза 119
анизотропный 132
анимация 131
анимированный GIF 129

анимированный GIF рисунок 129
аннотация 134
аннотированный признак 133
анонимное дигитализирование 361
антарктический полярный круг 135
антиалиасинг 136
антиподы 137
анти-спуфинг 138
антропогенный объект 2243
аппарат для исследования космического пространства 3356
аппаратная платформа 1805
аппаратное обеспечение 1804
аппаратное обеспечение дигитализирования 1088
аппаратные средства 1804
аппаратура 1314
аппликата 4039
аппроксимация 144
аппроксимация поверхности 3540
аппроксимация сплайна 3444
аппроксимирование 144
ареал 158
ареал обитания 1797
ареальный 168
ареоларный 168
арифметика изображений 1892
арктический полярный круг 156
Артиллерийская съёмка 2651
архив 151
архивация 152
архивирование 152
архивирование карт 2254
архивное хранение 152
архитектура 150
архитектура ГИС 1685
асимметричное распространение информации 213
асимметрия 214
асинхронная передача 217
асинхронный 215
асинхронный запрос 216
асинхронный перенос 217
ассоциация 199
ассоциация географической информации 200
ассоциированное число 198
астролябия 202
астрономическая долгота 206
астрономическая единица 210
астрономическая параллель 209
астрономическая широта 205
астрономические таблицы 2640

астрономический 203
астрономический азимут 204
астрономический меридиан 208
астрономическое зенитное расстояние 211
астрономическое картографирование 207
астрономия 212
астрономо-геодезическая сеть 201
атлас 1575
атлас в Web 3983
атмосфера 224
атмосферная абсорбция 225
атмосферная коррекция 227
атмосферное окно 232
атмосферное поглощение 225
атмосферные эффекты 228
атмосферный зонд миллиметрового диапазона 2422
атмосферный пробег 231
атомизация 235
атомное время 233
атрибут 244
атрибутивная таблица 259
атрибутивные данные 250
атрибутивный запрос 265
атрибутирование 262
атрибутная дефиниция 251
атрибутный тег 260
атрибут объекта 1374
атрибут отношения 255
атрибут спектра 3427
атрибут-характеристика 1374
атрибуты контура 2671
атрибуты, определяемые пользователем 3887
атрибуты остановок 3496
атрибуты перерывов 3496
атрибуты трафика 3784
атрибут элемента рельефа 1374
аффинная трансформация 77
аффинное отображение 76
аэронавигационная карта 73
аэронавигационная карта в разрезе 3249
аэронавигационная топографическая карта 75
аэронавигационные данные 74
аэроспектрометрирование 69
аэросъёмка 70
аэротриангуляция 71
аэрофотограмметрия 64
аэрофотоплан 68
аэрофотоснимок 65
аэрофототопография 64

база 924
база вневременных данных 220

база геометрических данных 1656

база данных 860

база данных авиасимулятора 1437

база данных географических регистров 1604

база данных географических средств 1563

база данных ГИС 1691

база данных для системы принятия решений 936

база данных из съёмок 2738

база данных об окружающей среде 1288

база данных растительного покрова 2078

база данных системы принятия решений 936

база данных уличной сети 3510

база знаний 2058

база картографических данных 442

база метаданных 2402

базисная линия 317

базовая пространственная единица 323

базовая станция 318

базовая таблица 319

базовый слой ГИС 1689

байесовские методы 328

байт 420

баланс массы 2362

баллистическая фотокамера 305

банк данных 858

банк данных изображений 1899

барабанный графопостроитель 1186

барабанный плоттер 1186

барабанный сканер 1187

барьер 400

батиметрическая карта 324

батиметрическая съёмка 325

батиметрические данные 326

батиметрия 326

БД 860

безбумажный 2696

безразмерное представление 1103

безрастровый снимок 724

безрастровый тон 723

безусловная классификация 3876

береговая зона 592

береговая линия 594

береговая черта 594

бесслойный лист 3870

бесшовная база данных 3239

бесшовная карта 720

бесшовные данные 717

бесшовный слой 3240

бета-индекс 336

БЗ 2058

библиографические данные 341

библиографическое индексирование 342

библиотека 2146

библиотека карт 2264

библиотека цифровых карт 1060

бикубическая поверхность 344

бикубический интерполятор 343

билинейная интерполяция 345

билинейный интерполятор 346

бинарный растр 352

биом 355

биорегион 356

биотоп 1797

бит 357

битовая плоскость 358

битовый слой 358

бланк 1460

блеск 404

ближайшая соседняя запись 2506

ближайший сосед 2506

ближайший узел 2510

ближняя навигация 3291

близкий допуск 2928

близлежащие точки 2514

близость 2929

"блоб" 349

блок 585

блок-диаграмма 1443

блок-диаграмма временного разреза 3690

блок закрашивания многоугольников 2840

блок заполнения полигонов 2840

блокирование 2222

блокировка 2222

блокировка координат 1482

блок переписи 488

блок сплошного закрашивания многоугольников 2840

блок-схема 1443

БнД 858

боковая проекция 3293

боковая связь 2110

боковое отклонение 816

боковое перекрытие 2111

болото 368

большой атлас 2107

большой двоичный объект 349

большой круг 1761

бортовая радиолокационная станция бокового обзора 3294

бортовая система сбора данных 85

бортовой самолётный радиопрофилометр 86

бортовой сканер 87

браузер 3963

браузер данных для вертикальной наводки 1248

бродилка 3963

броузинг 406

буквенно-цифровая сетка 100

буквенно-цифровой 98

буквенно-цифровые данные 99

буксировка 1178

булево выражение 369

булево изображение 370

булев оператор 372

булев поиск 373

булевы операции 371

бумажная карта 2697

буссоль 632

бутон захвата изображения 1724

бутон захвата кадра 1724

буфер 412

буферизация 410

буферизация с взвешиванием 3990

буферирование 410

буферная зона 412

"буферное" расстояние 409

быстрое оцифрование 2976

быстрый вызов 3288

валентность узла 3896

вариантность 3904

вариантность кригинга 2060

вариация данных 3905

вариограмма 3906

вверх по течению 3879

ввод аналоговой информации 1054

ввод в действие 1915

ввод данных 886

ведомство 973

ведущая линия 2127

ведущий край 2126

вектор 3907

векторизатор 3029

векторизация 3028

векторизированная ячейка 3916

векторная ГИС 3908

векторная графика 3914

векторная карта 3918

векторная модель данных 3911

векторная структура 3922

векторная функция 3913

векторное нетопологическое представление 3359

векторное поле 3912

векторное представление 3921
векторно-растровое
    преобразование 3923
векторно-топологическое
    представление 154
векторные данные 3909
векторный графопостроитель
    3919
векторный процессор 184
векторный слой 3917
векторный формат данных 3910
векторный формат
    пространственных данных
    3910
вектор расстояния 1136
вектор рельефа 1392
вектор-функция 3913
величина ординаты 2568
величина ординаты по карте
    2568
вероятная круговая
    погрешность 544
вероятность 2898
вероятность сферического
    отклонения 3435
вертикальная линия 3938
вертикальная невязка 3934
вертикальная невязка полигона
    3934
вертикальная отмывка
    возвышений 3936
вертикальная погрешность
    3933
вертикальная позиция 3939
вертикальная проекция 3940
вертикальная разрешающая
    способность 3941
вертикальная точность 3929
вертикальное начало отсчёта
    3931
вертикальное утрирование 3935
вертикальное утрирование
    размера или формы 3935
вертикальные углы 3930
вертикальный зазор 3934
вертикальный интерполятор
    3937
вертикальный разрез 3940
верх 2718
верхний слой 3701
вершина 2718, 3928
    от ~ы 1486
вершина фацета 1361
веха 2087
взаимная видимость двух точек
    2814
взаимная зависимость 778
взаимная корреляция 806
взаимное проникание 1997
взаимное проникновение 1997

взаимные азимуты 2481
взаимодействие 1996
взаимозаменяемость 1996
взаимосвязь между
    переписными районами 501
взвешенное среднее 3991
взвешивание обратных
    расстояний 2007
взгляд назад 304
взгляд с воздуха 72
видеокаска 1810
видимость 3953
видимость ребра 1224
видимый слой 3956
видимый топографический
    элемент 3955
видимый элемент 3955
вид крупным планом 581
видовой экран 3944
вид сбоку 3293
виды тематического
    картографирования 398
визуализатор 3963
визуализатор картографических
    данных 443
визуализация 3962
визуализация конфигурации
    местности 3618
визуализация реальности 3036
визуальная аккумуляция 3958
визуальная линия наземной
    позиции 3964
визуальная симуляция
    естественных явлений 3966
визуальное дешифрирование
    изображений 3960
визуальное представление 3965
визуальный атрибут 3959
визуальный интерфейс 3961
виртуальная ГИС 3950
виртуальная реальность 3951
виртуальные данные 3949
виртуальный пут 3952
висячая вершина 850
висячая дуга 848
висячая линия 849
висячий полигонный ход 3843
висячий узел 850
вкладывать 1961
включать 1961
включение 1923
вложение 1259
вложенность 1259
вневременные данные 219
внедрение 1259, 1915
внесение 1916
внешние устройства 2727
внешний ключ 1457
внешний объект 1351
внешний полигон 2675

внешний растр 1355
внешний слой 2670
внешний файл 1353
внешность 1354
вниз по течению 1174
внутреннее ориентирование
    1984
внутреннее преобразование
    1988
внутренний номер 1989
внутренний полигон 1956
внутренний район 1838
внутренний узел 1955
внутренний указатель 1991
внутренняя зона 1983
внутренняя область 1983
внутренняя ориентация 1984
внутренняя текстура 1957
внутренняя точка 1990
внутренняя точка полигона
    2068
вогнутость 673
водораздел 3975
военная система
    прямоугольных координат
    2421
военный атлас 2420
возвращаться 3159
возвышенность 3878
воздействие атмосферных
    условий 3981
воздержание 9
воздушная перспектива 63
возрастающий 188
возраст данных 79
возрастная пирамида 963
возрастно-половая пирамида
    963
возрастно-половая пирамида
    населения 963
воксел 1243
воспроизводимый атрибут
    1127
восток 3994
восточное положение в
    координатной или
    географической сетке 1210
восточное склонение 1210
восходящий 188
"вошь" 413
впадина 2757
вполне аналитическая
    аэротриангуляция 1489
вполне определённая точка
    3993
вращение 3176
врезка 1962
временная база данных 3597
временная колонка 3602
временная размерность 3598

временная размерность данных 3599
временная топология 3603
временная точность 3596
временная эталонная система 3601
временное расширение 3600
временной канал 3691
временные аспекты данных 917
время выборки 3227
время регистрации 531
время сбора 21
всемирная геодезическая система 4010
всемирная система географических координат 4011
всемирное время 758
всплывающее меню 2859
всплывающее окно 2860
вспомогательная горизонталь 290
вспомогательные данные 121
вспомогательный файл 4003
вставлять 1961
встраивание 1973
встроенный язык структурированных запросов 1258
вторая нормальная форма 3245
вторая проекция 3940
вторая разность 1170
вторичный эффект 3246
вуаль 1451
входные данные 1958
выбираемый слой 3254
выбор 3257
выборка 3191
выборка по зонам 4028
выборка по слоям 4028
выбор маршрута 2715
выбор масштаба 3221
выбор области 179
выбор объектов темы через объектов другой темы 3656
выбор оптимального маршрута 3259
выбор оптимального пути 3259
выборочная характеристика 2855
выборочное среднее 3192
выбор признака 1385
выбор пути 3182
выбор решения 934
выбор формы 1385
выбранная область 3255
выбранный объект 3256
выброс 3441
выброс сигнала 2683
выведенная колонка 984

выветривание 3981
вывод данных 903
вывод на графопостроитель 2794
выдвижное меню 2953
выделение 653
выделение данных 890
выделение диапазонов 3002
выделение краев 1219
выделение линейных элементов изображения 2172
выделение признаков 1379
выделение составляющих 1117
выездная проверка 1397
выклинивание 3667
выноска 428
выпадание 1185
выполненная вручную карта 1801
выпрямитель 3055
выпрямление 3054
выпуклая оболочка 751
выпуклость 752
выравнивание кривой 830
выражение 1347
вырезание 571
вырезанная область 804
вырезка 3317
высокая вода 1834
высокий уровень воды 1834
высокочастотный фильтр 1831
высота 1811
высота репера, отнесена к началу отсчёта 1813
высота сечения рельефа 730
высота ячейки 482
высотная отметка 1245
высотная съёмка 103
высотомер 102
высотометрия 104
высшая геодезия 1830
вытягивание контраста 737
выход 1340
выходные данные 1918
вычерчивание карты 2308
вычерчивание контуров 728
вычерчивание кривой 830
вычерчивание кривой по точкам 830
вычисление буфера 408
вычисление массы 2363
вычисление наклона 3320
вычисление площади 659
вычисление площади и периметра 174
вычисление сети 2518
вычислительная ошибка 658
вычислитель угла возвышения 1246
вычитаемые основные цвета

3529
вычитание изображений 1912
выше по течению 3879
выявление изменений 522
вьювер 3963
вьюер 3963

габарит 2674
газеттир 1515
галерея карт 2257
галочка 3674
гамильтонова цепь 1800
гамильтонов контур 1800
гарнитура 526
гарнитура шрифта 526
гауссова аппроксимация 1512
гауссово распределение 2556
гауссовские координаты 1513
гауссовский алгоритм 1511
гашение 238
ГВС 3996
гексотомическое дерево 1819
гелиографическая долгота 1814
генеалогия 2151
генерализационные операторы 1523
генерализация 1520
генерализация кусков поверхности 173
генерализация пространственных данных 3372
генератор отчётов 3132
генератор покрытия 796
генерация отчётов 3133
генерирование кривых 831
генерирование многоугольников 2841
генерирование полигонов 2841
генерирование сетки 1769
генетические алгоритмы 1528
географ 1565
географическая база данных 1585
географическая взаимосвязь 1621
географическая визуализация в реальном времени 3038
географическая единица 1633
географическая защита 1626
географическая зона 1582
географическая зона наблюдения 1582
географическая идентичность 1602
географическая иерархия 1599
географическая информационная система 1608

географическая информация 1584
географическая карта 1611
географическая картография 1577
географическая категория 1579
географическая метафора 1612
географическая область 1564
географическая область покрытия 1582
географическая основа карты 3706
географическая параллель 2701
географическая связь 1621
географическая сетка 2274
географическая ссылка 1619
географическая статистика 1628
географическая сфера деятельности 1564
географическая тема 1630
географические данные 1584
географические измерения 1561
географические координаты 1581
географические наименования 1613
географические науки 1624
географические операторы 1616
географический 1569
географический адрес 1568
географический атлас 1575
географический вид 1634
географический дисплей 1590
географический доступ 1566
географический запрос 1609
географический каталог политических и статистических регионов 1578
географический код 1601
географический маркетинг 1650
географический объект 1615
географический поиск 1625
географический полюс 1618
географический примитив 1595
географический процессор 1594
географический регистр 1603
географический ресурс 1623
географический север 3834
географический центр 1580
географический цикл 1583
географический эквивалент 1597
географически относимое адресуемое отображение 1672

географически относимые данные 1573
географически относимый 1572
географически упоминаемый 1572
географическое кодирование и стандартизация 1593
географическое представление 1622
географическое применение 1574
географическое разделение 1591
географическое распределение 1617
география адресного пространства Интернет 1636
география переписи 495
география транспорта 3805
геодезическая высота 1548
геодезическая долгота 1550
геодезическая засечка 3145
геодезическая кривизна 1546
геодезическая линия 1541
геодезическая окружность 1544
геодезическая параллель 1554
геодезическая референцная система 1555
геодезическая сеть 2530
геодезическая сеть высокой точностью 1832
геодезическая сеть стереоскопического вычерчивания 1477
геодезическая съёмка 1556
геодезическая широта 205
геодезические данные 1547
геодезические измерения 1552
геодезические координаты 1545
геодезический 1540
геодезический азимут 1543
геодезический меридиан 1553
геодезический прибор 1549
геодезический треугольник 1559
геодезическое зенитное расстояние 1560
геодезическое кручение 1558
геодезическое обследование 1557
геодезическое отображение 1551
геодезия 1542
геодезия поверхности 3541
геодемография 1539
геодинамика 1562
геоид 1637
геоизображение 1639
геоинформатика 1607
геоинформационная концепция

1641
геоинформационная система 1608
геоинформационная технология 1701
геоинформационное картографирование 1642
геоинформационное образование 1640
геоинформационный анализ 1688
геоинформационный проект 1698
геокод 1601
геокодирование 1592
геокодированная перепись 1535
геокодированная перепись населения 1535
геологическая карта 1646
геологическая местная сеть 1643
Геологическая съёмка США 3857
геологические данные 1644
геологическое картирование 1647
геологическое образование 1645
геомагнитная съёмка 1649
геомагнитное поле 1648
геомаркетинг 1650
геоматика 1607
геометрическая графика 1660
геометрическая коррекция 1653
геометрическая модель 1661
геометрическая операция 1662
геометрическая поверхность 1664
геометрическая поправка 1653
геометрическая сложность 1652
геометрическая точность 1651
геометрическая точность карты 2298
геометрическая трансформация 1665
геометрические данные 1655
геометрические зоны действия 1654
геометрический примитив 1663
геометрический фактор точности 1657
геометрическое искажение 1658
геометрическое преобразование 1665
геометрия иллюминации 1887
геометрия объекта 2589
геометрия пласта 524
геометрия покрытия 797

геометрия форм местности 3612
геомоделирование 455
геоморфологическая карта 2082
геоморфометрия 1666
геопортал 1669
геопотенциальное число 1670
геопространственные данные 1584
геореляционная концепция 1673
геореляционная модель данных 1674
геосинхронная орбита 1682
геостатистика 1628
геостационарная оверлейная навигационная система 1680
геостационарная орбита 1681
геостационарный 1679
геофизические события 1668
геоцентрическая гринвичская прямоугольная система координат 1202
геоцентрическая долгота 1532
геоцентрическая параллель 1534
геоцентрическая широта 1531
геоцентрические координаты 1529
геоцентрический меридиан 1533
геоцентрический репер 1530
герметизация 1263
гетерогенность 1816
гибридная ГИС 1859
гибридный редактор 1858
гидрографическая карта 1862
гидрографическая съёмка 1864
гидрографический 1861
гидрография 1865
гидрологическая информационная система 1870
гидрологический атлас 1867
гидрологический репер 1868
гидронимы 1863
гидротермическая диаграмма 566
гиперспектральный датчик изображения 1873
гиперспектральный формирователь изображений 1873
гипертекстовая карта 1874
гипотетическая горизонталь 1463
гипсографическая карта 1875
гипсометрическая карта 1876
гипсометрическая окраска 1878
гипсометрическая шкала 1879

гипсометрический способ 1877
гипсометрия 1880
ГИС 1608
ГИС в Web 3984
ГИС в Интернет 1695
ГИС-приложение 1684
ГИС-продукт 1697
ГИС-проект 1698
ГИС-технология 1701
гистограмма 315
главный город 512
главный масштаб 2892
глобализация 1708
глобальная вычислительная сеть 3996
глобальная ГИС 1706
глобальная локационная система 1710
глобальная позиционирующая система 1710
глобальная природно-ресурсная база данных 1711
глобальная сеть 3996
глобальная система индексирования 1707
глобальная система координат 4008
глобальная система мониторинга окружающей среды 1704
глобальная система окружающей среды 1705
глобальная система пространственных данных 1712
глобальная система рекогносцировки 1710
глобальное пространство координат 1703
глобально-однозначный идентификатор 1709
глобус 1209
глубина 978
гномоническая картографическая проекция 1715
гномоническая проекция 1715
головной экран 1810
горизонт 1841
горизонталь 732
горизонтальная линия отсчёта 1846
горизонтальная линия приведения 1846
горизонтальная ось координат 1846
горизонтальная плоскость 1849
горизонтальная позиция 1850
горизонтальная проекция 1851
горизонтальная разрешающая

способность 1852
горизонтальная съёмка 2778
горизонтальные координаты 1845
горизонтальный интерполятор 1848
горизонтальный разрез 1851
горизонтальный угол 1843
горизонталь с отрицательной отметкой 977
горизонт воды 3976
горная область 2452
городская ГИС 3881
городская планировка 3885
городское население 3883
городской участок 547
городской участок земли 547
город с пригородами 2416
город-центр 512
государственная единица 1717
государственная информационная система пунктов плановой сети 2489
государственная кадастровая информационная система 2488
государственная топографо-картографическая информационная система 423
ГПС 1710
ГПС-констелляция 1719
гравиметрическая сеть 1755
гравиметрическая съёмка 1757
гравиметрические исследования 1757
гравиметрия 1757
гравиразведочные работы 1757
градация 1725
градиент 1726
градиентная плотность 1728
градиентный анализ 1727
градуированная линия 425
градуированные условные знаки 1732
градуированные шкалы значков 1732
градус 949
гражданская информационная система на основе ГИС 1686
граница 377, 1216
    за ~ей 2616
граница административного района 1149
граница городского административного района 1149
граница застройки 415
граница земельного участка 1487

граница землепользования 2098
граница зоны 387
граница квартала 363
граница места 2763
граница метропольного района 2417
граница области 387
граница округа 787
граница отдельной области 389
граница переписного района 500
граница поверхностей 1216
граница поверхности 388
граница полигона 2834
граница полосы отвода 2084
граница почтового индекса 4023
граница почтового кода 4023
граница присоединения 237
граница территории 387
граница урбанизации 3882
границы карты 2280
границы листа карты 2334
границы подведомственной области 2052
границы сетки 1763
границы сферы полномочий 2052
граничная точка 390
граничное ребро 380
граничное условие 1218
граничные дуги 375
граничный геодезический знак 386
граничный грунтовой репер 386
гранулированный шум 3425
гранулярность 1734
грань 1216, 1360
грань области 164
граф 1735
графа для галочки 530
график 1018
графика для построения геометрических фигур 1660
графика координатами и линиями 761
графическая накладка 1742
графическая переменная 1749
графическая страница 1746
графическая суперпозиция 1747
графические элементы 1738
графический 1736
графический интерфейс пользователя 1748
графический масштаб 1744
графический образ 1739
графический объект, неограниченный окном 1508

графический оверлей 1742
графический планшет 1071
графический пользовательский интерфейс 1748
графический примитив 1274
графическое воспроизведение 3962
графическое изображение 1739
графическое наложение 1747
графическое представление 1743
графическое устройство ввода данных 1071
графоаналитические приёмы 1737
графопостроитель 2795
графство 786
гребень 3161
ГРИД 1711
гринвичское гражданское время 758
грубая погрешность 367
грунт 1784
группа 584, 585
группа блоков переписи 489
группа городов 3884
группа данных о генерализации 1521
группа по управлению объектами 2593
группирование 587
групповое кодирование 3189
групповой поиск 178
грязь 368
ГСМОС 1704
гуманитарная карта 1857
густота 969
густота распределения 1147

давление 2886
дальномер 1134
дальномерная аппаратура 1134
дальномерная рейка 3459
дальномерная съёмка 3459
дальномерное оборудование 1134
дальномер с окулярной сеткой 3459
дальность 2998
дальность видимости 3003
дамба 1095
данная величина 922
данное 922
данные 852
данные аэрокосмического зондирования 3121
данные в узлах координатной сетки 1775
данные в цифровой форме 1077

данные деловой графики 418
данные дистанционного зондирования 3121
данные для вертикальной наводки 1247
данные-заменители 3547
данные, не относящиеся к определённому времени 219
данные об изображении 1898
данные об окружающей среде 107
данные о полигональных ходах 3810
данные о рельефе 3608
данные о состоянии окружающей среды 107
данные переписей 492
данные, преобразованные в цифровую форму 1077
данные "спагетти" 3358
данные типа "спагетти" 3358
данные топологических объектов 3762
данные центральной линии 508
данные центровой линии 508
дата 920
датум 2811
датчик 3265
движение на восток 1210
двоичная линейная генерализация 350
двоичная сетка 348
двоичное пространственное разбиение 354
двоичные растровые данные 353
двоичный код 347
двоичный оверлей 351
двоичный растр 352
двойное дигитализирование 1171
двойное оцифрование 1171
двумерная ГИС 1012
двумерная поверхность 2773
двумерная топология 1190
двумерное изображение 1098
двухволновой приёмник 1192
двухточечная равнопромежуточная проекция 3839
двухточечная эквидистантная проекция 3839
двухчастотный приёмник 1192
деактивация 931
девиационный круг компаса 636
девиация 1005
дегеокодирование 3853
дезагрегирование 1117
дезагрегировать 1116

действительная мгновенная
    зона обзора 1229
действительность 26
действительный мгновенный
    сектор обзора 1229
декартова система координат
    434
декартовы координаты 433
декодирование 941
декомпозиция 942
декомпозиция области 162
декомпрессировать 943
делёж 1152
деление 1154, 3674
деление на округа 1150
деление на округа годографом
    3688
делительный циркуль 1153
деловая графика 417
дельта 1322
делянка 2089
демаркационная линия 385
демаркация 957
демографические данные 960
демографический анализ 959
демографический прогноз 962
демографический тренд 2858
демографическое
    прогнозирование 961
демография 965
денивелирование 1812
денормализация 967
департамент 973
депрессия 976
дерево 3811
    4-~ 2965
    Q-~ 2965
    R-~ 3185
дерево квадрантов 2965
дерево квадратов 2965
дерево октантов 2613
дерево решений 938
дерево-свидетель 4002
дескрипция 989
десятичный градус 933
детализация 3076
детализировать 1116
детектор 1000
дефект 944
деформирование 3974
децентрализованная
    картографическая база
    данных 1139
дешифрирование 941
дешифровочный признак 3295
джойн 2044
джойстик 2050
ДЗ 3120
диагональ 1017
диагональное направление 1016

диаграмма 1018
диаграмма взаимосвязи
    примитивов 1279
диаграмма последовательности
    1443
диаграмма разброса 3228
диаграмма рассеивания 3228
диаграмма с областями 160
диаграмма топологии 3757
диаграмма цветности 606
диаграммы Вороного 3665
диазотипия 1021
диазотипный процесс 1021
диалоговая обработка 1980
диалоговое геокодирование
    1976
диалоговое дигитализирование
    1975
диалоговое картографирование
    2625
диалоговое слежение
    горизонтали 1978
диалоговый доступ 2623
диалоговый запрос 2626
диалоговый каталог 2624
диалоговый режим 1979
диаметр 1020
диапазон 2999
диапазон адресов 42
дигитайзер 1078
дигитайзер, базированный на
    триангуляции 3822
дигитализирование 1054
    3D ~ 1189
дигитализирование 3D
    поверхностей 1189
дигитализирование
    изображения 1901
дигитализировать 1055
дизайн 992
динамическая генерализация
    1199
динамические представления
    1853
динамический обмен данными
    1198
динамическое геоизображение
    1200
динамическое сегментирование
    1201
директивный сетевой график
    1107
дирекционный угол 329
дискретизация 1123
дискретизирование 1123
дискретизированная
    фотография 1064
дискретное распределение 1121
дискретность 1734
дискретные данные 1120

дискретные явления 1122
дискриминантный анализ 1124
диспетчеризация 1125
дисплей 3231
дистанционная генерализация
    3122
дистанционная обработка
    данных 3117
дистанционная оценка
    объектов 3118
дистанционная таблица 3125
дистанционное зондирование
    3120
дистанционное
    позиционирующее
    устройство 3119
дистанционное присваивание
    адреса 1132
дистанционные измерения 3593
дистанционный зонд 3124
дистанционный орбитальный
    зонд 2641
дистанционный сбор данных
    3116
дистанция затухания 932
дистанция затухания волн 932
дисторсия 1137
дифференциальная
    позиционная точность 1024
дифференциальное
    выпрямление 1027
дифференциальное
    нивелирование 1023
дифференциальное
    определение дальности
    1026
дифференциальное
    определение дистанции
    1026
дифференциальное
    позиционирование 1025
дифференциальные поправки
    1022
дифференциальные промеры
    1026
дифференциальный замер
    дальности 1026
дифференцирование
    изображений 1900
диффузная отражательная
    способность 1029
диффузный цвет 1028
длина дуги 2136
длина отрезка 3188
длина поля 1401
длина пути 2711
длина пути в атмосфере 231
длина траектории 2711
длина траектории в атмосфере
    231

дневное картографирование 927
добавить 28
добавлять 28
доверительный интервал 684
дождемер 2995
документ 1158
документирование 3133
долгота 206
долина разлива 1441
доля 2089
домен 1159
домен атрибута 252
дополнительные данные 121
допуск 3698
допуск зерна 1733
допуски обработки 2900
допуски редактирования 1227
допуск отсчётов 3675
допуск разрядки 3989
допуск совмещения узлов 2539
допуск совпадения 2366
допуск совпадения узлов 2539
допустимая погрешность 58
допустимый диапазон адресов 2728
допустимый предел 3698
дорога 3177
"дорожки" 2125
дорожная карта 3168
дорожная символика 3169
дорожно-транспортная планировка 3804
дорожные символические обозначения 3169
дорожный атлас 3165
достоверные данные 562
доступ 13
доступ к данным 853
доступность 14
доступность данных 857
драйвер 1009
драйвер дигитайзера 1083
драйвер устройства 1009
дрань 2112
драпировка 1179
древовидная сеть 3813
древовидная схема 966
древовидная топология 3812
дробление 3445
дублирование 1196
дублированная линия 1195
дублированное окно 573
дублированные данные 1194
дублирующий ключ 101
дуга 146
дуги границы 375

евклидова геометрия 1324
евклидова плоскость 1325

евклидово пространство 1326
евклидово расстояние 1323
Европейское космическое агентство 1327
единица 2041
единица площади 3858
единица рельефа 1391
единицы покрытия 800
единицы расстояния 1135
единичная регрессия 3303
естественная среда 2498
естественное окружение 2498
естественные сети 2499

жёсткая классификация 1803
жёсткое тело 3339
жилая единица 1855
жилая зона 3149
жилищная единица 1855
жилищный массив 3150
жилой квартал 3150
жилой район 3149
жители или страны противоположных частей света 137

зависимая кнопка 2984
зависимая переменная 974
заголовок карты 2344
заграждающая линия 1395
загружаемый драйвер 2205
загущение 968
задание начальных условий 1952
задание размеров 756
задача коммивояжера 3808
задача об изменяемой единице площади 2442
задача о микселах 2436
задача о раскрашивании карт 2258
задача о четырёх красках 1466
задачи до цифрования 2882
задачи после дигитализирования 2873
задержка 952, 2125
задержка в ионосфере 2012
задержка в пути 2125
задержка сигналов в ионосфере 2012
задний отсчёт 304
задний план 300
зазор 1510
заказ 3140
заказная цветная таблица 836
закартированная таблица 2293
закартированные явления 2295
закрашенная область 1416
закрашенные области 610
закрашенный мозаичный

фрагмент 611
закрашенный мозаичный шаблон 611
закрашивание 1417
закрашивание многоугольников 2839
закрашивание округа 1418
закрашивание района 1418
закрашивание регионов 3086
закреплять 1430
закрывать 574
закрытие 582
заливание 1439
заливка 1417
заливная терраса 1441
заливной район 1441
замедление 952
замер дальности 3004
замерный пункт 2384
замкнутая горизонталь 578
замкнутая ломаная 580
замкнутая последовательность непересекающихся цепей или дуг 3162
замкнутая фигура 577
замкнутая форма 577
замкнутые объекты 576
замкнутые элементы 576
замкнутый полигон 579
замкнутый полигональный ход 579
замкнутый теодолитный ход 579
замок базы данных 866
замораживание координат 1482
замораживание слоя 1483
замощение 403
замыкание 582
замыкать 574
замыкающая понижения местности 977
занятие 2609
занятость 2609
заокеанский 2616
запаздывание 952
запирание 2222
запись 3184
запись меток 2066
заполнение 1417
заполнение полигонов 2839
заполненная зона 1416
запоминающее устройство 3500
запрос 3140
    SQL~ 3454
запрос на информацию 3141
запрос на проектирование 3142
запрос на утверждение 3142
запросная таблица 2975
запрос по шаблону 2973
запруда 1095

зарамочное оформление 2349
зарамочное оформление карты 2349
зарамочные данные 2350
зарамочные данные карты 2350
засечка 3145
затенение 3274
затененное изображение рельефа 3272
затопление 1439
затухание 238
захват 429
защёлка 3328
защита доступа к данным 855
звено 3248
земельная единица 2096
земельная информационная система 2083
земельная регистрация 2090
землемер 2094
землемерная съёмка 3723
землепользование 2097
землеустройство 2086
земля 1784, 2072
земляная система 2095
земная волна 1795
земная радиоволна 1795
земная сфера 1209
земное картографирование 3620
земной глобус 1209
земной импеданс 1788
земной эллипсоид 1203
зенит 4019
зенитное расстояние 4020
зеркальное отражение 3433
зернистость 1734
ЗИС 2083
знак 525
знаки-символы аппликаты 4040
знак пересечения 3298
значение атрибута 264
значение цветности 617
значение ячейки 485
значок 1881
зона 158
зона агломерации 81
зона анализа трафика 3781
зона близости 2932
зона видимости 3945
зона влияния 4032
зона градостроительства 3880
зона Делоне 950
зона доступности 142
зона землепользования 2105
зона координатной сетки 1783
зональная карта 4025
зональное распределение 4033
зональные операторы 4026

зональный индекс 4024
зональный многочлен 4027
зона нумерации 2576
зона нумерации переписного участка 366
зона обзора 1403
зона обслуживания 521
зона описательных данных 883
зона перехода 3796
зона почтового кода 4022
зона растительности 3925
зона решётки 1783
зона совместного пользования 2049
зонд 3343
зонирование 4033
зонная структура 4029
зоны картодиаграмм 541

иглообразный полигон 3319
идентификатор 1883
идентификатор записи 3050
идентификатор класса 551
идентификатор конечного узла 1267
идентификатор маркировочного знака 2088
идентификатор начального узла 3469
идентификатор оболочки 798
идентификатор объекта 2590
идентификатор ориентира на местности 2088
идентификатор покрытия 798
идентификатор полигона 2842
идентификатор системы географических координат 1620
идентификатор узла 2537
идентификатор формы 1382
идентификационный номер 1882
идентификационный номер признака 1382
идентификация местоположения 2765
идентификация полосы 3779
идентификация полосы пространства 3779
идентификация пользователя 3889
идентичность 1884
иерархическая база данных 1825
иерархическая картографическая база данных 1823
иерархическая классификация 1824
иерархическая модель 1826

иерархическая модель данных 1826
иерархическое географическое представление 1827
иерархия 1828
изаритмическая карта 2019
избирательный округ 3972
избыточность 3062
извлекать атрибутную информацию 1356
извлечение границы 382
извлечение данных 890
извлечение линий 2173
извлечение ребер 1221
извлечение цифровой модели высотных точек 958
извлечённая карта 983
изгиб 1939
изготовление карты 2310
издание карт 2317
издательский оригинал 1363
излучение 2983
измельчение 3076
изменение интонации цвета 612
изменение климата 567
изменение масштаба изображения 4036
изменение размера 3151
изменение формата 3077
изменения границы 378
изменения цвета 609
измененное изображение 523
измерение 2389
измерение высоты 104
измерение глубин 326
измерение псевдодальности 2943
измерение трёхмерных изображений 1100
измерительная линейка 3187
измерительное дешифрирование изображений 1905
измерительный пункт 3778
изнашивание 975
износ 1345
изобар 2021
изобара 2021
изобарическая форма 2022
изобата 979
изображение 1888
  2,5-мерное ~ 1099
изображение в виде диаграммы 1751
изображение картушки компаса 636
изображение координат 757
изображение кривой 832
изображение местоположения 2867

изображение нескольких
    датчиков 2472
изображение отмывки 1836
изображение оттенками серого
    цвета 1760
изображение, преобразованное
    в цифровую форму 1079
изображение рельефа
    горизонталями 3136
изображение рельефа
    штрихами 3137
изображения, генерированные
    компьютером 668
изогипса 732
изогнутая поверхность 828
изогона 2024
изогональная карта 2025
изодемографическая карта 2023
изоколы 1138
изолинейная блок-диаграмма
    2029
изолинии равных искажений
    1138
изолиния 2031
изолированная система 2027
изолированное расположение
    2026
изолированный многоугольник
    2020
изолированный участок 2020
изоляционная лента 2361
изоляция 2028
изометрическая проекция 2032
изометрическая сеть 2030
изометрическое представление
    2033
изоплета 2031
изотерма 2036
изотермическая карта 2037
изотермический слой 2035
изотропия 2040
изотропная кривая 2038
изотропность 2040
изотропный слой 2039
ИИ 187
икона 1881
иллюминация 1886
именование 2485
именование домена 1160
именованный масштаб 1343
имеющий одинаковое
    протяжение во времени или
    пространстве 597
имитационное моделирование
    3305
имитация 3305
имитация данных 3446
импеданс 1914
импеданс перерывов 3497
импеданс перехода 3837

импеданс поворота 3837
импортирование 1916
имя 2483
имя блока 364
имя домена 1160
имя заливки 1419
имя карты 2287
имя квадратного участка 2958
имя маршрута 2712
имя массива данных 913
имя переписного участка 364
имя покрытия 799
имя поля 1402
имя примитива 1278
имя пути 2712
инвариантность 2004
инвентаризационная карта 2005
индекс 1927
индекс TIGER улиц
    переписных районов 3678
индексация 1931
индекс достижимости 15
индекс доступности 15
индексирование 1931
индексирование базы данных
    863
индексирование элемента 2042
индексирование элемента
    данных 2042
индексирование элементов
    мозаичного изображения
    3683
индекс колонки 622
индексная карта фотографий
    2747
индексное перекрытые 1930
индекс слоя 2120
индекс улиц переписных
    районов 504
индекс чувствительности
    окружающей среды к
    внешнему воздействию
    1297
индивидуальное
    картографирование 2731
индикаторный кригинг 1935
индикатор состояния
    спутников 3205
индикатриса Тиссо 1253
индикационные признаки 1938
индикация 1934
инженерная геодезия 141
инициализация 1952
инкапсуляция 1263
инклинатор 1922
инклинометр 1922
инкремент 1925
инсталлировать 1963
инструмент 1967
инструментальная лента 3699

инструментальная линейка
    3699
инструментальное
    дешифрирование
    изображений 1905
инструментарии
    картографирования 2305
инструментарии разработчика
    1003
инструментарий измерения
    2380
инструменты базы данных 872
интегратор баз данных 864
интегратор данных 895
интегратор изображений 1904
интеграция 1973
интегрирование 1973
интегрированная ГИС 1969
интегрированная региональная
    база данных 1970
интегрированная съёмка 1971
интегрированное
    радиолокационное
    картографирование
    поверхности Земли 1972
интеллектуальный анализ
    данных 900
интеллигентная ГИС 1974
интенсивность цвета 542
интерактивная обработка 1980
интерактивное геокодирование
    1976
интерактивный режим 1979
интервал 2002
интервал возврата 3056
интервал временного канала
    3691
интервал группирования 559
интервал группировки 559
интервал между горизонталями
    730
интервал между изолиниями
    730
интервал привязки 3329
интервал сетки 1771
интервальная арифметика 2003
интервальные вычисления 2003
интероперабельность 1996
интерполирование 1998
интерполирование с равными
    интервалами 1305
интерполятор пикселов 2760
интерполяция 1998
интерполяция линий
    поверхности земли 3613
интерполяция обратных
    взвешенных расстояний
    2006
интерполяция рельефа
    местности 3613

интерпретация 2000
интерфейс 1982
интерфейс ODBC 2629
интерфейс машинной графики 670
интерфейс формуляров 1464
информативность карты 2276
информационная карта 1943
информационная карта набора данных 912
информационная надёжность 909
информационная поддержка 1944
информационная система 1945
информационная система аэрофотоснимков 66
информационная система водотоков 3977
информационная система географических названий 1614
информационная система для проточных вод 3190
информационная система окружающей среды 1293
информационная система, основанная на анализе изображений 1893
информационная система по планированию 2792
информационная система принятия решений 935
информационная система ресурсов окружающей среды 1296
информационное обеспечение 1944
информационный обмен 3788
информационный поток 892
информация о проездах и эстакадах 2677
информация о пространственных ссылках 3410
информация о точке маршрута 3980
инфракрасный сканер 1946
инфраструктура 1948
инфраструктура географической информации 1677
инфраструктура геопространственных данных 1677
инфраструктура пространственных данных 3373
инфраструктура сети 2525
инцидентность 1919

инцидентность сети-ребра 2398
инцидентность узла-ребра 2536
ионосферные погрешности 2013
иррегулярность 2014
ИС 1945
искажение 1137, 3974
искажение картографической проекции 2315
искажение направления 1111
искажение области 163
искажение примитива 1276
искажение расстояний 1133
искажение формы 3281
искание 3241
искать 1424
искусственная граница 186
искусственные данные о рельефе 3566
искусственный интеллект 187
искусственный объект 2243
искусственный спутник для экологического мониторинга 1298
использование 139
использование земель 2097
использование карт 2346
использование угодий 2097
исправление 3054
исправление геометрических искажений 1659
исправление контраста 736
исправления в реальном времени 3037
исправленная ширина колонки 775
исправленный формат строки 775
исследование операций 2635
истинная широта 3833
истинное трёхмерное изображение 3832
истинный север 3834
источник 2657
источники пространственных данных 3376
источник света 2148
источник событий 1332
источник явлений 1332
исходная геодезическая система 3462
исходная карта 3345
исходная таблица 3347
исходная точка 2811
исходная широта 2114
исходные данные 2673
исходный материал 3346
исходный пункт 2811
исчисление объёма 3967
исчисление эквипотенциальной

поверхности 2875

кадастр 422
кадастровая карта 2652
кадастровый класс 421
кадастровый план 2652
кадастр природных ресурсов 2501
кадр 1128
кадрирование 1478, 3999
кадрирование изображения 3999
кадровка 1478
кадр объектов 395
кайма 374
календарные данные 96
калиброванная линия 425
калиброванное адресуемое отображение 424
калибровка 426
калибровка данных 874
калибровка размера ячейки 484
канал информации 2756
каппа Коэна 599
капсулирование 1263
кардинальная точка 431
кардинальное направление 430
каркас 1476
каркасное изображение 4000
каркас сети 2525
карманный компьютер 2689
карта 1611
карта близкого обращения 3292
карта в Web 3985
карта видимости 3954
карта видимости/невидимости 3954
карта в равновеликой проекции 1840
карта-врезка 1962
карта, вычерченная в горизонталях 727
карта гидро-изоплет 1866
карта глубин 982
карта густоты населения 2856
карта дальностей 982
карта дикой природы 3997
карта дикой флоры и фауны 3997
карта дорог 3168
карта земельного кадастра 2652
карта землепользования 2101
карта земных полушарий 2288
карта значков 2755
карта зон затопления 1440
карта зоны обслуживания 3268
карта изолиний 727
карта изоплет 2034
карта изохрон 727
карта инцидентов 1920

карта киберпространства 843
карта классификации земель 2075
карта кластеров 589
карта леса 1459
карта лесного хозяйства 1459
карта локальной ориентации 2289
карта мира 2288
карта облачности 583
карта обслуживаемой области 3268
карта озёр и болот 2069
карта окрестностей 2930
карта ориентации 2289
карта осадков 2996
карта переписей 496
карта подземного рельефа 3527
карта поземельного кадастра 2652
карта покрытия 793
карта пригодности земли 2076
карта признаков 1383
карта разговоров 747
карта растительного покрова 2079
карта растительной жизни 2793
карта с горизонталями 727
карта сеансов 747
карта сети 2528
карта сети Usenet 3886
карта склонов 569
карта, созданная на базе дигитализированной информации 2312
карта состояния земель 2076
карта специального назначения 3424
карта с равными склонениями 2025
карта-схема 3229
карта торговой области 3780
карта удельной проводимости 683
карта узлов 3312
карта флоры 2793
карта характерных признаков 1383
картирование 2297
картировать 2290
картный файл 2272
картобиблиография 2255
картоведение 466
картограмма 539
картограмма безинтервальных шкал 540
картограф 435
картографирование 2297
картографирование аварий 1118

картографирование берега 590
картографирование в Web 3986
картографирование изолиний 733
картографирование имён доменов 1161
картографирование Интернет-связей 1993
картографирование информации 1940
картографирование киберпространства 844
картографирование контуров 733
картографирование морского берега 590
картографирование планиметрической основы 2785
картографирование утилит 3893
картографировать 2249
картографическая алгебра 2252
картографическая аппликация 438
картографическая база данных 442
картографическая библиография 2255
картографическая генерализация 450
картографическая единица 2306
картографическая запись 2275
картографическая изученность 2262
картографическая иллюстративная вставка 438
картографическая информационно-поисковая система 452
картографическая информация 451
картографическая коммуникация 439
картографическая надпись 453
картографическая прагматика 2309
картографическая проекция 2314
картографическая рабочая станция 464
картографическая разрешающая способность 2303
картографическая сводка 2338
картографическая семантика 2329
картографическая семиотика 2330
картографическая сетка 2274

картографическая символика 2340
картографическая синтактика 2341
картографическая система 2304
Картографическая система географической съёмки 1606
картографическая среда 447
картографическая стилистика 2337
картографическая тема 2343
картографическая техника 460
картографическая топонимика 461
картографическая трапеция 2318
картографическая функция редактирования 2268
картографические данные 2263
картографические обозначения 459
картографический 436
картографический банк данных 441
картографический браузер 2348
картографический вид 2347
картографический визуализатор 2348
картографический дизайн 444
картографический запрос 2320
картографический инструмент 2302
картографический каталог 437
картографический масштаб 2326
картографический метод исследования 454
картографический образ 456
картографический отсекающий объект 2256
картографический проектор 2316
картографический репер 529
картографический сегмент 2328
картографический сервер 1994
картографический сканер 458
картографический скрипт 2327
картографический слой 2291
картографический фонд 3495
картографический чертёж 445
картографический шаблон для рисунка 438
картографический элемент 448
картографическое агентство 2299
картографическое визуализирование 462
картографическое деление 2300

картографическое извлечение 2328

картографическое оборудование 1315

картографическое образование 446

картографическое печатание 2311

картографическое представление 457, 2267

картографическое производство 2901

картографическое резюме 2338

картографическое управление 440

Картографическое управление Министерства обороны США 946

картографическое черчение 445

картография 465

картодиаграмма 539

картометрические операции 468

картометрический показатель 467

картометрия 469

картопостроитель с классификацией геологических районов 3647

картосоставление 2297

картосхема 3229

картохранилище 2264

карты взаимодействия природы и общества 2335

касательная 3582

касательный конус 3580

каскадное меню 470

каскадно-расположенные окна 472

каскадный 471

каталог базы данных 862

каталог данных об окружающей среде 1289

каталог изображений 1894

каталог объектных ключей 2592

категорийное покрытие 474

категорийные данные 475

категория 476

качество данных 906

качество карт 2319

квадрант 2960

квадрат 3455

квадратический полином 2962

квадратический сплайн 2963

квадратический тренд 2964

квадратная сетка 3456

квадратная сеть 3456

квадратный блок 2960

квадратный участок 2960

квадратомическое дерево 2965

квадратура 3457

квадродерево 2965

квазигеоид 2972

квазигеоцентрические координаты 2971

квант времени 3691

квантование 1123

квантовать 2969

квартал 3150

квартал города 3150

кибергеография 842

километровая сетка 3456

кинематика 2057

кинематика в реальном времени 3039

кинематическая ГПС 2056

кипрегель 3594

КЛА 3356

клавиши движения курсора 824

клавиши курсора 824

клавиши управления курсором 824

класс 549

класс атрибута 247

класс-интервал 559

классирование 552

классификатор 558

классификатор мультиспектральных изображений с разделением смешанных пикселов 2761

классификатор, построенный по критерию минимального расстояния 2425

классификатор с разделением смешанных пикселов 2761

классификация 552

классификация блоков узкополосных фильтров 397

классификация данных 876

классификация дорог 3166

классификация земель 2074

классификация изображений 1895

классификация интеграторов с узкополосным фильтром 397

классификация методом параллелепипеда 2700

классификация пикселов 2759

классификация сегментированных изображений 3252

классификация спутниковых изображений 3201

классификация Сютса-Вагнера 3532

классифицированное изображение 556

класс объектов 2585

классовый интервал 559

класс признаков 1377

класс примитива 1275

кластер 585

кластеризация 587

кластерный анализ 586

клетка растра 1764

климадиаграмма 566

климатическая зона 568

климатическая кривая 566

клинометрическая карта 569

клиппирование 571

ключ 2055

ключ индекса 3103

ключ отношения 3103

кнопка настройки 2638

ковариантность 792

ковёр-самолёт 1450

код 595

код атрибута 248

код гидрологической единицы 1871

код длины линии 2180

код зоны 161

код зоны нумерации 2577

кодирование 1264

кодирование адреса 33

кодирование группами отрезков 3189

кодирование картографических элементов 2271

код класса 551

код класса признаков переписи 494

код конца строки 1269

код места 2764

код объекта 2586

кодовое обозначение 993

код переписи 490

код признака 1378

код типа топологии 3768

код управления признаком 249

код цвета 617

код элемента рельефа 1378

код юрисдикции 2053

ко-кригинг 600

колебание 1449

количественная геоморфология 1666

количественный анализ рельефа 2968

коллапс 602

коллектор полевых данных 1400

коллекция карт 2257

коллекция отбора примитивов 1282

коллинеарное преобразование 2918

колонка 620
колышки для разбивки откосов 3323
кольцо 3162
командная строка 626
комбинационная табуляция 815
комбинация 623
комбинирование объектов 2587
комбинированное изображение 1172
комбинировать 624
коммуникационная надёжность 628
компактность 629
компас 632
компас наклонения 1922
компасный азимут 634
компасный пеленг 634
компасный румб 2810
компиляция 638
комплексная карта 647
комплексная поверхность 651
комплексное картографирование 648
комплексный атлас 645
комплект шрифта 526
компонентная модель 652
компоненты карты 2260
компоновка 2453
компоновка карты 2285
компрессировать 654
компрессия 656
компьютерная графика 669
компьютерная карта 1239
компьютерная картография 660
компьютерный атлас 667
конвейер 2756
конвенциональная проекция 744
конверсия 748
конвертер 750
конвертирование форматов 1461
конволюция 753
конгломерат 689
конец строки 1268
конечная вершина 1266
конечная точка 1270
конечная точка дуги 1271
конечно-элементная модель 1427
конечные координаты 1265
конечный узел 1266
коническая проекция 690
конкатенация 672
конкретизация 1966
консолидация 702
консолидированный город 701
константа 703
констелляция 705

конструктивная блочная геометрия 708
контекстное меню 3289
контекстно-зависимое меню 3289
континентальная карта 714
континентальный шельф 715
контрастные фильтры 1731
контролируемая классификация 3536
контролируемая область 3535
контроллер пространственных объектов 3400
контроль 532, 3548
контроль вдоль дорожек 2231
контрольная точка 739
контрольное устройство ГПС 1720
контрольный пост 2445
контроль топологии 3753
контур 726
контурная карта 727
контурная линия 731
контурная обработка 2903
контурный объект 166
контур покрытия 794
контур распространения шума 2541
конусность 2606
конфигурация 685
конфигурация дигитайзера 1081
конфигурация местности 3605
конфигурация окон 3998
конфигурация триады 3817
конформная проекция 686
концентратор 1856
концентрические буферы 674
концептуальная карта 678
концептуальная модель 679
концептуальная основа 677
концептуальная структура 677
концептуальная точность 675
концептуальное описание 676
концептуальные рамки 677
конъектор 699
координата 754
  UTM ~ы 3895
  X~ 4013
  Y~ 4016
  Z~ 4018
координатная геометрия 760
координатная ручка 2050
координатная сетка 762
координатная сетка для трассировки 3183
координатная система 765
координатная система карты 2261
координатная система

устройства 1008
координатная трансформация 767
координатные метки 1396
координатный считыватель 764
координатный шар 3777
координатограф 768
координаты конца 1265
координаты на плоскости 2775
координаты отсчёта 3672
координаты проекции 2910
координаты устройства 1007
копировальная рама 2896
копирование 770
копия 769
копия штрихового оригинала 2171
корегистрирование 772
корень дерева 3174
коридор 782
коридор подхода 782
корневой узел 3173
коробка условных обозначений 2133
корпоративная ГИС 416
корпус 1476
корпус картодиаграммы 537
корректура карты 528
коррекция контраста 736
корреляционные точки 781
корреляционный анализ 779
корреляция 778
кортеж 3184
косая отмывка возвышений 2602
косая проекция 2603
косая производная 1109
косвенное измерение 1937
косвенные признаки 1938
космическая геодезия 3198
космические карты 3353
космический летательный аппарат 3356
космический сегмент 3354
космокарты 3353
косое направление 2606
косой вид 2605
косоугольная проекция 2603
коэффициент компактности 630
коэффициент контрастности 596
коэффициент местоположения 2220
коэффициент неплотности 3666
коэффициент помех 3296
коэффициент уменьшения масштаба 3220
краевая точка 390
краевое условие 1218
краевой эффект 381

край 1216, 2672
край поля 164
краткая форма 3288
крест нитей 808
кривая 826
кривая Безье 337
кривая клетка 3729
кривая перемещения 3798
кривая Пиано 2721
кривая регрессии 3093
кривая сдвига 3798
кривизна земной поверхности 825
криволинейные координаты 833
кригинг 2059
криджинг 2059
криптография 817
критерии фильтрирования 1423
критическая зона 803
критические точки 802
критический анализ 801
критический разбор 801
критический регион 803
кромка 1216
круг 543
круговая диаграмма 2753
круговая погрешность позиции 545
крупномасштабная вертикальная конвекция 2223
крупномасштабная карта 2108
крупномасштабные восходящие течения 2223
крупномасштабный вертикальный перенос 2223
крупномасштабный вид 581
крутизна 3482
крутизна местности 3617
крутость 3482
кубическая свёртка 818
кубический сплайн 820
кубический тренд 821
кубическое картирование окружения 819
курвиметр 834
курс 791, 3177
курсовая система 240
курсор 823, 1082
кусок поверхности 3545
кусочно-линейная аппроксимация 2825

"ладошка" 1724
лазерный плоттер 2109
ландшафтное районирование территории 2095
ЛВС 2206
легенда 2278

легенда карты 2278
лента 3511
лента прокручивания 3234
лесной район 1458
лимб картушки компаса 636
линейка 3187
линейка прокрутки 3234
линейка просмотра 3234
линейная диаграмма 2176
линейная интерполяция 2157
линейная карта 2160
линейная оптимизация 2163
линейная ось синфазности 2155
линейная погрешность 2154
линейная регрессия 3303
линейная сеть 2161
линейная структура 2168
линейная триангуляция 2170
линейное вытягивание 2167
линейное проективное отображение 2918
линейно-связанная структура 2159
линейно-угловая сеть 2153
линейно-угловой 2152
линейно-узловая структура картографических данных 153
линейно-узловая топология 155
линейно-узловое представление 154
линейные знаки 2193
линейные условные знаки 2193
линейный интерполятор 2158
линейный масштаб 1744
линейный масштабный коэффициент 2166
линейный объект 2162
линейный признак 2177
линейный сегмент 2189
линейный тренд 2169
линейный фильтр 2156
линии штриховки 1808
линия 2150
линия девиации 1006
линия долготы 2185
линия из нескольких параллельных линий 2464
линия картограммы 538
линия картодиаграммы 538
линия наибольшего отлива 2238
линия отклонения 1006
линия отрыва 400
линия паводка 1835
линия перемены даты 921
линия положения, определяемая азимутом 2982
линия потока 1445

линия равных высот 732
линия равных глубин 979
линия разъёма 400
линия растра 3018
линия резки на листы 3285
линия роста 3814
линия сетки 3018
линия сращения 3238
линия тренда 3814
линия узлов 2186
линия уровня 732
линия центров 507
линия широты 2184
листинговый файл 2272
лист карты 2333
лист кривых 1018
литера 526
литерал 2201
литеральная константа 2201
лифт 1252, 3234
лицевая поверхность 1488
лицевая сторона 1488
лицо 1488
логарифмическая шкала 2224
логическая взаимосвязь 2229
логическая связь 2229
логическая точность 2226
логический выбор 2230
логический запрос 2228
логический соединитель 2227
ложная величина абсциссы 1367
ложная величина ординаты 1368
ложная величина ординаты по карте 1368
ложное отшествие на восток 1367
ложное отшествие на север 1368
ложноцветный снимок 1366
ложный полигон 3452
локализация 2215
локальная вычислительная сеть 2206
локальная ГИС 2212
локальная совместимость 2208
локальное направление дренажа 2211
локальное эластичное соединение 2207
локальность 2213
локальный банк данных 2210
локальный направление 2211
локсодрома 3160
локсодромия 3160
локсодромная спираль 3160
ломаная 2852
ломаная дуга 2826
ломаная линия 2852
лофтинг 2223

лупа дигитайзера 1084
льготный тариф 1119

магазин 3458
магистраль 2756
магнитное направление
    северного меридиана 2240
магнитное склонение 635
магнитный азимут 634
магнитный север 2240
макет 3595
макет атласа 221
макет местности 3614
макет форм местности 3614
макро 2239
макроинструкция 2239
макрокоманда 2239
макрос 2239
максимальное правдоподобие
    2373
максимально правдоподобная
    классификация 2374
максимизированное окно 2372
маркер 2352
маркер границ 386
маркирование 2122
маркирование кластеров 588
маркировочный знак 2087
маркшейдерское дело 2427
маршрут 3177
маршрут движения с
    минимальными издержками
    2128
маршрутизация 3182
маршрутизация
    географических объектов
    1631
маршрутный монтаж 62
маска 2359
маска высотных точек 1250
маска данных 899
маскирование 2360
маскирующая лента 2361
массив 183
массив базы географических
    данных 1576
массив данных 911
масштаб 3210
масштаб генерализации 1524
масштаб издания 3218
масштабирование 3221, 4036
масштаб карты 2326
масштабная линия 3214
масштабная сетка 3068
масштабно-независимая ГИС
    2469
масштабный коэффициент 3216
масштабный коэффициент
    сетки 1778
масштабный множитель 3216

масштаб обобщения 1524
масштаб проекции 3217
масштаб сетки 1778
масштаб составления 640
математическая картография
    2368
математическая модель
    местности 1072
математическая основа карт
    2367
математическая функция 2369
математическое моделирование
    2370
материальная точка 2364
материковая отмель 715
матрица 2371
матрица высот 105
матрица классификации 554
матрица Мортона 2450
матрица неточностей 688
матрица ошибок
    классификации 2430
матрица перекрёстных
    соотношений 807
матрица поведения 332
матрица соотношений 780
матрица тени 3278
матричный процессор 184
МГИС 3881
мгновенная зона обзора 1965
мгновенный сектор обзора 1965
меандровая линия 2376
меандрообразная линия 2376
меандрообразная съёмка 2377
медиана 2386
медианный фильтр 2387
международная карта 1992
межевание 393
межевой знак 2087
межень 2382
межмаршрутное расстояние
    2191
межсетевой интерфейс
    географических ресурсов
    1676
мелкий круг 3326
мелкомасштабная карта 3327
мензула 3550
мензульная топографическая
    съёмка 2780
ментальный атрибут 2390
меню 2391
меню дигитайзера 1085
меню каскада 470
мера маршрута 3179
меридиан 2395
меридиональная проекция 3806
мерная решётка 2383
мерная цепь 516
местная ГИС 2212

местная сеть 2206
местная система координат
    2209
местничество 3083
местное географическое имя
    1163
местность 784
место 1627, 2762
местоназначение переписи 493
место назначения 997
место обитания 1797
местоположение 1627
место соединения 3238
метаданные 2401
метаинформационная система
    2404
метакартография 2400
метаморфизм 2405
метафайл 2403
метахронная блок-диаграмма
    3690
метеорологическая карта 3982
метеорологическая модель 2406
метка 2061, 3676
метка-идентификатор 428
метка на видеоэкране 823
метод интерполяции 1999
метод классификации 555
метод компрессирования 657
метод конечных элементов
    1426
метод линейного
    эталонирования 2164
метод наименьших квадратов
    2131
методология каталогизации
    объектов 1376
метод описания пределов
    земельного участка в
    терминах протяженности и
    векторов границ 2407
метод поиска допустимого
    решения 707
метод пространственного
    доступа 3362
метод рекурсивного спуска
    3058
метод трёхмерного
    дигитализирования 930
методы дистанционного
    зондирования 3123
метрическая система 2415
метропольный ареал 2416
микроволновый несущий
    сигнал 2419
микротриангуляция 181
микрофильм-плоттер 2418
миксел 2435
миллиметровая бумага 1750
минимальное расстояние 2424

минимальный ограничительный прямоугольник 2423
минута 2428
мировая аэронавигационная карта 4006
мировая дифференциальная ГПС 3995
мировая координата 4007
мировая координатная система 4008
мировая сеть 3996
мировое время 758
мишень 3583
МММ 1072
многогранная проекция 2851
многозначная поверхность 2479
многозональный 2455
многозональный приёмник 2456
многоканальный 2455
многоканальный приёмник 2456
многократная экспозиция 1172
многолучевость 2462
многомерное масштабирование 2459
многомерное пространство 2460
многопоисковое изображение 2978
многопоисковое радарное изображение 2978
многопроводная линия 2464
многопутность 2462
многосегментная траектория 2471
многосегментный пут 2471
многосенсорное изображение 2472
многослойное представление 2119
многоспектральная пиксельная классификация 2476
многоспектральное опробирующее устройство 2477
многоспектральное сканирующее устройство 2477
многоспектральные данные 2473
многоспектральный снимок 2475
многоспектральный электрооптический датчик 2474
многоугольная сеть 2846
многоугольник 2823
    3D ~ 1176

многоугольник Безье 338
многоугольники близости 3665
многоугольники Фурье 1468
многоуровневое представление 2121
многофункциональный кадастр 2467
многоцелевая система географических данных 2468
многоцелевой кадастр 2467
множественная регрессия 2465
множественное представление 2470
множественный элемент мозаичного изображения 2478
множество граней 392
множество данных 911
множество матричных и растровых данных 1765
мобильная ГИС 2437
моделирование 3305
моделирование взаимосвязи примитивов 1281
моделирование естественных явлений 2500
моделирование объёмов 2441
моделирование окружающей среды 1294
моделирование поверхности объекта с помощью многоугольников 2833
моделирование с помощью многоугольников 2833
модель 3595
модель атмосферного шума 229
модель атмосферных помех 229
модель базы данных 868
модель взаимосвязи примитивов 1280
модель высотных точек 1251
модель геопространственных данных 1678
модель глобальной циркуляции 1518
модель данных 901
модель данных растра 3009
модель индексного перекрытия 1932
модель компонентов 652
модель места обитания 1798
модельная площадь 2439
модель неперекрывающихся форм местности 3682
модель нечёткой логики 1501
модельное пространство 2439
модельно-познавательная концепция 2440
модель общей циркуляции 1518

модель общей циркуляции атмосферы 1518
модель объекта 2594
модель перекрёстной штриховки 810
модель пространственного взаимодействия 3392
модель риска к обвалу 2092
модель сетевых данных 2521
модель "спагетти" 3359
модифицированная проекция 2443
модовый фильтр 3978
модовый шум 3425
модуль дигитализирования 1089
модульная сетка 3068
мозаика 3681
мозаика данных 902
мозаичное размещение 3684
мозаичный шаблон 3679
моментальный снимок 3330
монолитная линия 3340
монолитное тело 3339
монохромный аэрофотоснимок 359
монтаж 2453
монтаж последовательных аэросъёмок 62
морское картографирование 2351
морской план 1802
морфологический фильтр 2446
морфология 2447
морфометрический показатель 2448
морфометрия 2449
мост 402
мотив 2716
мотив закраски 1421
МОЦ 1518
МСУ 2477
мультилиния 2464
мультимедийная ГИС 2461
мультиплексный приёмник 2466
мультиплицированное изображение 1172
муниципальная ГИС 3881

наблюдение 2607
наблюдение Земли 1205
набор данных 911
набор значений 3899
набор карт 2332
навигатор 2505, 3963
навигационная система 2504
навигационная система с помощи ГПС 1718
навигация 2503

навигация по псевдодальности 2944
навигация с помощи ГПС 1721
наводка на резкость 4036
нагорье 3878
надёжность данных 909
надёжность исследований по картам 3113
надёжность картографического метода исследования 3112
надёжность карты 2324
надел 2089
надир 2482
надпечатание 2681
надпись 3297
наезд 4035
название домена 1160
название карты 2344
наземная контрольная точка 1786
наземная линия 2084
наземная проверка 1793
наземные координаты 3619
наземный контроль 1785
наземный контроль данных 1793
наземный контрольный пункт 1786
наземный ориентир 2087
наземный профиль 2084
назначение 997
наикратчайший пут 3290
наилучшая линейная несмещённая функция оценки 335
наименование 2485
найти 1424
накапливание 80
накладывать 3533
наклон 1921
наклонение 1921, 2606
наклонная буссоль 1922
наклонная поверхность 3316
наклонная проекция 2603
накопитель 3500
налагать 3533
наложение 2122, 3534
наложение двух полигональных слоев 2843
наложение символов 3558
намывная равнина 1441
намывной вал реки 2137
нанесение полос 306
наплыв 1130
направление 791, 1108
направление взгляда 3943
направление на восток 1210
направление потока 1444
направленная связь 1106
направленный граф 1105

направляющий угол 329
направляющий фильтр 1110
напряжённость поля 1406
народосчисление 497
нарушение баланса детектора 1001
нарушение порядка 2014
населённый пункт, включённый в список городов 1924
наследование 1950
наследство 1950
наслоение 3504
настенная карта 3973
настольная ГИС 995
настольное картографирование 996
настольный плоттер 3572
настраивание 54
настройка 3269
настройки на требования пользователя 837
насып 2137
насыщение 3207
насыщенность цвета 617
науки о Земле 1675
научно-справочный атлас 3230
национальная база геопространственных данных 2491
национальная система координат 2492
национальное начало отсчёта высоты 2490
национальное планирование 785
Национальное управление по освоению океана и атмосферы США 2494
национальные стандарты точности картографирования 2493
национальный атлас 2487
национальный британский формат обмена 2495
национальный британский формат обмена векторными данными 2495
национальный исходный горизонт 2490
начало координат 2660
начало координатной системы 2660
начало отсчёта 2811
начальная дуга 331
начальная плоскость 925
начальная точка 3468
начальный меридиан 2889
начальный узел 3468
небесная карта 478

небесный глобус 480
небесный экватор 479
небольшой участок земли 2089
невидимая линия 1820
невидимый слой 2011
невыверенные данные об изображении 3874
невязка 1319
невязка полигона 1319
недвижимое имущество 2073
недокументированные данные 3849
недоступная выборка 1104
недоход до заданной координаты 2870
нежёсткое соединение 2235
независимая переменная 1926
независимый 2615
незаконченная линия 3850
незамкнутая горизонталь 3841
незамкнутость 1319
незамкнутый полигон 3842
неиерархический файл 1436
неизменность 2004
неизменность во времени 3472
неизометрическая линия 2550
неизотропный 132
неисправленные данные об изображении 3874
нейтральная точка 2532
некоммерсиальный ГИС-пакет 1699
неконтролируемая классификация 3876
нелинейный кригинг 2551
немая карта 3551
неметропольный ареал 2552
необнаруженная ошибка 3848
необработанные данные 3030
необрезанный 3840
неоднозначность 108
неоднородное искажение 3974
неорганизованная территория 3873
непараллельное несогласие 3506
неперекрывающееся размещение 3684
неперекрывающиеся карты 3680
неперекрывающийся образец 3679
непланарная сеть 2554
непланарная схема 2554
непланарная цепь 2554
неплоское пересечение 2553
непозиционной 190
непомеченный 3869
непомеченный полигон 2432

неправильная классификация 2429
неправильный многоугольник 2018
непрерывная картограмма 540
непрерывная логика 1500
непрерывная переменная 725
непрерывная поверхность 722
непрерывное отображение 720
непрерывное распространение 718
непрерывное явление 721
непрерывность 716
непрерывные данные 717
непрерывный оттенок 723
непрерывный характер местности 719
непроезжая улица 1913
непространственные данные 192
непространственный 190
непространственный атрибут 191
непространственный объект 193
непротиворечивость 700
непротиворечивость данных 878
непроходимая улица 1913
неравномерно распределённые данные 2017
неравномерно распределённые точки 2016
неравномерность 2014
нерезидентный драйвер 2205
нерезкий 1495
нерезкий объект 1504
нерезкость 1494
неровная линия 2015
неровность 92, 3851
несвязанный текст 3872
несвязная вершина 3845
несвязная дуга 3844
несвязный узел 3845
несимметрия 214
несогласие 3506
нестационарность 2555
неструктурная программа 3357
неточность 1917
неточность из-за недохода до заданной координаты 3847
неточность из-за перерегулирования 2682
неточный интерполятор 2549
неустойчивость детектора 1001
неуточнённые координаты 2927
неучастие 9
неучастие в голосовании 9
неформатированные данные 3852

нефтепровод 2756
нечёткая граница 1496
нечёткая классификация 1497
нечёткая кластеризация 1498
нечёткая логика 1500
нечёткий 1495
нечёткий граф 1499
нечёткий допуск 1507
нечёткий объект 1504
нечёткий узел 1503
нечёткое множество 1505
нечёткое число членов 1502
нечёткость 1494
неясность 108
неясный 1495
нивелир 2139
нивелирная геодезическая сеть 2140
нивелирный пункт 333
нивелирование 2141
низкая вода 2237
низкочастотный фильтр 2236
ниспадающее меню 2953
нисходящая сортировка 988
нисходящий индекс 986
номенклатура 2543
номенклатура карт 3287
номер зоны 4031
номер листа 3286
номер маршрута 3181
номер метки 2067
номер сегмента 3253
номер строки 2183
номинал 2544
номинальная плоскость орбиты 2545
номограмма 2546
нордовая разность широт 2568
нормализация 2559
нормализирование 2559
нормаль 2560
нормаль к фацету 1359
нормальная высота 2558
нормальная коническая проекция 2071
нормальная проекция 2562
нормальная форма 2557
нормальное отражение 3433
нормальное разделение на слои 2563
нормальное распределение 2556
нормальный вид 2557
нормирование 2559
нормы содержания 710
нормы содержания пространственных метаданных 711
носитель цепи 517
НП-полная задача 2547

нулевая высотная отметка 1787
нулевая гипотеза 2573
нулевая клетка 2572
нулевая плоскость 925
нулевое значение 2574
нулевой уровень 924
нуль государственных нивелировок 3237
нумерация 2575
нумерация квартала 365
нумерация переписного участка 365
нумерование 2575
нумерование квартала 365

обвал 2091
обвал земли 2091
обесценивание 975
обзор 2684, 3548
обзор заднего вида 304
обзорная карта 3551
обзорная съёмка 644
области исключения 1338
области квадродеревьев 2966
области проекции 177
область 158, 1159
область вычерчивания 2796
область значений 2999
область карты 3087
область меню планшета 3574
область операций 172
область отсекания 804
область переписи 499
область покрытия 1582
область почтового индекса 4022
область просмотра 3944
облачная карта 583
обмен данными 888
обнаружение знаний в базах данных 900
обнаружение краёв 1219
обновление 3877
обновление карты 2325
обновлять 3078
обобщённое расстояние 2241
обобщённые координаты 1526
обобщённый 1525
обозначение 993
обозначение листа 3284
обозначение полосами 306
обозначение репера 1474
обозримость 653
оболочка текста 3627
оборудование 1314
обработка адресных сведений 36
обработка данных, связанных с науками о Земле 1671
обработка изображений 1907

обработка на копировальном
    станке 2903
обработка по копиру 2903
обработка полигонов 2848
обработка растровых
    изображений 3016
обработка цифрового
    изображения 1053
образ 1888
образец 2926, 3191
образование складок 1179
обрамление 728
обратная поверхность 3158
обратная полярность 3157
обратная трансформация Фурье
    2008
обратное визирование 304
обратное отображение 2009
обратное рассеяние 303
обратный азимут 299
обращение 3064
обрезанный регион 570
обследование 3548
обследование домашних
    хозяйств 1854
обследование ресурсов Земли
    1207
обслуживаемая зона 521
обходный путь 419
обшивка 2693
общая батиметрическая карта
    океанов 1517
общая вариантность 3772
общая граница 691
общая топология 1713
общегеографический атлас
    1516
общедоступная запись 2952
общедоступная информация
    2949
общеземной эллипсоид 4009
общий объект 627
объединение 2044
объединение множество
    вариантов 877
объединение множество
    вариантов данных 877
объединение смежных
    полигонов 2838
объединение смежных
    полигонов с уничтожением
    границ между ними 2838
объединённый город 701
объезд 419, 3773
объездная дорога 419
объект 2584
    3D ~ 1157
объектно-ориентированная база
    данных 2596
объектно-ориентированная

обработка данных,
    связанных с науками о
    Земле 2598
объектно-ориентированное
    моделирование данных 2597
объект типа атрибута 254
объёмная столбчатая
    диаграмма 928
объёмное геоизображение 1011
объёмное изображение 3969
объёмные примитивы 3968
объёмный дисплей
    непосредственной
    трёхмерной визуализации
    1114
объёмный объект 1157
объёмный элемент 1243
объёмный элемент
    изображения 1243
обычный кригинг 2649
овал 2676
оверлей 2680
ограничение 546, 3155
ограничения данных 873
ограничивающая область
    объектов 395
ограничивающие координаты
    394
ограничивающий
    прямоугольник 395
ограничивающий ящик 395
один из результатов 905
один из результатов обработки
    данных 905
однозначная поверхность 3310
однозначное отображение 3309
однозначный ссылочный номер
    парцеллы 3856
одномерный датум 2619
одномерный объект 2162
однообразность 3854
однопоисковое изображение
    2620
однопроводная линия 3307
одноразмерное начало отсчёта
    2619
одноразмерный 2618
одноразмерный репер 2619
однородность 3854
однородные регионы 3855
одноточковое пересечение 3308
оживление 131
оживлённая карта 130
окрестность 1286
окаймление 728
окантовка 728
океанографическая модель
    2611
океанографический атлас 2610
океанская съёмка 2612

окно браузера 405
окно диаграммы 1752
окно дигитализирования 1093
окно легенды 2135
окно макета 2124
окно оцифровки 1093
окно прозрачности атмосферы
    232
окно просмотра 3944
окно статистики 3481
оконтуривание 728
окрестности 2213
окрестность точки 2808
окрестность узла 2540
округ 786, 1148
окружение 1286
окружность 543
окружность большого круга
    1761
октарное дерево 2613
октодерево 2613
октотомическое дерево 2613
опасная зона 1809
оперативная совместимость
    1996
оперативное нанесение на
    карту 2977
оперативный дистанционный
    доступ к СУБД 2204
оператор преобразования 3794
оператор присваивания 196
оператор формирования
    окрестностей 2513
оператор формирования
    окрестностей элемента
    изображения 2513
операторы генерализации 1523
операции анализа 114
операционная система 2633
операционная совокупность
    2634
операционная среда символов
    3554
операция наложения 3534
операция определения
    принадлежности линии
    полигону 2178
операция определения
    принадлежности точки
    полигону 2805
операция проецирования 2908
описание 989
описание данных 881
описание кривой 832
описательные данные 990
оползень 2091
опора 120, 3063
опорная геодезическая сеть
    2530
опорная контурная линия 3537

опорная плоскость 925
опорная станция 318
опорная точка 2811
опорный контур 1929
опорный край 3066
опорный уровень 924
определение граничного
    геодезического знака 2358
определение дальности 3004
определение данных 881
определение дистанции 3004
определение местности 2765
определение местоположения
    явления 1330
определение позиции
    спутником 3204
определение размеров 2389
определённость 2878
определитель листов карты
    2282
опробирование 3194
опробирование пространства
    3415
опрос 3140
оптимальный маршрут 2636
оптимальный пут 2636
оптимизация маршрута 2713
опускающееся меню 2953
опции дигитализирования
    1090
опция атрибута 256
опытный образец 2926
оразмерение 1101
орбита 2639
орбитальные данные 2640
организационная надёжность
    2654
организация землепользования
    2086
организация окон 3999
организация по сбору,
    классификации и
    распространению
    информации или услуг 563
орграф 1105
ординал 2648
ординальное число 2648
ординальный класс 2647
ордината 2650
оригинал географической
    основы 3707
оригинал карты 2658
оригинал надписей 2484
ориентация 2655
ориентация объекта 2595
ориентир 2656
ориентир на местности 2087
ориентированная сеть 1107
ориентированный граф 1105
ориентировка 2655

ориентировочная точка 1112
ортогональная проекция 2663
ортогональное проектирование
    2663
ортогональное пространство
    2664
ортогональный меридиан 2662
ортогональный оператор
    проектирования 2668
ортодрома 2661
ортодромия 2661
ортометрическая высота 1638
ортоморфная проекция 686
орторектификация 2669
ортоснимок 2666
ортотрансформирование 2669
ортофотокарта 2667
ортофотоплан 2667
осадкомер 2995
освежать 3078
освещённость 404
осевая линия 507
осевой меридиан 3070
оседание 3525
ослабление 238, 3525
оснащение карты 1315
основание 330
основа сферических координат
    1493
основная горизонталь 1929
основная земельная единица
    322
основная картографическая
    единица 320
основная линия 317
основная утолщённая
    горизонталь 1929
основные геодезические
    работы 321
остановка 2125
остов 1476
острие 3441
островок 2020
ось X 4012
ось Y 4015
ось Z 4017
ось долины 3561
ось проектирования 292
ось проекции 292
ось центров 507
отбор 2752
отбор выборок 3194
отвод земель 20
отдел 973
отделение 973
отделка филёнками 2693
отдельная линия 3307
отдельный предмет 2041
отдельный факт 1964
откат 3170

отклонение 1005, 3904
отклонение от заданного
    значения 2683
отклонение по дальности 3001
отклоняться от курса 4014
откосные лекала 3323
открытая площадь 2628
открытая связь с базами
    данных 2629
открытая система 2632
открытый ГИС Консорциум
    2631
открытый участок 2628
отлив 2237
отмена 954, 3170
отмена выделения 991
отмена геокодирования 3853
отмена отбора 991
отмена сглаживания 3875
отмена селекции 991
отметка 3674
отметка глубин 3447
отметка о прибытии 531
отметка уровня 1245
отметки географической сетки
    1753
отметки сетки 1753
отмеченная область 3258
отмеченный 2065
отмывка 3274
отмывка при боковом
    освещении 2604
отмывка при комбинированном
    освещении 625
относимая точка 2811
относительная высота 3105
относительная навигация 3107
относительная точность 3104
относительное
    позиционирование 3109
относительное расположение
    3106
относительное указательное
    устройство 3108
отношение основания к высоте
    316
отношение сигнал-помеха 3296
отношение сигнал-шум 3296
отображать 2248
отображение 2908, 3962
отображение Бельтрами 1551
отображение граней 1222
отображение минимального
    расстояния 2426
отображение окрестности 1300
отображение признаков 1384
отображение ребер 1222
отображение, сохраняющее
    траектории 1551
отображение трафика 3785

отображение характеристических признаков 1384
отображение элементов рельефа 1384
отображённая переменная 2296
отпечаток 1918
отправка 1345
отраслевая карта 3644
отрасли тематического картографирования 398
отрезок 3250
отрезок времени 3691
отроги 2672
отсек 3248
отсекание 571
отсекающее окно 572
отсек линии 3248
отсечение 571
отслеживание лучей 3031
отсутствующие полосы сканирования 2434
отсутствующий узел 2433
отсчёт 3191, 3676
отсылка 1345
оттенение 3274
оттенение рельефа 3115
оттененный символ 3273
отчёт качества данных 907
отчётливость 2878
отшествие на восток 1210
отъезд 4038
отъезд от изображения 4038
отыскание путей 2710
офлайн 2615
оффшорный 2616
охват 2999
охват гравиметрической съёмкой 1754
охват индекса 1930
охват сети 2520
оцененная стоимость 1320
оценивание 1321
оценка 1321
оценка градиента 1729
оценка карты 2269
оценка качества окружающей среды 1287
оценка крутости 1729
оценочная карта 1328
оценочная стоимость 1320
оцифрование 1054
оцифрование карты 2266
оцифрованная карта 1080
оцифрованное изображение 1079
оцифрованный снимок 1064
оцифрованные данные 1077
оцифровка 1054
оцифровывать горизонталь

1928
очертание 726
очерчивание 955
очистка 3236
ошибка 1317
ошибка в вычислении 658
ошибка дальности 3001
ошибка инверсии 2010
ошибка классификации 2429
ошибка наблюдения 2608
ошибка по дальности 3001
ошибка при проектировании 413
ошибочная классификация 2429

паводок 1834
пад 1364
падающее меню 2953
падение 1365, 1919
палетка 2383
палитра 2687
палитра закраски 1420
палитра маркера 2353
пампасы 2690
память 3500
панель инструментов 3699
панель контроля 530
панель легенды 2133
панель перечня 2200
панорамирование 2694
пантограф 2695
панхроматическая лента 2692
панхроматический 2691
паразитное изображение 1172
паразитный иглообразный полигон 3319
паразитный полигон 3452
пара изображений 1906
пара карт 2294
пара координат 763
параллакс 2698
параллель 2699
параллельная проекция 2702
параметризированное изображение 2703
параметрическая поверхность 2704
параметрические критерии 2705
параметрические тесты 2705
параметрический анализ 2706
параметры проекции 2914
парцелла 2089
паспортизация земель 2074
пеленгование 275
первая нормальная форма 1428
первая разность 3306
первичная цифровая модель поверхности 2887

первичный ключ 2888
перебор в ширину 399
перевёрнутое изображение 3175
перевод 3797
перевод в цифровую форму 1054
передаточная функция 3790
передатчик 3799
передача данных 918
передача размера единицы времени 3687
передача сигналов точного времени 3687
передача точек 2815
передвижение 2503
передний план 1456
передняя грань 1488
переименование 3126
переклассификация 3046
переклассификация изображений 1909
переклассифицированное изображение 3047
переключатель 530
перекрёстная классификация 805
перекрёстная табуляция 815
перекрёстная штриховка 809
перекрёстное оттенение 814
перекрёстное соотношение 806
перекрёстное табулирование 815
перекрёстное штрихование 809
перекрёстный 3807
перекрестье 808
перекрестье нитей 808
перекрытие 2679, 2680
перемасштабировать 3144
переменная 3900
переменная длина 3902
переменная окружения 1301
переменная поверхности 3546
переменный атрибут 3901
перемещаемость 2861
перемещение 1126
перемещение знака 2454
перемычка 1095
переназначение 3042
переносимость 2861
переносимость базы данных 869
перенос файлов 1414
перенумеровывать 3128
переписной район 499
переписной участок 499
переписной участок округа 491
перепись 497
перепись населения 497

перепись пространственной занятости 3402
перепуск 419
перераспределение населения 2857
перерегулирование 2683
перерисовывать 3060
пересекать 3809
пересечение 2001
пересечение в сетке 1770
пересечение зон 167
перестройка 3129
пересчёт 487, 3221
переупорядочение 3129
переупорядочение слоев 3130
переформатирование 3077
переход 3836
переходная зона 3796
переходный период естественного движения населения 964
перечень переписных районов с указанием их границ 503
перечерчивать 3060
перечисление 487
перешеек 403
перигелий 2725
периметр 2726
периодическое обновление 845
периферийное оборудование 2727
периферийные устройства 2727
периферийные устройства ввода и вывода 1960
периферия 546, 2727
периферный узел 1266
персональный компьютер 2730
перспектива 690
перспективная проекция 690
перспективное вычерчивание 2733
перспективный аэрофотоснимок 2601
перспективный рендеринг 2734
перспективный трёхмерный вид 1175
перспектограф 2732
перьевая палитра 2723
перьевой плоттер 2724
петля 2234
печатание 2895
печатающее устройство 2894
печатная карта 2697
пик 2718
пиксел 2758
пиктограмма 1881
пилот-проект 2754
пирамидальная диаграмма 2954
пирамидные слои 2955
ПК 2730

П-код 2879
плавная смена изображений 1130
плавное изменение 136
план 992
планарная карта 2769
планарная сеть 2771
планарные картографические проекции 2770
планарный граф 2767
планарный многоугольник 2772
план города 548
планета 2779
планетарная ГИС 1706
планетное картографирование 2782
планетный глобус 2781
планиметр 2783
планиметрическая карта 2788
планиметрическая основа 2784
планиметрическая съёмка 2789
планиметрические данные 2786
планиметрический вид 2790
планиметрический объект 1156
планиметрическое изображение 2787
планиметрическое представление 2790
планиметрия 2791
планирование землепользования 2102
планирование критических ситуаций 1261
планирование окружающей среды 1295
планировка 2123
планировка инфраструктуры 1949
планка инструментов 3699
плановая геодезическая сеть 2530
плановые координаты 1845
планшет 3550
планшетный графопостроитель 1432
планшетный сканер 1433
пласт 330
плато 3571
плеяда 705
плитка 3679
плоская возвышенность 3571
плоская координатная система 2776
плоская модель Земли 1435
плоская поверхность 2773
плоские координаты 1845
плоские сплайны 3668
плоский граф 2767
плоский графопостроитель 1432

плоский многоугольник 2772
плоский наклон 1434
плоский объект 1156
плоский сканер 1433
плоский файл 1436
плоскогорье 3571
плоское изображение 1098
плоское пересечение 2768
плоскостные координаты 2775
плоскость 2774
плоскость карты 2307
плоскость напластования 2777
плоскость начала отсчёта 925
плоскость нулевого меридиана 4021
плоскость орбиты 2642
плоскость относимости 925
плоскость уровня 2144
плоскость экватора 1308
плоскость элементов отображения 358
плотина 1095
плотная линия 3340
плотность 969
плотность наблюдений 970
плотность растра 3010
плотность точек 971
плоттер 2795
плоттер Келша 2054
площадная диаграмма 160
площадная триангуляция 181
площадь 157
площадь города с пригородами 2416
площадь застройки 414
площадь поверхности 171
плювиограф 2995
побитовое отображение 3015
поблочное присваивание адреса 362
поверхностный слой 3543
поверхность 3538
2D ~ 2773
поверхность Безье 339
поверхность гравитации 1756
поверхность карты 2339
поверхность напластования 2777
поверхность откоса 3544
поверхность, ощущаемая электронными приборами 1237
поверхность проекции 2915
поверхность сетки 1781
поверхность тренда 3815
поворот 3836
поворотная точка 3838
повторитель 1452
повторитель кривых 2174
повторная выборка 3143

повторная раскраска 3049
повторное деление на округа 3059
повторное изображение 1172
повторное позиционирование 3134
повторное раскрашивание 3049
повторно трассируемая линия 3156
повторный выбор 3146
повторный отбор 3146
повторяемость 3131
поглощение 8
погрешность 1317, 3698
погрешность, вносимая шкалой 3215
погрешность геокодирования 1536
погрешность данных 887
погрешность дигитализирования 1087
погрешность инструмента 1968
погрешность многолучевости 2463
погрешность обмена данными 889
погрешность позиционирования 2864
погрешность при наблюдении 2608
погрешность сканирования 3195
погрешность формата 1462
подблок 3517
подбор 2752
подбор методом наименьших квадратов 2130
подбор состава 2924
подвыборки 3523
поддерево 3530
подквадрант 3522
подкласс 3518
подмножество 3524
поднимание 1244
подобласть 3516
подобласть переписи 498
подобъект 3520
подокно 3531
подпись 3297
подплоскость 3521
подразбиение 3519
подразделение 3519
подразделение округа 789
подразделение охраны 835
подсистема аппаратуры пользователей 3892
подсистема наземного контроля и управления 742
подсказка атрибута 257
подслаивание 3846

подстановка 3526
подстрочное примечание 1454
подсчёт 783
подтверждение 1966
подъёмник 1252
поземельная база данных 2081
поземельная книга 422
позиционирование 2869
позиционирование в реальном времени 3040
позиционная погрешность 2864
позиционная точность 2863
позиционная эталонная система 2218
позиционный анализ 3311
позиционный допуск 2865
позиционный символ 2219
позиционный эталон 2217
позиция 239, 2862
познавательная карта 598
поиск 3241
поиск ближайшего соседа 2508
поиск близости 2931
поиск в глубину 981
поиск в определённой области 178
поиск в ширину 399
поиск информации 1942
поиск линии склона 3322
поиск оптимального маршрута 3259
поиск преимущественно в глубину 981
поиск преимущественно в ширину 399
покадровый просмотр 406
показатель снижения точности 1097
показатель снижения точности, обусловленный геометрическими факторами 1657
показатель снижения точности определения положения в пространстве 2866
покрытие 1582, 3543
покрытие гравиметрической съёмкой 1754
покрытие дуг 148
покрытие плоскости многоугольниками 3684
покрытие полигона 2837
поле атрибутного тега 261
полевая съёмка 3723
полевой блок 1409
полевые данные 1398
полевые измерения 1407
поле данных 891
поле для галочки 530
полезный коэффициент

кривизны земной поверхности 1228
поле зрения 1403
поле списка 2200
поле текста 3626
полигон 2823
полигональная погрешность 2828
полигональная сеть 2829
полигональное моделирование 2833
полигональное отсекание 2827
полигональное перекрытие 2847
полигональное соединение 2835
полигональный конвертор 2836
полигональный объект 166
полигональный ход 2830
полигон в форме выпуклой оболочки 751
полигональный ход 2830
полигоны Вороного 3665
полигоны Тиссена 3665
поликоническая проекция 2822
полилиния 2852
полилиния, сглаженная кривой 829
полимасштабная ГИС 2469
полимасштабное представление 2470
полиморфизм 2853
политические границы 2821
полиэдрическая проекция 2851
полная вариантность 3772
полная пространственная беспорядочность 643
полная пространственная случайность 643
полное множество 1339
полномочия географического доступа 1567
полный репер 642
половодье 1834
положение 239
полоса 374, 3511
полоса захвата 3553
полоса сканирования 3226
полоса сканирования квадранта 2970
полоса спектра 3428
полуавтоматизированное оцифрование 3261
полуавтоматическое цифрование 3261
полу-вариограмма 3262
полукраска 1799
полусфера 1815
полутон 1799
полутоновая карта 1762

полутоновая шкала 1759
полутоновое изображение рельефа 1837
получатель 3043
получение контрольных характеристик 334
получение точные размеры скважины 3549
полушар 1815
польза в месте спроса 2766
польза при широком распределении 2766
пользователь ГИС 1702
пользовательская метка 838
пользовательский интерфейс 3890
пользовательский символ 839
полюс 1618
полярная координатная система 2817
полярная проекция 2820
полярное расстояние 2818
полярность 2819
полярные координаты 2816
полярные координаты в пространстве 3437
помехи приёмника 3045
помеченный 2065
помеченный полигон 2062
помеченный регион 2063
понижение 3061
понимание 653
поперечная проекция 3806
поперечное сечение 812
поперечный 3807
поперечный разрез 812
поправка 776
поправка в карте 2253
популяция 2854
поразрядное отображение 3015
порог 3671
пороговое значение 3671
порог приёма 12
портовый план 1802
порядковая статистика 2646
порядковое числительное 2648
порядковое число 2648
порядок 2643
порядок вычерчивания 1184
порядок Мортона 2451
порядок сортировки примитивов 1283
последовательный доступ 3266
последовательный файл 3267
последующая классификация 2872
последующая обработка 2874
послойное представление 2119
послойно-организованная ГИС 2117

поставщик данных 904
постоянная 703
постоянная геометрическая точность 704
постоянная ошибка 3567
постоянные потери 2729
построение карты 2265
построитель SQL запросов 3453
построитель колонки 621
потерянные строки развёртки 2434
поточечное оцифрование 2807
поточное накопление 1442
почва 1784
почвенная карта 3338
почтовый адрес 2242
почтовый индекс 2871, 4034
почтовый код 4034
пошаговая линейная классификация 3483
пояснительные надписи 1342
права географического действия 1567
правила написания объектов на картах 2665
правило Боудитча 396
правильно расположенные данные 3095
правильность 777
правильные данные 562
превращение 748
предварительная обработка растра 3021
предварительные данные 2933
предел 546
    за ~ами морской границы 2616
предел видимости 3003
предельная точность масштаба 3213
предназначение 997
предположительное исчисление населения 962
предпроектное исследование 1373
предсказание 1455
представление банка данных 859
представление в виде символов 3556
представление границ 391
представление данных в виде таблиц 3577
представление динамически связанных данных 1197
представление землепользования 2104
представление координат 757
представление местности 3138
представление объекта 2599

представление пространственных данных 3375
представление табличных данных 3576
представление топографических поверхностей 3138
предшествование 2876
преобразование 748, 3793
преобразование в картографическом представлении 749
преобразование выборок 3196
преобразование данных 879
преобразование карт 2345
преобразование координат 755
преобразование примитива 1284
преобразование пространственных данных 3380
преобразование репера 923
преобразование секретных данных 2897
преобразование срединной оси 2385
преобразование точки в регион 2813
преобразование форматов 1461
преобразование Фурье 1469
преобразователь 750
преобразовывать в цифровую форму 1055
прерывание линии 2188
прибавлять 28
приближение 4035
приближение изображения 4035
приближённая интерполяция 143
приближённые координаты 2927
прибор для измерения угла наклона 1922
прибор для черчения 1183
прибор, записывающий длину и профиль пройдённого пути 956
прибрежная зона 592
прибрежный марш 591
приведение 3061
приведение данных 908
привлекательность 243
привязка 3074, 3328
привязка к квадранту 2961
привязка к растру 3025
привязывание 3328
привязывать к узлу 3332
пригородный ареал 2552

придавать новую форму 3148
придание квадратной формы 3457
приёмник 3043
приёмник позиционирования 1722
приёмочное испытание 11
приёмочный тест 11
приёмы анализа карт 2342
признак 3295
признаки весомости 3992
признак конца поля 1408
признак, снабжённый комментариями 133
прикладная геодезия 141
прикладное программное обеспечение 140
прилегание 45
прилежащий угол 47
приложение 139, 140
применение 139
примитив 1274, 2890
принтер 2894
принцип наименьшего сопротивления 2893
принятие решений 934
принятие решений по множеству критериев 2458
приобретать новую форму 3148
приобретение земельной собственности 20
приобретение земли 20
приподнятый берег реки 2137
приращение 1925
природоохранная ГИС 1290
прирост 1925
присваивание адреса 31
присваивание имён 2485
присваивание кластерам меток 588
присваивание меток 2066
присваивание объектам меток 2066
присваивание атрибута пространственным объектам 262
присоединение 53, 236
присоединение атрибута 246
притягательная сила 243
проба 3191
пробный чертёж 2919
проведение линию на карте 2181
проверка 532
проверка в процессе эксплуатации 1397
проверка на месте 1397
проволочно-каркасная модель 4001
проволочно-каркасное

изображение 4000
прогнозирование 1455
прогнозирование транспорта 3803
прогнозирование численности населения 962
прогнозная карта 2905
прогнозная карта абсолютной топографии 2904
прогнозная контурная карта 2904
программа-векторизатор 3029
программа географической статистики 1629
программа интерполяции горизонталей 729
программа отображения 2313
программа просмотра 3963
программа трассировки топологии географического расположения 1632
программа управления растра 3014
программная ошибка 413
программное обеспечение 3336
программное обеспечение ГИС 1700
программное обеспечение ГПС 1723
программное обеспечение диалогового графического проектирования 1977
программное обеспечение дигитализирования 1092
программное обеспечение для сохранения информации в карте 3337
программный пакет трансформации географических координат 1519
продолжительная транзакция 2232
продольное перекрытие 1465
продольный контроль 2231
проект 992
проективная алгебра 2916
проективная геометрия 2917
проективное преобразование 2918
проективный допуск 2906
проектирование 994. 2908
проектирование базы данных 861
проектирование карт 444
проектирование размера 2924
проектирование с помощью компьютера 661
проектировщик 956
проектирующая линия 2187

проектор 2911
проекция 2907
проекция UTM 3867
проекция Гала-Питерса 1509
проекция Гаусса-Крюгера 1514
проекция Ламберта 2071
проекция Меркатора 3867
проекция Молвайда 2444
проекция окружения 1300
проекция, определяемая пользователем 3888
проекция с числовыми отметками 3720
проецирование 2908
прозрачность 3800
прозрачный регион 3801
производить съёмку местности 2249
производная единица 985
производная карта 983
производная поверхности 3539
производная по направлению 1109
производная по оси 1109
производная путевой устойчивости 1109
производное поле 984
производные данные 3244
произвольная проекция 145
прокрутка 3235
прокручивание 3235
проксимальный допуск 2928
пролистывание 406
промежуток 1510
промежуточная горизонталь 1985
промежуточная посадка 3498
промежуточная сумма 3528
промежуточный узел 1986
промер 1864
промеры 3004
пропадание 1185
пропорциональное переразмерение объекта 2922
пропорциональность 2921
пропорциональные символы 2923
пропуск линии 2188
просмотровая таблица 2233
просмотрщик 3963
простая планарная сеть 2396
простая разность 3306
простая регрессия 3303
простое изображение 2620
простое спрямление 3302
простой кригинг 3299
простой объект 3300
простой полигон 3301
простой список 516

пространственная автокорреляция 3366
пространственная база данных 3371
пространственная взаимосвязь 3413
пространственная геодезическая сеть 3367
пространственная единица 3419
пространственная интерполяция 170
пространственная информационная система 3390
пространственная картина 3406
пространственная корреляция 3369
пространственная организация 3404
пространственная привязка 3412
пространственная ссылка 3409
пространственная статистика 3417
пространственная структура 3418
пространственная частота 3386
пространственно-временная база данных 3421
пространственно-временная ГИС 3422
пространственно-временная конвергенция 3693
пространственно-временная концептуализация 3692
пространственно-временное концептуальное представление 3692
пространственно-временные данные 3420
пространственное взаимодействие 3391
пространственное изображение 3406
пространственное индексирование 3388
пространственное моделирование 3397
пространственное мышление 3408
пространственное перекрытие 3405
пространственное пересечение 3393
пространственное положение 3395
пространственное разрешение 3414
пространственное расположение 3395

пространственное сглаживание 3416
пространственное соединение 3394
пространственные данные 3370
пространственные координаты 3368
пространственные метаданные 3396
пространственные окрестности 3398
пространственные примитивы 3968
пространственный адрес 3363
пространственный анализ 3364
пространственный атрибут 3365
пространственный вектор 3355
пространственный доступ 3361
пространственный запрос 3407
пространственный индекс 3387
пространственный объект 3399
пространственный охват данных 880
пространственный порядок 3403
пространственный репер 3382
пространство 3352
пространство признака 1388
пространство траекторий 3787
пространство устройства 1010
пространство формы 1388
противовес 2362
против течения 3879
проток 1446
протокол 2925
прототип 2926
протяжение 3360
протяжённость в определённом направлении 1487
протяжённость трассы 2711
профилограф 2902
профилометр 2902
профиль 812
профильная блок-диаграмма 813
проход 782
проходить 3809
процедура присваивания меток 2064
процедура расположения/размещения 2216
процесс выборки 3194
процессор векторно-растрового преобразования 3924
процессор выделения признаков 1380
процессор интерфейса сканера 3224

процессор переноса пространственных координат 3378
процессор растровых изображений 3017
прыжок 3328
прямая 3501
прямая линия 3501
прямая полярность 2561
прямая, пополненная бесконечными точками 1348
прямолинейная навигация 3502
прямоугольная координатная система универсального пространства 3866
прямоугольная проекция 2663
прямоугольная сетка 3052
прямоугольные координаты 433
прямоугольные фацеты 3051
прямые признаки 1113
псевдодальность 2942
псевдо-избирательный округ 2947
псевдоконическая проекция 2938
псевдо-перспектива 1370
псевдоскопический вид 2945
псевдослучайная последовательность 2941
псевдослучайный код 2941
псевдослучайный шум 2941
псевдостатика 2946
псевдоузел 2940
псевдоцвет 2934
псевдоцветное добавление 2935
псевдоцветное изображение 2936
псевдоцилиндрическая проекция 2939
психомоторная погрешность 2948
птичка 3674
пузырьковая диаграмма 407
пункт 2762
пунктир 2722
пунктирная линия 2722
пункт наземного контроля 1794
пункт плановой сети 740
пункт слежения 3778
пустая клетка 1262
пустая строка 360
путешествие 3773
путь 3177
путь в атмосфере 230
пятнистый шум 3425

рабочая область 4004
рабочая станция 4005
рабочее поле 23

рабочее пространство 4004
рабочий слой 24
рабочий файл 4003
равновеликая азимутальная проекция Ламберта 2070
равновеликая коническая проекция Алберса 89
равновеликая проекция 1316
равновеликий 1304
равновесие материалов 2362
равномерность 3854
равномощная поверхность 2145
равнопромежуточная проекция 1312
равнопромежуточные линии 1311
равнораспределение 1313
равносторонняя прямоугольная сетка 3053
равноугольная карта 2025
равноугольная проекция 686
равноугольник 2024
радарное дистанционное зондирование 2979
радиальная линия 2982
радиальная погрешность 2981
радиальное искажение 2980
радиовысотомер с самописцем 86
радиокнопка 2984
радиолокатор с синтезированной апертурой 3562
радиолокационная система обзорно-сравнительного метода наведения 3606
радиолокационная станция с искусственным раскрывом антенны 3562
радиометр 2985
радиометрическая калибровка 2986
радиометрическая коррекция 2987
радиометрическая чувствительность 2990
радиометрическое искажение 2988
радиометрическое разрешение 2989
радиометрия 2991
радиопозиционирование 2992
радиосигнал 2993
радиус 2994
радиус Экватора 1309
разбивать на подгруппы 2969
разбиение 942, 3445
разбиение карты на квадраты 2336
разбиение на области 3084

разбиение на регионы 3084
разбиение на слои 2121
разбиение на треугольники 3821
разбиение области 162
разбиение плоскости на многоугольники 3684
развёрнутое меню 2953
развёртывающаяся поверхность 1002
развитие 1004
развитие географической информации 1605
разграничение 957
разграфка 2575
раздел 1152, 3247
разделение изображений 1902
разделение на многоугольники 2844
разделение на полигоны 2844
разделение полосы частот 309
разделение признаков 1386
разделение форм 1386
разделение цветов 616
разделитель полей 1404
раздел окружающей среды 1299
разделять на многоугольники 2845
разделять на составные части 1116
различение 3048
различение изображений 1900
различие 1151
разложение 942
размах 3360
размельчать 1116
размер 1101
размер буфера 411
размер маркера 2355
размерность 1102
размер рабочего поля 3315
размер рабочего поля плоттера 2796
размер ячейки 483
разметка 334
разметочный циркуль 1153
разметчик 3648
размещение 94, 2215
размещение меток точечных объектов 2802
размещение ресурсов 95
размещение события 1330
размыкать 3871
размытая классификация 1497
размытая логика 1500
размытое множество 1505
размытость 1494
размытый 1495
размытый объект 1504
разновременность 3686

разнообразие 1151
разнородность 1816
разность 1151
разность времён 3686
разработка 1004
разработка заказного варианта 837
разрез 840, 3317
разрешающая способность 3152
разрешающая способность по времени 2149
разрешение 3152
разрешение деталей земной поверхности 1790
разрешение на местности 1790
разрешение неоднозначности 3153
разрешение по времени 2149
разрушение в атмосферных условиях 3981
разрядка 3988
разрядная матрица 358
разукрупнение 1117
разъединять 3871
район 1148, 3079
район вглубь от прибрежной полосы или границы 1838
районирование 1150
районная планировка 175
район под застройку 414
рама картодиаграммы 537
рамка 374
рамка листа карты 376
ранжирование областей 176
ранжированная карта 3000
раскопка данных 900
раскрывающееся меню 2859
расплывчатость 1494
расплывчатый 1495
распознавание 3048
распознавание изображений 2717
распознавание карты 2321
распознавание образов 2717
распознавание форм 3283
расположение 2215
расположение данных в виде таблицы 3577
расположенные с равными интервалами данные 3095
распределение 94, 552
распределение Гаусса 2556
распределение населения по возрастным группам и полу 963
распределение по поясам 4033
распределение хи-квадрат 533
распределённая база данных 1142

распределённая географическая информация 1144
распределённая информационная система 1145
распределённая картографическая база данных 1139
распределённая обработка 1146
распределённые данные 1140
распределённый банк данных 1141
распространение 2920
рассеяние 1129
рассеяние частоты 1484
расслоение 2121
расслоение масштабирования 4037
расслоение плотности 972
расслоённый кригинг 3505
рассредоточенная база данных 3371
рассредоточенный источник загрязнения 165
расстояние 1131
расстояние Бхаттачарья 340
расстояние видимости 3003
расстояние Махаланобиса 2241
расстояние невязки 2431
расстояние незамкнутости 2431
растеризатор 3924
растеризация 3923
растительный покров 2077
растр 3005
растровая карта 1772
растровая линия 3018
растровая модель 3009
растровая структура 3027
растровая тематическая карта 1782
растрово-векторное преобразование 3028
растрово-векторный процессор 3029
растрово-векторный редактор 1858
растровое изображение 3015
растровое представление 3023
растровое пространство 3026
растровое сканирование 3024
растровые данные 3008
растровый графопостроитель 3020
растровый объект 3019
растровый плоттер 3020
растровый файл 3012
растровый фон 3007
растровый формат 3013
растр протока 1447
расчёт маршрута движения с

минимальными издержками 2129
расчёт поверхности/периметра 174
расчёт сети 2518
расширение отображения 2270
расширение полосы частот 313
расширение псевдоцвета 2935
расширённая прямая 1348
расширённый объект 1349
расшифровка 941
расщепление 3445
расщепление мод 2438
расщепление плотности 972
расщеплять 3871
рафинированная генерализация 1425
РБД 1142
реализация 1915, 1966
реализм 3035
реальность 26
ребро 1215
ребро Вороного 3970
ребро многоугольника 164
ребро полигона 164
регенерация растра 3022
регион 3079
регионализм 3083
региональная база данных 3081
региональная ГИС 3082
региональная карта 3085
региональное покрытие 169
региональный атлас 3080
регистр 473
регистрационный файл 2225
регистрация адресов 41
регистрация земельных участков 2090
регистрация изображений 1910
регистрирование 3091
регистрированная таблица 3090
регистровая карта 2652
регрессия 3092
регулярная ошибка 3567
регулярная сетка 3094
редактирование 1225
редактирование значений пикселов 3011
редактирующая программа 1226
редактор 1226
редактор классифицированных изображений 557
редукция 3061
режим асинхронной передачи 218
режим работы в собственной системе команд 2496
резерв 3062
резервирование 3147

рейка 2112
реквизит 244
ректификация 3054
рекурсия 3057
релаксация 3110
релевантность 3111
рельеф 3605
рельефная карта 3114
рельефная поверхность 3542
реляционная база данных 3099
реляционная картографическая база данных 3098
реляционная связь 3102
реляционное индексирование 3101
рендеринг 3127
рендеринг форм местности 3615
реорганизация данных 910
репер 333, 1475
реперная отметка 924
реперная точка 740
репрезентативные единицы рельефа 3135
рестрикция 3155
ресурс 194
ретрансляция 3795
ретроспективный взгляд 1839
ретуширование 3274
ретушь 3276
референцная система 3073
референц-станция 318
референц-эллипсоид 3067
реципиент 3043
речная сеть 3163
решётка 2115
рисунок 2716
ров 1095
роликовый графопостроитель 3171
РСА 3562
рубеж 1152
рулонный графопостроитель 3171
рулонный плоттер 3171
руль высоты 1252
русло 330
ручное дигитализирование 2245
рынок географических информационных систем 1696
рычажный указатель 2050
ряд треугольников 520

самая внутренняя горизонталь 1954
самый низкий уровень 2482
сателлитная система навигации транспортных средств 3926

сближение 745
сближение меридианов 746
сбой счёта времени 3689
сбор данных 875
сбор и регистрация данных с борта самолёта 84
сборная таблица 1933
сборный лист 1933
сборный лист карты 1933
сведение для инструментальной кнопки 3700
свёртка 602, 753
свёртывание 753
светлота 404
световой источник 2148
светодиодный плоттер 2147
свечение 1886
свидетельство правооснования 515
свободная пространственная кривая спроса 1481
свободный узел 1480
сводка 1223
свойство топологии 3765
связанность 698
связанные данные 197
связанные дуги 693
связанные земные структуры 1791
связанные линейные сегменты 696
связанные таблицы 2198
связанный граф 695
связка 2199
связная поверхность 697
связное множество граней 790
связной граф 695
связность 698
связующая точка 3677
связываемость 698
связывание 199, 2199
связывание атрибутов 253
связывание карт 2281
связывание объектов с атрибутами 262
связывающее ребро 403
связь 2199
связь между приложениями 1981
связь меньше критической 2235
сгибание 1939
сглаживание 136
сглаживание границ кривых, наклонных линий и шрифтов 136
сглаживание данных 915
сглаживание кривой 830
сглаживание линии 2190
сглаживание поверхности 3540

сглаживание сплайна 3444
сдвиг 1126
сдвоённая разность 1170
сдвоённое изображение 1172
сдвоённое имя квадратного участка 1173
сдерживание 709
сеанс дигитализирования 1091
сеанс оцифрования 1091
север 2564
север картографической сетки 1774
северный 2567
северный магнитный полюс 2240
Северный полюс 2569
Северный полярный круг 156
северо-восток 2566
северо-запад 2570
сегмент 3250
сегментация 3251
сегментация изображения 1911
сегмент выделения признаков 1381
сегмент дороги 2714
сегментирование 3251
сегмент маршрута 2714
сегмент пользователей 3892
сегмент потребителей 3892
сегмент пути 2714
седловина 2708
секанс 3242
секторная диаграмма 2753
секторный 168
сектор обзора 1403
секунда 3243
секущая 3242
секционированное оразмерение 1473
секция 3248
селективная наличность 3260
селекция мод 2438
сельская местность 784
сельская территория 3947
сельский район 784
семейная единица 1371
семейство 584
сенсор 3265
сенсор замера положения 242
сенсор измерения позиции 242
"серая" выборка 1104
сервер баз данных 870
сервер географических данных 1538
сервер географических информационных систем 1694
серийный номер признака 1387
серия карт 2331
сертификат титула 515

сетевая информационная система 2526
сетевая карта 2528
сетевая картографическая база данных 2519
сетевая связь 2527
сетевая файловая система 2524
сетевой адрес 2515
сетевой анализ 2517
сетевой доступ 2623
сетевой каталог 2624
сетевой элемент 2523
сетка 2115, 2396
сетка адресов 37
сетка для трассировки 3183
сетка координат 762
сетка на карте 2274
сетка нитей 808
сетка полигонов 2829
сетка-указательница 2214
сеточная модель улиц 1780
сеточная поверхность 1781
сеточная система 1598
сеточный режим 1773
сеточный север 1774
сеточный юг 1779
сеть 2396
сеть Вороного 3971
сеть географических данных 1537
сеть гидроклиматических данных 1860
сеть гидрологических реперов 1869
сеть неравносторонних треугольников 3820
сеть опорных точек 2530
сеть трафика 3783
сеть утилит 3894
сечение 840
сечение горизонталей на карте 730
сжатие 656
сжатие полосы частот 312
сжатие спектра 312
сжатие цвета 608
сжать 654
сильная связь 575
сильное изменение картины рождаемости и смертности 964
символ 525
символ зоны 4030
символизация 3556
символические обозначения на карте 459
символически отображённые точки 3557
символ маркера 2356
символ мишени 3587

символ с тенью 3273
символ цели 3587
символы высотных отметок 3448
символы клифов 565
символы обрывов 565
символьная строка 527
символьное перекрытие 3558
символьный файл 3555
симулятор Земли 1208
симуляция 3305
симуляция потока 1448
симуляция трафика 1448
синклинальная линия 3561
синтаксический анализ 2706
синтаксический разбор 2706
синтез изображения 1897
синтетическая карта 3563
синтетическая разрешающая способность 3565
синтетическая тематическая карта 3646
синтетические показатели 82
синтетическое картографирование 3564
система автоматизированной картографии и ГИС в управлении сетями предприятий коммунального хозяйства 278
система адресации 38
система адресов 38
система ввода географических данных 1596
система визуализации конфигурации местности 3609
система вычерчивания в универсальной проекции 3864
система географических знаний 1610
система географических координат 1598
система государственной земли 2951
система государственной триангуляционной съёмки 2950
система диспетчерского контроля оптимальной маршрутизации 2637
система землепользования 2095
система и формат представления данных о пространственных объектах 1193
система кадрирования 1479
система картографирования в

Web 3987
система картографирования ложе моря с высокой разрешающей способностью 1833
система классификации земель по характеру их использования 2099
система координат 765
система координатных осей Земли 1204
система линейного эталонирования 2165
система маршрута 3180
система навигации транспортных средств 3926
система негеографических координат 2548
система обработки атмосферных и океанографических снимок 226
система обработки изображений 1908
система общедоступного топографического измерения 2950
система ориентации 241
система открытой съёмки местности 2950
система отображения поверхности Земли в цифровом коде 1056
система отслеживания рельефа местности 3606
система отсчёта 3073
система отсчёта курса и положения 240
система оцифровки дуг 149
система плоскостных координат США 3470
система понятий 677
система пространственной стабилизации 241
система прямоугольных координат 1777
система распознавания карт 2322
система редактирования конфигурации местности 3610
система, следящая рельефа местности 3611
систематизация областей 176
систематическая ошибка 3567
систематическая погрешность 3567
система управления базами данных 867
система управления

землепользования 2100
система управления ориентацией 241
система управления распределённых баз данных 1143
система управления реляционных баз данных 3100
система управления с помощью ГИС 1687
системное картографирование 3568
системные таблицы 3569
ситус 3314
скалярное поле 3209
сканер с полистовой подачей 1394
сканирование 3225
сканирование ближайшего соседа 2509
сканирование в поперечном направлении 22
сканирование в продольном направлении 97
сканированная карта 3223
сканированные данные 3222
складывание вперёд с захватом 2043
складывать 28
склонение 939
склонение компаса 634
сколка 1071
скомпилированный набор карт 641
скопление 80
скорость 3927
скорость в бод 327
скорость передачи данных в бод 327
скорость прорисовки 2797
скос 1921
скошенная проекция 2603
скошенность 214
скриншот 3232
скроллинг 3235
скрытая линия 1820
скрыть 1822
слабая связь 2235
след 1918, 3775
следящий элемент 1452
слежение за фазой несущей 432
слепое оцифрование 361
слияние 2051, 2393
слияние рек или пластов 2051
словарь базы данных 885
словарь данных 885
сложная поверхность 651
сложность 646
сложность кривой 827

слоистое представление 2119
слой 2116, 3317
слой данных 897
слой оперативным доступом 2202
служба исследования пространственной информации 3389
служба определения стандартного местоположения 3467
служба точного определения местоположения 2877
служебная карта 2614
случайная погрешность 16
случайные данные 2997
смежная грань 50
смежная сторона 50
смежная структура данных 713
смежное ребро 49
смежность 45, 712
смежность полигонов 2824
смежность сетей 2397
смежность узлов 2534
смежные дуги 48
смежные фацеты 51
смежный лист карты 52
смежный угол 47
смесь разнородных элементов 689
смешивание 2399
смещение 1126
смещение начала отсчёта 926
смещение репера 926
смыкание 2044
снижение точности определения положения в вертикальной плоскости 3932
снижение точности определения положения в горизонтальной плоскости 1847
снимок 2742
сноска 1454
событие 1329
совместимость 700
совместимость геоизображений 637
совместимость данных 878
совместная работа 2047
совместное пользование данных 914
совмещение 2365, 2679
совмещение узлов 2538
совокупность 83, 705, 2854
совокупность географических данных 1588
совокупность данных 875
совокупность полевых данных 1399
совокупность символов или знаков 3556
совокупность слабых источников загрязнения, рассредоточенных по большой площади 165
совокупность слабых источников, рассредоточенных по большой площади 165
совокупность точек 2800
совокупные показатели 82
совпадение 2365
совпадение рёбер 1223
совпадение узлов 2538
согласие 687
согласование 2365
согласование карт 2250
согласование признаками 2365
согласование цифровых изображений 1052
соглашение о наименовании 2486
соединение 2044, 2051
соединение множество вариантов 877
соединение резиновой нитью 3186
соединённые линии 2045
соединённые узлы 2046
соединитель 699
созвездие 705
создание конфигурации местности 3607
сокращение 3061
сокращение данных 908
сокращённое меню 3289
сокращённое наименование 3288
солнечная долгота 1814
сооружения 1314
соответствие 17, 687
соответствие границ 379
соответствие рёбер 1223
соотнесённость границ 379
соотношение многие-ко-многим 2246
соотношение один-к-одному 2621
соотношение один-ко-многим 2247
сопоставимость 631
сопоставление 2365
сопоставление карт 2283
сопровождение данных признаками 2066
сопряжение 53
сопряжённые точки 692
сортирование 3342
сортировать 3341
сортировка 3342
сортировка в восходящем порядке 189
сортировка по убыванию 988
соседние точки 2514
соседний угол 47
соседство 2929
составительский оригинал карты 2659
составление 638
составление карт 2297
составление схемы уличного движения 3804
составление таблицы 3577
составлять карту 2249
составляющая цвета 607
составная часть карты 448
составной объект 649
составной полигон 650
сохранение 709
сохранение площади 2885
сохранять 3208
социально-экономическая зона 3335
социально-экономическая карта 3334
социально-экономические данные 3333
сочетание 623
сочленение 672
спектральная коррекция 2987
спектральная надпись 3432
спектральная полоса 3428
спектральная разрешающая способность 3431
спектральная яркость 3429
спектральный класс 3430
спектрозональный аэрофотоснимок 1366
специальная граница 3423
специальная карта 3424
спецификация базы данных 871
спецификация открытой интероперабельности географических данных 2630
спецификация региона 3088
спецификация экологического воздействия 1291
спиннер 3442
списковое поле 2200
список атрибутов 3898
список классов 560
список контрольных точек 741
список переменной длины 3903
список свойств 3898
список топологических схем 3760
сплайн 3443

сплинтер 3319
сплошные данные 717
способ значков 2409
способ изолиний 2410
способ картографического
    изображения 2244
способ картографического
    изображения ареалов 2408
способ картографического
    изображения знаков
    движения 2412
способ картографического
    изображения значков 2409
способ картографического
    изображения изолиний 2410
способ качественного фона
    2413
способ количественного фона
    2414
способ линейных знаков 2411
спот-карта 3449
справка 3140
справочная система по
    информационным ресурсам
    3154
справочная таблица 2233
справочная таблица цветов 614
спрямление 3054
спускающееся меню 2953
спуск вниз 1365
спутник наблюдения Земли
    1206
спутниковая геодезия 3198
спутниковая геометрия 3199
спутниковая навигационная
    система 3203
спутниковая система
    позиционирования 1710
спутниковая триангуляция 3206
спутниковое изображение 3200
спутниковые данные 3197
спутниковые изображения 3202
сравнение изображений 1896
сравнимость 631
среда 1286
среда обитания 1797
среда символов 3554
средневзвешенная во времени
    величина 3694
средневзвешенная по площади
    разрешающая способность
    182
среднее 2379
среднее значение 2379
среднее отклонение 3904
среднеквадратическая
    погрешность 3172
среднеквадратическая
    погрешность единицы веса
    3164

среднемасштабная карта 2388
средний масштаб 1987
средний уровень моря 2378
средняя величина 2379
средняя площадь 2375
средняя тенденция 514
средство передачи 3791
срезание 571
ссылка 3064
ссылка атрибута 258
ссылочная целостность 3075
ссылочный эллипсоид 3067
стандарт CGI 670
стандарт DGM 1049
стандарт DIGEST 1047
стандарт DOQ 1062
стандарт IGES 1951
стандарт OMG 2617
стандартизация представления
    географических данных
    1593
стандартизованная растровая
    графика 3464
стандарт на передачу
    географических данных
    3379
стандартная линия 3465
стандартная параллель 3466
стандартное отклонение 3461
стандартный лист
    топографической карты
    2318
стандартный формат обмена
    3463
стандарт обмена цифровыми
    пространственными
    данными 3379
стандарт передачи 3792
стандарт передачи
    пространственных данных
    3379
стандарты на обработку
    информации 1393
стандарты содержания 710
стандарты точности 19
станция слежения 3778
старая, центральная часть
    города 512
старение данных 856
старение карты 2251
статика 3471
статистика географических
    ссылок 1628
статистика растительного
    покрова 2080
статистика хи-квадрат 534
статистическая графика 3476
статистическая зона 3473
статистическая категория 3475
статистическая поверхность

3480
статистический момент 3479
статистический параметр
    класса 561
статистически эквивалентная
    категория 3477
статистическое моделирование
    3478
статистическое среднее 3474
статистическое среднее
    значение 3474
стационарность 3472
стек 3458
стенная карта 3973
степень генерализации 1522
стерадиан 3484
стереовысота 3487
стереограф 2746
стереографическая проекция
    3488
стереодигитайзер 3486
стереомодель 3489
стереообрабатывающий
    графопостроитель 3492
стерео-оригинал 3485
стереопара 3491
стереоплоттер 3492
стереоскопическая проекция
    3494
стереоскопический
    аэрофотоснимок 3491
стереоскопический
    графопостроитель 3492
стереоскопическое
    вычерчивание 3493
стереофотограмметрия 3490
стиль линии 2192
стиль регионов 3089
стиль символов 3559
стирание 954
стирать 953
столбец 620
столбиковая диаграмма 315
столбчатая диаграмма 315
сточная канава 1095
стратегическое принятие
    решений 3503
стратиграфическое несогласие
    3506
стратиграфия 3507
стратификация 3504
стрелка 185
стрелка-указатель "север-юг"
    2565
стрелочный глиф 1714
строение 3512
строенная разность 3827
строка 527, 3184
строка развёртки 3226
строка сканирования 3226

струйный плоттер 1953
структура 3512
структура графических данных 1745
структура данных 916
структура данных сети неравносторонних треугольников 3696
структура дефектов 945
структура домена 1162
структура квадродерева 2967
структура объектов 2588
структура пространственных данных 3377
структура пространственных объектов 3401
структура сети 2525
структура таблицы 3573
структура тематических данных 3640
структурированная модель высотных точек 3514
структурированные данные 3513
структурная карта 3527
структуры связанных покрытий 694
ступенчатость 92
СУБД 867
субституция 3526
СУМ 2378
суммарная съёмка 644
суммирование изображений 1889
суммировать 28
суперпозиция 3534
СУРБД 1143
сферические координаты 3437
сферические проекции 3438
сферическое искажение 3434
сферическое отображение 3436
сфероид 3439
сфероидическая геодезия 3440
схема 1018
схема географического кодирования 1600
схема дорог 3168
схема затенения 3277
схема, изображаемая кружочками и стрелками 407
схема последовательности 1443
схема процесса, изображаемая кружочками и стрелками 407
схема расположения 2123
схема расположения профилей 1933
схема расположения соседних листов карт 1933

схема сближения меридианов 940
схематическое изображение 1019
схематическое представление 1019
сходящийся объект 2394
сценарный язык 3233
сцепление 672
сцеплённый ключ 671
счёт 783
счётный участок 499
сшивание 2044
сшивка карт 2286
съёмка 1556
съёмка землепользования 2103
съёмочный масштаб 3219
съёмочный планшет 3550

таблет 1071
таблица 3570
таблица атрибутов 259
таблица атрибутов дуг 147
таблица атрибутов маршрута 3178
таблица атрибутов объектов 1375
таблица атрибутов полигонов 2832
таблица атрибутов узлов 2535
таблица валентностей 3897
таблица выходных уровней 2142
таблица координат 766
таблица оперативным доступом 2203
таблица, представленная с помощью карты 2293
таблица просмотра 2233
таблица символов 3560
таблица события 1333
таблица текстовых атрибутов 3625
таблица цветности 614
таблица цветов 614
таблица явления 1333
табличные данные 3575
табулирование 3577
табуляция 3577
таксономическая классификация 3588
таксономия 3589
тальвег 3561
тангенциальная коническая проекция 3581
таргетирование 3585
таскание 1178
тасселяция 3684
тасселяция Дирихле 1115
тахеометр 3578

твёрдое тело 3339
тег атрибута 260
тегирование 2066
тегирование горизонталей 735
тегирование контуров 735
тексел 3634
текст 3623
текстовая верность 3631
текстовая зона 3626
текстовая информация 3632
текстовая метка 3628
текстовая панель 3626
текстовая точность 3631
текстовое сопровождение 453
текстовой символ 3630
текстовой стиль 3629
текстовые данные 3632
текстовый атрибут 3624
текстура 3633
текущий контроль 3548
телеизмерение 3593
телекоммуникационное картографирование 3592
телеметрия 3593
тело 3339
тема 1630
тема события 1334
тематическая географическая область 3641
тематическая карта 3644
тематическая карта на базе множества переменных 2480
тематическая карта одной переменной 2622
тематическая картография 3649
тематическая переменная 3655
тематическая символика 3657
тематическая точность 3635
тематическая фотокарта 3651
тематические данные 3639
тематический атлас 3636
тематический диапазон 3652
тематический класс 3637
тематический слой 3643
тематический слой явления 1334
тематическое картографирование 3649
тематическое ключевое слово 3642
тематическое моделирование 3650
тематическое оттенение 3654
тематическое разрешение 3653
тематическое содержание 3638
тенденция изменения структуры и численности населения 2858
тензор 3604

тень 3276
теодолит 3658
теоретическая невязка
    контуров 3659
теория картографии 3661
теория картографической
    коммуникации 3660
теория нечётких множеств 1506
теория размытых множеств
    1506
теория решения 937
тепловой многозональный
    сканер ИК-диапазона 3662
термический графопостроитель
    3664
термо-изолинейная карта 3663
территориально
    распределённая система
    1571
территориально
    распределённый 1570
территориально
    рассредоточенный 1570
территория 3621
тестовой участок 3622
тестовый ящик 530
техническая документация
    3590
техническая карта 1272
техническая надёжность 3591
технические средства 1804
техническое задание 3142
технология базы
    геопространственных
    данных 1692
технология географических
    данных 1589
технология картографирования
    информации 1941
течение 791
тип данных 919
типичные инструменты
    картографирования 1527
тип легенды 2134
тип линии 2196
тип признака 1390
тип примитива 1285
тип рельефа 1389
тип склона 3324
тип условных обозначений
    2134
тире 314
тождественное отображение
    1885
тождественность 1884
тон 3276
тонирование 3127
топографическая база данных
    3709
топографическая граница 3710

топографическая изученность
    территории 3715
топографическая карта 3714
топографическая контурная
    карта 3714
топографическая модель
    рельефа 3721
топографическая основа карты
    3706
топографическая планиметрия
    3717
топографическая плоскость
    3716
топографическая поверхность
    3722
топографическая проекция
    3720
топографическая съёмка 3723
топографическая точка 3718
топографические данные 3708
топографический 3705
топографический вид 3724
топографический лист 1405
топографический план 3714
топографический профиль 3719
топографический рисунок 3711
топографический чертёж 3711
топографический
    четырёхугольник 2318
топографический элемент 3712
топографическое деление 3710
топографическое отображение
    текстуры 3726
топография 3725
топография местности 3605
топогридная интерполяция
    3727
топологизация 3754
топологическая векторная
    решётка 3752
топологическая группа 3738
топологическая клетка 3729
топологическая окружность
    3730
топологическая операция
    замыкания 3732
топологическая очистка 3731
топологическая переменная
    3769
топологическая погрешность
    3735
топологическая ревизия 3753
топологическая связь 3747
топологическая сложность 3733
топологическая
    согласованность 3734
топологическая сортировка
    3748
топологическая статистика
    3767

топологическая структура 3750
топологическая схема 2123
топологическая трансформация
    3751
топологическая функция 3736
топологический анализ 3728
топологический граф 3737
топологический запрос 3766
топологический оверлей 3745
топологический примитив 3746
топологически полный
    четырёхугольник 3739
топологически сложные
    объекты 3740
топологически
    структурированные данные
    3742
топологическое имя 3761
топологическое моделирование
    3744
топологическое описание 3756
топологическое отображение
    3743
топологическое
    преобразование 3751
топологическое пространство
    3749
топология 3755
топология дуг и узлов 155
топология Интернета 1995
топология многоугольников и
    дуг 2831
топология полигона 2849
топология связей левых и
    правых сторон полигонов
    2132
топология сети 3764
топология сходимости по мере
    3763
топология типа "дерево" 3812
топонимика 3770
топонимия 3770
топонимы 1613
топооснова 3706
топоцентрическая основа 3704
топоцентрические координаты
    3703
топоцентрический репер 3704
топоцентрический угол 3702
точек на дюйм 3701
точечная диаграмма 3228
точечная карта 1167
точечная тематическая карта
    1167
точечное отображение 2806
точечный объект 2809
точечный способ 1168
точка 1166, 2798
точка географического
    кодирования 2323

точка замера 2384
точка измерения 2384
точка изображения 2758
точка маршрута 3979
точка направления наблюдения 1112
точка обзора 3957
точка отсчёта 1475
точка перегиба 811
точка перспективы 3957
точка поворота 3838
точка прерывания 401
точка приведения 2811
точка привязки 2811
точка признака 1936
точка сетки 2529
точка скрепления 3677
точка соединения или пересечения 3677
точка указателя 1936
точка центра 511
точное дигитализирование 2880
точномасштабное представление 3835
точность 17
точность атрибута 245
точность времени 3685
точность задания времени 3685
точность измерений 2382
точность измерений по картам 2284
точность картографирования 2298
точность карты 2298
точность классификации 553
точность масштаба 3212
точность масштаба карты 3212
точность определения местоположения 1842
точность плановой привязки 1842
точность съёмки 18
точный интерполятор 1335
точный позиционный сервис 2877
точный цифровой планшет 2881
траектория 3786
траектория в атмосфере 230
транзакция 3788
транзит 3795
трансект 3789
трансляция 3797
трансляция сетевого адреса 2516
транспортная развязка 3167
трансфокация 4036
трансформация 3793
трансформация проекций 2909
трансформация псевдоцвета 2937
трасса 3775
трасса орбиты 1792
трассирование 3776
трассировка 3182, 3776
трассировка координатной сетки 1766
трассировка линий 2195
трассировка лучей 3033
трассировка сети 2531
трассировка топологии географического размещения 1631
трафарет 3595
трафик
    по направлению ~ а 1174
    по ходу ~а 1174
    по ходу основного ~а 1174
требования пользователя 3891
трекбол 3777
тренд-линия 3814
тренд-поверхность 3815
треножные данные 3829
третья нормальная форма 3669
третья разность 3827
треугольная декомпозиция 3821
треугольная сетка 3819
треугольник 3818
треугольное подразделение, основанное на реберном представлении 1217
трёхмерная ГИС 1013
трёхмерная глобальная географическая модель 1014
трёхмерная многоугольная сеть 1177
трёхмерная отметка уровня 929
трёхмерная сейсмическая съёмка 1188
трёхмерное геоизображение 1011
трёхмерное дигитализирование 1189
трёхмерное изображение 3969
трёхмерный 3670
трёхмерный базис 929
трёхмерный гиперболический граф 1015
трёхмерный дигитайзер 3383
трёхмерный многоугольник 1176
трёхмерный объект 1157
трёхмерный репер 929
триангулирование 3821
триангуляционная марка 3823
триангуляционный маркер 3823
триангуляция 3821
триангуляция Делоне 951
триангуляция крупного масштаба 2106
триангуляция полигонов 2850
триангуляция элементарной площадки 3325
триггерная кнопка 530
тригонометрия 3824
трилатерация 3825
трилинейный интерполятор 3826
триплет 3828
трихотомическое дерево 3830
тройка 3828
тройка координат 3828
тропосферные задержки 3831
трубопровод 2756
трясина 368
туман 1451
тупик 1913
тур 3773
туристский атлас 3774

УАПП 3859
увеличенный масштаб 1273
увеличивание 4035
увеличивание изображения 4035
угловая высота 123
угловая мера 2381
угловое несогласие 3506
угловое расстояние 128
угловое соединение 774
угловое сочленение 774
угломер 3658
угломестные данные 1247
углы при вершине 3930
угол 122, 773
угол взгляда 127
угол возвышения 123
угол градиента 125
угол зрения 127
угол наклона 126
угол откоса 126
угол поля зрения 124
угол ската 126
угол уклона 126
удаление невидимых линий 1821
удалённое присваивание адреса 1132
удалённый от промышленного центра район 1838
удвоение 1196
удобство места 2766
узел 1627, 2533
узел назначения 998
узел-потомок 987
узел сети 2529
узкая зона 782
узловой регион 1490
узлы координатной сетки 1776

узлы объекта 2600
узор 2716
указание 2752
указатель 993, 2801
указатель географических
    названий 1515
указательная сетка 2214
указательное устройство 2804
указующее устройство 2804
укладка 1259
уклон 1921
уклонение 947
уклонение отвесной линии 948
укорачивание 3288
укрупнение 702
улавливание 429
уличная сеть 3509
улучшение изображения 1903
улучшение качества
    изображения 1903
улучшение псевдоцвета 2935
улучшенная картографическая
    система 60
улучшенное картографическое
    оборудование 59
улучшенный формат DLG 1155
уменьшение 3061, 4038
уменьшение изображения 4038
уменьшение объёма 968
уменьшение ступенчатости 136
уменьшение точности
    положения в
    горизонтальной плоскости
    1847
унаследование свойств 1950
универсальная географическая
    идентичность 3861
универсальная ГИС 3862
универсальная десятичная
    классификация 3860
универсальная полярная
    стереографическая
    проекция 3863
универсальная поперечно-
    цилиндрическая проекция
    Меркатора 3867
универсальная
    пространственная
    прямоугольная система
    координат 3866
универсальное время 758
универсальное
    координированное время по
    Гринвичу 758
универсальный асинхронный
    приёмопередатчик 3859
универсальный выпрямитель
    3865
универсальный полигон 3868
уникальный идентификатор

1709
упаковка 1259, 2685
уплотнение 656, 968
уплотнение сети 2522
уплотнённые цифровые данные
    для вертикальной наводки
    655
уплотнять 654
упорядочение 2645
упорядочение Пиано-Гилтона
    2719
упорядоченный список 2644
управление географических
    данных 1587
управление информации об
    окружающей среде 1292
управление параллельных
    процессов 680
управление природных
    ресурсов 2502
управление слоями 2118
управление средствами 1362
управление топографической
    информации 3713
управляющая станция 743
управляющий символ 738
упрощение 3304
уравнение проекции 2912
уравнивание 54
уравнивание наземной линии
    2085
уравнивание полигонального
    хода 55
урез воды 594
уровен 2138
уровень алидады 1844
уровень воды 3976
уровень вырезки 3318
уровень моря 3237
уровень отображения 2143
уровень серого 1758
уровень слоя 3318
уровень яркости 1758
уровневая линия 732
усечение 571
условная линия 681
условная проекция 744
условное направление
    северного меридиана 1774
условное направление южного
    меридиана 1779
условное обозначение 2278
условное цифровое
    обозначение 1882
условные внемасштабные
    знаки 2812
условные картографические
    знаки 459
условные площадные знаки 180

условный горизонт 924
условный знак местоположения
    2219
условный оператор 682
усовершенствование 3076
усовершенствованный
    радиометр очень высокой
    разрешающей способностью
    61
успокоение 238
усредняющий фильтр 291
устанавливать отношение 3097
устанавливать связь 3097
устанавливать соответствие
    2248
устанавливать соотношение
    3097
установка 3269
установка точности стандарта
    3460
установление 3269
установление границ 957
установление контрольных
    точек 334
устаревшие данные 78
устойчивое развитие 3552
устойчивые потери 2729
устранение избыточных
    промежуточных точек в
    цифровой записи линий
    2197
устранение искажения 2542
устранение скрытых линий
    1821
устройство ввода 1959
устройство ввода позиции 2221
устройство графического ввода
    1740
устройство графического
    вывода 1741
устройство изготовления
    тематических карт 3647
устройство оцифровки 1078
устройство оцифровки
    видеоизображений 3942
устройство термического
    картографирования 1947
устройство управления
    позицией 2804
устье 1322
утонение 3667
утончение 3667
утончение линий 2194
уточнение адреса 35
уточнение контура 728
утрирование 1336
утрирование размера или
    формы 1336
ухватываться за сетку 3331
ухватываться за узел 3332

участок 1148, 3248
участок земли 2089
участок земли клином 1716
участок между зданием и
  дорогой 1487
учёбный атлас 222
учёт земель 2090
учёбный курс 1119
учётный процент 1119
учёт трафика 3782
учёт численности 487

фазовая автоподстройка
  частоты 2735
фазовая синхронизация 2735
фазовый метод 2736
фазовый метод измерения
  дальностей 2736
файл 1411
файл CFF 449
файл аэрофотоснимков
  элементарных квадратных
  участков 67
файл взаимосвязей между
  переписными районами 502
файл внешней базы данных
  1352
файл графических форм 3282
файл данных граничных
  условий 383
файл карты 2272
файл координат 759
файл меток 3673
файл описательных данных 884
файл остановок 3499
файл палитры 2688
файл площадей и значений
  участков 1412
файл проекции 2913
файл точек 2803
файл учёта основного капитала
  195
файл формата CFF 449
файл четырёхугольных
  участков 2957
файл элементарных квадратных
  участков 2957
фактор внешнего порядка 1354
фактор местоположения 2220
фальшивое начало 1369
фамилия 584
ФАПЧ 2735
фасад 1488
фацет 1360
фацеты сети неравносторонних
  треугольников 3697
феномен 2737
фигура 3279
физико-географическая карта
  2751

фиксирование позиции 2868
фиксирование позиции
  спутником 3204
фиксированное расстояние
  1431
фиксировать 1430
фильтр 1422
фильтрация пространства 3385
фильтр верхних частот 1831
фильтрирование градиента
  1730
фильтрирование крутости 1730
фильтр мод 3978
фильтр низких частот 2236
фильтр обнаружения краёв
  1220
фильтр пятнистого шума 3426
фильтр растяжения 1096
фильтр типов волн 3978
фильтр типов колебаний 3978
фильтр частотного интервала
  1485
флажок 530
флажок события 1331
флуктуация 1449
фон 300
фонетический код 3344
фоновая карта 302
фоновая фигура 1410
форма 3279
форма маркера 2354
формат DLG 1058
формат DXF 1182
формат GIF 1683
формат JPEG 2048
формат JPG 2048
формат MOSS 2292
формат NTF 2495
формат RINEX 3044
формат TIF 3579
формат TIGER 3741
формат TRIM 3616
формат атласа 223
формат векторного
  произведения 3920
формат векторных
  изображений 3915
формат данных 893
формат данных для
  вертикальной наводки 1249
формат для передачи
  инженерной графики 1951
формат изображений BSQ 310
форматирование карты 2273
формат картографического
  произведения 463
формат обмена 1337
файл цифровых
  картографических данных
  1586

формат обмена чертежей 1182
формат передачи изображений
  с построчным хранением
  данных 307
формат передачи изображений
  последовательностью
  значений яркости каждого
  пиксела 308
формат переноса
  кадастровых планов 2653
формат строки 2179
формат топологического
  интегрированного
  географического
  кодирования и
  стандартизации 3741
формат файла 1413
формат файла с разметкой для
  хранения изображений 3579
формирование кадра 1478
формулировка воздействия на
  природную среду 1291
формуляр 1460
формы местности 3605
фотограмметрически
  оцифрованный 2740
фотограмметрия 2741
фотокамера для внешних
  траекторных измерений 305
фотокарта 2745
фотокопирующее устройство
  2739
фотоплан 2748
фотоплёнок 2743
фотоплоттер 2418
фоторельеф 2744
фоторепродукционная камера
  2739
фотоснимок 2742
фотосфера 2749
фотосъёмка 2742
фототеодолит 305
фототеодолитная съёмка 2750
фрагмент 3545, 3679
фрагмент земли 1789
фрагментирование 3684
фрагмент мозаичного
  изображения 3679
фрактал 1470
фрактальная поверхность 1472
фрактальное изображение 1471
фракционный размер 1473
функциональная поверхность
  1492
функциональная связь 1491
функциональные возможности
  ГИС 1693
функция группирования 1796
функция отображения 2301
функция передачи 3790

функция распределения
вероятности 2899
функция цели 2591

характеристика
народонаселения 2855
характеристика совокупности
2855
характеристики округа 788
характеристическая карта 1383
ход 3773
ход лучей 3032
хозяйство 1371
холмистая местность 3878
холодный старт 601
хорда 535
хорология 536
хранилище пространственных
данных 3381
хранитель 835
хребет 3161
хронология данных 894
художественное
проектирование карт 444

ЦАП 1073
цвет 619
цвет заливки 1415
цвет заполнения 1415
цветной аэрофотоснимок 603
цветной тест 606
цветовая гамма 2687
цветовая компрессия 608
цветовая палитра 615
цветовая панель 605
цветовая плотность 542
цветовая шкала 618
цветовое инфракрасное
изображение 613
цветовой сканер прибрежной
зоны 593
цветовой фон 604
цветоотделение 616
цветоотдельный оригинал
карты 2259
цвет фона 301
целевая карта 3586
целевая функция 2591
целевой район 3584
целеуказание 3585
целостность базы данных 865
целостность данных 896
целостность на уровне ссылок
3075
целостность примитива 1277
цель 997, 3583
цельнокроеный слой 3240
центр 505
централизованный
картографический банк

данных 513
центральная линия 507
центральная проекция 690
центральная точка 511
центральная часть города 512
центральный регион 1490
центр геофизических данных
1667
центр гомологии 510
центр инерции 509
центрирование 506
центр класса 550
центр массы 509
центр обмена информацией 563
центровая линия 507
центроид 159
центр перспективы 510
центр проектирования 510
центр проекции 510
центр распределения 514
цепная линия 477
цепное кодирование 519
цепной код 518
цепочка 516
цепочка символов 527
цепь 516
цепь треугольников 520
цилиндрическая проекция 847
цилиндрическая равновеликая
проекция 846
цилиндрическая
равнопромежуточная
проекция 1310
циркуль 1153
циркуль-измеритель 1153
цифрование 1054
цифрование потоковым вводом
3508
цифрование топологии 3758
цифрователь 1078
цифровая геотопографическая
информационная система
1050
цифровая геохимия 1046
цифровая кадастровая карта
1032
цифровая камера 1033
цифровая карта 1239
цифровая карта-основа мира
1036
цифровая картографическая
модель 1034
цифровая картография 1035
цифровая классификация 1037
цифровая матрица высотных
точек 1042
цифровая модель высотных
точек 1043
цифровая модель местности
1072

цифровая модель рельефа 1072
цифровая модель ситуации
1067
цифровая надпись 1066
цифровая объектно-
структурированная модель
местности 1057
цифровая отметка горизонтали
734
цифровая поверхность 1070
цифровая таксономия 2583
цифровая топография 1075
цифровая фотограмметрия 1063
цифрово-аналоговый
преобразователь 1073
цифровое изображение 1051
цифровое покрытие 1039
цифровое пространство 1069
цифровое цветоделение 1038
цифровой 1031
цифровой векторный
географический формат
1048
цифровой графопостроитель
1065
цифровой источник 1068
цифровой монтаж снимков
облачности 1076
цифровой ортоснимок 1061
цифровой планшет 1071
цифровой плоттер 1065
цифровой снимок 1064
цифровой формат 1045
цифровой формат обмена 1044
цифровые данные 1040
цифровые данные высотных
точек 1041
цифровые картографические
данные 1059
цифровые оси 2582
цифровые топографические
данные 1074
цифры сетки 1768
ЦММ 1072
ЦМР 1072

часовой пояс 3695
частичная сумма 3528
частное значение 905
частный масштаб 2707
частный масштаб карты 2707
частота штриховки 1807
часть карты вне рамки 2678
часть кода, занимаемая адресом
34
часть кода, занимаемая
номером памяти 34
чёрно-белый аэрофотоснимок
359
черновая карта 2883

черновая топология 2884
черта 314
чертёжное устройство 1181
чертёжный инструмент 1183
чертить карту 1180
чётко определённая точка 3993
чёткость 2878
чётность адреса 40
четырёхвершинник 2959
четырёхмерная ГИС 1467
четырёхугольник 2959
четырёхугольный участок 2956
числа Пиано 2720
численный 1031
численный масштаб 3139
число 783
числовой 1031
числовой атрибут 2581
число контрольных точек 2578
число регионов 2579
число строк 2580
чистая кайма 2511
чистая окантовка 2511
чистка 3236
чистое обрамление 2511
чистый бордюр 2511
чтение карты 2321
чувствительность 3263
чувствительность к обвалу
    2093

шаблон 3595
шаблон дигитайзера 1086
шар 3777
шаровой указатель 3777
шатер 3258
шероховатость 3851
шестая часть окружности 3270
ширина колонки 2179
ширина полосы 311
ширина полосы ленты 311
широкое устье 1322
широкое устье реки 1322
широкополосный спектр 3451
широта 205
широта-долгота 2113
широта ячейки 486
шкала 3211
шкала гипсометрической
    окраски 1879
шкала заложений 3321
шкала оттенков серого цвета
    1759
шкала полутонов 1759
шкала цветов 618
шкала цветового охвата 606
шкала яркостей 1759
шлем-дисплей 1810
шлюз географических ресурсов
    1676

шрифт 526
шрифтовая палитра 1453
штамп 1918
штрих 314
штрихи 3271
штрихи для обозначения
    профилей местности 3271
штрих-линия 2722
штрихование 3275
штриховая линия 2722
штриховка 3271
штриховое маркирование 851
штриховой негатив 2182
штриховой оригинал 999
штриховой оригинал карты 999
штриховой фон 1806
штрихпунктирная линия 2722
штукатурка 2112

щель 1510

эвристика 1818
эвристический 1817
экватор 1306
экваториальная плоскость 1308
экваториальные координаты
    1307
эквидистантные данные 3095
эквидистантные линии 1311
эквидистантные профили 3096
эквипотенциальная
    поверхность 2145
экземпляр 769, 1964
экзогенный фактор 1354
экологическая оценка 1287
экологическая экспертиза 1287
экологическое планирование
    1295
экономика картографического
    производства 1214
экономическая география 1213
экран 3231
экранное оцифрование 2627
экранный дамп 3232
экран состояния спутников
    3205
экспериментальная точка 905
экспертная система 1341
экспликация 2278
эксплуатационные данные 1398
экспозиция 633
экспозиция склона 633
экспонирование 633
экспортирование 1345
экспортирование векторных
    объектов 1346
экстенсивное измерение 1350
экстент отображения 2270
экстент покрытия 795
экстент сетки 1767

экстент страницы 2686
экстраполирование 1357
экстраполяция 1357
экстремальные точки 1358
эксцентрицитет 1211
эксцентричность 1211
эластичная линия 1231
эластичная трансформация
    1232
эластичное соединение 3186
эластичный круг 1230
элеватор 1252
электромагнитная разведка
    1235
электромагнитная съёмка 1235
электромагнитное излучение
    1233
электромагнитные
    исследования 1235
электромагнитный спектр 1234
электромеханический сенсор
    1236
электронная карта 1239
электронная таблица 3450
электронное измерение
    расстояний 1238
электронный атлас 667
электронный планиметр 1240
электронный тахеометр 3771
электростатический плоттер
    1241
элемент 2041
элементарная величина 234
элемент данных 922
элемент изображения 2758
элемент карты 448
элемент меню 2392
элемент мозаичного
    изображения 3679
элемент объёма 1243
элемент площади 1242
элемент поверхности 1242
элемент разрешения 3545
элемент растра 1764
элемент сети 2523
элемент текстуры 3634
элемент топологии 3759
элемент трёхмерного
    изображения 1243
элементы маркировки 2357
эллипс искажений 1253
эллипсоид 1254
эллипсоидальные координаты
    1255
эллипсоид вращения 1256
эллиптическая проекция 1257
эллиптическая проекция
    Дональда 1164
эпиполярная ось 1303
эпиполярный 1302

эстуарий 1322
эталон 3063
эталонирование 3074
эталонная карта 3069
эталонная область 3065
эталонная пространственная
　　система 3411
эталонная сеть 3071
эталонная система 3073
эталонное тестирование 334
эталонное тестирование ГИС
　　1690
эталонные сети высокой
　　точностью 1829
эталонный приёмник 3072
эталонный эллипсоид 3067
этикет 2061
эфемериды 2640
эффект глубины 980
эффект Доплера 1165
эффект первого порядка 1429
эффект рельефа 1260
эхо 1212
эхо-изображение 1172

юг 3348
юг картографической сетки
　　1779
юго-восток 3349
юго-запад 3351
Южный полюс 3350
Южный полярный круг 135

явление 2737
явления реального мира 3041
ядро 1856
ядро популяции 2571
язык доступа к данным 854
язык запросов 2974
язык картографических
　　надписей 2279
язык карты 2277
язык манипулирования данных
　　898
язык обработки
　　пространственных данных
　　3374
язык описания данных 882
язык описания сценариев 3233
язык определения данных 882
язык подготовки сценариев
　　3233
язык разметки для географии
　　1635
язык структурированных
　　запросов 3515
язык сценариев 3233
якорь 120
яркость 404
ячейка 481

ячейки Вигнера-Зейтца 3665